# Studies in Computational Intelligence 459

**Editor-in-Chief**

Prof. Janusz Kacprzyk
Systems Research Institute
Polish Academy of Sciences
ul. Newelska 6
01-447 Warsaw
Poland
E-mail: kacprzyk@ibspan.waw.pl

For further volumes:
http://www.springer.com/series/7092

# Preface

The last decade has witnessed a tremendous attention devoted to both modeling and simulation of nonlinear dynamical systems. This high interest is due to the growing awareness that nonlinear dynamics is inherent in a vast class of systems, phenomena and events: natural biological systems, physical systems, engineering systems, etc. A study of the relevant state-of-the-art reveals various scientific contributions based on both modeling, simulation and control of nonlinear dynamical systems. Various interesting and striking states of these systems are of particular interest: periodic, quasi-periodic, stable, unstable, deterministic, stochastic, synchronized, torus and chaotic dynamics, etc. A good understanding and mastering of these states does lead to various potential interesting applications.

Model approximation appears to be a very common modeling technique in engineering since accurate models are generally very complex and tough to solve either analytically or even numerically. Further, it is well known that numerical simulation is prone to accumulation of round-off errors during computation, which could result into non-realistic outcomes especially in case of extremely stiff systems. Numerical simulation is also eventually very time-consuming and simulation convergence of is not well tractable systematically. Thus, some trade-off between the potential analysis techniques appears necessary. Approximate models are easily solvable either analytically or numerically although the results obtained are of less accuracy. Accurate modeling while mastering related computational concerns is indeed a serious challenge for modern computational engineering.

This book is a collection of recent advanced contributions in the field of nonlinear dynamics and synchronization, with special applications in the area of theoretical electrical engineering. The book contains twenty-one chapters grouped in five parts.

The first part consisting of five chapters does focus on theoretical issues related to chaos and synchronization and their potential applications in mechanics, transportation, communication and security. The second part having six chapters handles dynamic systems modeling and simulation with special applications to real physical systems and phenomena. The third part consisting of four chapters discusses some fundamentals of electromagnetics (EM) and addresses the modeling and simulation in some real physical electromagnetic scenarios. The fourth part of three chapters does mainly address stability concerns. Finally, the last part having three chapters assembles some sample applications in the area of optimization, data mining, pattern recognition and image processing.

In the following the contents of twenty-one chapters of this books is respectively briefly presented.

The first chapter, **Synchronization of two nonidentical clocks: What Huygens was able to observe?**, by Krzysztof Czolczynski, Przemysław Perlikowski, Andrzej Stefanski, and Tomasz Kapitaniak, addresses the synchronization of two nonidentical clocks. An in-depth analytical study is carried out to demonstrate the occurrence of both almost complete (in-phase) and almost anti-phase synchronizations. These types of synchronization are defined in terms of the phase shift. Evidence is provided to explain the phenomenon of almost anti-phase synchronization of two pendulum clocks that Huygens was unable to observe experimentally in the XVIIth century.

The second chapter, **On the synchronization of 1D and 2D multi-scroll chaotic oscillators**, by J.M. Muñoz-Pacheco, E. Zambrano-Serrano, O.G. Félix-Beltrán, E. Tlelo-Cuautle, L.C. Gómez-Pavón, R. Trejo-Guerra, A. Luis-Ramos, C. Sánchez-López, investigates synchronization issues in 1D and 2D multi-scroll chaotic oscillators. The applicability of these issues in communication and security systems is demonstrated. The 1D and 2D multi-scroll chaotic oscillators are simulated using the analog computing paradigm. A chaotic synchronization scheme for multi-directional multi-scroll chaos generators is introduced and the Hamiltonian theory is used to derive synchronization criteria. Finally, two schemes are set-up to transmit encrypted binary and analog signals by applying chaotic switching technique and additive chaotic masking, respectively. Both schemes are also implemented using analog computing. This computing paradigm is simulated/emulated on

The third chapter, **Nonlinear filtering of chaos for real-time applications**, by V. Kontorovich and Z. Lovtchikova investigates the nonlinear filtering of chaotic signals in presence of additive white Gaussian noise (AWGN). Specifically, nonlinear filtering and chaotic phenomena are investigated with concrete applications in electrical engineering, basically in communications, control, etc. It is demonstrated that both nonlinear and chaotic filtering are processes leading to various potential applications including those which are related to real-time regime conditions. Examples of these applications include chaos-based communications systems synchronization, real-time control of chaos, radio-frequency interference filtering and mitigation, and chaotic system identification, just to name a few.

The fourth chapter, **Time-of-flight estimation using synchronized chaotic systems**, by Christian F. Wallinger and Markus Brandner develops a theoretical ranging concept based on time-of-flight (ToF) estimation. It is demonstrated that the concept involving chaotic signal leads to various potential metrological applications. A series of performance criteria for these applications are defined as fundamental parameters of the delay estimation process. An experimental demonstration is described showing the applicability of synchronized chaotic systems in a ToF measurement system. Many time-delays are evaluated analytically and experimentally and a comparison is performed to validate the concepts developed.

The fifth chapter, **Binary synchronization of complex dynamics in cellular automata and its applications in compressed sensing and cryptography**, by Radu Dogaru and Ioana Dogaru applies a concept involving cellular automata to both modeling and synchronization of complex dynamics. A specific focus is devoted to the binary synchronization of the transmitted bit stream. It is demonstrated that the decoding of this stream is possible only when the structure of the cellular automata (encryption key) is known. Finally, some applications of the concept developed are considered in cryptography, spread spectrum communications, and compressed sensing. Finally, an implementation of the concept on FPGA is presented.

The sixth chapter, **Fast switching behavior in nonlinear electronic circuits: a geometric approach**, by Tina Thiessen, Sören Plönnigs and Wolfgang Mathis develops a geometric concept to detect some striking and unexpected behavior of a specific class of nonlinear electronic circuits. An example of this behavior is the fast switching behavior of circuits which is characterized by the jumps in their state space. An analytical study is developed and appropriate criteria are derived for the occurrence of jump conditions. The approach developed in this work is essential as it allows a separation between the transient states of the electronic circuits and their permanent states. Finally, it is demonstrated that the geometric concept developed can easily be adapted to a system of equations based on MNA, allowing an implementation on circuit simulators such as SPICE. Numerical simulations are considered to validate the analytical results developed.

The seventh chapter, **Dynamics of Liénard optoelectronic oscillators**, by Bruno Romeira, José Figueiredo, Charles N. Ironside and Julien Javaloyes investigates the behavior a nonlinear optoelectronic oscillators (OEO) of Liénard types. It is demonstrated that the effects due to negative differential resistance across the tunnel diode leads to oscillations within the system. The concept developed in this chapter is validated by some recent experimental results presented by the relevant state-of-the-art. A specific focus is devoted to both electrical and optical systems, likely to exhibit complex dynamics ranging from self-sustained relaxation oscillations, to injection locking and chaotic dynamics. Potential applications of the concept in this chapter are found in the development of secure communication systems and highly stabilized OEO devices for microwave-photonic systems.

The eighth chapter, **Application of coupled dynamical systems for communities detection in complex networks**, by Nikolai Nefedov, models the stable communities' detection and links predictions processes in complex networks by coupled ordinary differential equations. The dynamics of the local states is described by the well-known Kuramoto model. The method is based on the dynamical formulation of modularity using a random walk and then is extended to coupled dynamical systems to detect communities at different hierarchical levels. Both attractive and repulsive coupling are considered and their effects on the dynamics of the global network are analyzed. As proof of concepts of the approach developed in the chapter, practical datasets recorded during the Nokia mobile-data collection campaign are considered and the concept developed is applied to derive the potential structures of the social community and to perform link predictions or recommendations.

The ninth chapter, **Infinite networks of hubs, spirals, and zig-zag patterns in self-sustained oscillations of a tunnel diode and of an erbium-doped fiber-ring laser**, by Ricardo E. Francke, Thorsten Pöschel, and Jason A.C. Gallas, investigates the dynamics of a nonlinear circuit containing diodes as analog devices. Some strange and oscillating patterns are depicted analytically which cannot be explicitly observed experimentally due to their appearance in a compressed form. To overcome this problem a circuit using a tunnel diode is designed. This circuit is likely to display two large spiral cascades over a wide range of parameter settings. It is demonstrated that such a display can be easily observed experimentally.

The tenth chapter, **Study of the dynamics of atmospheric pollution and its association with environmental parameters**, by Siwek Krzysztof, Osowski Stanislaw, and Swiderski Bartosz investigates the dynamic processes of creation of atmospheric pollution due to SO2, NO2 and ozone propagation. The interdependence between the pollution and the environmental atmospheric parameters (e.g. temperature, wing, humidity, insolation, etc) is investigated. It is demonstrated that the interdependence is nonlinear and could lead to chaotic dynamics. Finally, the results of these studies are exploited in the development of prediction models of the concentration of particular pollutants.

The eleventh chapter, System dynamics modeling of intelligent transportation systems – Human and social requirements for the construction of dynamic hypotheses, by Oana Mitrea, structures the knowledge and requirements for the (system dynamics) modeling of distributed actions in the intelligent transportation systems from the perspective of the sociology of technology. This chapter concentrates on the potential of the theory of the distributiveness of actions in heterogeneous constellations (sociology of technology) to enhance the formulation of dynamic hypotheses for system dynamics modeling of ITS. It is demonstrated that the performance of ITS systems depend not only on the interactivity between the vehicle-human systems and the environment, but also on the impact of socio-political and other contextual factors from a broader socio-technical constellation.

The twelfth chapter, **Fundamentals of electrodynamics**, by Branko Miškovic, develops some fundamental basics of the electromagnetics (EM) theory. Some lights are shed on important basic aspects of the EM theory which were not clearly explained by the relevant literature. This is achieved considering three classes of EM quantities namely, electric charges, currents, and kinetic magnetic fields. A new mathematical expression of the general kinetic central law is derived using the concept developed in this chapter. Further, the secondary field transformations are also completed. The relativistic postulates and respective kinematical transformations, based on the incomplete field transformations, are finally disqualified. On the other hand, Einstein's equation and Lorentz' mass function are explained and validated. The circular spatial axes are also explained and validated by the wave cosmic process, which determines by itself gravitational attraction, with unique and uniform lapse of time, in agreement with the Galilean view.

The thirteenth chapter, **Advanced adaptive algorithms in 2D finite elements method of higher order of accuracy**, by Pavel Karban, Ivo Doležel, František Mach, and Bohuš Ulrych develops a new method for automatic adaptivity in finite element methods of higher order of accuracy. The main attention is devoted to hp-adaptivity techniques that exhibit the highest level of flexibility and exponential convergence of results. The technologies are implemented in the adaptive FEM codes and finally both Agros2D and Hermes2D based on a higher-order finite element method are illustrated by three typical examples.

The fourteenth chapter, **SPICE model for fast time domain simulation of power transformers, exploiting the ferromagnetic hysteresis and eddy-currents**, by Lucian Mandache, Dumitru Topan, Mihai Iordache, and Ioana Gabriela Sirbu, considers both modeling and simulation (using SPICE) of power transformers. Specific states are depicted, namely saturation regime, static hysteresis and eddy currents. The principle of magnetic circuit modeling is based on analog lumped equivalent circuits and the SPICE implementation uses the principle of modularity. It is demonstrated that the method developed in this chapter allows simulation of normal operation modes, as well as critical transients and faulty conditions. A case study is presented to prove the feasibility, usefulness and accuracy of the proposed modeling and simulation approach, where the SPICE results are validated by experimental ones.

The fifteenth chapter, **Hard-coupled modeling of induction shrink fit of gas-turbine active wheel**, by Václav Kotlan, Pavel Karban, Bohuš Ulrych, Ivo Doležel, and Pavel Kus, develops a hard-coupled model of induction heating of a ferromagnetic disk. The mathematical model is derived from three coupled partial differential equations (for the distribution of electromagnetic field, temperature field and field of thermo-elastic displacements) whose coefficients are temperature-dependent functions. The finite element method is implemented into the Hermes and Agros codes and numerical solutions are obtained. The methodology is illustrated by a typical example, i.e., heating of an active wheel of a high-speed gas turbine that is to be hot-pressed on a shaft with the aim of obtaining a shrink fit allowing transferring the given torque at the nominal revolutions. Finally the parameters of the heating process in transverse and longitudinal magnetic fields are evaluated and discussed.

The sixteenth chapter, **Stability analysis and limit cycles of high order sigma-delta modulators**, by Valeri Mladenov, considers the stability analysis of limit cycle oscillations in a one bit high order sigma-delta modulators. An approach is developed and applied to sigma-delta modulators as a general analysis framework. The approach developed is based on a parallel decomposition of the modulator and a direct nonlinear systems analysis. In this representation, the general N-th order modulator is transformed into a decomposition of low order, generally complex modulators, which interact only through the quantizer function.

The seventeenth chapter, **Stability analysis of vector equalization based on recurrent neural networks**, by Mohamad Mostafa, Werner G. Teich, and Jürgen Lindner, focuses on an application of RNNs in communications engineering, namely the vector equalization. The importance of this procedure arises from the fact that there is no need for training. The parameters of the RNN to act as vector equalizer can be obtained by investigating the stability properties of these networks and by choosing a suitable activation function, which is the core of this work. Both global and local stability of the discrete-time and continuous-time RNN in the context of a vector equalizer are analyzed and the impact of the stability conditions on the equalization process is depicted. It is shown that the global stability is interesting if the weight matrix is not symmetric. Finally, it is demonstrated that a vector equalizer based on RNNs with increasing slope of the activation function during the iteration process can be interpreted as a scheme with a transition from global stability to local stability.

The eighteenth chapter, **Stability of linear circuits with interval data: a case study**, by Zygmunt A. Garczarczyk analyses the stability of a linear lumped electric circuit with interval data that model uncertainties of their element parameters (passive element values R, L, C and controlled source coefficients k). An approach is developed based on the checking of the stability of a symmetric interval matrix associated with the state matrix and on some interval analysis results. The method is demonstrated to be appropriate for the stability analysis. It's depicted from both numerical and experiment investigations that the procedure works effectively for some interval state matrix.

The nineteenth chapter, **Data reconciliation and bias estimation in on-line optimization**, by Moufid Mansour, develops techniques for bias estimation (BE) and data reconciliation (DR) for the detection, estimation and elimination of biases and random errors. It is demonstrated that these techniques can be successfully employed within an online Integrated System Optimization and Parameter Estimation (ISOPE) scheme for the determination of the process optimum, despite the existence of model-reality differences and measurement errors. The performance of the resulting scheme is demonstrated by applying it to the optimization of a two tank CSTR system. The robustness of the optimization scheme is confirmed when applying this scheme to particular process data with multiple biases and noise. Finally, it is demonstrated that the BE and DR techniques are suitable for performing robust online optimization event in cases of noisy and/or errors data.

The twentieth chapter, **Image edge detection and orientation selection with coupled nonlinear excitable elements**, by Atsushi Nomura, Yoshiki Mizukami, Koichi Okada and Makoto Ichikawa, develops an algorithm for edge detection and orientation selection with a grid system consisting of coupled nonlinear excitable elements. The model uses the well-known Fitz-Hugh-Nagumo coupled equations. Two types of coupling are considered i.e. strong inhibitory coupling and strong nonlinearity induced stationary pulses. Using this latter type of coupling the algorithm for edge detection is developed based on a technique where the grid system self-organizes a pulse pattern at edges in an initial condition. This algorithm is further experimentally applied (with a great success) to artificial binary and real images. Finally a benchmarking is performed by comparing the results provided by the algorithm developed in this wok with some results proposed by the relevant literature.

The chapter twenty-first, **Consecutive repeating state cycles determined periodic points in a Turing Machine**, by Michael Stephen Fiske studies the Turing machine with new methods motivated by the notion of recurrence in classical dynamical systems theory. The state cycle of a Turing machine is introduced. It is proven that each consecutive repeating state cycle in a Turing machine determines a unique periodic configuration (point) and vice versa. This characterization is a periodic point theorem for Turing machines. Using the notion of a prime directed edge and a mathematical operation called edge pattern substitution, a search (or optimization) procedure finds consecutive repeating state cycles. Both periodic and aperiodic states are depicted and these states alternate together when monitoring a control parameter.

Klagenfurt,
August 2012

The Guest Editors

# Contents

# Part I

# Theory of Chaos and Synchronization and Applications in Mechanics, Transportation, Communication and Security Related System Concepts

# Chapter 1
# Synchronization of Two Nonidentical Clocks: What Huygens was Able to Observe?

Krzysztof Czolczynski, Przemysaw Perlikowski,
Andrzej Stefanski, and Tomasz Kapitaniak

**Abstract.** We consider the synchronization of two clocks which are accurate (show the same time) but have pendulums with different masses. We show that such clocks hanging on the same beam can show the almost complete (in-phase) and almost antiphase synchronizations. By almost complete and almost antiphase synchronization we defined the periodic motion of the pendulums in which the phase shift between the displacements of the pendulums is respectively close (but not equal) to $0^\circ$ or $180^\circ$ . We give evidence that almost antiphase synchronization was the phenomenon observed by Huygens in XVII century. We support our numerical studies by considering the energy balance in the system and showing how the energy is transferred between the pendulums via oscillating beam allowing the pendulums' synchronization.

## 1.1 Introduction

In the 60-ties of XVII century the longitude problem, i.e., finding a robust, accurate method of the longitude determination for marine navigation was the outstanding challenge. Huygens believed that pendulum clocks, suitably modified to withstand the rigors of the sea, could be sufficiently accurate to reliably determine the longitude (The discrepancy in the clock rate equal to one oscillation of the second pendulum (pendulum with a period of oscillations equal to 1 [s] in a day corresponds to the error in the longitude determination, approximately equal to 500 m in a day (at the equator).). In a letter to the Royal Society of London of 27 February 1665 Huygens described his famous experiment which showed the tendency of two pendulums (of the clocks) to synchronize, or anti-synchronize when mounted together on the same beam (*Huygens, 1665*). Originally, he used the phrase "an odd kind of sympathy" to

Krzysztof Czolczynski · Przemysław Perlikowski · Andrzej Stefanski · Tomasz Kapitaniak
Division of Dynamics, Technical University of Lodz,
Stefanowskiego 1/15, 90-924 Lodz, Poland
e-mail: `tomasz.kapitaniak@p.lodz.pl`

K. Kyamakya et al. (Eds.): Selected Topics in Nonlinear Dynamics, SCI 459, pp. 3–18.
DOI: 10.1007/ 978-3-642-34560-9_1 

describe the observed behavior in two maritime clocks. The original drawing showing this experiment is shown in Figure 1.1. Two pendulums, mounted together, will always end up swinging in exactly opposite directions, regardless of their respective individual motion. This was one of the first observations of the phenomenon of the coupled harmonic oscillators, which have many applications in physics (*Pikovsky et al., 2001, Blekham, 1988; Strogatz & Steward, 1993, Golubitsky et al. 1999*). Huygens originally believed the synchronization occurs due to air currents shared between two pendulums, but later after performing several simple tests he dismissed this and attributed the sympathetic motion of pendulums to imperceptible movement in the beam from which both pendulums are suspended.

Huygens' study of two clocks operating simultaneously arose from the very practical requirement of the redundancy: if one clock stopped, had to be cleaned or winded up, then the other one provided the proper timekeeping (*Huygens 1669*). Ultimately, the innovation of the pendulum did not solve the longitude problem, since slight and almost insensible motion was able to cause an alteration in their going (*Birch, 1756; Britten, 1973*).

Recently, this idea has been validated by a few groups of researchers which tested Huygens' idea (*Bennet et al., 2002, Pogromsky et al. (2003), Kanunnikov & Lamper, 2003, Senator, 2006, Dilao, 2009, Kumon et al., 2002, Fradkov, and B. Andrievsky, 2007, J. Pantaleone, 2002, Ulrichs et al., 2009, Czolczynski et al., 2009a,b, 2011a,b*). These studies do not give the definite answer to the question; what Huygens was able to observe, e.g., *Bennet et al. 2002* stated that to repeat Huygens' results, the high precision (the precision that Huygens certainly could not achieve) is necessary and *Kanunnikov & Lamper, 2003* showed that the precise antiphase motion of different pendulums noted by Huygens cannot occur. Different pendulums (possibly with different masses) were definitely used by Huygens as can be seen in his drawing shown in Figure 1.2. In this paper we consider the synchronization of two clocks which have pendulums with the same length but different masses. Such clocks are accurate, i.e., show the same time as both pendulums have the same length. We show that two such clocks hanging on the same beam can show the almost complete (in-phase) and almost antiphase synchronizations. By almost

**Fig. 1.1** An original drawing of Huygens illustrating his expermiments with pendulum clocks

complete and almost antiphase synchronization we defined the periodic motion of the pendulums in which the phase shift between the pendulums displacements is respectively close (but not equal) to 0° or 180°. We give evidence that the almost antiphase synchronization of the clocks was the phenomenon which Huygens observed in XVII century. This paper is organized as follows. Sec. 1.2 describes the model of the clocks which has been used. In Sec. 1.3 we derive the energy balance of the synchronized pendulums. Section 1.4 presents the results of our numerical simulations and describes the observed synchronizations states. Finally, we discuss the energy balance of the synchronized pendulums and summarize our results in Sec. 1.5.

## 1.2 Model

The analyzed system is shown in Figure 1.3. It consists of the rigid beam and two pendulum clocks suspended on it. The beam of mass $M$ can move in a horizontal direction, its movement is described by coordinate $x$. The mass of the beam is connected to the refuge of a linear spring and linear damper $k_x$ and $c_x$. The clocks' pendulum consists of the light beam of the length $l$ and mass mounted at its end. We consider the pendulums with the same length $l$ but different masses $m_1$ and $m_2$. The same length of both pendulums guarantees that the clocks are accurate, i.e., both show the same time. The motion of the pendulums is described by angles $\varphi_1$and$\varphi_2$ and is damped by the dampers(not shown in Figure 1.3) with the same damping coefficients $c_\varphi$.

The pendulums are driven by the escapement mechanism described in details in (*Rawlings, 194, Lepschy et al., 1993, Roup et al., 2003, Moon and Stiefel, 2006, Czolczynski et al. 2009b*). Notice that when the swingng pendulums do not exceed

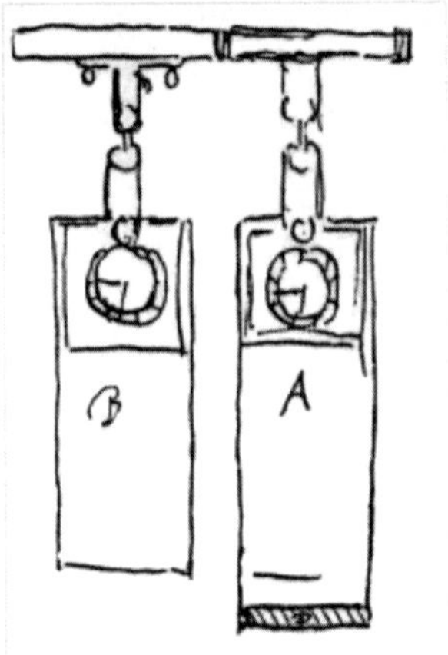


**Fig. 1.2** Details of the Huygens' experiment; it is clearly visible that the clocks used in the experiment have not been identical, D denotes the weight used to stabilize the clock case

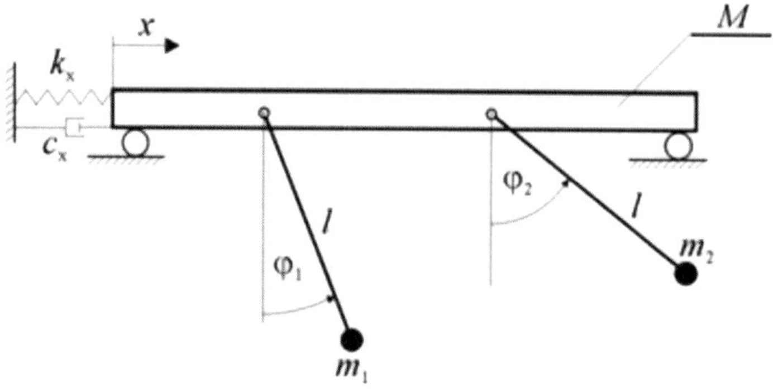


**Fig. 1.3** The model of the system – two pendulum clocks are mounted to the beam which can move horizontally

certain angle $\gamma_N$, the escapement mechanisms generate the constant moments $M_N$ (the same for both pendulums). This mechanism acts in two successive steps i.e., the first step is followed by the second on and the second one by the first one (detailed description of the escapement mechanism has been given in our previous work (*Czolczynski el al., 2009b*)). In the first step if $0 < \varphi_i < \gamma_N$ (i=1,2) then $M_D = M_N$ and when $\varphi_i < 0$ then $M_D = 0$. For the second stage one has for $-\gamma_N < \varphi i < 0 M_D = -M_N$ and for $\varphi i > 0 M_D = 0$. The energy supplied by the escapement mechanism balance the energy dissipated due to the damping. The parameters of this mechanics have been chosen in the way that for the beam $M$ at the rest both pendulums perform the oscillations with the same amplitude. Typically pendulum clocks oscillate with amplitude smaller then $2\pi/36$ and for the clocks with long pendulums like marine clocks this amplitude is even smaller (*Rawlings, 1994*). Since the damping coefficients of the two pendulums are identical, in the case of unmovable beam the pendulums oscillate with the same amplitude. the movement of the beam may change both the period and the amplitude of pendulums' oscillations.

The equations of motion of considered systems are as follow:

$$m_1 l^2 \ddot{\varphi} + m_1 \ddot{x} l \cos\varphi_1 + c_\varphi \dot{\varphi}_1 + m_1 g l \sin\varphi_1 = M_D$$
$$m_2 l^2 \ddot{\varphi} + m_2 \ddot{x} l \cos\varphi_2 + c_\varphi \dot{\varphi}_2 + m_2 g l \sin\varphi_2 = M_D \qquad (1.1)$$

$$\left(M + \sum_{i=1}^{2} m_i\right) \ddot{x} + c_x \dot{x} + k_x x + \sum_{i=1}^{2} m_i l (\ddot{\varphi}_i \cos\varphi_i - \dot{\varphi}_i^2 \sin\varphi_i) = 0 \qquad (1.2)$$

Note that the escapement mechanism acts only when the amplitude of the pendulum oscillations is larger than $\gamma_N$.

## 1.3 Energy Balance

### 1.3.1 Energy Balance of the Pendulum

Multiplication of both sides of eq.(1.1) by the angular velocity of the $i$-th pendulum gives:

$$m_i l^2 \ddot{\varphi}_i \dot{\varphi}_i + m_i g l \dot{\varphi}_i \sin\varphi_i = M_D \dot{\varphi_i^2} - m_i \ddot{x} l \cos\varphi_i \dot{\varphi}_i, \quad i = 1,2 \tag{1.3}$$

In the case of the periodic motion of the pendulums after integration eq.(1.3) gives the energy balance of the $i$-th pendulum in one period of motion:

$$\begin{aligned} &\int_0^T m_i l^2 \ddot{\varphi}_i \dot{\varphi}_i dt + \int_0^T m_i g l \dot{\varphi}_i \sin\varphi_i dt = \\ &\int_0^T M_D \dot{\varphi}_i dt - \int_0^T c_\varphi \dot{\varphi_i^2} dt - \int_0^T m_i \ddot{x} l \cos\varphi_i \dot{\varphi}_i dt, \quad i = 1,2 \end{aligned} \tag{1.4}$$

Left hand side of eq.(1.4) represents the decrease of the total energy of the $i$-th pendulum. In the case of the periodic behavior of the system(1,2) this decrease is equal to zero, so

$$\int_0^T m_i l^2 \ddot{\varphi}_i \dot{\varphi}_i dt + \int_0^T m_i g l \dot{\varphi i} \sin\varphi_i dt = 0, \quad i = 1,2 \tag{1.5}$$

The work done by the escapement mechanism during one period of pendulum's oscillations can be expressed as

$$W^{DRIVE} = \int_0^T M_D \dot{\varphi}_i dt = 2\int_0^{\gamma_N} M_N d\varphi_i = 2M_N \gamma_N, \quad i = 1,2 \tag{1.6}$$

Note that $W^{DRIVE}$ does not depend on the pendulum's masses, displacement $\varphi_{1,2}(\varphi_{1,2} > \gamma_N)$ and velocity. Energy dissipated in the damper is given by

$$W_i^{DAMP} = \int_0^T c_\phi \dot{\phi}_i^2 dt, \quad i = 1,2 \tag{1.7}$$

The last component of eq.(1.7) represents the energy transferred from the $i$-th pendulum to the beam $M$ (the pendulum loses part of its energy to force the beam to oscillate), so we have:

$$W_i^{SYN} = \int_0^T m_i \ddot{x} l \cos\phi_i \dot{\phi}_i dt, \quad i = 1,2 \tag{1.8}$$

Note that in the case of pendulums in the state of complete synchronization ($\phi_1(t) = \phi_2(t)$), energies dissipated by dampers are equal and energies transmitted to the beam are proportional to pendulums' masses. Substituting eqs. ((1.5)-(1.8)) into eq.(1.4),

$$W^{DRIVE} - W_1^{DAMP} - W_1^{SYN} = 0, W^{DRIVE} - W_2^{DAMP} - W_2^{SYN} = 0, \quad (1.9)$$

one gets energy balances for pendulums.

### *1.3.2 Energy Balance of the Beam and Whole System (1,2)*

Multiplying equation of the beam motion (2) by beam velocity $\dot{x}$ one gets:

$$\left(M + \sum_{i=1}^{2} m_i\right)\ddot{x}\dot{x} + c_x\dot{x}^2 + k_x x\dot{x} + \left(\sum_{i=1}^{2} m_i l(\ddot{\varphi}_i \cos\varphi_i - \dot{\varphi}_i^2 \sin\varphi_i\right)\dot{x} = 0 \quad (1.10)$$

Integrating eq.(1.10)) over the period of oscillations we obtain the following energy balance:

$$\begin{aligned} &\int_0^2 \left(M + \sum_{i=1}^{2} m_i\right)\ddot{x}\dot{x}dt + \int_0^2 k_x x\dot{x}dt + = \\ &-\int_0^T \left(\sum_{i=1}^{2} m_i l(\ddot{\varphi}_i \cos\varphi_i - \dot{\varphi}_i^2 \sin\varphi_i\right)\dot{x}dt = -\int_0^T c_x\dot{x}^2 dt \end{aligned} \quad (1.11)$$

Left hand side of eq.(1.11) represents the increase of the total energy of the beam which for the periodic oscillations is equal to zero:

$$\int_0^2 \left(M + \sum_{i=1}^{2} m_i\right)\ddot{x}\dot{x}dt + \int_0^2 k_x x\dot{x}dt + = 0 \quad (1.12)$$

The first component on the right-hand side of eq.(1.11) represents the work performed by the horizontal component of the force with which the pendulums act on the beam causing its motion:

$$W_{beam}^{DRIVE} = -\int_0^T \left(\sum_{i=1} 2m_i l(\ddot{\varphi}_i \cos\varphi_i - \dot{\varphi}_i^2 \sin\varphi_i)\right)\dot{x}dt. \quad (1.13)$$

The second component on the right hand side of eq.(1.11) represents the energy dissipated by the damper $c_x$:

$$W_{beam}^{DAMP} = \int_0^T c_x\dot{x}^2 dt. \quad (1.14)$$

Substituting eqs.((1.12)-(1.14)) into eq.(1.11) one gets energy balance in the following form

$$W_{beam}^{DRIVE} - W_{beam}^{DAMP} = 0. \quad (1.15)$$

In the case of the periodic oscillations it is possible to prove that

$$W_1^{SYN} + W_2^{SYN} = W_{beam}^{DRIVE}. \tag{1.16}$$

Adding eqs.(1.9) and (1.15)

$$2W^{DRIVE} - W_1^{DAMP} - W_2^{DAMP} - W_1^{SYN} - W_2^{SYN} + W_{beam}^{DRIVE} - W_{beam}^{DAMP} = 0,$$

and considering eq.(1.16) one obtains

$$2W^{DRIVE} - W_1^{DAMP} - W_2^{DAMP} - W_{beam}^{DAMP} = 0, \tag{1.17}$$

Eq.(1.17) represents the energy balance of the whole system (1,2).

## 1.4 Numerical Results

In our numerical simulations eqs ((1.1),(1.2)) have been integrated by the Runge-Kutta method. The stability of the obtained synchronous states has been investigated using the variational equations as described in (Czolczynski et. al. 2009(a,b)).

We used the following system parameters: mass of pendulum 1 $m_1 = 1.0$[kg]; the length of the pendulums $l = g/4\pi^2$=0.2485 [m] (it has been selected in such a way when the beam $M$ is at rest, the period of pendulum oscillations is equal to $T$=1.0 [s] and oscillations frequency to $\alpha = 2\pi[s^{-1}]$), $g$=9.81[m/s$^2$] is an acceleration due to the gravity, beam mass $M$=10.0[kg], damping coefficients $c_\phi$=0.01 [Ns] and $c_x$=1.53 [Ns/m] and stiffness coefficient $k_x$=4.0 [N/m]. When the displacements of the pendulums are smaller than $\gamma_N$=5.0°, the escapement mechanisms generate driving moments $M_N = 0.075$[Nm] and allow the pendulums to oscillate with amplitude $\Phi_1 = \Phi_2 = \Phi = 0.2575$ ($\approx 14.75°$) when beam $M$ is at rest. The mass $m_2$ of the pendulum 2 has been taken as a control parameter. It varied in the wide interval to examine the influence of its changes on the type of the observed synchronization.

### *1.4.1 From Complete to (Almost) Antiphase Synchronization*

The evolution of system (1,2) behavior starting from complete synchronization of identical clocks ($m_1$=$m_2$=1.0 [kg]) and the increase of the values of control parameter $m_2$ is illustrated in Figure 1.4(a-d). In Figure 1.4(a) we present the bifurcation diagram of the system (1,2). The mass of pendulum 2 – $m_2$ has been taken as a control parameter and it increases in the interval [1.0, 41.0]. In the initial state ($m_1 = m_2$) the pendulums exhibit complete synchronization ($\phi_1 = \phi_2$) and the beam moves in contrary phase to the pendulums. The increase of the bifurcation parameter leads to the reduction of the oscillations amplitudes of both pendulums which are in the state almost complete synchronization; their movements are not identical, but very close to identical as shown in Figure 1.4(b). Figure 1.4(b) shows the displacements of the pendulums $\phi_1$, $\phi_2$ and the beam $x$ (for better visibility enlarged 10 times) for $m_1$=1.0 [kg] and $m_2$=3.0 [kg]. Time on the horizontal axis is given in the following way $t = NT$, where $N$=1,2,3,... and $T$ is a period of pendulum's oscillations when the beam is at rest. Notice that $\phi_1 \approx \phi_2$ as the differences are hardy visible.

Further increase of the mass $m_2$ results in the further reduction of pendulums' amplitudes and the increase of the beam amplitude as can be seen in Figure 1.4(c) for the $m_2$=20.0 [kg]. The period of pendulums' oscillations decreases (in Figure 1.4(b) we observe 11.5 periods while in Figure 1.4(c) 17.5 in the same time interval). This reduction is due to the fact that while increasing mass of pendulum 2, the center of the mass moves towards the ends of the pendulums, i.e., towards the material points with masses $m_1$ and $m_2$ and moves away from the beam with constant mass $M$. For $m_2$=30.3 [kg] the amplitude of pendulums' oscillations decreases to limit $\Phi \approx \gamma_N = 5°$, below which the escapement mechanism is turned off. For larger values of $m_2$, we observe oscillations of pendulum 1 with amplitude $\Phi \approx 15°$, and small oscillations of pendulum 2 (whose escapement mechanism is turned off as can be seen in Figure 1.4(d). The pendulum moves due to the energy supplied to it by pendulum 1 via the beam. The bifurcation diagram of Figure 1.4(a) shows the existence of: (i) complete synchronization for $m_1$=$m_2$=1.0 [kg], (ii) almost complete synchronization for 1.0 [kg] $< m_2 <$ 30.3 [kg], (iii) almost antiphase synchronization, with one pendulum with turned off escapement mechanism (in Figure 1.4(d) pendulum 2 has turned off mechanism) for $m_2 >$ 30.3 [kg].

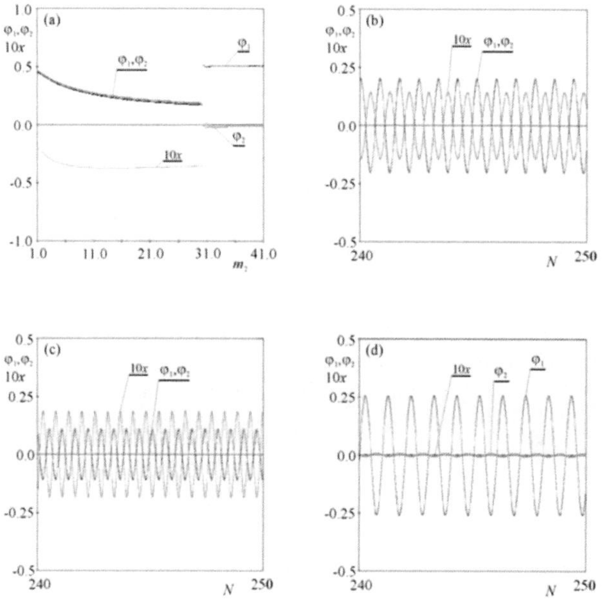


**Fig. 1.4** Evolution from complete to almost antiphase synchronization; (a) bifurcation diagram for increasing values of $m_2$, (b) time series of almost complete synchronization $m_1$=1.0 [kg] and $m_2$=3.0 [kg]; (c) time series of almost complete synchronization $m_1$=1.0 [kg] and $m_2$=20.0 [kg]; (d) time series of almost antiphase synchronization for $m_1$=1.0 [kg] and $m_2$=35.0 [kg]– the escapement mechanism of pendulum 2 is turned off

### 1.4.2 From Complete Synchronization to Quasiperiodic Oscillations

Evolution of system (1,2) behavior starting from complete synchronization of identical pendulums ($m_1$=$m_2$=1.0 [kg]) and the decrease of the values of the control parameter $m_2$ is illustrated in Figure 1.5(a-d). Figure 1.5(a) shows the bifurcation diagram for decreasing values of mass $m_2$ ($m_2 \in$[0.01,1.00]). In the interval 1.0 [kg] > $m_2$ >0.25 [kg], both pendulums are in a state of almost-complete synchronization. Their oscillations are "almost identical" as can be seen in Figure 1.5(b) for $m_1$=1.0 [kg] and $m_2$=0.3 [kg], the differences between the amplitudes and phases of $\phi_1$ and $\phi_2$ are close to zero, both pendulums remain in the (almost) antiphase to the oscillations of the beam. Further reduction of mass $m_2$ leads to the loss of synchronization and the motion of the system becomes quasiperiodic as shown in Figure 1.5(c). Figure 1.5(d) presents the Poincare map (the displacements and velocities of the pendulums has been taken at the moments of the greatest positive displacement of the first pendulum) for $m_2$=0.2 [kg]. In summary, the bifurcation diagram of Figure 1.5(a) shows the existence of: (i) complete synchronization for $m_1$=$m_2$=1.0 [kg], (ii) almost complete synchronization for 1.0 [kg] > $m_2$ >0.25 [kg], (iii) the lack of synchronization and a quasiperiodic oscillations for $m_2$ <0.25 [kg].

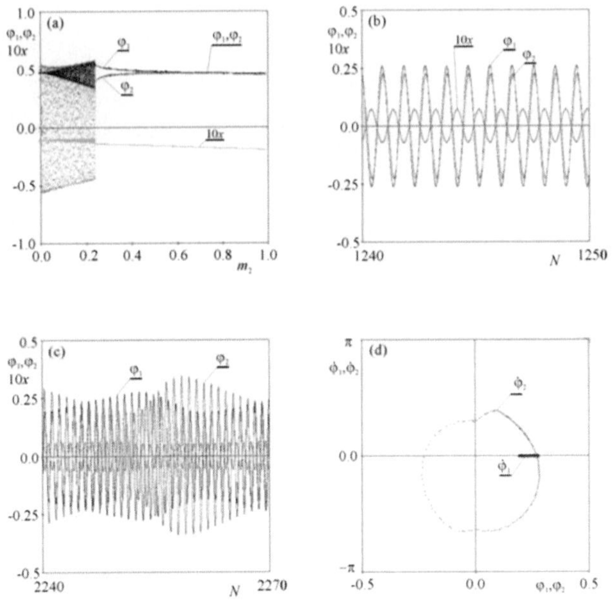

**Fig. 1.5** Evolution from complete synchronization to quasiperiodic oscillations; (a) bifurcation diagram for increasing values of $m_2$, (b) time series of almost complete synchronization for $m_1$=1.0 [kg] and $m_2$=0.3 [kg], (c) time series of quasiperiod oscillations for $m_1$=1.0 [kg] and $m_2$=0.2 [kg], (d) Poincare map showing quasiperiodic oscillations for $m_1$=1.0 [kg] and $m_2$=0.2 [kg]

### 1.4.3 From Antiphase to Almost-Antiphase Synchronization

The evolution of system (1,2) behavior starting from antiphase synchronization of identical pendulums ($m_1$=$m_2$=1.0) and the increase of the values of the control parameter $m_2$ is illustrated in Figure 1.6(a-c). Figure 1.6(a) presents another bifurcation diagram for the increasing values of $m_2$ ($m_2 \in$[1.0,41.0]). This time we start with a state of antiphase synchronization of pendulums with masses $m_1$=$m_2$=1.0 [kg], during which two pendulums are moving in the opposite way, such that $\phi_1$(t)=$-\phi_2$(t) and the beam is at rest. The increase of bifurcation parameter $m_2$ leads to the increase of the oscillations amplitude of pendulum 1 (with constant mass ($m_1$=1.0 [kg]) and decrease of the amplitude of the pendulum 2 (with increasing mass $m_2$). The pendulums remain in the state of almost-antiphase synchronization as phase shift between their displacements is close to 180° as shown in Figure 1.6(b) for $m_1$=1.0 [kg] and $m_2$=2.0 [kg]. In Figure 1.6(b) displacements $\phi_1$ and $\phi_2$ are almost in antiphase (difference between antiphase and almost antiphase is hardly visible). The beam is oscillating and its displacement $x$ is shifted by approximately 90$deg$ (respectively forward and backward) in relation to the pendulums' displacements. The beam oscillations are caused by the energy transmitted to the beam by pendulum 2. Part of this energy is dissipated in the beam damper $c_x$ and

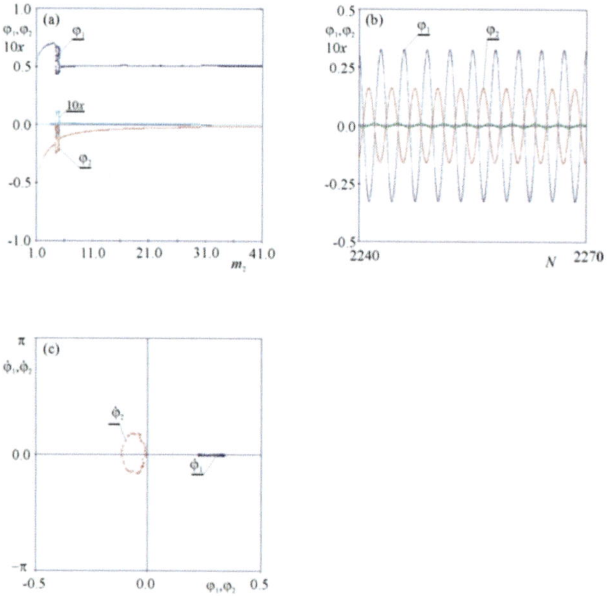


**Fig. 1.6** Evolution from antiphase to almost antiphase synchronization; (a) bifurcation diagram for increasing values of $m_2$, (b) time series of almost antiphase synchronization for $m_1$=1.0 [kg] and $m_2$=2.0 [kg], (c) Poincare map of quasiperiodic oscillations for $m_1$=1.0 [kg] and $m_2$=5.0 [kg]

part is transferred to pendulum 1. As the results of this transfer pendulum 1 oscillates with the amplitude larger than initial $\Phi \approx 15^{\circ}$ (see discussion in Sec.1.3). When mass $m_2$ reaches a value equal to 4.45 [kg] the amplitude of pendulum 2 oscillations decreases below the critical value $\Phi \approx \gamma_N = 5^{\circ}$. In the interval 4.45 [kg] $< m_2 <$5.3 [kg] we observe a quasiperiodic oscillations of the system (1,2), in this range the escapement mechanism of pendulum 2 is turned off. The example of quasiperidic oscillations is shown in Figure 1.6(c). There are two irregular phases of these oscillations: (i) after turning off the escapement mechanism of pendulum 2, it does not provide energy to pendulum 1, the amplitude of pendulum 1 oscillations decreases and simultaneously the amplitude of pendulum 2 oscillations increases and its escapement mechanism is turned on, (ii) after turning on of the escapement mechanism of pendulum 2, it provides the energy to pendulum 1, which causes the reduction of pendulum 2 amplitude and leads to the turn off of the escapement mechanism. For $m_2 >$5.3 [kg] the escapement mechanism of pendulum 2 is still turned off and the system tends to almost antiphase synchronization. The bifurcation diagram of Figure 1.6(a) shows the existence of: (i) antiphase synchronization for $m_1$=$m_2$=1.0 [kg], (ii) almost-antiphase synchronization for 1.0 [kg] $< m_2 <$4.45 [kg], (iii) the lack of synchronization and quasi-periodic oscillations for 4.45 [kg]$< m_2 <$5.3 [kg].(iv) almost-antiphase synchronization with the turn off of the escapement mechanism of pendulum 2 for $m_2 >$5.3 [kg].

## 1.4.4 From Antiphase Synchronization to Quasiperiodic Oscillations

The evolution of system (1,2) behavior starting from antiphase synchronization of identical pendulums ($m_1$=$m_2$=1.0 [kg]) and the decrease of the values of the control parameter $m_2$ is illustrated in Figure 1.7(a,b)). Figure 1.7(a) shows the bifurcation diagram of the system (1,2) for decreasing values of $m_2$ ($m_2$ decreases from an initial value 1.0 [kg] up to 0.01 [kg]). We start from the state of antiphase synchronization observed for $m_1$=$m_2$=1.0 [kg]. In the interval 1.0 [kg]$> m_2 >$0.38 [kg]

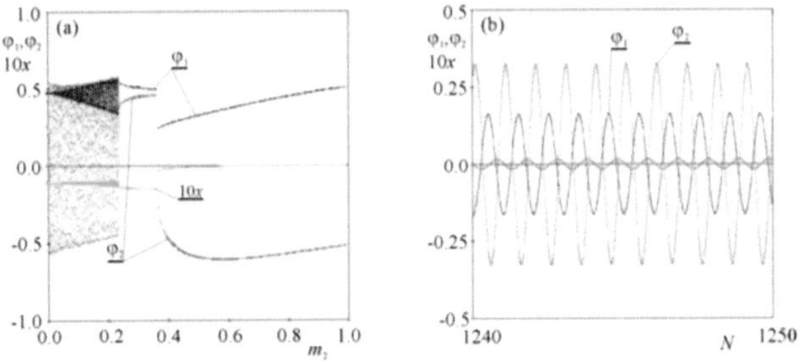


**Fig. 1.7** Evolution from antiphase synchronization to quasiperiodic oscillations; (a) bifurcation diagram of system (1,2) for decreasing m2, (b) time series of almost antiphase synchronization for $m_1$=1.0 [kg], $m_2$=0.5 [kg]

both pendulums are in the state of almost antiphase synchronization. Its displacements are out of phase by an angle close to 180° as shown in Figure 1.7(b) showing the time series of $\phi_1$, $\phi_2$ and $x$ for $m_1$=1.0 [kg] and $m_2$=0.5 [kg]. When the mass $m_2$ decreases from 1.0 [kg] to 0.56 [kg], the amplitude of oscillations of pendulum 1 decreases and the amplitude of oscillations of pendulum 2 increases. In the interval 0.56 [kg]> $m_2$ >0.38 [kg] we observe fast decrease of pendulum 2 amplitude up to the limit value $\Phi \approx \gamma = 5°$. For $m_2$=0.38 [kg] system (1,2) changes the type of the synchronization to almost complete (previously observed in Figure 1.5(b) and described in Sec.1.4.3). Further reduction of $m_2$ leads to the quasi-periodic motion of the system (as described in Sec.1.4.3). The bifurcation diagram of Figure 1.7(a) shows the existence of: (i) antiphase synchronization for $m_1 = m_2$=1.0 [kg], (ii) almost antiphase synchronization for 1.0 [kg]> $m_2$ > 0.38 [kg], (iii) almost complete synchronization for 0.38 [kg]> $m_2$ >0.25 [kg], (iv) the lack of synchronization and quasiperiodic oscillations for $m_2$ <0.25 [kg].

## 1.5 From Complete to (Almost) Antiphase Synchronization

In the considered system of two clocks suspended on the horizontally movable beam we identified the following types of synchronizations: (i) the complete synchronization during which the pendulums' displacements fulfill the relation $\phi_1(t) = \phi_2(t)$ and the phase shift between pendulums displacements $\phi_1(t)$ and $\phi_2(t)$ is equal to zero, (ii) the almost complete synchronization during which the pendulums' displacements fulfill the relation $\phi_1(t) \approx \phi_2(t)$ and the phase shift between pendulums' displacements $\phi_1(t)$ and $\phi_2(t)$ is close to zero, (iii) antiphase synchronization during which the pendulums' displacements fulfill the relation $\phi_1(t) = \phi_2(t)$ and the phase shift between the pendulum displacements is equal to 180°, (iv) almost antiphase synchronization during which the phase shift between the pendulum displacements is close to 180°. Additionally, in our previous work we find the possibility of long period synchronization during which the difference of the pendulums' displacements $\phi_1 - \phi_2$ is a periodic function of time and chaotic behavior of the clocks' pendulums (Czolczynski et al., 2011b). Note that types (i) and (iii) are possible only for non robust case of identical pendulum masses ($m_1 = m_2$). For cases (ii) and (iv) pendulums' energy balance looks differently as can be shown in Figure 1.8(a-c).

*(i) Energy balance in the state of almost complete synchronization*

In the state of complete synchronization pendulums' displacements fulfill the relation $\phi_1(t) = \phi_2(t)$, both dampers dissipate the same amount of energy and both pendulums transfer the same amount of energy to the beam:

$$\begin{aligned} W_1^{DAMP} &= \int_0^T c_\phi \dot{\phi}_1^2 dt = \int_0^T c_\phi \dot{\phi}_2^2 dt = W_2^{DAMP}, \\ m_2 W_1^{SYN} &= m_2 \int_0^T m_1 \ddot{x} l \cos\phi_1 \dot{\phi}_1 dt = m_1 \int_0^T m_2 \ddot{x} l \cos\phi_2 \dot{\phi}_2 dt = m_1 W_2^{SYN}. \end{aligned} \tag{1.18}$$

After substituting the energy values satisfying eqs.(1.18) into eqs.(1.9), eqs.(1.9) are not contradictory equations only in two cases: (i) masses of both pendulums are equal ($m_1 = m_2$) and both pendulums transmit the same amount of energy to the beam (see Figure 1.8(a)), (ii) synchronization energies $W_{1,2}^{SYN}$ are equal to zero, i.e., both pendulums dissipate the whole energy supplied by the escapement mechanism (see Figure 1.8(b)). In general case when $m_1 \neq m_2$ and $c_x \neq 0.0$, instead of complete synchronization we observe almost complete synchronizations during which the pendulums' displacements are not identical (but close to each other) and one gets the following expressions for considered energies

$$\begin{aligned} W_1^{DAMP} &= \int_0^T c_\phi \dot{\phi}_1^2 dt \approx \int_0^T c_\phi \dot{\phi}_2^2 dt = W_2^{DAMP}, \\ W_1^{SYN} &= \int_0^T m_1 \ddot{x} l \cos\phi_1 \dot{\phi}_1 dt \approx \int_0^T m_2 \ddot{x} l \cos\phi_2 \dot{\phi}_2 dt = W_2^{SYN}. \end{aligned} \tag{1.19}$$

which fulfills eqs.(1.9). The scheme of the energy balance is similar to the one in Figure 1.8(a).

*(ii). Energy balance in the state of almost antiphase synchronization*

In the state of antiphase synchronization the pendulums' displacements fulfill the relation $\phi_1(t) = -\phi_2(t)$ and both dampers dissipate the same amount of energy. The energies transmitted to the beam have the same absolute values but opposite signs, i.e.,

$$\begin{aligned} W_1^{DAMP} &= \int_0^T c_\phi \dot{\phi}_1^2 dt = \int_0^T c_\phi \dot{\phi}_2^2 dt = W_2^{DAMP}, \\ m_2 W_1^{SYN} &= m_2 \int_0^T m_1 \ddot{x} l \cos\phi_1 \dot{\phi}_1 dt = -m_1 \int_0^T m_2 \ddot{x} l \cos\phi_2 \dot{\phi}_2 dt = -m_1 W_2^{SYN} \end{aligned} \tag{1.20}$$

After substituting the energy values satisfying eqs.(1.20) into eqs.(1.9), eqs.(1.9) are not contradictory equations only when the beam acceleration is zero, which implies the zero value of its velocity and displacement (in the synchronization state of the behavior of the system is periodic). This condition requires a balancing of the forces which pendulums act on the beam, and this in turn requires that the pendulum have the same mass. The scheme of the energy balance is similar to the one in Figure 1.8(b). If the pendulums' masses are different, instead of antiphase synchronization we observe an almost-antiphase synchronization, during which the displacement pendulums have different amplitude and phase shift between these displacements is close, but not equal to 180°. Hence

$$W_1^{DAMP} \neq W_2^{DAMP}, W_1^{SYN} \neq W_2^{SYN} \tag{1.21}$$

The energy balance for the case of almost antiphase synchronization is shown in Figure 1.8(c). Part of the energy supplied by the escapement mechanism of pendulum 1 (let us assume that it has smaller mass) $W_1^{DRIVE}$ is dissipated by the damper of this pendulum ($W_1^{DAMP}$) and the rest ($W_1^{SYN}$) is transmitted to the beam. The damper of pendulum 2 dissipates the energy $W_2^{DRIVE}$ supplied by the escapements mechanism and energy ($W_2^{SYN}$) transmitted from the beam (mathematically

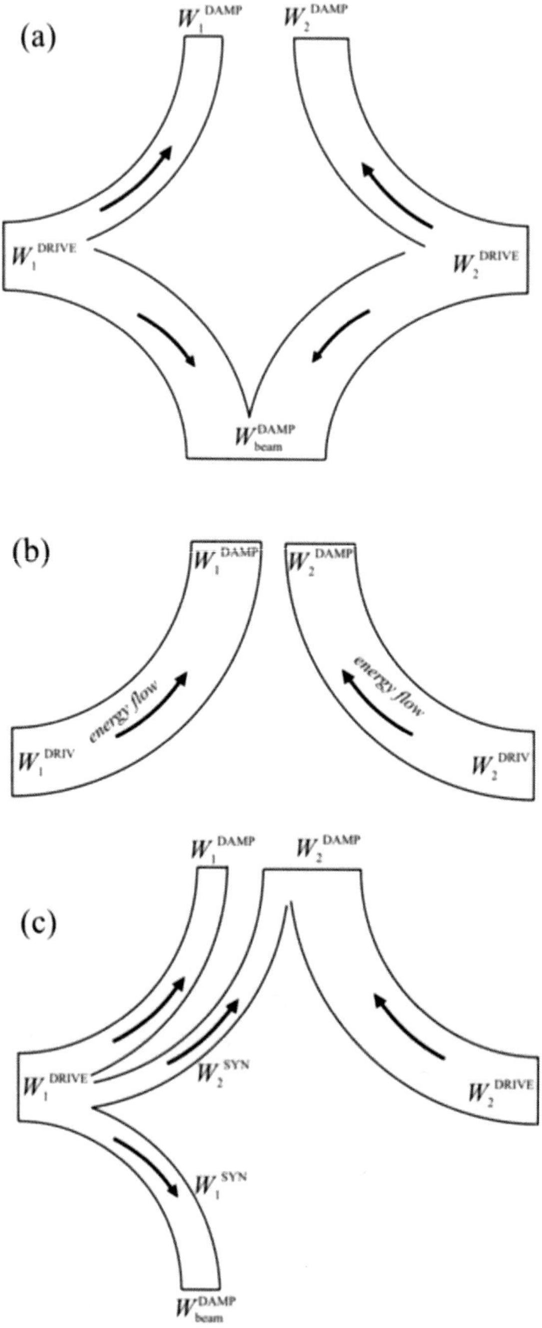


**Fig. 1.8** Energy balance schemes: (a) almost complete synchronization – both pendulums driver the beam, (b) antiphase synchronization – beam at rest, (c) almost antiphase synchronization – pendulum 1 drives the beam and supplies energy to pendulum 2

this energy is negative). The damper of the beam dissipates the rest of the energy $W_1^{SYN}$ : $W_{beam}^{SYN} = W_1^{SYN} - (-W_2^{SYN})$. Finally we give answer to the initial question (posed in the title): *our results give evidence that Huygens in his famous experiment was unable to observe antiphase synchronization as stated in his letters (Huygens, 1665) but observed almost antiphase synchronization of two pendulum clocks. In his times the distinction between antiphase and almost antiphase synchronization for the clock with similar masses of the pendulums was impossible.*

**Acknowledgements.** This work has been supported by the Foundation for Polish Science, Team Programme – Project No TEAM/2010/5/5. PP. acknowledges the support from Foundation for Polish Science (the START fellowship).

## References

1. Bennet, M., Schatz, M.F., Rockwood, H., Wiesenfeld, K.: Huygens's clocks. Proc. Roy. Soc. London, A 458, 563–579 (2002)
2. Blekham, I.I.: Synchronization in Science and Technology. ASME, New York (1988); Birch, T.: The history of The Royal Society of London for improving of natural knowledge, in which the most considerable of those papers communicated to the Society, which have hitherto not been published, are inserted in their proper order, as a supplement to the Philosophical Transactions, vol. 2, pp. 19, 21, 23–24. Johnson, London (1756) (reprint 1968)
3. Britten, F.J.: Britten's old clocks and watches and their makers; a historical and descriptive account of the different styles of clocks and watches of the past in England and abroad containing a list of nearly fourteen thousand makers. Methuen, London (1973)
4. Czolczynski, K., Perlikowski, P., Stefanski, A., Kapitaniak, T.: Clustering of Huygens' Clocks. Prog. Theor. Phys. 122, 1027–1033 (2009a)
5. Czolczynski, K., Perlikowski, P., Stefanski, A., Kapitaniak, T.: Clustering and Synchronization of Huygens' Clocks. Physica A 388, 5013–5023 (2009b)
6. Czolczynski, K., Perlikowski, P., Stefanski, A., Kapitaniak, T.: Huygens' odd sympathy experiment revisited. Int. J. Bifur. Chaos 21 (2011a)
7. Czolczynski, K., Perlikowski, P., Stefanski, A., Kapitaniak, T.: Why two clocks synchronize: Energy balance of the synchronized clocks. Chaos 21, 023129 (2011b)
8. Dilao, R.: Antiphase and in-phase synchronization of nonlinear oscillators: The Huygens's clocks system. Chaos 19, 023118 (2009)
9. Fradkov, A.L., Andrievsky, B.: Synchronization and phase relations in the motion of twopendulum system. Int. J. Non-linear Mech. 42, 895 (2007)
10. Huygens, C.: Letter to de Sluse. In: Oeuveres Completes de Christian Huygens (letters; no. 1333 of 24 February 1665, no. 1335 of 26 February 1665, no. 1345 of 6 March 1665). Societe Hollandaise Des Sciences, Martinus Nijhoff, La Haye (1893)
11. Huygens, C.: Instructions concerning the use of pendulum-watches for finding the longitude at sea. Phil. Trans. R. Soc. Lond. 4, 937 (1669)
12. Golubitsky, M., Stewart, I., Buono, P.L., Collins, J.J.: Symmetry in locomotor central pattern generators and animal gaits. Nature 401, 693–695 (1999)
13. Kanunnikov, A.Y., Lamper, R.E.: Synchronization of pendulum clocks suspended on an elastic beam. J. Appl. Mech. & Theor. Phys. 44, 748–752 (2003)

14. Kumon, M., Washizaki, R., Sato, J., Mizumoto, R.K.I., Iwai, Z.: Controlled synchronization of two 1-DOF coupled oscillators. In: Proceedings of the 15th IFAC World Congress, Barcelona (2002)
15. Lepschy, A.M., Mian, G.A., Viaro, U.: Feedback control in ancient water and mechanical clocks. IEEE Trans. Education 35, 3–10 (1993)
16. Moon, F.C., Stiefel, P.D.: Coexisting chaotic and periodic dynamics in clock escapements. Phil. Trans. R. Soc. A 364, 2539 (2006)
17. Pantaleone, J.: Synchronization of metronomes. Am. J. Phys. 70, 992 (2002)
18. Pikovsky, A., Roesenblum, M., Kurths, J.: Synchronization: An Universal Concept in Nonlinear Sciences. Cambridge University Press, Cambridge (2001); Pogromsky, A.Y., Belykh, V.N., Nijmeijer, H.: Controlled synchronization of pendula. In: Proceedings of the 42nd IEEE Conference on Design and Control, Maui, Hawaii, pp. 4381–4385 (2003)
19. Roup, A.V., Bernstein, D.S., Nersesov, S.G., Haddad, W.S., Chellaboina, V.: Limit cycle analysis of the verge and foliot clock escapement using impulsive differential equations and Poincare maps. Int. J. Control 76, 1685–1698 (2003)
20. Rawlings, A.L.: The science of clocks and watches. Pitman, New York (1944)
21. Senator, M.: Synchronization of two coupled escapement-driven pendulum clocks. Journal Sound and Vibration 291, 566–603 (2006)
22. Strogatz, S.H., Stewart, I.: Coupled oscillators and biological synchronization. Scientific American 269(6), 102–109 (1993)
23. Ulrichs, H., Mann, A., Parlitz, U.: Synchronization and chaotic dynamics of coupled mechanical metronomes. Chaos 19, 043120 (2009)

# Chapter 2
# On the Synchronization of 1D and 2D Multi-scroll Chaotic Oscillators

J.M. Muñoz-Pacheco, E. Zambrano-Serrano, O.G. Félix-Beltrán, E. Tlelo-Cuautle, L.C. Gómez-Pavón, R. Trejo-Guerra, A. Luis-Ramos, and C. Sánchez-López

**Abstract.** In this chapter, the guidelines to synchronize one-directional (1D) and two-directional (2D) multi-scroll chaos generators by means of Generalized Hamiltonian forms are presented. First, the multi-scroll chaotic oscillator is simulated at the electronic system level by applying state-variables and piecewise-linear approaches. Besides, we apply scaling procedures to modify the breaking points, slopes and frequency of the chaotic signals in order to reduce their excursion levels within practical values for electronic devices. Second, a chaotic synchronization scheme for multi-directional multi-scroll chaos generators is introduced. We use Generalized Hamiltonian forms approach to determine the synchronization conditions when one and two state-variables of the master system are sent to control the nonlinear functions in the slave system. Additionally, two schemes are set-up to transmit encrypted binary and analog signals by applying chaotic switching technique and additive chaotic masking, respectively. Both schemes are implemented

J.M. Muñoz-Pacheco
Department of Electronics and Telecommunications, UPPue, Puebla, MÉXICO
e-mail: jpacheco@uppuebla.edu.mx

E. Zambrano-Serrano · O.G. Félix-Beltrán · L.C. Gómez-Pavón, A. Luis-Ramos
School of Electronics Sciences, BUAP, Puebla, MÉXICO
e-mail: zambrano@ece.buap.mx, olga_flix@yahoo.com.mx, luz.gomezpavon@gmail.com

E. Tlelo-Cuautle
Department of Electronics, INAOE, Tonantzintla, MÉXICO
e-mail: etlelo@inaoep.mx

R. Trejo-Guerra
Snowbush Mexico Design Center, MÉXICO
e-mail: rodotg13@gmail.com

C. Sánchez-López
Department of Electronics, UAT and IMSE, Apizaco-MÉXICO and Sevilla-SPAIN
e-mail: carlsan@ieee.org

K. Kyamakya et al. (Eds.): Selected Topics in Nonlinear Dynamics, SCI 459, pp. 19–40.
DOI: 10.1007/978-3-642-34560-9_2 

by using traditional operational amplifiers. Finally, theoretical results are confirmed by performing numerical and SPICE simulation results.

## 2.1 Introduction

The vast discipline of nonlinear engineering divides into two complementary practices: one that pursues the elimination of undesired nonlinear effects, and another one that seeks to harness nonlinear effects for useful engineering purposes [1]. Focusing in the second practice, nonlinear science has had quite a triumph in all conceivable applications in science and technology [1–6]. In this manner, the most studied nonlinear phenomenon is the complex, random-like behavior called chaos. Chaos can occur widely on both natural and man-made systems and it possesses special features such as being extremely sensitive to tiny variations of its initial conditions, fractional topological dimension and a positive Lyapunov exponent [2].

In recent years, chaotic systems have been an attractive field for research in various areas, among them physics, communications, robotics, and electronics where there are potential applications involving true random number generators [7], liquid mixers [8], cooperative robotics and robot navigation [9, 10], high-performance electronics circuits like sigma-delta modulators [11], radar systems [12] and secure communications [13, 14]. Although remarkable research efforts have been invested in recent years, trying to export concepts from physics and mathematics into real-world engineering applications, the circuit implementation of reliable nonlinear circuits for generating various complex chaotic signals is a key issue for future applications of chaos [5]. In particular, creating various complex multi-directional multi-scroll chaotic attractors by using some simple electronic devices is a topic of both theoretical and practical interests [3]. However, it has been identified that it is quite difficult to synthesize multi-directional multi-scrolls by analog electronic circuits directly, because a wide dynamic range for the amplifiers is required for the physical realization of nonlinear resistors with multiple segments [3–5]. To cope with this challenge, in [5] is presented an approach based on behavioral modeling for synthesizing chaotic systems from mathematical level to electronic circuit level, which offers one possible way to abstract the features of interest into a circuit block. Individual blocks can be realized using diverse kinds of operational amplifiers [6], e.g. voltage operational amplifiers [3, 5], current-feedback operational amplifiers [15], current conveyors, etc. In this chapter, the 1D and 2D multi-scroll chaotic oscillator is numerically simulated at the electronic system level by applying state-variables and piecewise-linear approximations [5, 16]. Later, we apply a scaling procedure to modify the breaking points and slopes of the saturated functions in order to reduce the excursion levels of the chaotic signals within practical values for electronic devices and finally, current-saturated and voltage-saturated functions are synthesized using traditional voltage operational amplifiers [17].

In the applications side, some of the most pressing issues involve privacy and security for processing communication signals [1–5]. This new practice is evolving on many fronts and levels, reaching a state of maturity where it can be applied to real-world problems. However, a synchronization process is highly required in order to exploit chaos based secure communications [15]. The classical synchronization has been known since at least the seventeenth century; what was unexpected was that a similar phenomenon could be had with chaotic signals, as demonstrated by Pecora and Carroll in 1990 [18]. This issue generates a turning point in the investigation of chaos for communication systems, which allows chaos to be modulated and demodulated like in an analog carrier. We can count five basic chaotic synchronization techniques [19]:

- *Master-slave synchronization*: It is based on an autonomous system that unidirectionally drives a stable subsystem.
- *Non-autonomous synchronization*: It is based on a non-autonomous system that unidirectionally drives a stable identical non-autonomous system.
- *Inverse system synchronization*: It is based on a receiver that is a formal dynamical inverse of the transmitter, which will reproduce the latter's forcing function.
- *Adaptive control synchronization*: It is based on the numerous variants of adaptive control for chaotic systems.
- *Coupled synchronization*: It is based on bidirectionally coupled identical systems similar to the traditional classical form involving sinusoidal oscillators.

The first four forms of chaotic synchronization are suitable for standard communications purposes, while the fifth is suitable for network communications [19]. Focusing on the former technique, this chapter introduces the guidelines to synchronize one-directional (1D) and two-directional (2D) multi-scroll chaos generators by means of Generalized Hamiltonian forms because it has foremost advantages over other synchronization methods [20], namely: a) enables synchronization to be achieved in a systematic way and clarifies the issue of deciding on the nature of the output signal to be transmitted; b) it can be successfully applied to several well-known chaotic systems; c) it does not require the computations of any Lyapunov exponent and; d) it does not require initial conditions belonging to the same basin of attraction. Recently, multi-scroll chaotic oscillators have been addressed as a potential solution to improve the security in encrypted communications by using their native properties; i.e. the direction (1D or 2D) of the chaotic attractor; the effort to estimate chaotic dynamics according to the system parameters and the high sensitivity of chaotic systems to initial conditions. These characteristics imply strong cryptographic properties for data encryption, which makes them robust against attacks on the public channel [20, 21]. Because of the newness of these discoveries, many studies are still needed to face important engineering and operational issues [4].

The aim of this chapter is to present an approach for synchronizing 1D and 2D multi-scroll chaos generators by controlling the nonlinear functions in the nonlinear observer with a combination of the number of state-variables from the master system. The difference of this approach, compared with other papers on this subject [22], is the introduction of the synchronization conditions for multi-dimensional (1D, 2D, and 3D) chaos generators with multiple nonlinear functions, no matter the values of their Lyapunov exponents. Also, it is proposed a quasi-optimal region for synchronization gains.

The safeguard of information is a central and old problem of great interest for the humanity. Nowadays, data security plays an increasing role in common life; e.g. Internet, banking, commerce, industry, personal communications, etc. Different methods have been proposed for implementing and demonstrating analog or digital modulation of information using a chaotic carrier. In essence, the information is hidden in the noise in transmission and can be extracted or decoding in the receiver using the inherent determinism of chaos and subtracting or filtering operations. Major chaos-based modulation methods being investigated and developed internationally for communication applications, including [19]:

- *Additive chaotic masking*: Here, the information is added to the carrier as a small perturbation and usually demodulated using a cascaded form of master-slave synchronization.
- *Chaotic switching*: Here, an analog signal of finite duration represents a digital symbol consisting of one or more bits. In this case, the digital symbol is uniquely mapped to an analog waveform segment coming from a distinct region of a single strange attractor.
- *Forcing function modulation*: Here, a sinusoidal forcing function in a nonautonomous chaotic system is analog or digitally modulated with the information in a classical manner, with the transmitted signal being some other state variable. This modulation typically involves the nonautonomous or inverse synchronization methods and is the basis for the Aerospace development effort addressing high-data-rate, chaos-based communications.
- *Multiplicative chaotic mixing*: It is based on a chaos-based version of the traditional direct-sequence spread-spectrum approach, except in this case; the receiver actually divides by the chaotic carrier to extract the original information.
- *Parametric modulation*: In this case, the information directly modulates a circuit parameter value (such as resistance, capacitance, or inductance), and some state variable from the chaotic system is sent that contains the information in a complex manner.
- *Independent source modulation*: Here, the information becomes an independent voltage/current source that is inserted in the chaotic transmitter circuit.
- *Generalized modulation*: In this approach, the information and chaotic carrier are combined in a more general invertible manner.

Implications between theoretical postulates and engineering applications are being pursued by researches in order to exploit the deterministic, yet random-like behavior of chaos, particularly with regard to inherently secure signal processing and transmission. In this manner, this chapter also presents two chaos-based communication schemes to transmit encrypted information. First, a scheme is set-up to transmit binary signals by applying chaotic switching technique [19–21]. Both, the coupling signal and the confidential message are sent by only one transmission channel. Additionally, a second scheme is proposed to encrypted analog signals by using additive chaotic masking, which uses two transmission channels [19–21]. Theoretical results are confirmed by performing numerical and SPICE simulations to show the usefulness of the proposed synchronization technique. Finally, we use traditional operational amplifiers for designing the encryption schemes.

The chapter is organized as follows: the circuit synthesis of the 1D and 2D multi-scroll chaotic oscillators is covered in section 2.2. The Generalized Hamiltonian forms and synchronization conditions for 2D-4- and 3D-4-scroll chaos generators are discussed in section 2.3. Design tradeoffs of the proposed encrypted communication schemes as well as SPICE simulations are introduced in section 2.4. Finally, the conclusions are summarized in section 2.5.

## 2.2 General Aspects for the Amplifiers-Based Design of Chaotic Oscillators

The design methodology for chaotic systems is performed by three hierarchical levels as introduced in [5, 17]. The high-level descriptions capture the behavior of the chaotic attractor and include the number of scrolls, position of the scrolls, voltage or current level of the chaotic signals, and frequency of the attractor. Besides, the realization of a nonlinear characteristic with multi-segments is needed for implementing chaotic attractors with multi-directional orientation.

### 2.2.1 *High-Level Modeling*

The continuous-time chaos generator in (2.1) is taken as the core for generating chaotic behavior in 1D, 2D and 3D. The chaotic system is modeled by applying state variables approach, where $x,y,z$ are the state variables and $a,b,c$ positive real constants. To guide the linear system (2.1) to generate chaotic behavior in 1D, 2D and 3D; one, two or three nonlinear controllers should be added to stretch and fold the trajectories of the system repeatedly, respectively. The nonlinear functions are approximated by using the piecewise-linear (PWL) approximation defined by (2.2), where $k > 0$ is the slope and plateau of the saturated function series, $h > 2$ is the saturated delay time of the saturated function series, p and q are positive integers and $\alpha$ is the breakpoint; as depicted in Fig. 2.1. Details on the 1D n-scrolls attractor behavioral modeling are given in [5]. 2D chaotic attractors are obtained by using (2.3) [17].

$$\begin{aligned} \dot{x} &= y \\ \dot{y} &= z \\ \dot{z} &= -ax - by - cz + d_1 f(x;k_1,h_1,p_1,q_1), \end{aligned} \tag{2.1}$$

$$f(x;\alpha,k,h,p,q) = \begin{cases} (2q+1)k, & \text{if} \quad x > qh + \alpha \\ k/\alpha(x - ih) + 2ik, & \text{if} \quad |x - ih| \leq \alpha, -p \leq i \leq q \\ (2i+1)k, & \text{if} \quad ih + \alpha < x < (i+1)h - \alpha, -p \leq i \leq q - 1 \\ -(2p+1)k, & \text{if} \quad x < -ph - \alpha. \end{cases} \tag{2.2}$$

$$\begin{aligned}\dot{x} &= y - \tfrac{d_2}{b} f(y;k_2,h_2,p_2,q_2)\\ \dot{y} &= z\\ \dot{z} &= -ax - by - cz + d_1 f(x;k_1,h_1,p_1,q_1) + d_2 f(y;k_2,h_2,p_2,q_2),\end{aligned} \tag{2.3}$$

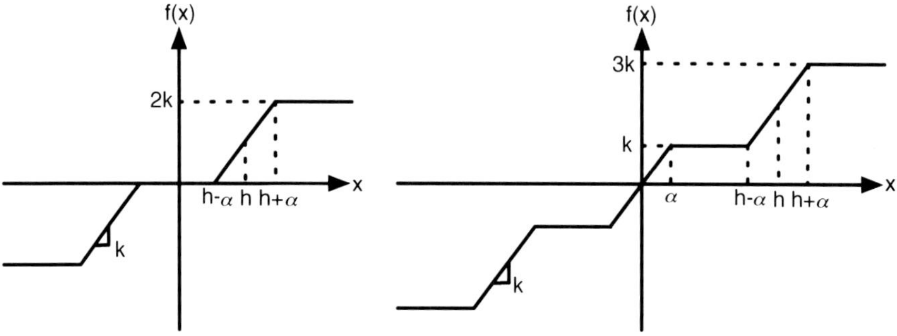


**Fig. 2.1** PWL function to generate odd and even $n$-scrolls

Note that for implementing multi-directional multi-scroll attractors using practical operational amplifiers (opamps), a scaling procedure should be applied on the excursion levels of the chaotic signals, to accomodate the maximum dynamic range of the opamps [17]. In this manner, $\alpha$ is used to change the breakpoints of the PWL function and consequently reduce the excursion of the signal. Similarly, by changing the value of the passive elements (capacitors and inductors), one can modify the frequency response of the attractor as shown in [23].

### 2.2.2 *Opamp-Based Circuit Synthesis*

A voltage-saturated function can emulate the behavior of the PWL function shown in Fig. 2.1. Therefore, the opamp finite-gain model is herein used to synthesize the nonlinear controller. In [17] is proposed a generic basic cell based in opamps devoted to generate the required plateaus and slopes. For instance, Fig. 2.2 shows the circuit synthesis of a PWL function to generate a 2D-4-scroll chaotic attractor, where $Rc$ realizes the current-voltage transformation. By parallel-connecting basic cells, one can obtain a higher number of scrolls [5, 17]. The value of the plateaus $k$, in voltage and current, the breakpoints $\alpha$, and the saturated delay $h$ are close related to the gain, bandwidth, slew rate and saturation of the opamps by using SPICE macro-models and Verilog-A behavioral models [5]. Moreover, the chaotic system in (2.3) can also be synthesized using opamps, as shown in Fig. 2.3. Note that, the blocks $SF(x_1)$ and $SF(x_2)$ represent the PWL functions in Fig. 2.2.

By selecting $Vsat = \pm 6.4V$, $R = 10K\Omega$, $Rca = 64K\Omega$, $Ria = 1K\Omega$ and $Rfa = 1M\Omega$, in Fig. 2.2; and $C = 2.2nF$, $Rx = Ry = Rz = 7K\Omega$, $Ri = Rf = 10K\Omega$ in Fig. 2.3 one gets a 2D-4-scrolls chaotic oscillator as shown in Fig. 2.4. The Lyapunov exponents are computed by using the method published in [24]. These are LE1= 0.155, LE2=0, LE3=-0.851.

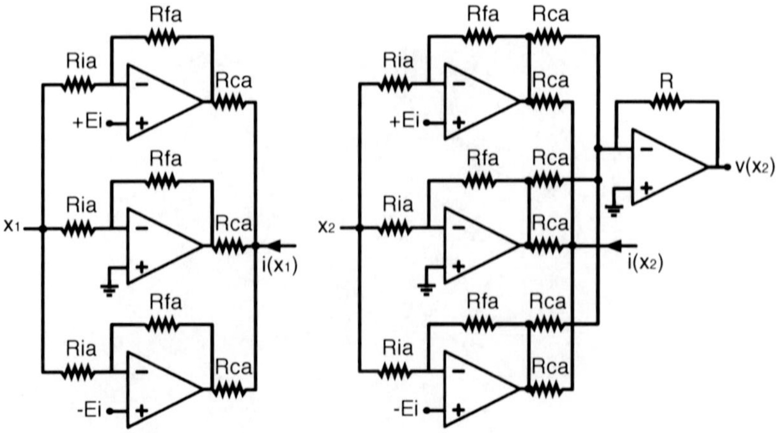


**Fig. 2.2** Opamp-based synthesis of PWL function in Fig. 2.1

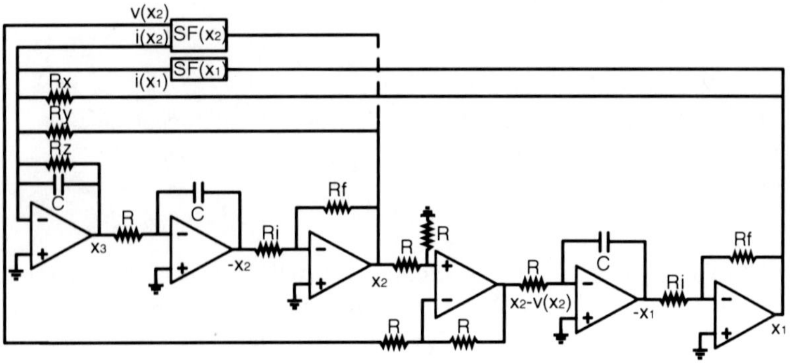


**Fig. 2.3** Opamp-based synthesis of chaotic system in (2.3)

## 2.3 Hamiltonian-Based Synchronization of Multi-directional Multi-scroll Chaotic Oscillators

Synchronizing chaotic oscillators implemented with electronic devices is quite important in engineering, such as in secure communications [10,13,21,25–27]. Robust applications require chaotic oscillators with more complex behavior than by simply using traditional double scroll attractors [23,25], e.g. chaotic oscillators generating multi-scrolls (1D) [3, 6, 15, 24, 28], and multi-directional behavior (2D) [5, 17, 29]. Because the multi-direction of a chaotic attractor (1D and 2D), is closely linked to an increased number of equilibrium points and consequently an increased effort to estimate chaotic dynamics according to the system parameters and the high sensitivity of the chaotic systems to initial conditions.

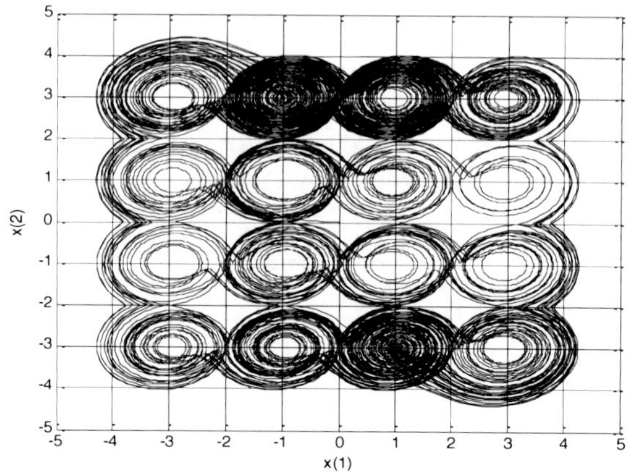


**Fig. 2.4** SPICE simulation results of a 2D-4-scrolls chaotic attractor

Lets us consider the dynamical system

$$\dot{x} = f(x) \tag{2.4}$$

where $x \in R^n$ is the state vector and $f : R^n \rightarrow R^n$ is a nonlinear function; so that (2.4) can be written in the following generalized canonical form [20]:

$$\dot{x} = J(x)\frac{\partial H}{\partial x} + S(x)\frac{\partial H}{\partial x}, \quad \mathbf{x} \in R^n, \tag{2.5}$$

where $H(x)$ describes an energy function which is globally positive-definite in $R^n$. The gradient vector of $H(x)$, denoted by $\partial H/\partial x$, is assumed that exists everywhere [20]. Frequently, quadratic energy functions of the form: $H(x) = \frac{1}{2}x^t Mx$, are used, with $M$ being a symmetric matrix positive-definite. In this case $\partial H/\partial x = Mx$. The squared matrices $J(x)$ and $S(x)$ in (2.5) satisfy: $J(x) + J^T(x) = 0$ and $S(x) = S^T(x)$ which represents the energy structure of the system [20–25]. Considered a special class of generalized Hamiltonian systems with destabilizing vector fields and linear output $y$, it is given by

$$\begin{aligned} \dot{x} &= J(y)\frac{\partial H}{\partial x} + S(y)\frac{\partial H}{\partial x} + F(y), \quad x \in R^n \\ y &= C\frac{\partial H}{\partial x}, \quad y \in R^m \end{aligned} \tag{2.6}$$

being $S$ a constant symmetric matrix, $y$ the output vector of the system and $C$ a constant matrix. By selecting $\xi$ as the estimated state vector of $x$, and $\eta$ as the estimated output in terms of $\xi$; a dynamic nonlinear state observer for (2.6) is given in (2.7); with $K$ being a constant vector, known as the observer gain [20].

$$\begin{aligned} \dot{\xi} &= J(y)\frac{\partial H}{\partial \xi} + S(y)\frac{\partial H}{\partial \xi} + F(y) + K(y - \eta), \\ \eta &= C\frac{\partial H}{\partial \xi}. \end{aligned} \tag{2.7}$$

The following theorems must be accomplished in order to demonstrate the synchronization by using generalized Hamiltonian forms [20–25].

**Theorem 2.1.** *The state $x$ of the nonlinear system (2.6) can be globally, exponentially, asymptotically estimated by the state $\xi$ of an observer of the form (2.7), if the pair of matrices $C, S$ are either observable or, at least, detectable.*

**Theorem 2.2.** *The state $x$ of the nonlinear system (2.6) can be globally, exponentially, asymptotically estimated, by the state $\xi$ of the observer (2.7) if and only if there is a constant matrix K; such as $[W-KC]+[W-KC]^T = 2[S-\frac{1}{2}(KC-C^TK^T)]$, and is negative definite.*

### 2.3.1 Synchronization of 2D-4-Scroll Chaos Generators

Lets us consider the 2D multi-scroll chaotic system defined by (2.3), we propose a Hamilton energy function and its gradient vector as given in (2.8) and (2.9), respectively.

$$H(x) = \frac{1}{2}\left[ax_1^2 + bx_2^2 + x_3^2\right], \tag{2.8}$$

$$\frac{\partial H}{\partial x} = \begin{bmatrix} ax_1 \\ bx_2 \\ x_3 \end{bmatrix}, \tag{2.9}$$

One can obtain the matrices $S$ and $J$ as shown in (2.10) and (2.11), respectively. Thay way, the 2D chaos generator in (2.3) can be described in generalized Hamiltonian forms given in (2.6) and (2.7), as shown in (2.12) and (2.13), where (2.13) is the nonlinear state observer of (2.12).

$$S(x) = \frac{1}{2}\left\{\begin{bmatrix} 0 & \frac{1}{b} & 0 \\ 0 & 0 & 1 \\ -1 & -1 & -c \end{bmatrix} + \begin{bmatrix} 0 & 0 & -1 \\ \frac{1}{b} & 0 & -1 \\ 0 & 1 & -c \end{bmatrix}\right\} = \begin{bmatrix} 0 & \frac{1}{2b} & -\frac{1}{2} \\ \frac{1}{2b} & 0 & 0 \\ -\frac{1}{2} & 0 & -c \end{bmatrix} \tag{2.10}$$

$$J(x) = \frac{1}{2}\left\{\begin{bmatrix} 0 & \frac{1}{b} & 0 \\ 0 & 0 & 1 \\ -1 & -1 & -c \end{bmatrix} - \begin{bmatrix} 0 & 0 & -1 \\ \frac{1}{b} & 0 & -1 \\ 0 & 1 & -c \end{bmatrix}\right\} = \begin{bmatrix} 0 & \frac{1}{2b} & \frac{1}{2} \\ -\frac{1}{2b} & 0 & 1 \\ -\frac{1}{2} & -1 & 0 \end{bmatrix}. \tag{2.11}$$

$$\begin{bmatrix} \dot{x}_1 \\ \dot{x}_2 \\ \dot{x}_3 \end{bmatrix} = \begin{bmatrix} 0 & \frac{1}{2b} & \frac{1}{2} \\ -\frac{1}{2b} & 0 & 1 \\ -\frac{1}{2} & -1 & 0 \end{bmatrix} \frac{\partial H}{\partial x} + \begin{bmatrix} 0 & \frac{1}{2b} & -\frac{1}{2} \\ \frac{1}{2b} & 0 & 0 \\ -\frac{1}{2} & 0 & -c \end{bmatrix} \frac{\partial H}{\partial x} + \begin{bmatrix} -\frac{d_2}{b} f(x_2) \\ 0 \\ d_1 f(x_1) + d_2 f(x_2) \end{bmatrix} \tag{2.12}$$

$$\begin{bmatrix} \dot{\xi}_1 \\ \dot{\xi}_2 \\ \dot{\xi}_3 \end{bmatrix} = \begin{bmatrix} 0 & \frac{1}{2b} & \frac{1}{2} \\ -\frac{1}{2b} & 0 & 1 \\ -\frac{1}{2} & -1 & 0 \end{bmatrix} \frac{\partial H}{\partial \xi} + \begin{bmatrix} 0 & \frac{1}{2b} & -\frac{1}{2} \\ \frac{1}{2b} & 0 & 0 \\ -\frac{1}{2} & 0 & -c \end{bmatrix} \frac{\partial H}{\partial \xi} \tag{2.13}$$

$$+\begin{bmatrix} -\frac{d_2}{b}f(x_2) \\ 0 \\ d_1 f(x_1) + d_2 f(x_2) \end{bmatrix} + \begin{bmatrix} k_1 & k_4 \\ k_2 & k_5 \\ k_3 & k_6 \end{bmatrix} (y - \eta).$$

with $\eta$ being

$$\eta = \begin{bmatrix} \frac{d_1}{a} & \frac{d_2}{b} & 0 \\ 0 & \frac{d_2}{b^2} & 0 \end{bmatrix} \frac{\partial H}{\partial \xi} \tag{2.14}$$

By evaluating the stability of the approach using Observability's criterion, one obtains

$$det \begin{vmatrix} \frac{d_1}{a} & \frac{d_2}{2b^2} & \frac{\left(\frac{1}{4b^2}+\frac{1}{4b}\right)d_1}{a} \\ \frac{d_2}{b} & \frac{d_1}{2ab} & \frac{d_2}{4b^3} \\ 0 & -\frac{d_1}{2a} & \frac{cd_1}{2a} - \frac{d_2}{4b^2} \end{vmatrix} \neq 0, \tag{2.15}$$

Besides, it is also necessary to demonstrate Theorem 2.2, in order to gain insight about the synchronization of two-directional chaos generators. So that matrices $S$, $C$ and $K$ are used to evaluate the equation in Theorem 2.2, resulting on (2.16).

$$2\left[\begin{bmatrix} 0 & \frac{1}{2b} & -\frac{1}{2} \\ \frac{1}{2b} & 0 & 0 \\ -\frac{1}{2} & 0 & -c \end{bmatrix} - \frac{1}{2}\left\{\begin{bmatrix} k_1 & k_4 \\ k_2 & k_5 \\ k_3 & k_6 \end{bmatrix}\begin{bmatrix} \frac{d_1}{a} & \frac{d_2}{b} & 0 \\ 0 & \frac{d_2}{b^2} & 0 \end{bmatrix} + \begin{bmatrix} \frac{d_1}{a} & 0 \\ \frac{d_2}{b} & \frac{d_2}{b^2} \\ 0 & 0 \end{bmatrix}\begin{bmatrix} k_1 & k_2 & k_3 \\ k_4 & k_5 & k_6 \end{bmatrix}\right\}\right]$$
$$= \begin{bmatrix} -\frac{2k_1d_1}{a} & \frac{1}{b} - \frac{k_1d_2}{b} - \frac{k_4d_2}{b^2} - \frac{k_2d_1}{a} & -1 - \frac{k_3d_1}{a} \\ \frac{1}{b} - \frac{k_1d_2}{b} - \frac{k_4d_2}{b^2} - \frac{k_2d_1}{a} & -\frac{2k_2d_2}{b} - \frac{k_5d_2}{b^2} & -\frac{k_3d_2}{b} - \frac{k_6d_2}{b^2} \\ -\frac{1}{b} - \frac{k_3d_1}{a} & -\frac{k_3d_2}{b} - \frac{k_6d_2}{b^2} & -2c \end{bmatrix} \tag{2.16}$$

The Sylvester's criterion [20–25], is used herein to demonstrate that the matrix in (2.16) is negative definite. Indeed, the values for the observer gain, matrix in (2.13), are also obtained by calculating the roots of the determinants in (2.16). For the first determinant, one obtains

$$-\frac{2k_1d_1}{a} < 0 \Longrightarrow k_1 > 0 \tag{2.17}$$

Equation (2.18) is obtained by solving for the minor of the matrix 2x2 in (2.16).

$$det = -\frac{1}{b^4a^2}\Big(-2k_1d_1d_2b^3ak_2 - 4k_1d_1d_2b^2ak_5 + b^2a^2 - 2b^2a^2k_1d_2 - 2ba^2k_4d_2 \tag{2.18}$$
$$-2b^3ak_2d_1 + k_1^2d_2^2b^2a^2 + 2k_1d_2^2ba^2k_4 + k_4^2d_2^2a^2 + 2k_4d_2ak_2d_1b^2 + k_2^2d_1^2b^4\Big) < 0$$

Considering that $k_1 = k_4$, $k_2 = k_5$ y $k_3 = k_6$, (2.18) can be updated by (2.19). This assumption is valid since the nonlinear functions $f(x_1)$ and $f(x_2)$ in (2.11) are the same.

$$det2 = -\frac{1}{b^4a^2}\Big(-2k_1d_1d_2b^3ak_2 - 2k_1d_1d_2b^2ak_2 + b^2a^2 - 2b^2a^2k_1d_2 - 2ba^2k_1d_2 \tag{2.19}$$

$$-2b^3ak_2d_1 + k_1^2d_2^2b^2a^2 + 2k_1^2d_2^2ba^2 + k_1^2d_2^2a^2 + k_2^2d_1^2b^4\Big) < 0$$

By solving (2.19), the interval values for $K_2$ are obtained as shown in (2.20).

$$\frac{\left(b + k_1d_2b + k_1d_2 - 2\sqrt{b^2k_1d_2 + k_1d_2b}\right)a}{d_1b^2} < k_2 < \frac{\left(b + k_1d_2b + k_1d_2 + 2\sqrt{b^2k_1d_2 + k_1d_2b}\right)a}{d_1b^2}. \tag{2.20}$$

Note that, $K_3$ since it has no influence on the observable eigenvalues of the nonconservative structure of the 2D chaos generator.

### 2.3.2 *Synchronization of 3D-4-Scroll Chaos Generators*

Lets us consider the 3D-multi-scroll chaotic system defined by (2.21), with $f(x_1)$, $f(x_2)$ and $f(x_3)$ being a PWL function.

$$\begin{aligned}\dot{x}_1 &= x_2 - \tfrac{d_2}{b}f(x_2)\\ \dot{x}_2 &= x_3 - \tfrac{d_3}{b}f(x_3)\\ \dot{x}_3 &= -ax_1 - bx_2 - cx_3 + d_1f(x_1) + d_2f(x_2) + d_3f(x_3),\end{aligned} \tag{2.21}$$

where $x_1, x_2, x_3$ are state-variables, and $a, b, c, d_1, d_2, d_3 = 0.7$ are positive real constants. By using (2.6), (2.7) and (2.9), the 3D chaos generator in (2.21) can be described in generalized Hamiltonian forms given in (2.5) as shown in (2.22). Consequently, the nonlinear state observer for 3D chaos generator in (2.21), according to (2.7), is shown in (2.23).

$$\begin{bmatrix}\dot{x}_1\\ \dot{x}_2\\ \dot{x}_3\end{bmatrix} = \begin{bmatrix}0 & \frac{1}{2b} & \frac{1}{2}\\ -\frac{1}{2b} & 0 & 1\\ -\frac{1}{2} & -1 & 0\end{bmatrix}\frac{\partial H}{\partial x} + \begin{bmatrix}0 & \frac{1}{2b} & -\frac{1}{2}\\ \frac{1}{2b} & 0 & 0\\ -\frac{1}{2} & 0 & -c\end{bmatrix}\frac{\partial H}{\partial x} + \begin{bmatrix}-\frac{d_2}{b}f(x_2)\\ -\frac{d_3}{b}f(x_3)\\ d_1f(x_1) + d_2f(x_2) + d_3f(x_3)\end{bmatrix} \tag{2.22}$$

$$\begin{bmatrix}\dot{\xi}_1\\ \dot{\xi}_2\\ \dot{\xi}_3\end{bmatrix} = \begin{bmatrix}0 & \frac{1}{2b} & \frac{1}{2}\\ -\frac{1}{2b} & 0 & 1\\ -\frac{1}{2} & -1 & 0\end{bmatrix}\frac{\partial H}{\partial \xi} + \begin{bmatrix}0 & \frac{1}{2b} & -\frac{1}{2}\\ \frac{1}{2b} & 0 & 0\\ -\frac{1}{2} & 0 & -c\end{bmatrix}\frac{\partial H}{\partial \xi} + \begin{bmatrix}-\frac{d_2}{b}f(x_2)\\ -\frac{d_3}{b}f(x_3)\\ d_1f(x_1) + d_2f(x_2) + d_3f(x_3)\end{bmatrix} + \begin{bmatrix}k_1 & k_4 & k_7\\ k_2 & k_5 & k_8\\ k_3 & k_6 & k_9\end{bmatrix}(y - \eta). \tag{2.23}$$

with $\eta$ being

$$\eta = \begin{bmatrix} \frac{d_1}{a} & \frac{d_2}{b} & \frac{d_3}{b} \\ 0 & \frac{d_2}{b^2} & 0 \\ 0 & 0 & \frac{d_3}{b} \end{bmatrix} \frac{\partial H}{\partial \xi} \tag{2.24}$$

According to (2.15), the approach is stable and the demonstration of Theorem 2.2 for the synchronization of three-directional chaos generators is given in (2.25), by using the matrices $S, K$, and $C$ in (2.10), (2.23) and (2.24), respectively.

$$\begin{bmatrix} -\frac{2k_1d_1}{a} & \frac{1}{b} - \frac{k_1d_2}{b} - \frac{k_4d_2}{b^2} - \frac{k_2d_1}{a} & -1 - \frac{k_3d_1}{a} - \frac{k_1d_3}{b} - \frac{k_7d_3}{b} \\ \frac{1}{b} - \frac{k_1d_2}{b} - \frac{k_4d_2}{b^2} - \frac{k_2d_1}{a} & -\frac{2k_2d_2}{b} - \frac{2k_5d_2}{b^2} & -\frac{k_2d_3}{b} - \frac{k_8d_3}{b} - \frac{k_3d_2}{b} - \frac{k_6d_2}{b^2} \\ -\frac{1}{b} - \frac{k_3d_1}{a} - \frac{k_1d_3}{b} - \frac{k_7d_3}{b} & -\frac{k_2d_3}{b} - \frac{k_8d_3}{b} - \frac{k_3d_2}{b} - \frac{k_6d_2}{b^2} & -2c - \frac{2k_3d_3}{b} - \frac{k_9d_3}{b} \end{bmatrix} \tag{2.25}$$

Similarly, the values of the observer gain are calculated by using the Sylvester's criterion as previously shown in subsection 2.3.1. Note that the first two determinants in (2.25) are identical to that on (2.16), since the extra nonlinear function $f(x_3)$ is only related to state-variable $x_3$. In this manner, the intervals for $k_1$ and $k_2$ are also given by (2.17) and (2.19), respectively. Now, considering that $k_1 = k_4 = k_7$, $k_2 = k_5 = k_8$ and $k_3 = k_6 = k_9$, since the nonlinear functions $f(x_1)$, $f(x_2)$ and $f(x_3)$ in (2.23) are the same. By evaluating (2.23), it is found the interval of values for $k_3$ shown in (2.27). For sake of simplicity, the results are shown by substituting the values of $a, b, c, d_1, d_2, d_3 = 0.7$.

$$\begin{aligned} det_{3\times3} = {} & 16.58k_3 - 9.71k_1 + 9.78k_2 + 0.18k_2k_3 - 3.45k_2^2 + 6.93k_3^2 + 16.42k_1k_2 \\ & -(2e-9)k_1k_3^2 - 28.20k_3k_1 + 8.25k_1^2 + (2e-9)k_1k_2k_3 + 2.85 < 0 \end{aligned} \tag{2.26}$$

$$\begin{aligned} k_3 = {} & \frac{4.14e9}{-3.46e9 + k_1} + \frac{4.59e7k_2}{-3.46e9 + k_1} - \frac{4.05e9k_1}{-3.46e9 + k_1} + \frac{0.50k_1k_2}{-3.46e9 + k_1} \\ & + \frac{1}{-3.46e9 + k_1} (0.50 (4.89e19 - 6.7e19k_2 - 1.66e20k_1 - 1.16e20k_1k_2 \\ & +2.39e19k_2^2 - 6.73e9k_1k_2^2 + 1.41e20k_1^2 4.64e9k_1^2k_2 + k_1^2k_2^2 + 1.65e10k_1^3\Big)^{1/2}) \end{aligned} \tag{2.27}$$

### 2.3.3 *Numerical Simulation Results*

This section is devoted to show the usefulness of the proposed approach. By selecting $k_1 = k_4 = 1$, $k_2 = k_5 = 2$ and $k_3 = k_6 = 0$ for the observer in (2.13), one obtains the synchronization of 2D-4-scroll chaos generators, the coincidence of their states is represented by a straight line, with slope equal to unity, in the phase plane for each state and its error $e_x$, which is the difference between the observed state $x$ and the estimated state $\xi$, as shown in Fig. 2.5 and Fig. 2.6, respectively.

By selecting $k_1 = k_4 = k_7 = 1$, $k_2 = k_5 = k_8 = 2$ and $k_3 = k_6 = k_9 = 0$ for the observer in (2.23); one obtains Fig. 2.7 and Fig. 2.8.

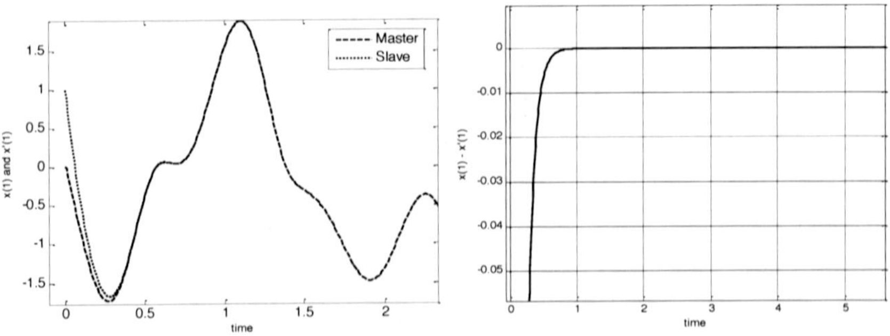


**Fig. 2.5** (a) Observed and estimated state in time-domain for two 2D-4-scroll chaos generators, (b) Error between the synchronized 2D-4-scroll chaos generators

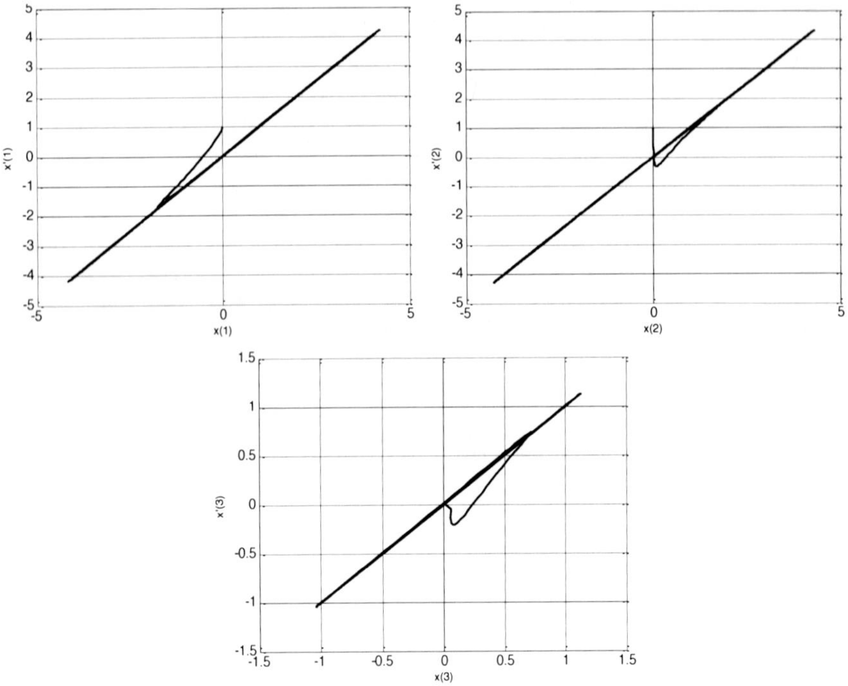


**Fig. 2.6** Phase plane diagrams for the state-variables in (2.12) and (2.13)

A prediction of the values for the observer gain is shown in Fig. 2.9. The solution is calculated from (2.20) and any value selected between solution (+) and solution (-) as shown in Fig. 2.9 leads to the synchronization for 2D and 3D-multi-scroll chaos generators given in (2.3) and (2.21).

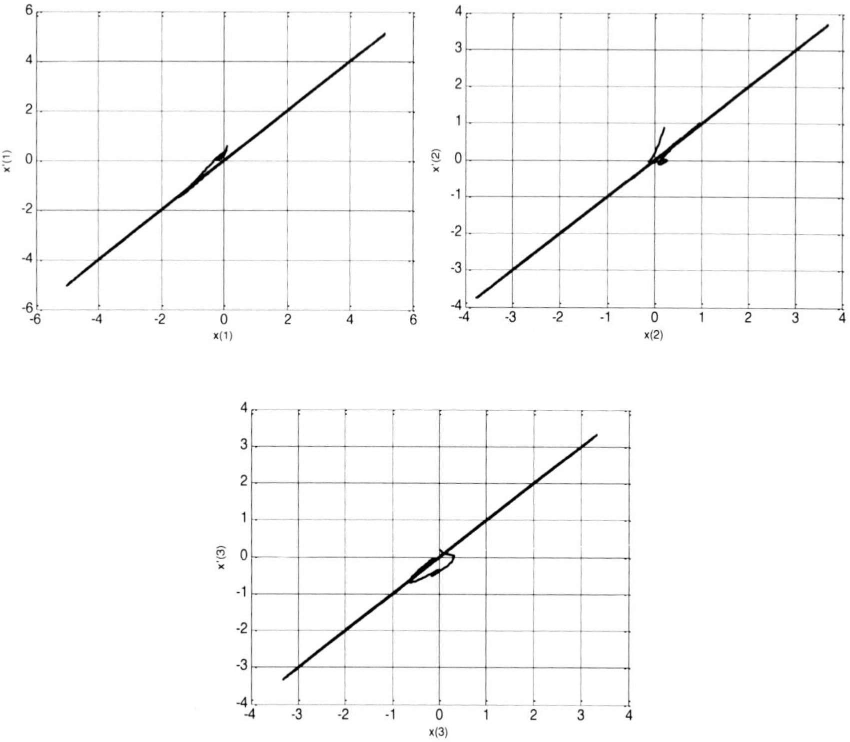


**Fig. 2.7** Phase plane diagrams for the state-variables in (2.22) and (2.23)

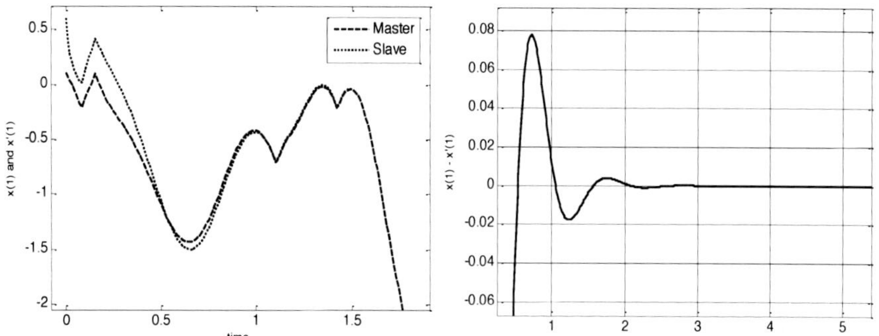


**Fig. 2.8** (a) Observed and estimated state in time-domain for two 3D-4-scroll chaos generators, (b) Error between the synchronized 3D-4-scroll chaos generators

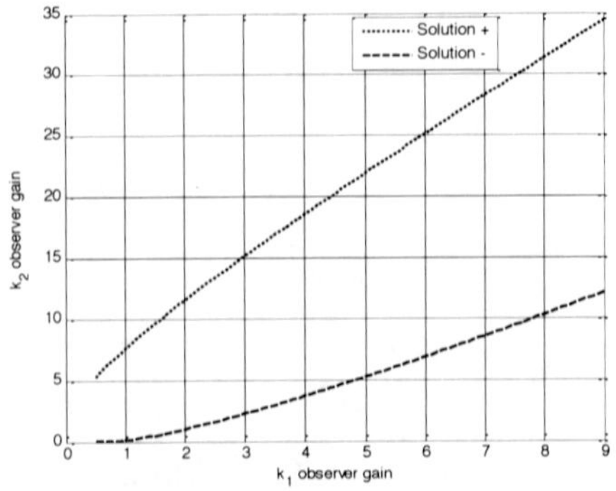


**Fig. 2.9** Synchronization region according to the observer gain

## 2.4 Design of Chaos-Based Encrypted Communication Schemes

The multi-directional multi-scroll chaotic system introduced in section 2.2 and previously synchronized is used to design two synchronization schemes in order to transmit confidential information [21]. As a result, the proposed scheme is shown in Fig. 2.10 when 1D-4-scroll chaotic oscillator is used. It can be seen from Fig. 2.10 that $x_1$ will be the coupling signal $f(y)$ which is transmitted from the master circuit to the observer. As one can infer, the nonlinear component SF in Fig. 2.10, is expected to be controlled directly by the master circuit.

The following subsections introduce two schemes for sending encrypted information using the synchronized circuit shown above by selecting $Rix = 10K\Omega$, $C = 2.2nf$, $R = 7K\Omega$, $Rx = Ry = Rz = 10K\Omega$, $Ri = 10K\Omega$, $Rf = 10K\Omega$, $Rhio = 10k\Omega$, $Rhfo = 3.9M\Omega$, $Rhko = 18\Omega$. Resistor $Rhko$ transforms the output voltage from the opamp in a current feedback that is injected to the slave system. Note that, the difference between the master signal and slave signal is compared at the differential amplifier and its error signal must be sent to the slave in order to progressively minimize the error and reach the synchronization. On the other hand, the necessary and sufficient condition, for master-slave synchronization to occur, is that the non-driven slave subsystem must be asymptotically stable. This is proved by adopting the generalized Hamiltonian forms as demonstrated in this chapter, because it does not require initial conditions belonging to the same basin of attraction. Therefore, the time-series generated from the slave and master circuits will converge as time increases. This is valid if the necessary and sufficient conditions are satisfied. In this context, Hamilton offers a good method for synchronizing chaotic systems, independently of their initial conditions. However, if the parameters of the circuits suffer a several variation, the chaotic behavior disappears. Advanced circuit analysis such as corner analysis [28] could be used to estimate the impact of these non-idealities.

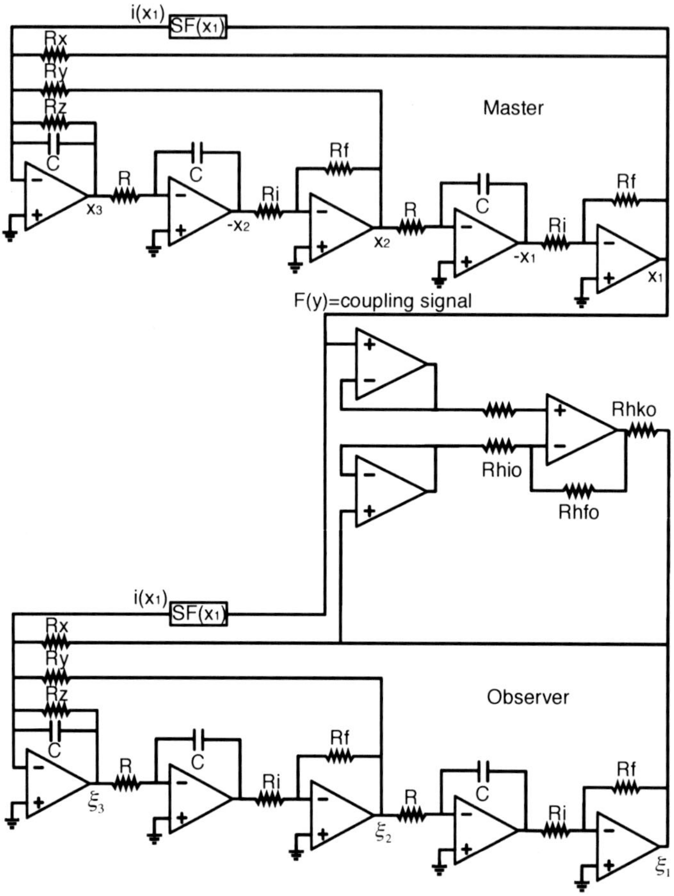


**Fig. 2.10** Circuit diagram to synchronize 1D multi-scrolls chaotic oscillators

## 2.4.1 *Binary Transmission*

A state from the original chaotic system is usually pointed by the Hamiltonian approach as the signal to be transmitted. This makes possible the synchronization in the receiver circuit and also to notice the differences (error signal) experienced when the master circuit has been perturbed regarding the information to be transmitted [19, 29]. From Fig. 2.10, the coupling signal (data carrier) is $x1$. In order to transmit binary data using the chaotic switching technique [19], six components in the transmitter of Fig. 2.11(a) are chosen to switch between two different values for experimenting the synchronizing and no-synchronizing of the receiver to the signal $x_1$. Therefore, the consequent changes in the synchronization error are available at the receiver as a voltage [13, 14]. The basic idea of this scheme is as follows: A private message is transmitted over an insecure communication channel. To avoid any unauthorized intruder located at the mentioned channel; the message is encrypted prior to transmission to generate an encrypted message by using two n-chaotic oscillators that generate, in this case, 1D-4-scrolls.

The encrypted binary message is sent to the receiver, where the message is recovered by the synchronization error. The chaotic switching technique is based on one transmission channel, which is a remarkable feature, since a single channel is usually available for communication applications [19].

A secure channel (bold line in Fig.2.11(a)) is used for transmission of data and the synchronization keys, which are obtained by switching the parameters that determine the dynamics of the chaotic oscillators; such as: $Rx = Ry = Rz = 10K\Omega$, $Rca = 100K\Omega$, $Rfa = 1M\Omega$, $Ria = 10K\Omega$, $E1 = 2V$; and $Rxb = Ryb = Rzb = 11K\Omega$, $Rcb = 105K\Omega$, $Rf = 10K\Omega$, $Ri = 10K\Omega$ $E1 = 2V$, $R = 7K\Omega$, at transmitter and receiver, respectively. With these values, the transmitter exhibits two different but qualitatively similar chaotic attractors. Therefore, to encoding "1" or "0" the private message is used to drive switches in Fig. 2.4. SPICE simulation results for the encrypted transmission of 8 bits are shown in Fig. 2.12(a).

Although chaotic switching is more robust and simplest form of chaotic parameter modulation, it suffers from a lower information transmission rate because the receiver has to wait until synchronization is achieved [19–21]. Nevertheless, given the observer-based synchronization scheme, the convergence rate of synchronization can be assigned by appropriately selecting the observer gain given by **k** [25]. High information transmission rate may cause big synchronization error which used to decode binary signals.

### 2.4.2 Analog Transmission

The experimental set-up to transmit private analog signals by chaotic additive masking [19–21] is shown in Fig. 2.11(b). The basic idea is similar to the previously proposed one to transmit binary data. With respect to the observer equation, the system requires the injection of a current proportional to the error signal, between state $x_1$ and $\xi_1$. Therefore, the transmitter and receiver circuits are synchronized when the chaotic signal $x_1(t)$ is sent by the first channel; while the confidential message $m(t)$ is encrypted in the chaotic signal $x_2(t)$ by means of a additive process with $Rs = 10K\Omega$. In this manner, the confidential information is sent by the second channel. To recover the original message, it is only necessary to apply the reverse operation as shown in Fig. 2.12(b).

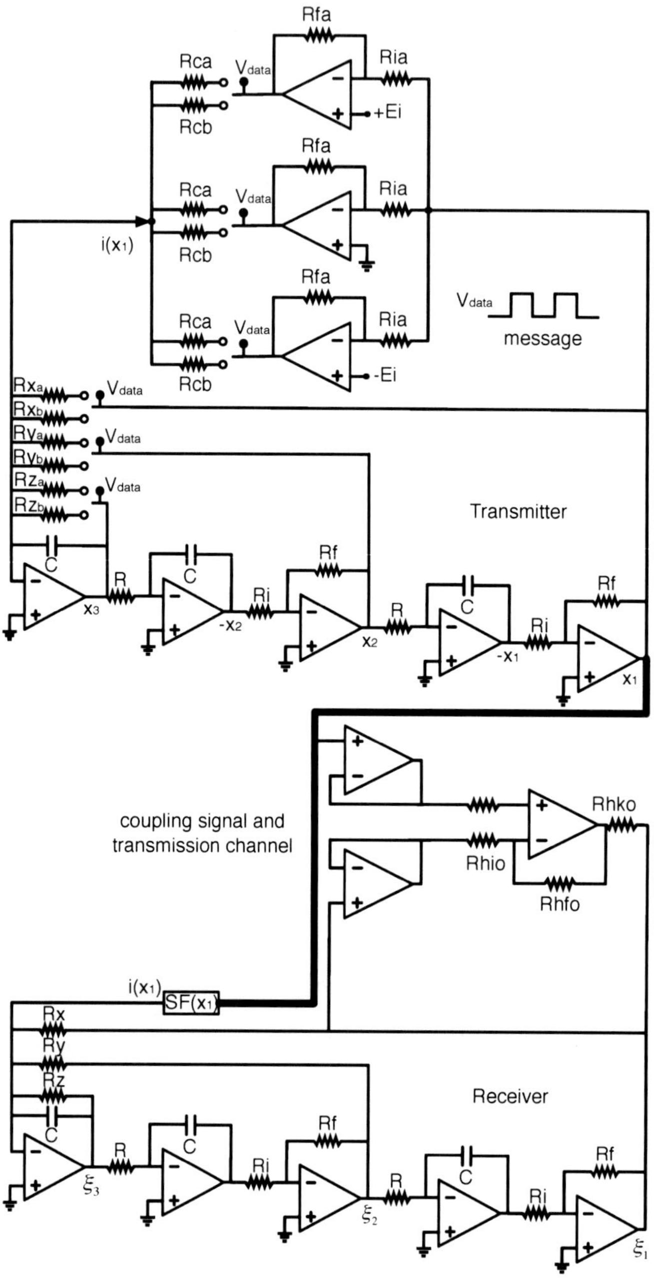


**Fig. 2.11** (a) Circuit diagram to transmit encrypted (a) binary information

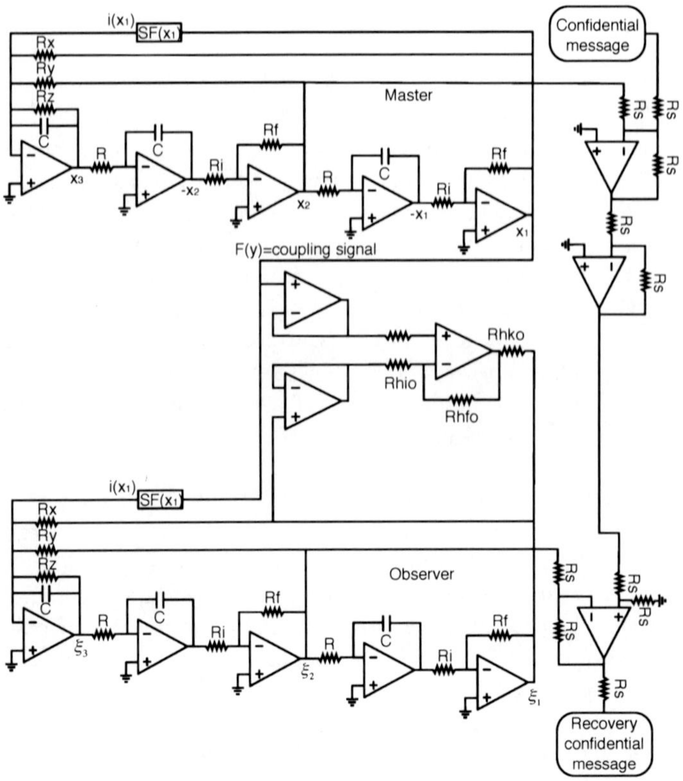


**Fig. 11** (b) Circuit diagram to transmit encrypted (b) analog information

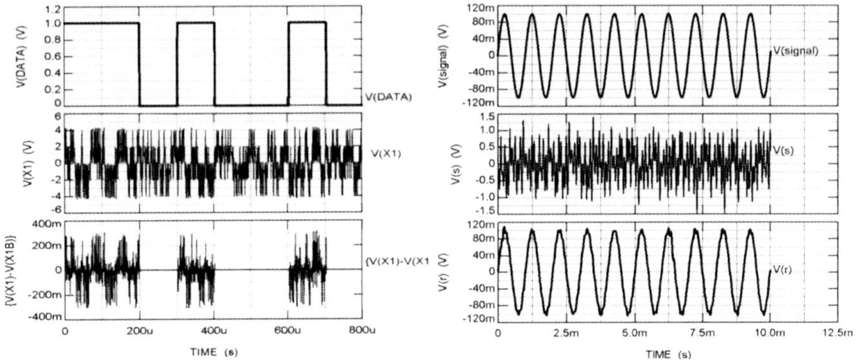


**Fig. 2.12** (a) Private message transmitted (11010010), chaotic signal in the public channel and recovered binary signal at the receiver; (b) Analog message transmitted, chaotic signal in the public channel and recovered analog signal at the receiver

## 2.5 Conclusions

It has been shown the synthesis of 2D-4-scrolls attractors by behavioral modeling. The 2D multi-scroll chaotic oscillator was modeled by state-variables and PWL approximations, and the synthesis process was focused on the implementation of PWL approximations by scaling the excursion levels of the chaotic signals to implement those ones using opamps. In this manner, it was shown that voltage and current saturated functions can be synthesized with opamps by controlling the break points and slopes. As a result, guidelines to synchronize multi-directional multi-scroll chaos generators were introduced. In particular, it was shown the synchronization of 2D-4-scroll and 3D-4 scroll chaotic attractors by using the generalized Hamiltonian forms. The synchronization was achieved by controlling the nonlinear functions in the slave chaos generator with the state-variables from the master chaos generator. Furthermore, it has also been presented a prediction for the values of the observer gain.

To demonstrate the potential application of multi-directional multi-scroll chaotic systems in secure communications, two communication schemes of practical realization to transmit encrypted confidential information, in particular, binary and analog messages, were presented. The proposed communication schemes are based on chaotic switching and chaotic additive masking, respectively. Since SPICE simulations are in good agreement with theoretical results, we can conclude on the usefulness of the proposed synchronization approach. Finally, we use traditional operational amplifiers for designing the encryption schemes. This is an important part of the chapter since it opens new lines for future research on the implementations using different kinds of electronic devices.

**Acknowledgements.** Zambrano-Serrano and Tlelo-Cuautle acknowledge CONACyT for the financial support given trough the scolarship 350385 and the project 131839-Y, respectively; to the PROMEP project UPPUE- PTC-033. Author Sánchez-López thanks the support of the JAE-Doc program of CSIC, co-funded by FSE.

## References

1. He, J.-H.: Nonlinear science as a fluctuating research frontier. Chaos, Solitons & Fractals 41(5), 2533–2537 (2009)
2. Strogatz, S.H.: Nonlinear Dynamics And Chaos: With Applications To Physics, Biology, Chemistry, and Engineering. Westview Press (2001)
3. Lü, J., Chen, G.: Generating multiscroll chaotic attractors: theories, methods and applications. Int. J. Bifurcat. Chaos 16(4), 775–858 (2006)
4. Zeraoulia, E., Sprott, J.C.: Some Open Problems in Chaos Theory and Dynamics. International Journal of Open Problems in Computer Science and Mathematics 4, 1–10 (2011)
5. Muñoz-Pacheco, J.M., Tlelo-Cuautle, E.: Electronic design automation of multi-scroll chaos generators. Bentham Sciences Publishers, Dubai (2010)

6. Trejo-Guerra, R., Tlelo-Cuautle, E., Jiménez-Fuentes, M., Muñoz-Pacheco, J.M., Sánchez-López, C.: Multiscroll Floating Gate based Integrated Chaotic Oscillator. International Journal of Circuit Theory and Applications (2011), doi:10.1002/cta.821
7. Tang, K.W., Kwok, H.S., Tang, W.K.S., Man, K.F.: A Chaos-Based Random Number Generator for Eight-Bit Micro-Controller System. International Journal of Bifurcation and Chaos 18(03), 851–867 (2008)
8. Zhang, Z., Chen, G.: Chaotic motion generation with applications to liquid mixing. In: European Conference on Circuit Theory and Design, pp. 225–228 (September 2005)
9. Sooraksa, P., Klomkarn, K.: No-CPU Chaotic Robots: from classroom to commerce. IEEE Circuits and Systems Magazine, 46–53 (2010)
10. Arena, P., De Fiore, S., Fortuna, L., Frasca, M., Patan, L., Vagliasindi, G.: Reactive Navigation through multiscroll systems: from theory to real-time implementation. Auton. Robot 25, 123–146 (2008)
11. Reiss, J.D., Sandler, M.B.: The benefits of multibit chaotic sigma delta modulation. Chaos 11, 377–383 (2001)
12. Liu, Z., Zhu, X., Hu, W., Jiang, F.: Principles of Chaotic Signal Radar. International Journal of Bifurcation and Chaos 17(5), 1735–1739 (2007)
13. Kwon, O.M., Park, J.H., Lee, S.M.: Secure Communication based on Chaotic Synchronization via Interval Time-varying Delay Feedback Control. Nonlinear Dynamics 63(1-2), 239–252 (2010)
14. Reza Naseh, M., Haer, M.: An optimal approach to synchronize non-identical chaotic circuits: An experimental study. International Journal of Circuit Theory and Applications 39(9), 947–962 (2011)
15. Carbajal-Gómez, V.H., Tlelo-Cuautle, E., Trejo-Guerra, R., Sánchez-López, C., Muñoz-Pacheco, J.M.: Experimental Synchronization of Multiscroll Chaotic Attractors using Current-Feedback Operational Amplifiers. Nonlinear Science Letters B: Chaos, Fractal and Synchronization 1(1), 37–42 (2011)
16. Leenaerts, D., van Bokhoven, W.M.G.: Piecewise Linear Modeling and Analysis. Springer (1998)
17. Muñoz-Pacheco, J.M., Tlelo-Cuautle, E.: Automatic synthesis of 2D-n-scrolls chaotic systems by behavioral modeling. Journal of Applied Research and Technology 7(1), 5–14 (2009)
18. Pecora, L.M., Carroll, T.L.: Synchronization in chaotic systems. Phys. Rev. Lett. 64, 821–824 (1990)
19. Jovic, B.: Synchronization Techniques for Chaotic Communication Systems. Springer, Heidelberg (2011)
20. Sira-Ramirez, H., Cruz-Hernandez, C.: Synchronization of chaotic systems: A generalized Hamiltonian systems approach. Int. J. Bifurcat. Chaos 11(5), 1381–1395 (2001)
21. Muñoz-Pacheco, J.M., Tlelo-Cuautle, E., Trejo-Guerra, R., Sánchez-López, C., Camacho-Pernas, V.M.: Chaos-based Communication Systems by applying Hamiltonian Synchronization. In: IEEE International 53rd Midwest Symposium on Circuits and Systems (MWSCAS 2010), Seattle, Washington, August 1-4, pp. 343–346 (2010)
22. Pikovsky, A., Rosenblum, M., Kurths, J.,, S.: A universal concept in nonlinear sciences. Cambridge University Press (2008)
23. Tlelo-Cuautle, E., Muñoz-Pacheco, J.M.: Numerical simulation of Chua's circuit oriented to circuit synthesis. International Journal of Nonlinear Sciences and Numerical Simulation 8(2), 249–256 (2007)
24. Trejo-Guerra, et al.: On the relation between the number of scrolls and the Lyapunov exponents in PWL-functions-based n-scroll chaotic oscillators. International Journal of Nonlinear Sciences and Numerical Simulation 11(11), 903–910 (2010)

25. Trejo-Guerra, R., Tlelo-Cuautle, E., Cruz-Hernandez, C., Sanchez-Lopez, C.: Chaotic com-munication system using Chua's oscillators realized with CCII+s. Int. J. Bifurcat. Chaos 19(12), 4217–4226 (2009)
26. Yu, S., Lü, J., Chen, G.: A novel multiscroll chaotic system and its realization. In: IEEE International Symposium on Circuits and Systems (ISCAS 2008), Seattle, Washington, May 18-21, 2008, pp. 2390–2393 (2008)
27. Chen, H., Ding, Q., Ding, L., Dong, X.: Experimental study on secure communication of different scroll chaotic systems with identical structure. ICIC Express Letters 2(2), 201–206 (2008)
28. Trejo-Guerra, R., Tlelo-Cuautle, E., Jiménez-Fuentes, J.M., Sánchez-López, C., Muñoz-Pacheco, J.M., Espinosa-Flores-Verdad, G., Rocha-Pérez, M.: Integrated circuit generating 3- and 5- scroll attractors. Communications in Nonlinear Sciences and Numerical Simulation (2012), doi:10.1016/j.cnsns.2012.01.029
29. Zambrano-Serrano, E., Muñoz-Pacheco, J.M., Félix-Beltran, O.G., Trejo-Guerra, R., Gómez-Pavón, L.C., Tlelo-Cuautle, E., Sánchez-López, C., Luis-Ramos, A.: Synchronization of multi-directional multi-scroll chaos generators: A Hamiltonian approach. In: Third International Workshop on Nonlinear Dynamics and Synchronization (INDS 2011), Klagenfurt, Austria, July 25-27, pp. 87–91 (2011)

# Chapter 3
# Nonlinear Filtering of Chaos for Real Time Applications

V. Kontorovich and Z. Lovtchikova

**Abstract.** Nowadays chaotic modelling of real phenomena in electrical engineering problems, particularly in communications, control, etc, is a topic of growing interest. Therefore, the filtering of chaos is one of the most important techniques for many applications including those which are related to real- time regime conditions. These applications include, but are not limited to chaos-based communication system synchronization, real-time control of chaos, radio-frequency interference filtering and mitigation, chaotic system identification, etc. This chapter presents among well known original results of a study of nonlinear filtering of chaotic signals in presence of additive white Gaussian noise (AWGN) and related topics.

## 3.1 Introduction

Chaos filtering is one of the most important techniques for many applications of chaos: chaos-based communications, control systems, interference modelling, chaotic system identification, etc. In fact, the complete list of references related to this topic is enormous and it is hardly possible to provide it. However, to get familiar to some of them, it is possible to make search at [18–20] and references therein. Despite of the growing interest of researches, practical specialists and academics in the topics of chaos filtering, some of the main problems related to real time applications of nonlinear algorithms are still not solved. For instance: why chaos filtering (even for quasi-optimum cases) provide with higher accuracy of filtering compared with traditional models?, which is the reasonable compromise between accuracy

V. Kontorovich
CINVESTAV-IPN, Av. IPN 2508 C.P. 07360 Mexico D.F.,
e-mail: valeri@cinvestav.mx

Z. Lovtchikova
UPIITA-IPN, Av. IPN 2580 C.P. 07340 Mexico D.F.
e-mail: alovtchikova@ipn.mx

K. Kyamakya et al. (Eds.): Selected Topics in Nonlinear Dynamics, SCI 459, pp. 41–59.
DOI: 10.1007/ 978-3-642-34560-9_3 

requirements for filtering and complexity of algorithms?, how to improve the accuracy of filtering for low Signal/Noise scenarios? and many, many others. Sure, in the framework of the short chapter it is hardly possible to give the complete answers on those questions and many others about filter synchronization, chaos quantification, etc immensely important for this topic.Due to the lack of space, authors decided to concentrate on the presentation of the original material of quasi-optimum algorithms of nonlinear filtering for low-dimensional chaotic signals and of the trends of their improvement in order to encourage readers to pay attention to this new trends.

The first part of this chapter is a brief extraction from a previous publication [12] not related to filtering and dedicated to the cumulants analysis of chaos It is shown that chaotic signals generated by strange attractors and applied as models for real time phenomena can be considered as Non-Gaussian stochastic process. Moreover, it is explained that these chaotic signals can be seen as a degenerated or quasi-deterministic Markov processes described by Stochastic Differential Equations (SDE) with external white noise force tending to zero. This material is introductory and helps for better understanding of the ongoing material. Then, the problem of optimum filtering of chaos as a degenerated Markov process is described following the fundamental ideas of R. Stratonovich and H. Kushner.

Chaotic signals considered in this chapter for modelling real phenomena in the filtering problem are those generated from the well known attractors: Lorentz, Chua and Rössler.

Optimum filtering algorithms are difficult to implement in real-time applications. In this regard, the approximate nonlinear algorithms are considered: Extended Kalman Filter(EKF), Functional, Integral approximation approaches for a-posteriori PDF, Unscented filter algorithms, Quadrature Kalman Filters, etc., and the results of the comparative study of these algorithms are presented. Then, one opportunistic approach for improving of the filtering accuracy is discussed as well.

## 3.2 Chaotic Modelling of Random Signals

It is well known [4], that each dissipative continuous time dynamic system (strange attractor) can be defined with the following equation:

$$\dot{x} = \mathbf{f}(x(t)), \quad x \in \mathbf{R}^n, \quad x(t_0) = x_0, \tag{3.1}$$

where $f(\bullet) = [f_1(x),...,f_1(x)]^T$

Chaotic attractors, described by (3.1) can be classified as hyperbolic, quasi-hyperbolic and non-hyperbolic [5]. All attractors, considered here (see, for example [4, 5]) are of the quasi-hyperbolic or non-hyperbolic type: (Lorenz, Chua, Rössler, etc.). The quasi-hyperbolic attractors do not actually differ from the hyperbolic (robust, ideal) attractors and the existence of an invariant measure for them is practically guaranteed (see details in [1, 2]). If the invariant measure is a stationary PDF of the chaos, then its existence is immensely important for the applications. In the following, we will take a physical measure [12] as an invariant measure applying the following idea of Kolmogorov: in (3.1) weak external noise $\xi(t)$ has to be considered, i.e

$$\dot{x} = \mathbf{f}(x(t) + \varepsilon\xi(t)) \tag{3.2}$$

where $\xi(t)$ is a vector of a weak external white noise with the related positive defined matrix of "intensities" $\varepsilon = \left[\varepsilon_{ij}\right]^{nxn}$

It can be seen that (3.2) is a Stochastic Differential Equation (SDE), generating the continuous Markov process with a physical measure (PDF) [4, 5, 20]. Note that relations between chaotic continuous processes and Markov processes were thoroughly discussed in [5]. We will only stress here, that the linear Kolmogorov-Fokker-Plank (KFP) operator for $W_\varepsilon(x,t)$ for SDE (3.2) has the property: $\lim_{t\to\infty} W_\varepsilon(x,t) = W_{st}^{\varepsilon}(x)$ when the vector function $f(\bullet)$ does not depend on time 't' (sufficient conditions, see [20]).

Thus, the physical measure $W_{st}(x,t)$ has to be taken as

$$W_{st}(x) = \lim{}_{\forall\varepsilon_{ij}\to 0} W_{st}^{\varepsilon}(x) \tag{3.3}$$

When noise intensities in SDE (3.2) tend to zero, the Markov process is a so-called 'quasi-deterministic' or 'degenerated' Markov process.

The existence of the stationary PDF for Rössler attractor was confirmed at [2]. The same was checked experimentally by the authors for Lorenz and Chua attractors.

As any PDF, the $W_{st}(x,t)$ as well as its characteristic function, is totally defined by the complete set of cumulants [18].

Since $W_{st}(x,t) = F^{-1}\{\theta(jv)\}$, where $F\{\bullet\}$ and $F^{-1}\{\bullet\}$ are direct and inverse Fourier transforms respectively, and $\theta(jv)$ is the characteristic function defined as:

$$\theta(jv) = \exp\left[\sum_{s=1}^{\infty} \frac{j^s}{s!} \sum_{m_1,m_2,\ldots,m_n}^{s} \kappa_{m_1,m_2,\ldots,m_n}^{\xi_1,\xi_2,\ldots,\xi_n} v_1^{m_1} \ldots v_n^{m_n}\right], \tag{3.4}$$

where $\{\xi_i\}_1^n$ are random variables, $m_1 + m_2 + \ldots + m_n = s, \kappa_{m_1,\ldots,m_n}^{\xi_1,\ldots,\xi_n}$ is the joint cumulant of the s-th order.

We will concentrate hereafter on the analytical representations of PDF's for Lorenz, Chua and Rössler attractors.

### 3.2.1 Approximations for PDF of Strange Attractors

In this section, the mathematical representations for the above considered strange attractors will be presented.

Lorenz Attractor

$$\begin{aligned}
\dot{x} &= \sigma(y - x), \\
\dot{y} &= Rx - y - z, \\
\dot{z} &= xy - Bz,
\end{aligned} \tag{3.5}$$

where $\sigma$, $B$ and $R$ are the parameters of the attractor , $x = [x,y,z]^T$.

Chua Attractor

$$\begin{aligned}
\dot{x} &= \beta_1 (y - x) - \alpha h(x), \\
\dot{y} &= \beta_2 (x - y) + \beta_4 z, \\
\dot{z} &= -\beta_3 y,
\end{aligned} \tag{3.6}$$

where $\beta_1 \div \beta_4$ and $\alpha$ are the parameters of the attractor; $x = [x,y,z]^T$ and an inertial non-linearity $h(x)$ has the following form:

$$h(x) = \begin{cases} -L & x < -L \\ x & |x| < L \\ L & x > L \end{cases} \tag{3.7}$$

Rössler Attractor

Equation (3.1) for the Rössler attractor has the following form:

$$\begin{aligned}
\dot{x} &= -y - z, \\
\dot{y} &= x + ay, \\
\dot{z} &= b + zx - zc,
\end{aligned} \tag{3.8}$$

where $a,b,c$ are the parameters of the attractor, $x = [x,y,z]^T$

All above mentioned attractors are represented by ordinary differential equations of third order with components "$x$", "$y$", and "$z$", so the number of dimensions of the chaotic models applied in the following is rather low. Now, some analytical approximations for the PDF's for components of attractors are presented.

It is worth to mention that different approaches were applied for analytical approximations. These approaches utilize significantly the values of cumulants of chaos (Gramm-Charlie method, etc.) and they were explained in details at [12, 18, 20]

The PDF $W(x)$ (in normalized scale) for the component "$x$" for the Lorenz attractor as well as the experimental (simulation) histogram and the approximate PDF-Gaussian distribution is illustrated in Fig. 3.1.

Let us present an analytical approximation for the Chua attractor also for the $x$-component. An approximation of the PDF histogram that presents bimodal characteristics is chosen (Fig. 3.2) according to the following equation [20]:

$$W(x) = C(p,q)\exp\left(px^2 - qx^4\right), \tag{3.9}$$

where $C$ represents a normalization constant, while $p$ and $q$ are approximations parameters.

The parameters $p$ and $q$ can be obtained from the central moments equations for (3.9):

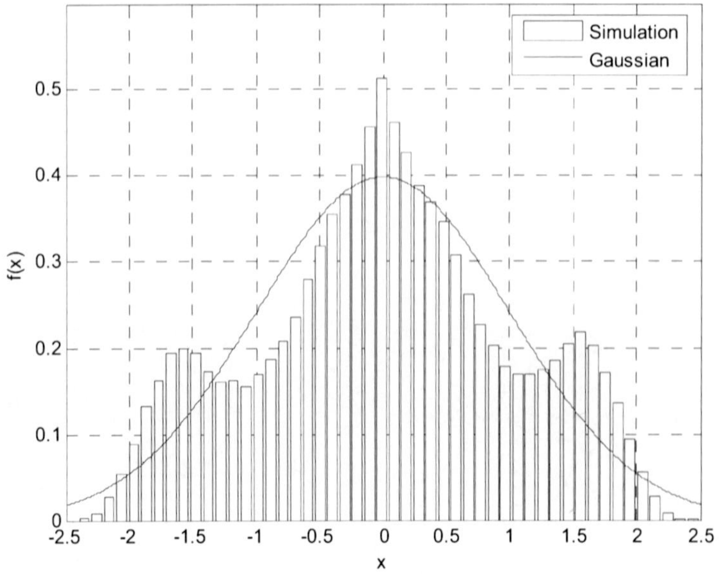


**Fig. 3.1** The Lorenz attractor and its approximations

$$\overline{x^{2n}} = \frac{\Gamma\left(n+\frac{1}{2}\right) D_{-n-\frac{1}{2}}(-\delta)}{\sqrt{\pi} D_{-\frac{1}{2}}(-\delta)\,(2q)^{n}/_{2}}, \tag{3.10}$$

where $n = 1, 2, \dots, \delta = \frac{p}{\sqrt{2q}}$ which can be found from $\kappa_2^x$ and $\kappa_4^x$.

Parameters "$p$" and "$q$" are $q = 1.5; p = 3.5$ while the normalization constant $C$ is 0.063.

In order to confirm that the approximation is valid, the comparison between the histogram of the PDF obtained from the attractor and the analytical approximation was made using the Kolmogorov-Smirnoff goodness of fit test (KST) (Fig. 3.3) with a level of significance: $\alpha = 0.05$. From that test, it can be concluded that the analytical approximation is valid.

For Rössler attractor, the PDF histograms for "$x$" and "$y$" components are described by means of the Laplace distribution defined as:

$$f(x) = \frac{1}{2\lambda} \exp\left\{-\frac{|x-\mu|}{\lambda}\right\}, \tag{3.11}$$

where $\mu$ and $\lambda$ are location and scale parameters, respectively. The use of Laplace distribution allows to make a right description for the "$x$" and "$y$" components of the Rössler attractor as it is observed in Figures 3.4 and 3.5.

Then, as it follows from Figures 3.4 and 3.5, the PDF for the "$x$" component of the Rössler attractor is approximated by a Laplace distribution with the local parameter $\mu = 0$ and scale parameter $\lambda = 1.1$ and for the "$y$" component with local

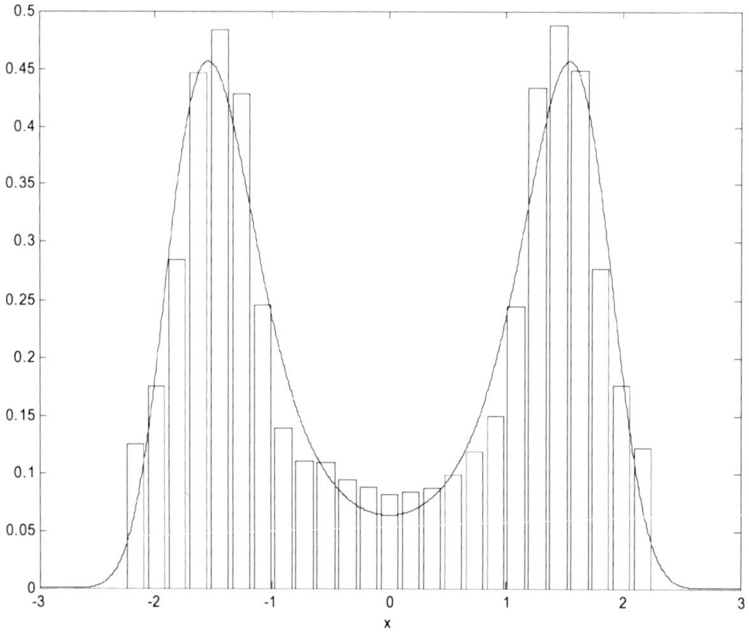


**Fig. 3.2** PDF histogram for $x$ component

parameter $\mu = 0$ and scale parameter $\lambda = 0.85$. Examples of the practical validity of such kind of approximations can be found at [12].

As it can be seen from the material presented above and [12], chaotic models can be interpreted as significantly Non-Gaussian random processes: let us stress here that these PDF's for instantaneous values are typical, at least, for the Radio-Frequency Interferences (RFI) Non-Gaussian statistical features. But for the rather complete characterization of the Non-Gaussian processes it is not enough to mention only one-moment statistical characteristics and it is reasonable to consider cumulants (cumulants functions) of the higher order and its multi-moment statistical description.

As it was mentioned in the Introduction, these characteristics provide with correct "choice" of the filtering strategy for chaos. Hereafter only few material of multi-moment cumulant functions description will be presented. The detailed analysis is rather cumbersome (see [14, 18]).

### *3.2.2 Degenerated Cumulant Equations for Two-Moment Cumulants*

Let us consider equations (3.1) and (3.2).Then, as it follows from [18, 20] to obtain the transient PDF $W(x, t|\, x_0, t_0)$ for solution of SDE (3.2), operating with the

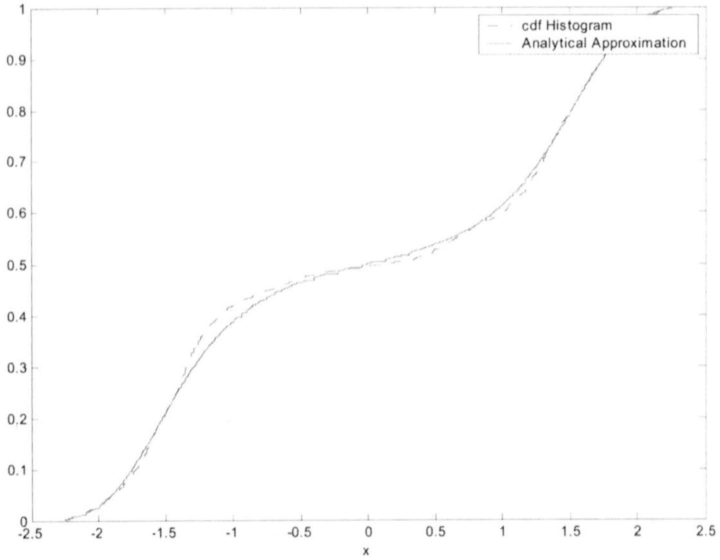


**Fig. 3.3** Illustration of the KST application of Chua attractor with approximation.

linear Kolmogorov-Fokker-Plank (FPK) operator and its adjoin inverse Kolmogorov operator $L^{+}(x)$ [2, 18, 20]:

$$L^{+}(x) = K_{1i}(x)\frac{\partial}{\partial x_i} + K_{2_{ij}}(x)\frac{\partial^2}{\partial x_i \partial x_j}, \tag{3.12}$$

where $K_{1i}(x) = f(x)$ and $K_{2_{ij}} = \left[\varepsilon_{ij}\right]^{n\times n}$ are kinetic coefficients [20] and $\epsilon_i j$ are the white noise intensities defined at (3.2). Taking into account that the two-moment PDF: $W(\mathbf{x}_0, t_0, \mathbf{x}, t) = W(\mathbf{x}_0, t_0)\, W(x, t|\, \mathbf{x}_0, t_0)$, satisfies the same FPK equation as $W(x, t|\, x_0, t_0)$ applying the same methodology to obtain degenerated cumulant equations as it was presented in [12].

For stationary conditions, which are assumed in the following, $t - t_0$ is defined as $\tau$ and all two-moment cumulants will be two-moment cumulant functions that depend on $\tau$. In the following, only final results for concrete attractors are presented and detailed analysis can be found at [14].

Lorenz Attractor

Let us make the following notations:

$< x_1, x_{1_\tau} >= \kappa_2^1(0,\tau)$, is a covariance function of the first component. Then using (3.4), it is possible to get:

$$\kappa_2^1(0,\tau) = \kappa_2^1 \exp\left[-\sigma\left(1 - \frac{\kappa_{1,1}^{1,2}(0,\tau)}{\kappa_2^1(0,\tau)}\right)|\tau|\right] \tag{3.13}$$

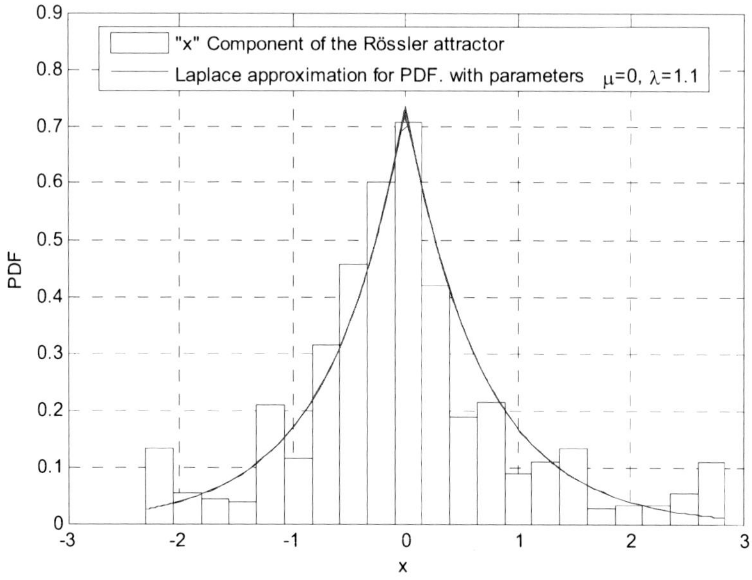


**Fig. 3.4** Laplace approximation of the $x$ component

For two asymptotic cases from (3.13) $\tau << \tau_{corr}$ and $\tau >> \tau_{corr}$ (here $\tau_{corr}$ is a correlation time) one can get:

When $\tau << \tau_{corr}$ , $\kappa_{1,1}^{1,2} \approx \kappa_2^1$ and $\kappa_2^1(0,\tau) \approx \kappa_2^1$ , so it is easy to show, that:

$$\kappa_2^1(0,\tau) = \kappa_{2_0}^1 \left(1 - \frac{(\Delta f_{eff}\tau)^2}{2}\right), \tag{3.14}$$

where $\Delta f_{eff}$ is the effective bandwidth for the first component of the statistically linearized attractor; $\tau_{corr} \sim \frac{1}{\Delta f_{eff}}$ (see [20] for details).

If $\tau >> \tau_{corr}$ and supposing that $\lim_{\tau\to\infty} \kappa_{1,1}^{1,2}(0,\tau) \sim 0$, with assumption that the tendency to zero, when $\tau \to \infty$, is "faster" for $\kappa_{1,1}^{1,2}(0,\tau)$ than for $\kappa_2^1(0,\tau)$. Then for $\tau \to \infty$:

$$\kappa_2^1(0,\tau) \to 0 \tag{3.15}$$

Chua Attractor

Almost in the same way as it was done above for the Lorenz case, it is possible to get: For $\tau << \tau_{corr}$ :

$$\kappa_2^1(0,\tau) \approx \kappa_{2_0}^1 \left(1 - \Delta f_{eff}|\tau|\right) \tag{3.16}$$

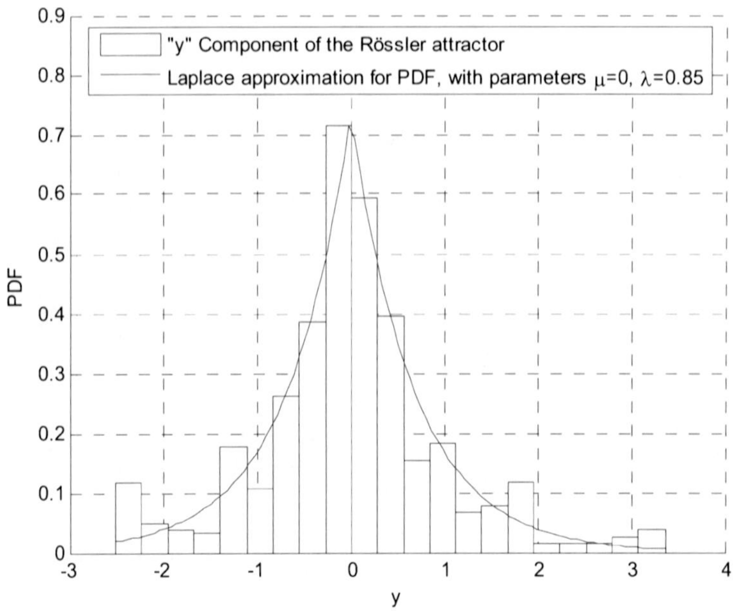


**Fig. 3.5** Laplace approximation of the $y$ component

For $\tau << \tau_{corr}$:

$$\kappa_2^1(0,\tau) \overset{\tau \to \infty}{\to} 0 \tag{3.17}$$

When $\tau << \tau_{corr}$ , the behavior of $\kappa_2^1(0,\tau)$ for Lorenz and Chua attractors is completely different. The approximate method for calculating $\kappa_{2_0}^1$ was presented at [12]. Examples of the higher order cumulants functions can be found at [14].

## 3.3 Filtering of Chaotic Signals in Presence of Additive Gaussian Noise

### *3.3.1 Markov Theory of Non-linear Filtering*

Let us consider the following filtering scenario where the received signal is:

$$y(t) = s(t, x(t)) + n_0(t) \tag{3.18}$$

where $y(t)$ is vector of the received signal with dimension "$m$", $s(\cdot)$ is a vector function of the desired signal of the same dimension "$m$", $n_0$ is the vector of the white additive noises with the intensity matrix $N_0(mxm)$. Here the signal $s(\cdot)$ depends on the "message" $x(t)$ (see SDE (3.1) and (3.2)) which is subject of filtering and is modeled by means of the following SDE as an n-dimensional Markov diffusion process:

$$\dot{x} = g\,(t,x) + \xi(t) \tag{3.19}$$

Formally, SDE coincides with (3.2) and the vector function $g\,(\cdot)$ is similar to $f\,(\cdot)$ in (3.2); the matrix of intensities in (3.19) is $D$. As it is well known (see, for example [20,22], etc.) with this assumption the a priori Probability Density Function, or a priori PDF for x(t) follows the so called Fokker-Plank-Kolmogorov (FPK) equation:

$$\frac{\partial W_{PR}(x,t)}{\partial t} = -\sum_{i=1}^{n}\frac{\partial}{\partial x_i}\left[g_i(t,x)W_{PR}(x,t)\right] + \frac{1}{2}\sum_{i=1}^{n}\sum_{j=1}^{n}\frac{\partial^2}{\partial x_i \partial x_j}\left[D_{ij}W_{PR}(x,t)\right], \tag{3.20}$$

where $W_{PR}\,(x,t_0) = W_0\,(x)$

The equation (3.20) can be rewritten in another form [10,21]:

$$\frac{\partial W_{PR}(x,t)}{\partial t} = -div\pi(x,t) \tag{3.21}$$

or

$$\frac{\partial W_{PR}(x,t)}{\partial t} = L_{PR}\left\{W_{PR}(x,t)\right\} \tag{3.22}$$

where $\pi\,(x,t)$ is a probabilistic "flow" with the components:

$$\pi(x,t) = g_i(x,t)W_{PR}(x,t) - \frac{1}{2}\sum_{j=1}^{n}\frac{\partial}{\partial x_j}\left[D_{ij}W_{PR}(x,t)\right] \tag{3.23}$$

In (3.20)-(3.23) $\{g_i(x,t)\}_1^n$ are drift coefficients and $\{D_{ij}\}$ are diffusion coefficients of the Markov process. Note that in the following, they are defined in the Stratonovich sense [20, 22]and $L_{PR}\{\bullet\}$ is a FPK linear operator. Then, as it was shown in [22] the integro-differential equation for the a posteriori PDF $W_{PS}\,(x,t)$ is given in the following equivalent forms:

$$\begin{aligned}\frac{\partial W_{PS}(x,t)}{\partial t} =& L_{PR}\left\{W_{PS}(x,t)\right\} + \\ & \frac{1}{2}\left[F(x,t) - \int_{-\infty}^{\infty} F(x,t)W_{PS}(x,t)dx\right]W_{PS}(x,t)\end{aligned} \tag{3.24}$$

or

$$\frac{\partial W_{PS}(x,t)}{\partial t} = -div\hat{\pi}(x,t) + \frac{1}{2}\left[F(x,t) - < F(x,t) >\right]W_{PS}(x,t), \tag{3.25}$$

where $< F(x,t) >= \int_{-\infty}^{\infty} F(x,t) W_{PS}(x,t) dx$, $\hat{\pi}(x,t)$ is (3.25), $W_{PR}(x,t)$ is substituted by $W_{PS}(x,t)$ and

$$F(x,t) = \left[ y(t) - \frac{1}{2} s(x,t) \right]^T N_0^{-1} \left[ y(t) - \frac{1}{2} s(x,t) \right] \tag{3.26}$$

Equations (3.24), (3.25) together with (3.26) are called Stratonovich-Kushner Equations (SKE) and have a following interpretation: the first summand in (3.26), (3.27) describes the dynamics of the a priori dates of the x(t) and the second summand depends on the innovation of the a priori dates from the analysis of observations(see details at [13]).

Note that the equation SKE fully describes the "evolution" of $W_{PS}(x,t)$ in time but actually does not provide exact analytical solutions. There are very few exceptions: linear SDE (3.4) [7, 10, 16, 17, 21–23]; the Zakai approach [24], etc. Due to this the nonlinear filtering algorithms are practically always approx-imate.

Supposing of the main practical interest of low SNR, it might be reasonable to consider the application of the high order cumulants (HOS) (see [11, 21, 22] for example).

As it follows from the material of the section 2, the filtering of chaos is equivalent to the filtering of quasi-deterministic (or degenerated) Markov process and for this special case it can be found some general features of the solutions of SKE equations (3.31)-(3.33): the $W_{PS}(x,t)$ at least for very low and very high signal to noise ratios (SNR) tends to a delta-function ,defined at the completely deterministic solution of (3.1). Such kinds of solutions are "singular" ones and show that the filtering algorithm is "tuned" to the attractor solution and asymptotically does not depend on the SNR value. This property of the optimum algorithms of chaos filtering was proofed in [13] and in the following it will be shown and proofed by simulations, that beginning from the rather low SNR, it is true for the approximate algorithms of nonlinear filtering as well.

### 3.3.2 *Approximate Algorithms of Non-linear Filtering of Chaos*

As it was commented in [3, 21], it is always "better" to approximate the a posteriori PDF $W_{PS}(x,t)$ than the nonlinearity at (3.21), (3.26). In this context, let us mention the following approximate approaches for $W_{PS}(x,t)$:

- Gaussian approximations: Extended Kalman Filter (EKF) [16,21,22]; Unscented Kalman Filter (UKF) [8]; Quadrature Kalman Filter (QKF) [3]; Integrated Kalman Filter (IKF), etc. It is the main stream "ideological" trend for the creation of the approximate algorithms (see references above).
- Functional approximations for $W_{PS}(x,t)$ [10, 21];
- Integral or Global approximations for $W_{PS}(x,t)$ [10];
- HOS approximations for $W_{PS}(x,t)$; etc.

Let us start with the Extended Kalman Filter (EKF). Considering $W_{PS}(x,t)$ as a three dimensional Gaussian PDF $\hat{W}_G(x,t)$ From (3.31) it is possible to obtain the following equations for per-component of the mean estimates $\{\hat{x}_i\}_1^3$ and for estimates of the elements of the a posteriori covariance matrix $\left\{\hat{R}_{ij}\right\}_{i,j=1}^{3}$:

$$\dot{\hat{x}}_i = \int_{-\infty}^{\infty} \left(\hat{\pi}^T(x,t)\,\mathrm{grad}x_i\right) dx + \frac{1}{2}\left[\int_{-\infty}^{\infty} x_i F(x,t)\,\widehat{W}_G(x,t)\,dx - \widehat{x}_i \int_{-\infty}^{\infty} F(x,t)\,\widehat{W}_G(x,t)\,dx\right] \tag{3.27}$$

or

$$\dot{\widehat{R}}_{ij} = \int_{-\infty}^{\infty} \left(\hat{\pi}^T(x,t)\,\mathrm{grad}\overset{\circ}{x}_i\overset{\circ}{x}_j\right) dx + \frac{1}{2}\left[\int_{-\infty}^{\infty} \overset{\circ}{x}_i\overset{\circ}{x}_j F(x,t)\,\widehat{W}_G(x,t)\,dx - \widehat{R}_{ij} \int_{-\infty}^{\infty} F(x,t)\,\widehat{W}_G(x,t)\,dx\right], \tag{3.28}$$

where $\overset{\circ}{x}_i = x_i - \hat{x}_i$, $\overset{\circ}{x}_j = x_j - \hat{x}_j$.

Practically it is possible to assume for $\forall \hat{R}_{ij}(t)$, (when $t \to \infty$) that they are converging to the stationary values $\overline{R_{ij}}$, and in consequence the second equation in (3.31) usually tends to the system of non-linear algebraic equations, which can be solved numerically.

$$W_{PS}(x,t) = \prod_{i=1}^{3} W_{PS}(x_i)\left[1 + \sum_{q=2}^{3}\sum_{j=1}^{q-1} \frac{R_{qj}}{R_{qq}R_{ji}}(x_q - \hat{x}_q)(x_j - \hat{x}_j)\right] \tag{3.29}$$

From (3.32), it is possible to observe that the Functional Approximation for the PDF is sufficiently non-Gaussian (marginal $W_{PS}(x_i,t)$ are arbitrary) but for "joint" characterization of the vector $\hat{x}$, only elements of the a posteriori covariance matrix are considered.

It can be shown that the equations for $\{\hat{x}_i\}_1^n$ and $\left\{\hat{R}_{ij}\right\}$ are the same as in (3.31), being the only difference that instead of $\hat{W}_G(x,t)$, the approximation (3.32) for $W_{PS}(x,t)$ has to substituted in (3.31). The corresponding integrals can be solved analytically or by the Gauss-Hermit quadrature formula (see [3, 8]).

When the SNR is low, it is necessary to look for approach, which is called Integral approximation. This approach was proposed for successful approximation of $W_{PS}(x,t)$ including the PDF's "tails", i.e. for the whole span of $x$. For the lack of space it will be omitted here; however, description of this approach is presented at [13].

Let us present an example of the analytical treatment of the approximate algorithm.

For simplicity let us consider the following special case of the one dimensional scenario:

$$y(t) = x_1(t) + n_0(t) \tag{3.30}$$

where $x_1(t)$ is the first (observable) component of any strange attractor (Lorenz, Chua, Rössler) and $n_0(t)$ is an scalar white noise.

Let us consider clearly Non-Gaussian case of the Chua attractor; additionally let us assume the low SNR scenario and applying the functional approximation (3.33) for $W_{PS}(x,t)$, it is possible to get:

$$\begin{aligned}\dot{\hat{x}}_1 = &- 2D\hat{x}_1(p_1+q_1) + 2Dq_1\hat{R}_{11} + \\ &2Dp_1\frac{2(y(t)-\hat{x}_1)\hat{R}_{11}}{N_0} - \frac{\hat{x}_1}{N_0}\left(y^2(t) - \hat{R}_{11}\right)\end{aligned} \tag{3.31}$$

If $D \to 0$ and the SNR is low, then from (3.34) and (3.31) it follows:

$$\dot{\hat{x}}_1 = -2D\hat{x}_1(p_1+q_1) + \frac{2(y(t)-\hat{x}_1)}{N_0} \tag{3.32}$$

and immediately obtain:

$$\dot{\hat{x}}_{11} = -\frac{D}{2} + \frac{\hat{R}^2_{11}}{N_0} + 4D(p_1+q_1)\hat{R}_{11} \tag{3.33}$$

It is easy to show, that (3.35), (3.36) coincides totally with the EKF for one component $x_1$. For $t \to \infty$ $\hat{R}_{11}(t)$ tends to its stationary value $\overline{R}_{11}$, which can be simply calculated as

$$\overline{R}_{11} = \frac{-4D(p_1+q_1) + \sqrt{16D^2(p_1+q_1)^2 + \frac{D}{N_0}}}{\frac{2}{N_0}} \geq 0 \tag{3.34}$$

Invoking $D \to 0$,

$$\overline{R}_{11} \cong 0.71 \cdot \sqrt{N_0 D} \tag{3.35}$$

Assuming that $N_0 \cong 1$, then

$$\overline{R}_{11} << 1 \tag{3.36}$$

Therefore, the EKF shows its adequacy for application to the case of Chua attractor. Unfortunately the analytical analysis of all approximate algorithms mentioned above is impossible to achieve. So in the next section the simulation results for some of them will be presented.

### 3.3.3 Comparative Analysis of Nonlinear Filtering Approach

In this section, extended Kalman filter (EKF), unscented Kalman filter (UKF) , Gauss-Hermite Quadrature filter (GHF), and Kalman Quadrature filter (KQF) are compared by simulations from the point of view MSE = $f$(SNR), where MSE is a minimum square error of filtering.

The detailed description of the algorithms for these filtered are presented in the above cited references. Hereafter only the simulation results of application of these algorithms for the filtering of components of Lorenz, Chua and Rössler at-tractors are presented.

It can be easily shown that UKF involves the bigger complexity, while EKF seems to be the simpler algorithm.

Chua Attractor

Fig. 3.6 shows the MSE vs. SNR for the Chua attractor for the different filters mentioned above. Even so, the MSE generated by the EKF is really small (SNR is about 0.5)

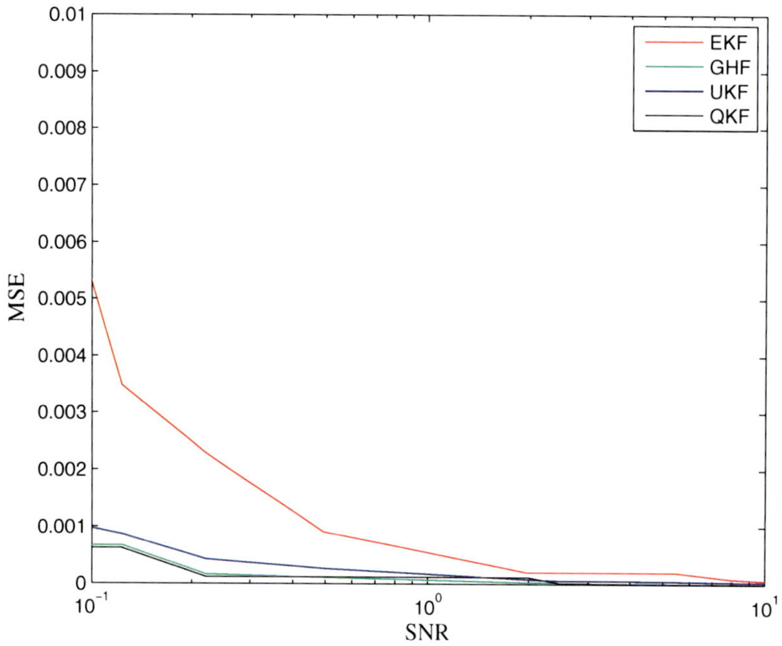


**Fig. 3.6** MSE vs.SNR for Chua attractor

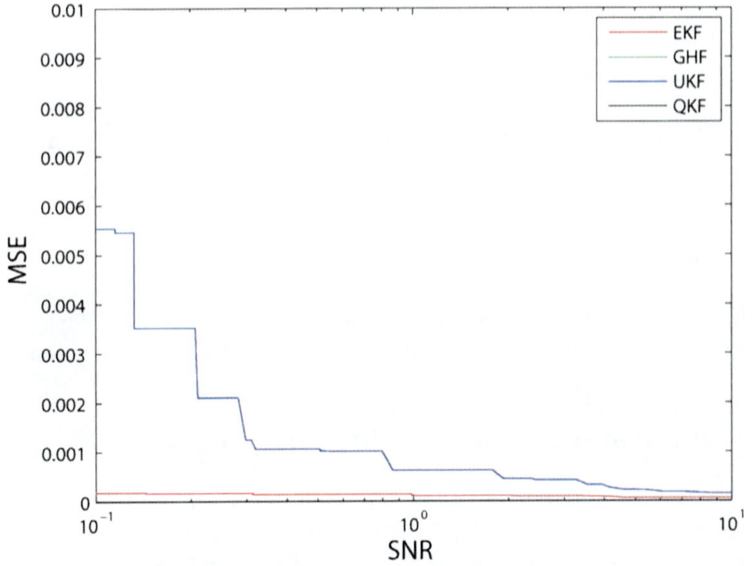


**Fig. 3.7** MSE vs.SNR for Lorenz attractor

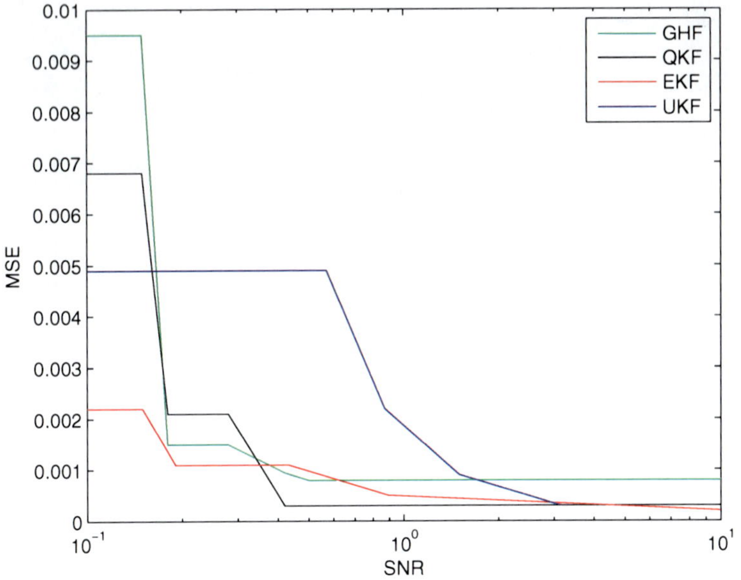


**Fig. 3.8** MSE vs.SNR for Rössler attractor

Lorenz Attractor

The effectiveness of the filtering methods working on the Lorenz attractor is analyzed by plotting the MSE versus SNR in. Simulation results appear in Fig. 3.7.

For the case of Lorenz attractor, GHF, QKF and EKF give better results than UKF. MSE for GHF and QKF are omitted here as they indistinguishable with x-axis.

Rössler Attractor

For Rössler system, shown in Fig 3.8, EKF works better than GHF, QKF, and UKF. It can be seen that as a compromise between complexity and filtering accuracy, the EKF is the best choice for the real time applications as it provides with the less than 0.1% MSE for SNR about 0.5(-3 db) and is rather simple for implementation.

## 3.4 "Multi-moment" Nonlinear Filtering of Chaos

From section 3, it is possible to observe that all algorithms tackled there are of "one-moment" type, i.e they are dealing instantaneously with the data of one moment of time. However, it might be challenging if it is possible to apply more a-priori information regarding statistical features of the filtered process to improve the filtering

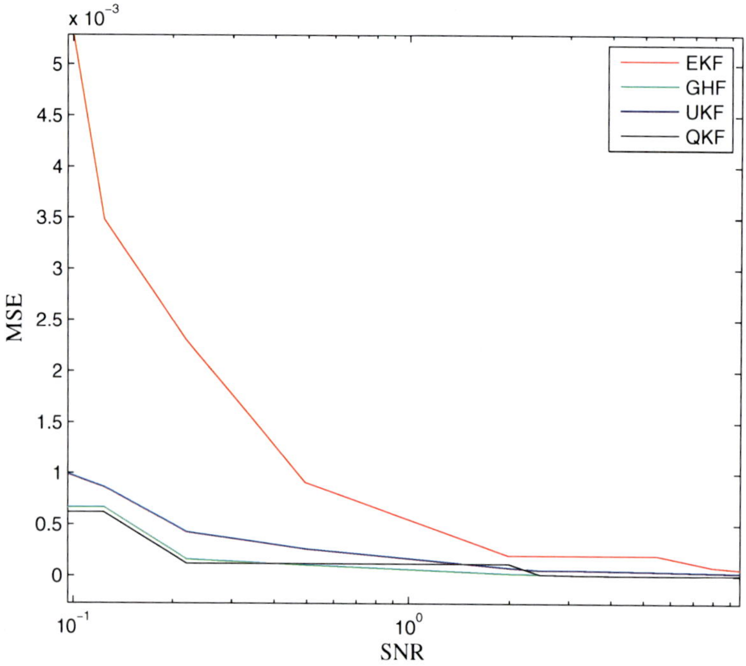


**Fig. 3.9** Chua attractor

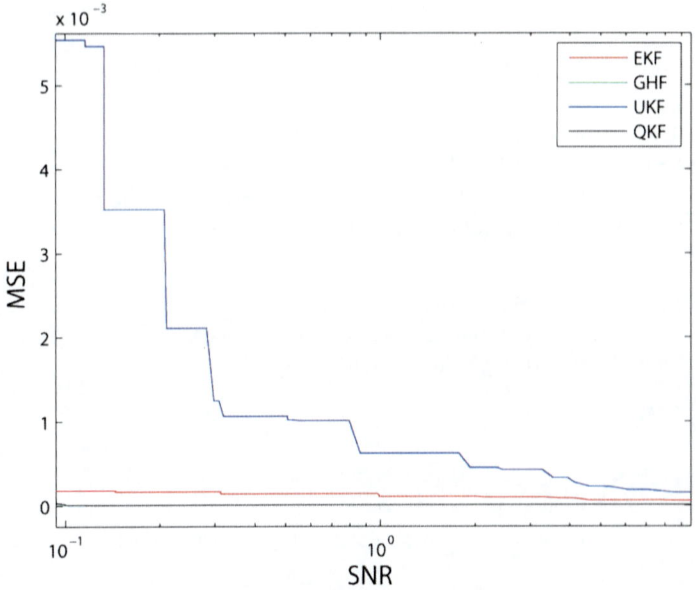


**Fig. 3.10** Lorenz attractor

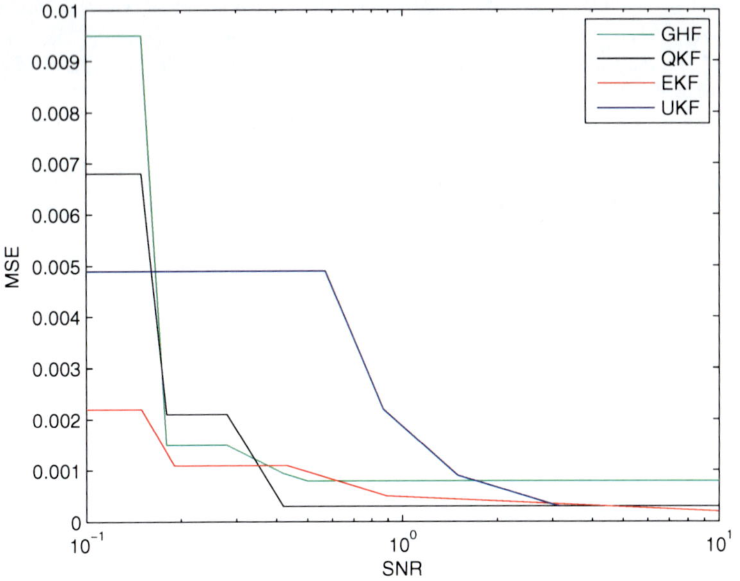


**Fig. 3.11** Rössler attractor

accuracy (see [15]). The latter is the main idea behind of the "multi-moment" algorithm design.

The heuristic quasi-optimum algorithm which for only two-moment filtering shows that practically without increasing the complexity of the EKF is possible to increase the filtering accuracy.

As for one moment algorithms but for two-moment case it is possible to expect the same MSE with almost 30% less SNR. This is presented at the Figs. 3.9- 3.11 for the cases of Lorenz, Chua and Rössler attractors.

**Acknowledgements.** This work was financially supported by Intel Corporation.

Authors would like to express their deep thanks to Dr. J. Meda, Dr. F. Ramos and M. Sc. B. Rodriguez for their help in the preparation of this material.

## References

1. Anischenko, V.S., et al.: Mixing and spectral-correlation properties of chaotic and stochastic systems: numerical and physical experiments. New Journal of Physics 7(76), 129 (2005)
2. Anischenko, V.S., et al.: Statistical properties of dynamical chaos. Physics-Uspehi 48(2), 151–166 (2005)
3. Arasaratnan, I., Haykin, S., Elliot, R.: Discrete-time non-linear filtering algorithms using Gauss-Hermite quadrature. Proceedings of the IEEE 96(5), 953–977 (2007)
4. Eckmann, J.P., Ruelle, D.: Ergodic Theory and Strange Attractors. Review of Modern Physics 57(3), 617–656 (1985)
5. Horton, W., Ichikawa, Y.H.: Chaos and Structures in Non-Linear Plasmas. World Scientific (1996)
6. Ito, K., Xiong, K.: Gaussian filters for non-linear filtering problems. IEEE Trans. on AC 45(5), 910–927 (2000)
7. Jazwinski, A.: Stochastic Processing and filtering theory. N.Y. Academic (1970)
8. Julier, S., Uhlman, J.: Unscented filtering and non-linear estimation. Proceedings of the IEEE 92(3), 401–422 (2000)
9. Kassam, S.: Signal detection in Non-Gaussian noise. Springer (1987)
10. Kazakov, I., Artemiev, V.: Optimization of dynamic systems with random structure. Nauka (1980)
11. Kontorovich, V.: Non-linear filtering for Markov stochastic processes using high-order statistics (HOS) approach. Non-linear Analysis; Theory, Methods and Applications 30(5), 3165–3170 (1997)
12. Kontorovich, V., Lovtchikova, Z.: Cumulant Analysis of Strange Attractors: Theory and Applications. In: Kyamakya, K., Halang, W.A., Unger, H., Chedjou, J.C., Rulkov, N.F., Li, Z. (eds.) Recent Advances in Nonlinear Dynamics and Synchronization. SCI, vol. 254, pp. 77–115. Springer, Heidelberg (2009)
13. Kontorovich, V., Lovtchikova, Z., Meda-Campaa, J.A., Tinsley, K.: Nonlinear Filtering Algorithms for Chaotic Signals. A comparative Study. ISAST Transactions on Computers and Intelligent Systems 2(1), 34–44 (2010)
14. Kontorovich, V., Lovtchikova, Z., Ramos-Alarcon, F.: Correlation Properties of Chaos: Cumulant approach. Mathematical and Computational Applications. Special Issue on Advanced Analytical Methods for Nonlinear Problems 15(5), 946–953 (2010)

15. Kontorovich, V., Lovtchikova, Z.: "Multi-moment" Nonlinear Filtering of Chaos. In: Proceedings of the Joint INDS 2011&ISTET 2011, pp. 289–294 (2011)
16. Kushner, H.: Dynamic equations for optimal non-linear filtering. J. Differ. Eq. 3, 179–190 (1971)
17. Kushner, H., Budhiraja, A.: A nonlinear filtering algorithm based on an approximation of the conditional distribution. IEEE Trans. on AC 45(3), 580–585 (2000)
18. Malakhov, A.N.: Cumulant Analysis of Random Non-Gaussian Process and their Transformation. Sovetskoe Radio, Moscow (1976)
19. Mijangos, M., Kontorovich, V., Aguilar-Torrentera, J.: Some statistical properties of strange attractors: Engineering view. Journal of Physics. Conference Series, 01214796, 6 (2008)
20. Primak, S., Kontorovich, V., Lyandres, V.: Stochastic Methods and their Applications to Communications. Stochastic Differential Equations Approach. John Wiley & Sons (2004)
21. Pugachev, V., Sinitsyn, I.: Stochastic Differential Systems. Analysis and Filtering. John Wiley & Sons (1987)
22. Stratonovich, R.: Topics of the theory of random noise, vol. 1-2. Gordon and Breach (1963)
23. Van Trees, H.: Detection, Estimation and Modulation theory. John Wiley & Sons (2001)
24. Zakai, M.: On the optimal filtering of diffusion processes. Wahrscheinlikenheitstheorie verngebiete 11, 230–243 (1969)

# Chapter 4
# Time-of-Flight Estimation Using Synchronized Chaotic Systems

Christian F. Wallinger and Markus Brandner

**Abstract.** Time-of-Flight (ToF) estimation is a basic building block in many metrological applications. Performance criteria for these applications are the variance and the bias of the derived delay estimate. From a signal processing point of view chaotic signals exhibit properties which make them well suited for metrological applications. In this chapter we experimentally investigate the applicability of synchronized chaotic systems in a ToF measurement system. In particular, we show that the choice of the numerical solver has a significant impact on the estimation performance. We further present a new delay estimator based on Poincaré intersections and compare the resultant estimation performance with the performance of a standard correlation-based delay estimator.

## 4.1 Introduction

Time-of-Flight (ToF) estimates are primary measurands in many metrological applications such as distance measurement, localization, and tracking. From the metrological point of view such applications are required to deliver estimates with small measurement uncertainties in the presence of bandwidth limitations, small signal-to-noise ratios (SNRs), and different kinds of disturbers. In the last two decades, the synchronization of chaotic systems has received a great deal of attention in the area of signal processing and communication engineering [16]. In this context, the beneficial properties of signals generated by chaotic systems are their unpredictability and their noise-like appearance.

For any discrete-time implementation of the chaotic system, the resolution of the measurand is inherently limited by the sampling interval. However, it is possible to increase the temporal resolution of the estimator using interpolation schemes. The

Christian F. Wallinger · Markus Brandner
Institute of Electrical Measurement and Measurement Signal Processing,
Graz University of Technology, Austria
e-mail: {christian.wallinger,brandner}@TUGraz.at

K. Kyamakya et al. (Eds.): Selected Topics in Nonlinear Dynamics, SCI 459, pp. 61–79.
DOI: 10.1007/ 978-3-642-34560-9_4 

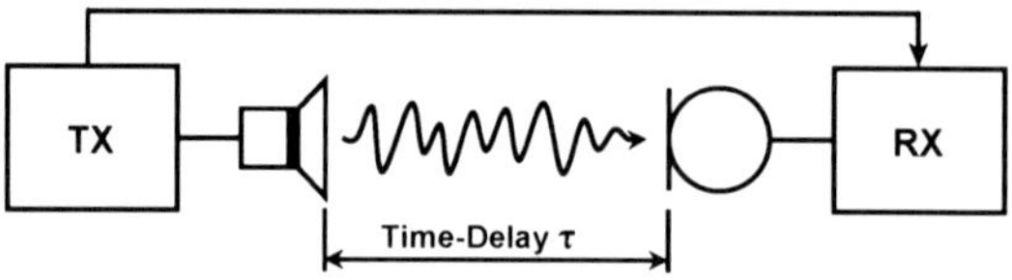


**Fig. 4.1** ToF measurements using chaotic systems. The delay parameter is estimated at the receiver (RX) based on the acquired acoustic signal and an undistorted reference signal provided by the transmitter (TX).

frequently applied polynomial interpolation is known to introduce an estimation bias [4]. Other approaches are based on the design of proper inverse systems using a synchronized receiver system as filter to reconstruct a transmitted information signal [5], [15]. Alonge *et al.* [3] report about a ToF measurement system using a chaotic modulated information signal as reference and pulse position modulation for transmission. In a recent work, Sorrentino and DeLellis [13] propose an adaptive method for estimating the ToF utilizing the self-synchronization process of chaotic systems itself.

In this chapter, we experimentally compare the performance of different ToF estimators using two synchronized chaotic systems. We modulate the amplitude of a carrier signal with the output of a Lorenz system. The demodulated signal is used to synchronize a second Lorenz system at the receiver side. In particular, we investigate the estimation bias and variance with respect to the use of different numerical solvers, interpolation schemes, and SNRs of Gaussian noise channels. The results are compared with a standard correlation-based ToF estimator.

### *4.1.1 Time-of-Flight Measurements*

The principle setup of a ToF measurement system is depicted in Figure 4.1. A transmitter (TX) generates a signal which is transmitted over a communication channel. Apart from other disturbers this channel adds a time-delay $\tau$ to the signal. In metrological terms the unknown quantity $\tau$ is referred to as the measurand. The subsequent receiver (RX) is now able to provide an estimate $\hat{\tau}$ of this delay given an undistorted reference signal transmitted by the TX using an ideal (e.g. undistorted and zero-delay) channel. A practical implementation of a ToF measurement system uses a modulated ultrasound signal which is transmitted over a variable distance to a receiver while the reference signal is passed directly from the TX to the RX. The estimated delay now corresponds to the time-of-flight of the ultrasound signal on the channel. The ideal ToF estimator $\hat{\tau}$ is unbiased, i.e. $b(\hat{\tau}) = E\{\hat{\tau}\} - \tau = 0$, and exhibits a minimum variance $var(\hat{\tau}) = E\{(\hat{\tau} - E\{\hat{\tau}\})^2\}$, where $E\{\cdot\}$ denotes the expectation operator. This combination of properties ensures that the resultant measurement system is able to deliver ToF estimates with a small measurement uncertainty.

### 4.1.2 Outline

The remainder of this chapter is structured as follows: The subsequent Section 4.2 introduces the notation of synchronized chaotic systems and presents different implementation details. Experimental results using different numerical solvers and channel parameters are presented in Section 4.3. We close our discussion with conclusions and an outlook to future work in Section 4.4.

## 4.2 Synchronized Chaotic Systems

In this work we use the self-synchronization property of coupled autonomous chaotic systems to estimate the ToF of a signal. We consider two identical unidirectionally coupled Lorenz systems using real state variables $x$, $y$, and $z$ as well as real parameters $\rho$, $\sigma$ and $\beta$. Commonly, the parameter $\rho$ is used as bifurcation parameter. The transmitter system is given by

$$\mathbf{F}(\mathbf{x}_T): \begin{aligned} \dot{x}_T &= \sigma(y_T - x_T) \\ \dot{y}_T &= \rho x_T - x_T z_T - y_T \\ \dot{z}_T &= x_T y_T - \beta z_T , \end{aligned} \tag{4.1}$$

where the transmitted signal is given by $s_T(t) = x_T(t)$. On the receiver side a signal $s_R(t) = \phi(s_T(t))$ is observed, where the real-valued mapping $\phi(\cdot)$ summarizes the influences of the channel including in particular the measurand $\tau$. In the simplest case, the received signal is a delayed version of the transmitted signal, i.e. $s_R(t) = s_T(t-\tau)$. The receiver system is given by

$$\mathbf{F}(\mathbf{x}_R) + C(\mathbf{x}_R, s_R - x_R): \begin{aligned} \dot{x}_R &= \sigma(y_R - x_R) \\ \dot{y}_R &= \rho x_R - x_R z_R - y_R + \alpha(\rho - z_R)(s_R - x_R) \\ \dot{z}_R &= x_R y_R - \beta z_R + \alpha y_R (s_R - x_R) , \end{aligned} \tag{4.2}$$

where a unidirectional coupling scheme with a coupling strength $\alpha$ is applied. In order to derive the according error system, we use the notation $x_{R,\tau^+} = x_R(t+\tau)$ to take care of the delayed receiver system. By introducing the transverse manifold $e_x := x_T - x_{R,\tau^+}$, $e_y := y_T - y_{R,\tau^+}$, and $e_z := z_T - z_{R,\tau^+}$, summarized in the state vector $\mathbf{e}^T = [e_x, e_y, e_z]^T$ the error system is given by

$$\dot{\mathbf{e}} = \mathbf{F}(\mathbf{x}_T) - \mathbf{F}(\mathbf{x}_{R,\tau^+}) - C\left(\mathbf{x}_{R,\tau^+}, s_{R,\tau^+} - x_{R,\tau^+}\right) \tag{4.3}$$

$$= \begin{matrix} \sigma(e_y - e_x) \\ (\rho - z_T)e_x - e_y - x_T e_z - \alpha(\rho - z_T)e_x + (1-\alpha)e_x e_z \\ y_T e_x + x_T e_y - \beta e_z - \alpha y_T e_x + (1-\alpha)e_x e_y \end{matrix}$$

$$= (D\mathbf{F}(\mathbf{x}_T) + DC(\mathbf{x}_T, 0))\,\mathbf{e} + g_{\mathbf{F}}(\mathbf{e}) + g_C(\mathbf{e}) .$$

The error system can be divided into a linear part $\mathbf{A} := D\mathbf{F}(\mathbf{x}_T) + DC(\mathbf{x}_T, 0)$ represented by the Jacobian matrix $D\mathbf{F}(\mathbf{x}_T)$ and the derivative of the coupling function $DC(\mathbf{x}_T, 0)$ determining the linear combinations of the error state variables for the coupling. All terms of higher order in $\mathbf{e}$ are summarized in the nonlinear driving terms $g_{\mathbf{F}}(\mathbf{e})$ and $g_C(\mathbf{e})$ vanishing in the case of $\mathbf{e} = 0$. Although the coupling is based on a time-dependent weighting between the receiver's state variable $x_R$ and the received signal through the factor $\alpha(\rho - z_R)$ in the differential equation $\dot{y}_R$ and the factor $\alpha y_R$ in the differential equation of the state variable $z_R$, it can be shown that the overall dynamical behavior of the synchronization framework is not affected by the time-delay $\tau$, i.e. the error system (Equation 4.3) has its equilibria in the origin.

In any practical implementation, the set of systems need to be solved numerically. From a metrological point of view this requires the consideration of the convergence behavior, numerical issues, and the derivation of estimators.

### *4.2.1 Convergence*

In this subsection we investigate the convergence rate within the range of complete synchronization. In principle the convergence rate is given by the maximum conditional Lyapunov coefficient $\lambda_{\max}$ of the according linearized error system [10], [7], i.e. the driving terms $g_{\mathbf{F}}(\mathbf{e})$ and $g_C(\mathbf{e})$ in Equation 4.3 will be neglected yielding $\lambda_{\max}\{\mathbf{A}\}$. Two identical systems coupled in the presented way will converge for $\lambda_{\max} < 0$. In contrast, if $\lambda_{\max} > 0$ the systems will diverge. This stability condition is necessary but does not guarantee that there are no areas of local unstability on the attractor [11] because of its averaging behavior characterizing the global stability over the whole chaotic attractor. The rate of convergence determined by $\lambda_{\max}$ is an important indicator for the time it takes to meet a certain measurement uncertainty threshold. In general, it will give a bound on the speed of tracking perturbations at the receiver.

From the multiplicative ergodic theorem of Oseledec [1] we know that, in the limit $t \to \infty$ the Lyapunov coefficients converge independently of the trajectory of the system or its initial values in the basin of attraction. For finite time intervals the estimates of the Lyapunov coefficients depend on both the observation time (i.e. the window length) and the trajectory of the system. In situations where a linearized error system is available the Lyapunov spectrum can be used to evaluate the local convergence behavior of the system [2], [8], [18], [12]. An alternative approach to the determination of the local convergence behavior is based on the fact that the convergence rate directly depends on $\lambda_{\max}$. Thus an estimator can be derived using the synchronization error of the coupled systems. Figure 4.2a illustrates our estimation principle from the error time series. Once the synchronization error measured as the Euclidean distance between the transmitter $\mathbf{x}_T^T = [x_T, y_T, z_T]^T$ and receiver $\mathbf{x}_R^T = [x_R, y_R, z_R]^T$ given as $\mathrm{SE} := \| \mathbf{x}_T - \mathbf{x}_R \|_2$ is transformed into the natural logarithmic domain, $\lambda_{\max}$ can be estimated as the slope of a regression line (black line in Figure 4.2a). This estimator is asymptotically unbiased and has a variance which decreases with increasing window length $L$. However, the estimation is limited because of the existing synchronization noise floor. Hence, $L$ cannot be made arbitrary large. Figure 4.2b shows histograms of estimated conditional $\lambda_{\max}$-values ($10^5$ estimates) for a coupling strength $\alpha = 1$ and three different window sizes $L$. The parameters of the according Lorenz systems are $\rho = 45.92$, $\beta = 4$ and $\sigma = 16$. The decrease of the standard deviation follows $L^{-\nu}$ with an estimated $\nu \cong 0.87$. This is in good agreement with [2]. For better illustration the normalized histograms are presented within a logarithmic ordinate.

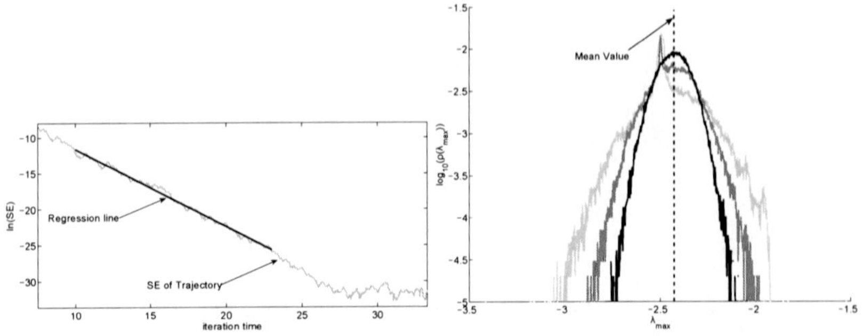


(a) (b)

**Fig. 4.2** (a) Principle of estimating the local conditional $\lambda_{\max}$ as slope of the regression line of the synchronization error between the transmitter and receiver, $\| \mathbf{x}_T - \mathbf{x}_R \|_2$. The length of the regression line indicates the estimation window length. (b) Normalized histogram $\rho(\lambda_{\max})$ showing the distribution $\lambda_{\max}$ estimated from different window length (light gray 300 iteration steps, dark grey 600 iteration steps and black 1200 iteration steps (iteration step-size = 0.01)) at a coupling strength $\alpha = 1$. The black dashed line indicates the mean value of the distributions. In order to illustrate the decreasing variance of the estimates with increasing window length, the histograms are illustrated within a logarithmic ordinate.

The convergence rate strongly depends on the coupling strength $\alpha$. Figure 4.3a shows the behavior of $\lambda_{\max}(\alpha)$ at a time-delay $\tau = 0$ calculated from the linearized error model (Equation 4.3). Whereas Figure 4.3b illustrates $\lambda_{\max}(\alpha)$ estimated from the error time series SE. As can be seen, the results are in good agreement. The Lorenz systems (TX and RX) are bounded in phase space [14]. Thus, even if no complete synchronization happens, the error trajectory is bounded in phase space, too. In general, this yields an upper bound on the ability of the estimator to detect positive Lyapunov coefficients, $\lambda_{max} \leq 0$ except for the case of close enough initial conditions of TX and RX.

## 4.2.2 Detection Range

An upper bound on the maximum detectable time-delay exists due to the synchronization mechanism of chaotic systems. The analysis of this bound is based on stability considerations of the underlying error system. In contrast to one-dimensional

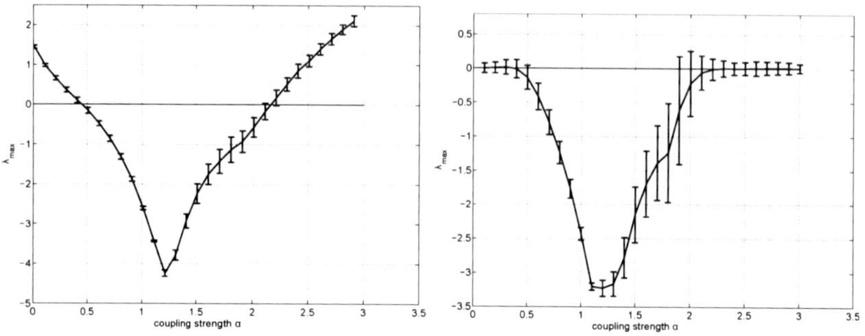


(a) (b)

**Fig. 4.3** Estimations of $\lambda_{\text{max}}$ as a function of the coupling strength $\alpha$ at a channel time-delay $\tau = 0$. In (a), $\lambda_{\text{max}}$ is calculated from the linearized synchronization error system, whereas in (b) $\lambda_{\text{max}}$ is estimated from the error time series.

equations, stability criteria are much harder to derive for systems of the form as given in Equation 4.3. For that reason stability is analyzed numerically. To study the stability of the synchronization manifold, we compute trajectories for different coupling strengths $\alpha$ and different channel time-delays $\tau$. As discussed in the previous subsection, the stability is given by a negative maximal Lyapunov coefficient $\lambda_{\text{max}}$. Thus, the region of stability is defined implicitly by its boundaries at which $\lambda_{\text{max}} > 0$. Since we want to apply the synchronization of chaotic systems in measurement science we are interested in complete synchronization of the transmitter and receiver. In order to detect the region of complete synchronization the higher order terms in Equation 4.3 will be neglected. As can be seen in Figure 4.4, the stability region for this coupling remains constant within an interval $0.5 < \alpha < 2.2$ over a range of different time-delays $\tau$. Of course, the results of this study represent a theoretically detection range given by the synchronization framework, i.e. whenever a real transmission channel exists and has an impact on the synchronization, the detection range will be limited to a certain ToF.

### *4.2.3 Discretization Algorithms and Numerical Issues*

Analytical error models as denoted in Equation 4.3 have stable equilibria at the origin, which allows for unbiased synchronization. Different numerical solvers exist to transform and iterate those continuous-time dynamical systems into the discrete-time domain. These solvers differ in their numerical behavior. In particular, when integrating a chaotic system, any errors associated with the solver in combination with the information generating property lead to trajectory hopping [9]. While this effect can be neglected for many applications, this is no longer true for synchronized chaotic systems: Trajectory hopping is no problem at the master system, but its appearance at the slave leads to synchronization errors. As a consequence of this biased synchronization will happen [17].

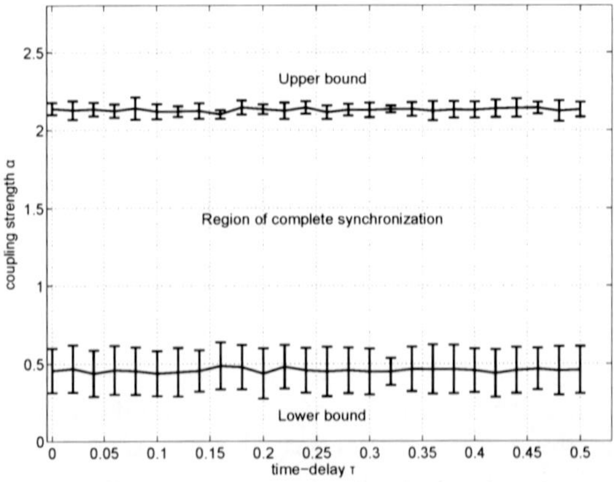


**Fig. 4.4** A numerically estimated stability diagram for the coupled Lorenz systems as a function of the channel time-delay $\tau$. Complete synchronization is achieved approximately within the lower bound of $\alpha > 0.5$ and the upper bound $\alpha < 2.2$. As one can see, the time-delay induced by the channel has no significant impact on the stability of the synchronization.

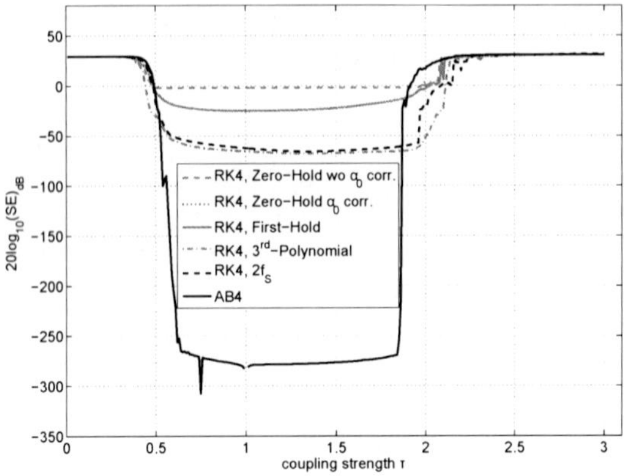


**Fig. 4.5** Synchronization error as a function of the coupling strength as well as different numerical integration schemes ($4^{th}$-order Runge-Kutta (RK4) and $4^{th}$-order Adams-Bashforth predictor (AB4)). Furthermore, different half-step generating filters for the RK4 method are compared at a step-size $\Delta = 0.01$.

Figure 4.5 compares different solver strategies in the context of a varying coupling strength at a step-size $\Delta = 0.01$. In particular, a comparison in terms of the averaged synchronization error defined in the text of subsection 4.2.1 of $4^{th}$-order Runge-Kutta (RK4) methods and explicit $4^{th}$-order Adams-Bashforth predictor (AB4) is given. At the RK4 method, the generation of the next iteration step is based on predicted gradients at half the step-size $\Delta$. Since the receiver is driven by the detected signal $s_R(t)$, we also need information of this signal at half the step-size at the receiver. In general, an analytical description of the detected signal $s_R(t)$ at the receiver does not exist. Thus, this information must be generated by a filtering process of the driving signal. Different filtering processes such as zero order- [6], first order-, and third order polynomial filtering are compared. Any group delay introduced by the filtering method needs to be properly considered in order not to misinterpret resultant synchronization errors. As an alternative method to filtering, the input signal at the receiver is sampled at half the step-size ($\equiv \frac{\Delta}{2}$). From Figure 4.5 we can see that the Runge-Kutta solver introduces a systematic bias depending on the applied filter. This systematic bias appears as a noise floor and will act as a bound on the achievable performance, i.e., in the context of metrology it will limit the achievable uncertainty. Moreover, this synchronization error induced by the RK4 methods strongly depend on the step-size. In contrast, the Adams-Bashforth solver does not require any interpolation and leads to unbiased synchronization errors which are dominated by the number representation of the applied system environment. In the present case the MATLAB simulation environment is used.

In general, the AB4 solver and the RK4 solver are based on different strategies [9]. The AB4 solver belongs to the class of multi-step methods, whereas the RK4 solver is a representative of single-step methods. Thus, we have to verify, if the global dynamical behavior of the systems and their synchronization behavior will be conserved. For this reason, we compare the numerically evaluated Lyapunov spectrum of the Lorenz system obtained from the RK4 and AB4 solver. In both cases $\lambda_{\text{max}} \approx 1.5$, $\lambda_{middle} \approx 0$ and $\lambda_{\text{min}} \approx -22.5$ being in good agreement with the estimates presented in [2]. Figure 4.6 represents estimated histograms of the conditional $\lambda_{\text{max}}$ describing the synchronization convergence behavior for the different solver strategies at a short window length and a coupling strength $\alpha = 1$. From this experiment we conclude, that the RK4 method using zero-order hold filtering causes a significant variation of $\lambda_{\text{max}}$, whereas the high-order RK4 methods ($3^{rd}$-order polynomial filtering, doubling the sampling frequency (2fs) at the receiver) and the AB4 method show the same behavior.

In the AB4 solver, the generation of the next iteration step is based on a weighted sum of the last four gradients including the current one. In principle, this only requires the calculation of the current gradient and the storage of the last three gradients. In contrast, the RK4 solver is based on a weighted sum of four gradients too, but these gradients are based on the current iteration step and therefore, they have to be calculated at every iteration step. Thus, the RK4 solver requires 4 times the evaluation of the vector field at every iteration step. Of course, it does not need to keep any gradients in a storage. A quite big difference between these two solver strategies relates to the problem of initialization. Whereas the RK4 method needs only an initialization point at time index 0, the AB4 method needs initialization information for time indices $= \{0, -1, -2, -3\}$. To overcome this problem, one can pursue the strategy of starting with a $1^{st}$-order, followed by a $2^{nd}$- and $3^{rd}$-order Adams-Bashforth solver and finally the proposed AB4 will be applied. Note that the $1^{st}$-order AB predictor represents the forward Euler algorithm.

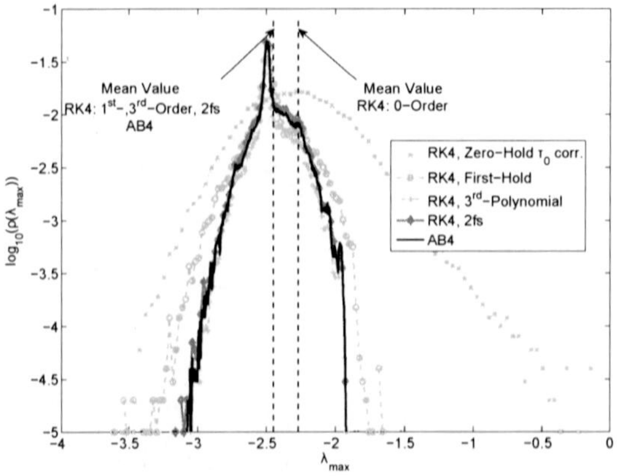


**Fig. 4.6** Comparison of normalized histograms $\rho(\lambda_{\max})$ of the max. conditional Lyapunov coefficient $\lambda_{\max}$ for different numerical integration schemes at a window size of 300 iteration steps and a coupling strength $\alpha = 1$. The histograms are based on $10^5$ estimates.

### 4.2.4 *Delay Estimators*

A classical method to estimate the time-delay is based on the cross-correlation of the transmitted and the received signal. We use this method as reference standard in the subsequent comparison of different estimators. The following discussion is restricted to coupled Lorenz systems with matched parameters as described by Equations 4.1 and 4.2. We consider three types of estimators as outlined in the signal model shown in Figure 4.7:

- $\hat{\tau}_{Signal}$: The ToF is estimated given the position of the maximum of the cross-correlation between the state variables $s_T(t)$ and $s_R(t)$.
- $\hat{\tau}_{Lorenz}$: The receiver is used as a filter so that the ToF is based on the position of the maximum of the cross-correlation between $s_T(t)$ and the synchronized state $x_R(t)$.
- $\hat{\tau}_{State}$: Again, the receiver is used as a filter. For each Lorenz system a sequence of dirac impulses is generated based on the intersection of the trajectory with a Poincaré plane. The ToF estimation is based on the position of the maximum of the cross-correlation between the derived impulse sequences.

#### 4.2.4.1 Poincaré Estimator Design

The Lorenz system has in its chaotic regime one unstable equilibrium at $EQ_1 = (0,0,0)$ and two symmetric, unstable equilibria at $EQ_{2,3} = (\pm\sqrt{\beta(\rho-1)}, \pm\sqrt{\beta(\rho-1)}, \rho-1)$. When the transmitter and receiver are in synchrony the trajectories of both systems are traveling around the same equilibria $EQ_2$ and $EQ_3$ in state space which have a distance $\rho - 1$ in $z$-direction from the origin. By inserting a 2-dimensional Poincaré hyperplane in the 3-dimensional state space positioned at a constant distance of $z = \rho - 1$, we are able to detect the

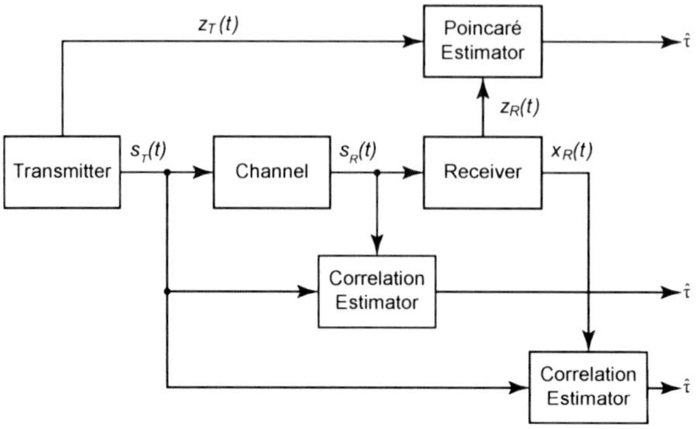


**Fig. 4.7** Overview of different ToF estimators based on the cross-correlation of a transmitter sequence and a receiver sequence

individual time instances of intersection of the trajectories with the according hyperplane. This hyperplane segments the state space into two distinct regions which are assigned the symbols $+1$ if the trajectory is above the Poincaré plane and $-1$ if the trajectory is below the described Poincaré plane. Consequently, deriving the generated symbol sequence we obtain an uncorrelated impulse sequence, which can be used for a computationally efficient and robust subsample ToF estimation.

For the Lorenz system, the symbol sequences can be generated by quantizing the $z$-component as depicted in Figure 4.8a. The Poincaré plane (dashed line) carries out a binary quantization of the state space marked by the symbols $+1$ and $-1$. This leads to a pseudo random sequence of rectangular pulses. By deriving the square wave we obtain the according impulse sequence as illustrated in Fig. 4.8b. The autocorrelation function of such a sequence as shown in Figure 4.8c exhibits a Dirac-like behavior indicating its applicability to ToF measurement.

#### 4.2.4.2 Subsampling Techniques

Both the transmitter and the receiver system are evaluated in discrete time using the step-size $\Delta$ as underlying quantization step. Without any further interpolation procedures the quantization error $Q$ of the correlation-based estimators is bounded by half the step-size, i.e., $|Q| \leq \frac{\Delta}{2}$. In order to be able to detect fractional delays subsample interpolation is applied. Unfortunately, the design of a minimum-variance-unbiased estimator is not possible because analytical correlation sequences of the used Lorenz state variables ($s_T(t)$,$s_R(t)$ and $x_R(t)$) are to the best of our knowledge unknown. Interpolators are used to be able to detect fractional delays. Due to the approximation of the true correlation sequences by using a proper interpolation filter, in general, we are not able to estimate the true position of the maximum of the cross-correlation unbiased, i.e., the estimates of the time-delay $\tau$ are biased. In Figure 4.9 the underlying basic problem in correlation analysis is depicted. The pins $y_i$ represent the discrete correlation sequence centered around the correlation shift zero. The true maximum $\tau$ is given by the dashed line but this is unknown. Hence, the

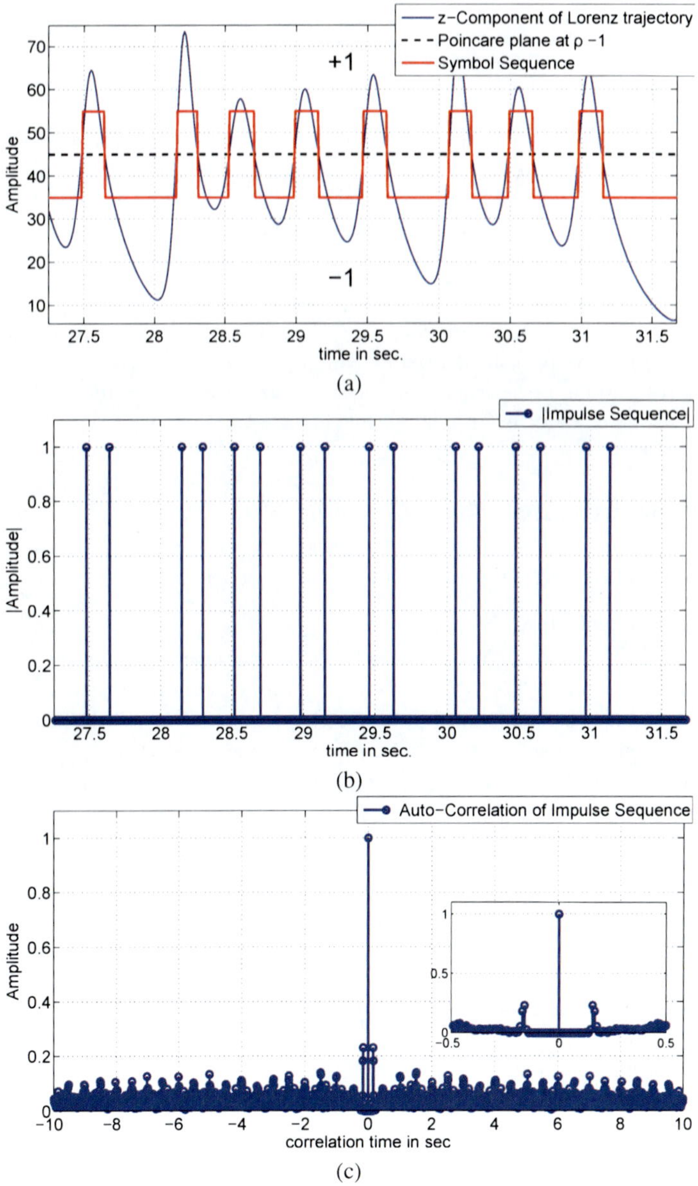


**Fig. 4.8** Derivation of symbol sequence from Poincaré plane. (a) Typical $z$-component of a Lorenz trajectory quantized by the Poincaré plane (dashed line) at $z = \rho - 1$ resulting in the symbol sequence (bold line). (b) Normalized impulse sequence. (c) represents a typical auto-correlation sequence of such an impulse sequence.

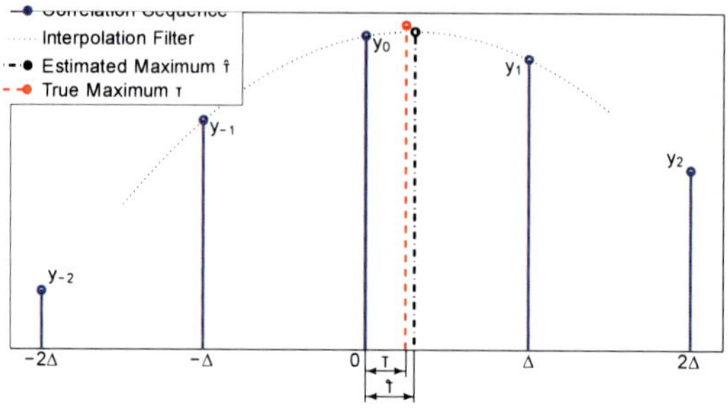


**Fig. 4.9** Illustration of the basic problem in correlation analysis. The analytical description of the correlation sequence (solid line with according maximum $\tau$ (dashed line)) is unknown. Estimating the position of the maximum a proper interpolation filter (dotted line) is used ending up in a biased estimate $\hat{\tau}$ (dashed-dotted line).

position of the maximum is estimated by using a proper interpolation filter (dotted line) ending up in an biased estimate $\hat{\tau}$ (dashed-dotted line). In the ongoing discussion, we give a brief overview of the investigated interpolators.

- **Cosine-Interpolation**
  One possibility of modeling the main lobe of the cross-correlation sequence is given by $y = A\cos(\omega x + \phi(\tau))$. Such a *cosine*-function consists of 3 degrees-of-freedom namely the amplitude $A$, the frequency $\omega$ and the time-delay dependent phase $\phi(\tau)$. Again, knowing the correlation samples $y_{-1}, y_0$ and $y_1$ we are able to determine the parameters as

$$\omega = \frac{1}{\Delta}\arccos\left(\frac{y_{-1}+y_1}{2y_0}\right) \tag{4.4}$$

$$\phi(\tau) = \arctan\left(\frac{y_{-1}-y_1}{2y_0\sin(\omega\Delta)}\right)$$

$$A = \frac{y_0}{\cos(\phi(\tau))}\,.$$

  The time-delay estimate $\hat{\tau}$ is given by the phase $\phi(\tau)$

$$\hat{\tau} = -\frac{\phi(\tau)}{\omega}\,. \tag{4.5}$$

- **Center of Gravity**
  For the time-delay estimation based on the cross-correlation of the impulse sequences we do not need an approximative interpolation. This is due to the fact that these sequences are unique. Hence, the time-delay can easily be determined by the center of gravity. The time-delay estimate by the center of gravity needs in an ideal case only 2 correlation samples. This is, those two correlation samples in the vicinity of the true position of the correlation maximum. Hence, referring to the notation given in Figure 4.9 the time-delay estimator $\hat{\tau}$ is given by

$$\hat{\tau} = \frac{y_0}{y_0 + y_1}\Delta. \tag{4.6}$$

Other influencing factors which are not covered in this work relate to practical implications of using finite non-rectangular windows.

## 4.3 Experiments

This section presents experiments related to the feasibility of synchronized chaotic systems for ToF measurements. Two coupled Lorenz systems with a parameter set $\rho = 45.92$, $\beta = 4$ and $\sigma = 16$ and a coupling strength $\alpha = 1$ are used in a discrete-time realization with an iteration step-size of $\Delta = 0.01$. Both systems are connected with a unit-gain additive white Gaussian noise (AWGN) channel introducing a known time-delay $\tau$. All the estimates are evaluated using averages of 100 pairs of time series uniformly initialized within the basin of attraction for each time-delay. The computations are done after the transient phase of the synchronization process, i.e. we start the measurements at 20 times $\lambda_{\max}$. The performance of the time-delay estimation is given in terms of the bias $b(\hat{\tau})$ and an according standard deviation obtained by fitting the estimates to a Gaussian distribution.

The experiments consist of three parts: *Part* I determines the influence of the window size of a rectangular window on the estimator performance of $\hat{\tau}_{Signal}$, $\hat{\tau}_{Lorenz}$ and $\hat{\tau}_{State}$ applying the discussed subsample interpolators. *Part* II determines the influence of the AWGN channel on the estimator performance for signal-to-noise ratios (SNR) ranging from 100 dB down to 20 dB. In order to explore the influence of the numerical solver routine all the experiments are executed for both the numerical solvers RK4 and AB4. *Part* III discusses ToF-estimates depending on the order of the numerical solver. We compare results obtained from the proposed AB4 predictor and results obtained from the AB3 and AB5 predictor.

### 4.3.1 *Different Window Lengths*

In correlation analysis, in general, the window length has a scaling impact on the estimation uncertainty, i.e., if the window length increases, the parameter uncertainty decreases. In order to verify this behavior, we compare time-delay estimates of the $\hat{\tau}_{Signal}$-, $\hat{\tau}_{Lorenz}$- and $\hat{\tau}_{State}$-estimators for a correlation window length of 10 s and 100 s and an ideal transmission channel. In Figure 4.10a the estimation performance of the $\hat{\tau}_{Signal}$- estimator is depicted as a function of the true time-delay $\tau$. Both RK4 and AB4 exhibit a similar performance in this experiment. Figure 4.10b shows the estimation performance when applying the $\hat{\tau}_{Lorenz}$-estimator (synchronization is used). In this case the ToF estimates obtained from the low-order interpolation methods at the receiver when applying the RK4 solver show a worse situation. For the other methods the behavior is roughly the same as without synchronization. Both results are obtained utilizing the *cosine*-subsample interpolation. When applying the Poincaré estimator $\hat{\tau}_{State}$ (Figure 4.10c) the situation is in this ideal case quite the

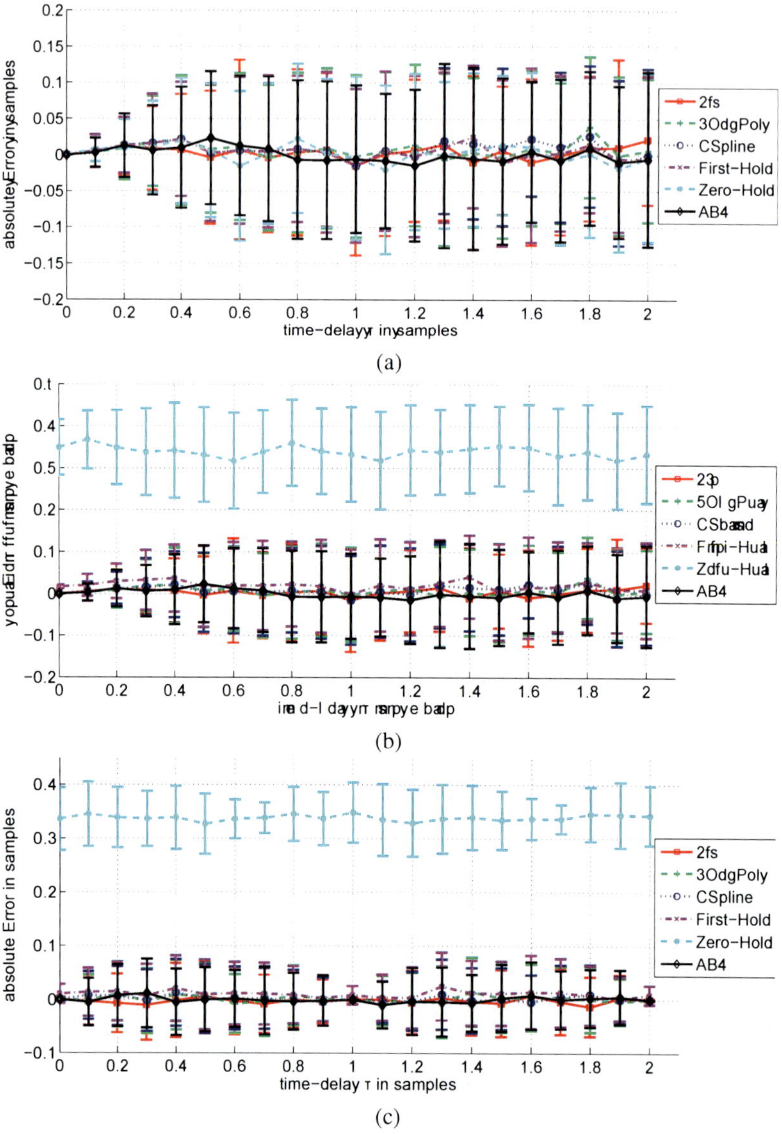


**Fig. 4.10** Bias and standard deviations of ToF estimations for (a) $\hat{\tau}_{Signal}$-, (b) $\hat{\tau}_{Lorenz}$- and (c) $\hat{\tau}_{State}$-estimators using a correlation window length of 10 s. *Cosine*-subsample interpolation is used in the first two estimators.

same except for the variances of the estimates, which become smaller. All the experiments are carried out for a correlation window length of 10 s.

In the next step, we increase the correlation window length to 100 s. As already mentioned, the uncertainty of the estimators should decrease. Figure 4.11a represents the results for the $\hat{\tau}_{Signal}$-estimator. In Figure 4.11b the estimation performance of the $\hat{\tau}_{Lorenz}$-estimator is depicted. Again *Cosine*-subsample interpolation is used for these estimators. Figure 4.11c represents the estimation performance obtained from the $\hat{\tau}_{State}$-estimator.

Estimation errors of all estimators are decreased by a factor of $\sim \frac{1}{10}$. As can be seen, the bias of the estimates obtained by the low-order interpolation methods from the RK4 solver (zero-order and first-order hold) does not change with the window length. We attribute this behavior to the limiting synchronization noise floor induced by the RK4 solver. For scaling reasons and to enable a closer look, we neglected the estimation error performance of the RK4 solver using zero-order hold interpolation at the receiver in Figure 4.11c. As expected, the RK4 method using higher-order interpolation methods ($3^{rd}$-order polynomial and cubic spline interpolation as well as 2fs) shows the same performance as the AB4 solver.

### 4.3.2 Different Noise Levels

In the second part of the presented experiments we investigate the influence of different noise levels of the transmission channel on the ToF estimation performance. Results of the $\hat{\tau}_{Lorenz}$- (Figure 4.12a) and $\hat{\tau}_{State}$- (Figure 4.12b) estimators based on the AB4 solver are presented for SNRs of 100 dB, 50 dB, 30 dB and 20dB and a correlation window length of 100 s. Again, *cosine*-subsample interpolation is applied in the $\hat{\tau}_{Lorenz}$-estimator. As one can see throughout those experiments, the error performance of the time-delay estimates decreases with decreasing SNR. Having a closer look at the statistics of the estimates, the variances of the $\hat{\tau}_{State}$-estimator are smaller when compared with the $\hat{\tau}_{Lorenz}$-estimator, except in the situations of integer time-delays of the step-size. Moreover, the estimation bias induced by the $\hat{\tau}_{State}$-estimator shows systematic behavior with decreasing SNR ending up in a correctable uncertainty.

### 4.3.3 Different Orders of Numerical Solver

In the previous subsections we compared different numerical solvers of $4^{th}$-order. Now, we investigate and briefly discuss the question about the dependency of the ToF estimates on the order of the numerical solver. In what follows, we compare ToF estimates obtained from the $\hat{\tau}_{Lorenz}$- and $\hat{\tau}_{State}$-estimators utilizing the AB3, AB4, and AB5 solver strategy. Figure 4.13 illustrates results obtained from the $\hat{\tau}_{Lorenz}$ estimator in the case of an ideal transmission channel (a) and SNR $=$ 20 dB (b). The correlation window length is 100 s. As one can see, the AB4 solver outperforms the AB3 and AB5 solver throughout these experiments. Figure 4.14 illustrates results obtained from the same experimental setup utilizing the $\hat{\tau}_{State}$ estimator. In this case, the estimates are based on cross-correlation sequences representing time information. Of course, this time information is again derived from amplitude information (Poincaré estimator), but it is more robust against amplitude variations within certain boundaries. Thus, the behavior of the different solver strategies are similar in

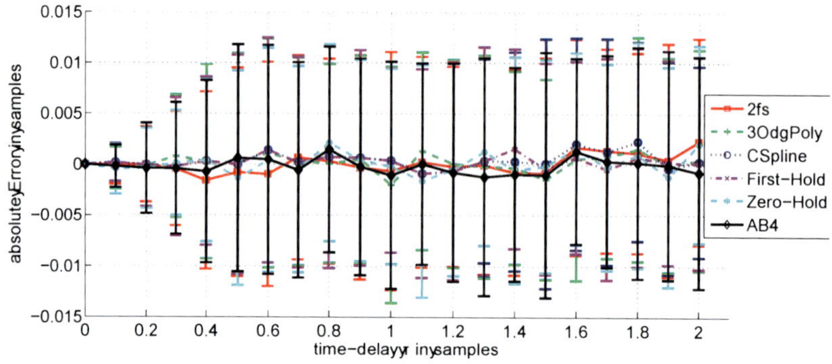


(a)

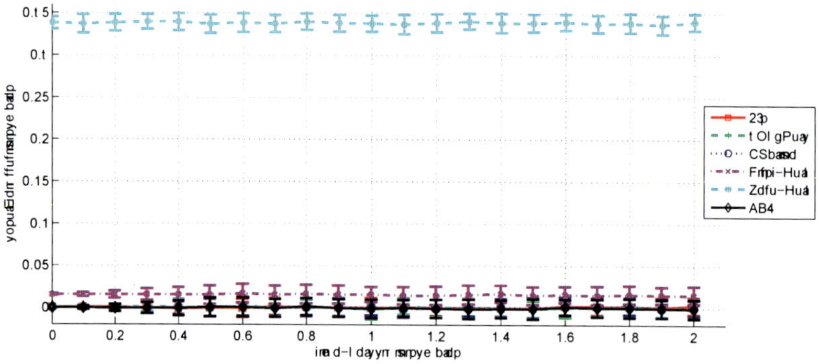


(b)

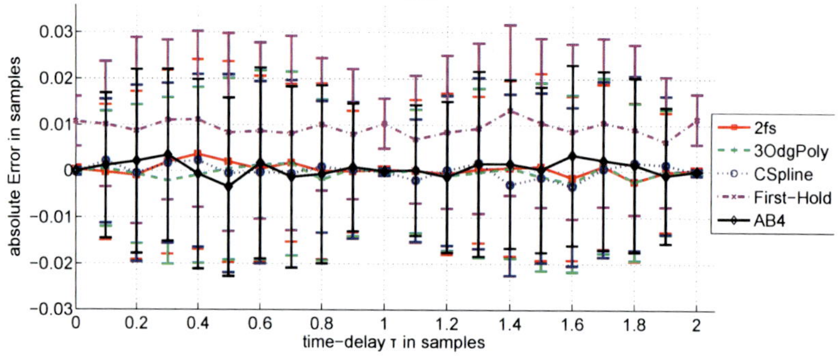


(c)

**Fig. 4.11** Bias and standard deviations of ToF estimations for (a) $\hat{\tau}_{Signal}$-, (b) $\hat{\tau}_{Lorenz}$- and (c) $\hat{\tau}_{State}$-estimators using a correlation window length of 100 s. *Cosine*-subsample interpolation is used in the first two cases. For scaling reasons the error performance of the RK4 solver using zero-order hold interpolation at the receiver is neglected in (c). Whereas the error performance of the RK4 method using higher-order interpolation at the receiver and the AB4 solver scales down with increasing correlation window length, the methods based on low-order interpolation remain constant compared to the results utilizing the short window length.

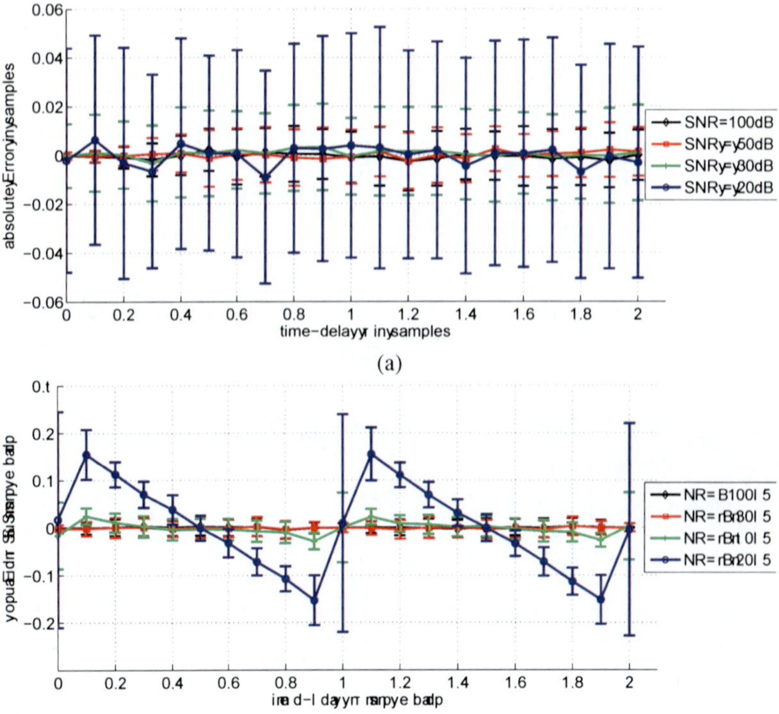


(a)

(b)

**Fig. 4.12** Bias and standard deviations of ToF estimations for (a) $\hat{\tau}_{Lorenz}$- and (b) $\hat{\tau}_{State}$-estimators for different SNRs: SNR = 100 dB, 50 dB, 30 dB and 20 dB. Results are based on the AB4 solver.

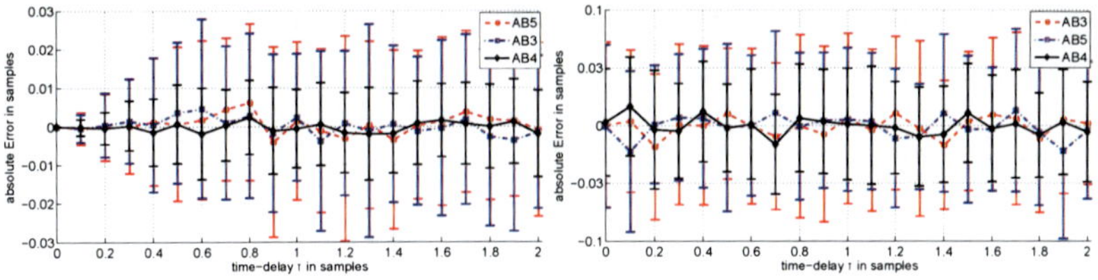


(a) (b)

**Fig. 4.13** Bias and standard deviations of ToF estimations for the $\hat{\tau}_{Lorenz}$-estimator applying the AB3, AB4 and AB5 solver strategy. (a) Error estimates for an ideal transmission channel (SNR $\rightarrow \infty$ dB). (b) Error estimates at SNR = 20 dB.

the undistorted case. In the case of a decreasing SNR (Figure 4.14b), the amplitude variations of the AB3 and AB5 become stronger ending up in a worse estimation performance. From this point of view we can conclude that the ToF estimation in principle is not an invariant of the order of the numerical solver, but utilizing our

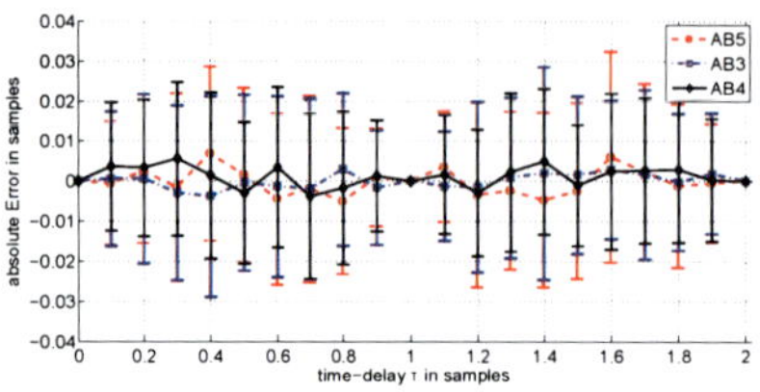


(a)

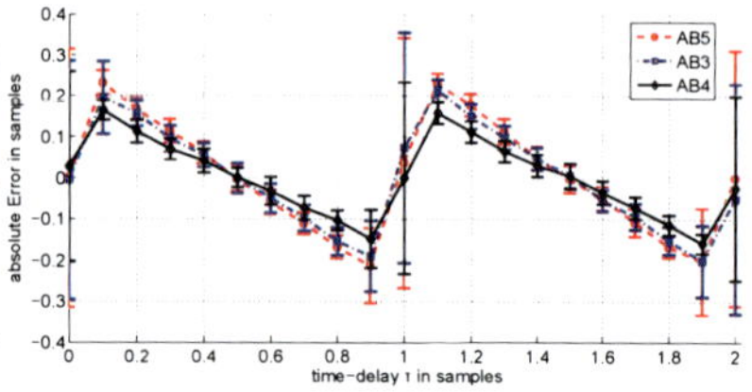


(b)

**Fig. 4.14** Bias and standard deviations of ToF estimations for the $\hat{\tau}_{State}$-estimator applying the AB3, AB4 and AB5 solver strategy. (a) Error estimates for an ideal transmission channel (SNR $\rightarrow \infty$ dB). (b) Error estimates at SNR = 20 dB.

proposed Poincaré estimator, we are able to keep the impact of the order of the numerical solver restricted.

## 4.4 Summary and Conclusions

This paper discusses the application of synchronized chaotic systems to ToF measurements. We show that two coupled Lorenz systems with identical parameters can be used to robustly estimate the ToF of an additive white Gaussian noise channel with unit-gain.

We present a novel delay estimator based on Poincaré intersections and compare its performance to a standard correlation-based estimator. The paper further covers the influence of numerical issues such as the choice of the solver strategy and of interpolation schemes. Of course, the proposed ToF-estimation framework is also applicable to other chaotic systems like the well known Roessler system or Chua circuit just to mention a few. Since the method is based on the self-synchronization property of chaotic systems, the main question is, how strong can the self-synchronization property of such systems be distorted by the transmission channel?

Future work is concerned with the investigation of additional channel models and the practical implementation of a ToF measurement system.

## References

1. Abarbanel, H.: Analysis of Observed Chaotic Data. Springer (1996)
2. Abarbanel, H., Brown, R., Kennel, M.: Lyapunov exponents in chaotic systems: Their importance and their evaluation using observed data. International Journal of Modern Physics B 5(9), 1347–1375 (1991)
3. Alonge, F., Branciforte, M., Motta, F.: A novel method of distance measurement based on pulse position modulation and synchronization of chaotic signals using ultrasonic radar systems. IEEE Transactions on Instrumentation and Measurement 58(2), 318–329 (2009)
4. Cespedes, I., Huang, Y., Ophir, J., Spratt, S.: Methods for estimation of subsample time delays of digitized echo signals. Ultrasonic Imaging 17, 142–171 (1995)

5. Feldmann, U., Hasler, M., Schwarz, W.: Communication by chaotic signals: The inverse system approach. International Journal of Circuit Theory and Applications 24, 551–579 (1996)
6. Freedman, A.E.: Transmission channel compensation in self-synchronizing chaotic systems (1995)
7. Fujisaka, H., Yamada, T.: Stability theory of synchronized motion in coupled-oscillator systems. Progress of Theoretical Physics 69(1), 32–47 (1983)
8. Geist, K., Parlitz, U., Lauterborn, W.: Comparison of different methods for computing lyapunov exponents. Progress of Theoretical Physics 83(5), 875–893 (1990)
9. Parker, T.S., Chua, L.O.: Practical Numerical Algorithms for Chaotic Systems. Springer (1989)
10. Pecora, L.M., Carroll, T.L.: Driving systems with chaotic signals. Physical Review A 44(4), 2374–2383 (1991)
11. Pecora, L.M., Carroll, T.L.: Master stability functions for synchronized coupled systems. Physical Review Letters 80(10), 2109–2112 (1998)
12. Pecora, L.M., Carroll, T.L., Johnson, G.A., Mar, D.J., Heagy, J.F.: Fundamentals of synchronization in chaotic systems, concepts and applications. Chaos 7(4), 520–543 (1997)
13. Sorrentino, F., DeLellis, P.: Estimation of communication-delays through adaptive synchronization of chaos. Chaos, Solitons and Fractals 45(1), 35–46 (2012)
14. Sparrow, C.: The Lorenz Equations: Bifurcation, Chaos, and Strange Attractors. Springer-Verlag New York Inc., New York 10010 (1982)
15. Spears, B.K., Tufillario, N.B.: A chaotic lock-in amplifier. American Journal of Physics 76(3), 213–217 (2008)
16. Stavroulakis, P.: Chaos Applications in Telecommunications. CRC Press, Inc., Boca Raton (2005)
17. Wallinger, C.F., Brandner, M.: Numerical aspects of the synchronization performance of discretized dynamical systems. In: 19th IEEE Workshop on Nonlinear Dynamics of Electronic Systems (March 2011)
18. Wolf, A., Swift, J.B., Swinney, H.L., Vastano, J.A.: Determining lyapunov exponents from a time series. Physica D 16, 285–317 (1984)

# Chapter 5
# Binary Synchronization of Complex Dynamics in Cellular Automata and its Applications in Compressed Sensing and Cryptography

Radu Dogaru and Ioana Dogaru

**Abstract.** Complex dynamics of the type used in random number generators may emerge in elementary cellular automata with properly designed structures and cells. This chapter reviews recent results in quantifying the complexity of the dynamics in cellular automata with an emphasis on the recently discovered phenomenon called *binary synchronization*. It allows that two cellular automata systems with the same structure will synchronize (the receiver will duplicate the n-dimensional state vector of the transmitter) receiving only a single bit stream, produced by the output of a single cell of the transmitter cellular automaton. The decoding of this stream is possible only when the structure of the cellular automata (encryption key) is known. It is shown how the key space may be increased using various methods (e.g. using hybrid models or perturbing the cellular network model into a small-worlds model). Applications in cryptography, spread spectrum communications, and compressed sensing are reviewed. Some particularities for the implementation of such cellular automata systems in FPGA technologies are provided.

## 5.1 Introduction and Motivation

In various applications (remote sensing, secure data transmission, compressive sensing [1]) some message must be passed from a transmitter system (abbreviated in the next as Tx) to a remote receiver (abbreviated in the next Rx), such that cannot be intercepted by an unauthorized intruder. This problem is usually approached by various methods form the mature area of cryptography. Yet, some applications require low power consumption and consequently a simple mechanism is needed for generating ciphering sequences, given a key. Correspondingly, a simple algorithm for

Radu Dogaru · Ioana Dogaru
University "Politehnica" of Bucharest, Dept. of Applied Electronics
and Information Engineering, Natural Computing Laboratory, Room B232,
Bvd. Iuliu Maniu 1-3, Sector 6, Bucharest, Romania
e-mail: radu_d@ieee.org

K. Kyamakya et al. (Eds.): Selected Topics in Nonlinear Dynamics, SCI 459, pp. 81–95.
DOI: 10.1007/ 978-3-642-34560-9_5 

deciphering will be implemented in the receiver. Starting with the work of Pecora and Caroll [2], proving that continuous-time chaotic signals generated by nonlinear dynamic system may be used to synchronize a receiver with a similar structure ("drive-response" method), such an approach was further considered and developed as an alternative to the traditional encryption methods, particularly for low power and low complexity applications.

The main idea of chaos synchronization is that a nonlinear dynamic system can be tuned in its vector of parameters **G** to produce a chaotic signal $s(t)$ capable to modulate the useful message $m(t)$ in the transmitter (Tx). The resulting signal $r(t)$ is sent over a channel and is received by a receiver (Rx) built around an *identical* nonlinear dynamic system (i.e. defined by the same equations and parameters **G**). A replica $\hat{s}(t)$ of the original chaotic signal $s(t)$ is extracted from $r(t)$ and used to synchronize the state of the receiver. For a properly chosen dynamic system and its parameters G, the Rx system will synchronize (after a certain amount of time) such that its entire state vector will duplicate the state vector in the Tx. Since the state vector was used to encrypt the message $m(t)$, the recovered version of the Tx state vector can now be used in receiver to decrypt the received message. The encryption key corresponds to set of parameters G and the equations defining the nonlinear systems with complex (chaotic) dynamics. Various modulation and demodulation schemes were proposed [3][4] but, although initially chaos synchronization for communications held much promise, its critics (e.g. [5][6]) pointed out on several potential drawbacks that must be carefully considered before successfully using it in applications:

i) The case of continuous time and state (often discussed in theory) corresponds in practice to implementations of two *identical* analog circuits (e.g. Chua's circuit and other similar). Such a requirement is extremely difficult to ensure in mass production and parameter mismatches result in loss of synchronization or at least in a degradation of the recovered message. To date, the above drawback is alleviated considering discrete-time and discrete-state (finite computing precision) nonlinear systems, with digital implementations. Common examples of such systems used in chaos-based cryptography are the logistic map [7][8], tent map [9] and other nonlinear maps. In order to ensure good quality of the encryption sequences, various techniques are used (sub-sampling, mixing, delays) but they usually increase the system implementation complexity. As shown in the next, good cryptographic properties at low implementation complexity can be achieved using the equivalent of the nonlinear map approach cast into the cellular automata framework (Section 2). Particularly, for the proposed cellular automata systems, there is no transient time, as is usually the case for various nonlinear maps [10]. ii) There are nonlinear dynamic system theorems [11] showing that the structure of the Tx nonlinear system (including all its hidden states) can be recovered based on output samples coming from the system, with the single condition to have at least 2n such consecutive samples (where $n$ is the number of state variables, that is usually unknown for the intruder). In other words the "key" (structure of the system) may be revealed by the apparently encrypted message sent over the channel. To alleviate this problem, the state vector dimension n must be very large. This requirement leads to difficulties in designing a nonlinear system, particularly when it is required to have in addition the synchronization property. On the other hand, the solution discussed herein considers the cellular automaton with $n$ cells as the nonlinear dynamic system. Since cellular automata are scalable, there is no problem to choose $n$ as large as desired to minimize the risk of cryptographic attack; iii) To our knowledge, most of systems proposed for chaos synchronization provide a continuous-state (or its finite precision representation) signal $s(t)$. Synchronization in such systems will suffer from low immunity to channel noise. Instead, a binary stream $s(t)$ will ensure the highest immunity to noise, as required in practical applications. Only a few exceptions were so far reported, but they are built around low-dimensional dynamical systems [12][13] thus suffering from drawbacks discussed in paragraph (ii). In [14] we proved for the first time that synchronization can be achieved between properly designed dynamical systems while sending a minimal quantity of information over the channel. Recently [15] this phenomenon was more carefully investigated in the context of cellular automata, providing that design solutions for building high-dimensional, yet having low complexity, nonlinear dynamic systems with synchronization capability do exists. This chapter reviews recent results in defining a novel class of chaos synchronization systems suitable for implementation in digital technologies. Such systems achieve good cryptographic properties while maintaining the low level of complexity required by certain type of applications (mostly in the area of remote sensing). Our synchronization model is provided in Fig. 5.1.

Both Tx and Rx systems are implemented as digital nonlinear maps F. The novelty of our approach is that F is implemented as the feedback loop of cellular

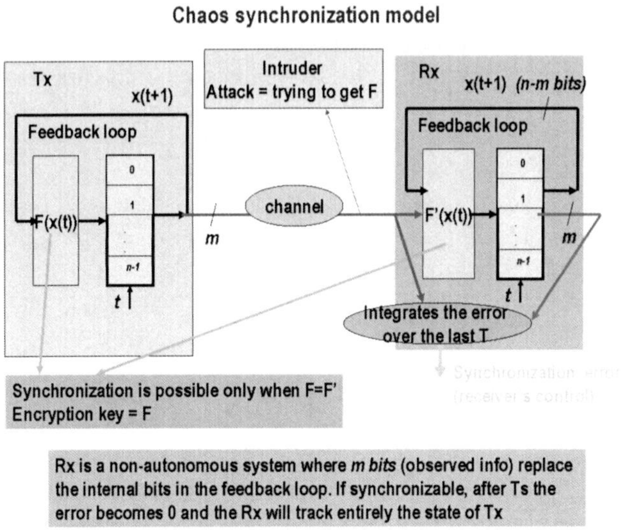


**Fig. 5.1** A general chaos synchronization model based on nonlinear maps represented in finite precision ($n$ bits). Both Tx and Rx systems implement nonlinear maps. For each time-step only $m$ bits are sent over the channel. In the limit $m=1$ binary synchronization is implemented.

automata or "small world" networks. The detailed architecture of such systems is presented in Section 2. We are particularly interested to optimize F such that:

i) Binary synchronization is achieved, i.e. *m=1* bit in the model represented in Fig. 5.1. In this case the information needed to resynchronize the Rx is minimal and consists of only 1 bit per iteration. Consequently it may be easily embedded and recovered in various forms of modulation/demodulation.

ii) Complex dynamics (we avoid the term "chaotic" since it is normally associated with continuous state systems) is achieved; In particular, some measures of complexity are defined in Section 3. In particular we are interested in implementing *conservative systems* (with no transients) with *maximal cycle length* and high *degree of chaos* In Section 5.5 some efficient methods to implement such systems in FPGA technologies are discussed. Finally some specific applications of the proposed complex signal synchronization scheme are briefly exposed in Section 5.

## 5.2 Automata Network Models and the Key Space

In order to implement the nonlinear dynamic system to be embedded in both Tx and Rx systems (see the model in Fig. 5.1) cellular automata (CA) systems and their variations were chosen. The choice is motivated by the following reasons: a) In [16] an in-depth analysis of the elementary cellular automata dynamics is provided. Particularly it is shown that CA with odd number $n$ of cells governed by rules ID=45 (and its 3 equivalents ID=75,89, and 101) or ID=154 (or its equiv-alents, ID=166, 180, 210) are the single *non-linear rules* leading to *conservative* dynamic behaviors. In other words, such automata have the state space organized such that no state is "ephemeral" i.e. there is no transient. All states are enclosed in one or more cycles, and the effort will be directed to further adjust the structure F such that there will be one very long cycle (almost $2^2$ states for an automaton with $n$ cells) with complex (chaotic) behavior. The existence of transients in usual maps (like the logistic, etc.) is well known and has a negative impact on designing communication systems based on chaos. Only recently a model is proposed to estimate the duration of such transients [10] but their existence demonstrates that a part of the state space is lost (they belong to the transient, not to the useful cycle providing the complex sequence). On the other hand, in [15] it is shown that among the conservative rules, only ID=45 (and its 3 equivalents ID=75,89, and 101) ensures the property of binary synchronization ($m = 1$ bits in the model presented in Fig. 5.1). None of the few linear rules (or their combinations in form of hybrid cellular automata, frequently cited in the literature [17][18]) ensuring conservative behaviors posses the binary synchronization property as follows from results in [19] which extensively discuss synchronization in linear CA. Note that in linear CA, Boolean functions describing the cell have the canonical form $y = a_0 \oplus a_1x_1 \oplus a_2x_2 \oplus a_3x_3$, where $\oplus$ denotes the exclusive-or (XOR) operator and $a_i$ are binary coefficients. b) Cellular systems, where each cell is in fact a Boolean function with only 3 inputs (to be connected to other cells in the network) are perfectly suited for low complexity implementations in digital technologies, as detailed in [20] and in Section 5.5. Figure 5.2 presents two automata structures that were successfully tested for having both binary synchronization and long chaotic cycle properties. Note that they can be operated in either autonomous mode (as is the case in the Tx) or with one input forced by the synchronization signal (as is the case in the Rx).

The discrete-time dynamics of the hybrid cellular automata (HCA) in Fig.5.2 is given by the next equation, which applies synchronously to all cells (a cell is identified by an index $i \in \{1,2,...,n\}$ ):

$$x_i^T(t+1) = m_i \oplus Cell(x_{i-1}^T(t), x_i^T(t), x_{i+1}^T(t), ID) \quad (5.1)$$

where the upper index "$T$" stands for the transmitting CA counter, $\oplus$ is the logical XOR operator and $Cell(u1, u2andu3)$ is a Boolean function with 3 binary inputs $(u1, u2, andu3)$, also called the CA (local) rule. The local CA rule is characterized by a decimal identifier (ID), which encodes the relationship between inputs and output. For instance ID=101, with its binary representation 01100101, provides the outputs for each of the 8 possible input codes. In its binary representation, the most significant bit of ID corresponds to the cell output when the input code $[u3, u2, u1] = [1,1,1]$. A periodic boundary condition is also assumed i.e. the leftmost cell $(i = 1)$ is connected to the rightmost one $(i = n)$. The binary *mask vector* $mathbfm = [m_1, m_2, ..., m_n]$ has to be optimized [21] for any odd counter size up to $n \leq 29$ to obtain a maximal cycle length ($r = N/2^2 \rightarrow 1$). Note that if all $m_i = 0$ (non-inverted cell outputs), the cellular automaton is a standard, homogenous one. The cells with inverted outputs correspond to ID=154 (the remaining non-linear rule credited for conservative dynamics [16] when used in cellular automata). As shown in [15], in the case of homogenous CA, a maximal cycle length is not always attainable for an arbitrary $n$. The optimization process of hybrid CA is a simple random

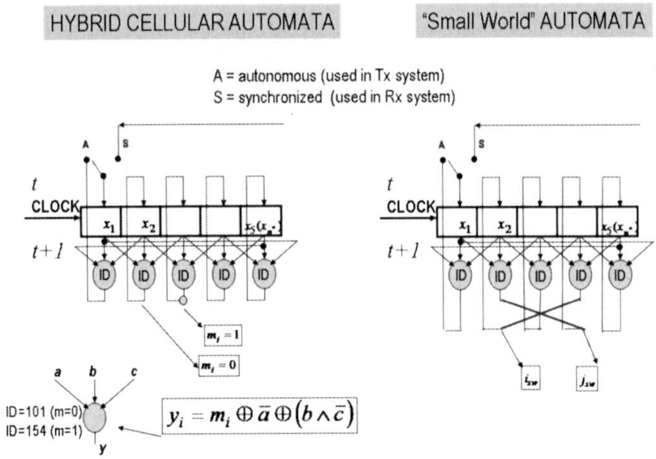


**Fig. 5.2** Two automata network structures using 3-input cells that have both binary synchronization property and complex (chaotic) dynamics

search and may be performed successfully on actual standard PC platforms only for $n \leq 29$ due to the exponential computational complexity. For instance, in the case $n = 21$ the optimal *mask vector* is [000000000010100000010] or 1282 in decimal. A table with optimal masks for $n \leq 29$ is provided in [21]. In the above example, the length of the maximal cycle is 2097151, i.e. $2^{21} - 1$. It is possible to have more different masks, providing the same maximal length.

The **key space** for this CA model is represented by the number of cells $n$ , the value of "mother cell" (it can be ID=101 or any of its 3 equivalents [16]) and the specified mask vector.

There are many possibilities to increase the key space. One possibility, recently investigated, is to consider an alteration of the regular cellular topology into a "small worlds" one, as shown in Fig.5.2 (right side). Essentially the model is described by the same equation ((5.1)) where the mask can be removed (i.e. all $m_i = 0$) but where the optimization of the maximal cycle length may now be achieved using a random search process in a space of permutations (one or more pairs of outputs are swapped as shown in Fig.5.2). The mask is now replaced by the set of swapping pairs $(i_{sw}, j_{sw})$. In addition one can consider the hybrid cellular model instead of the homogenous one, and consequently expand the key space furthermore.

## 5.3 Characterizing Complex Dynamics in Automata

Any automata, including the models discussed above, has a finite state space with $2^n$ states. A tableau of all possible dynamics in such a state space is given in Fig.5.3 for the case $n = 4$. In addition to the binary synchronization property we want to design the feedback logic F under the constraints of the chosen cellular models (described in Section 2) such that the maximal length of the major cycle is maximized (and consequently the number of cycles is reduced) and the dynamics on this cycle is complex. In order to describe the complexity of the dynamics we consider the following approach:

In [15] a simplified measure of dynamics complexity (chaos) was defined observing that in a "chaotic counting automata", unlike in a "normal counting automata" the average jumps (in terms of Hamming distance) between consecutive binary vector states (as given by the $n$ cell outputs) becomes $n/2$ instead of 1. Therefore for any arbitrary counting cycle $C_j$ of length $L_j$ a *scattering coefficient* $S_j$ is defined by averaging the Hamming distances between all consecutive binary vector states in that cycle:

$$S_j = \frac{1}{nL_j} \sum_{k=1}^{L_j} \sum_{i=1}^{n} |x_i(k) - x_i(k-1)| \tag{5.2}$$

where $k$ is the time index of consecutive states in the cycle $j$ A *degree of chaos* $\lambda_j$ is defined such that it becomes maximum if $S_j = 0.5$ and zero for the extreme, non-chaotic cases of both fixed points and period 2 cycles (with $S_j = 0$ and $S_j = 1$ respectively):

$$\lambda_j = 1 - |2S_j - 1| \tag{5.3}$$

The *degree of chaos* may be regarded as qualitatively similar to the Lyapunov exponent, usually used in continuous-state systems to characterize chaotic behaviors. In our case its largest value is $\lambda_j = 1$ indicating the highest degree of randomness in a finite-length cycle. It ranges between $\lambda_j = 0$ for low complexity dynamics (fixed points, periodic limit cycles where consecutive states are close to each other), to $\lambda_j = 1$ in good random number generators. This is actually the case for the architectures presented in Section 5.2, as an effect of the ID choice. It is important to note that rule ID=101 (and its equivalents) belong the most complex classes (Hyper-Bernoulli-shift) in Chua's taxonomy of elementary cellular automata dynamics [16]. The conservative property of homogenous CA with rule ID=101 guarantees the absence of transients and the effective use of all states in the state space. The modified models proposed in Section 5.2 to increase the key space and improve the maximal cycle length are also conservative, as resulted from our numerical simulations. Note that in comparison with digital implementation of the logistic map, the cellular models discussed in Section 2 have several clear advantages, as pointed out in [22]:

a) They have a very low implementation complexity growing linearly $(O(n))$ with the number of cells. Instead, implementation of F in the case of logistic map requires multiplication, an operation requiring an implementation complexity of $O(n^2)$;

b) There are no transients and the entire state space is efficiently used. Also there are no finite-precision effects as in the case of the logistic function. For instance,

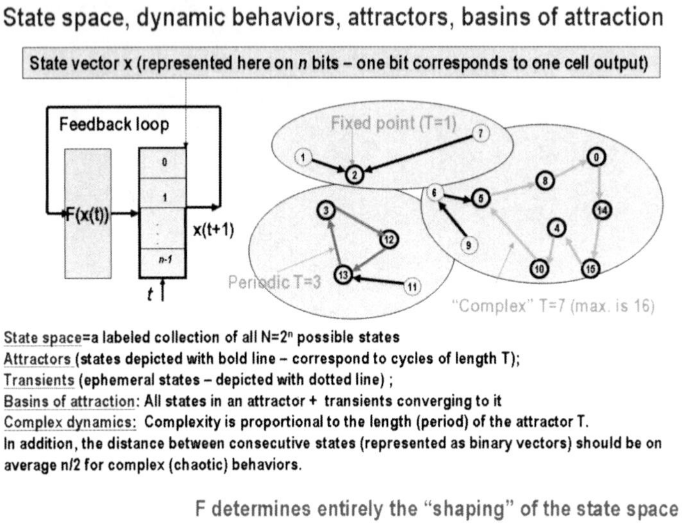


**Fig. 5.3** Profile of the state space in an automata network (here exemplified for $n = 4$ cells). A good random number generator must have no transient and a small number of cycles (ideally one cycle) with maximal length T and complex dynamic behavior.

a logistic map implementation with $n = 29$ bits converges towards cycles with a length of only T=16420 states. Instead, the hybrid CA with the same number of cells ensures the maximal cycle with a length T=2097151 (almost 128 times longer !!). Other usually employed nonlinear maps have similar problems.

c) They have the binary synchronization property, not common among usual nonlinear maps. In terms of synchronization time $T_S$ (i.e. average number of iterations required by a receiver initialized in a randomly chosen state) our experimental results shown in Fig.5.4 emphasize an exponential dependence on $n$ following the approximate formulae: $T_S \cong 1.6^n$.

In addition to a strict dynamical characterization of automata (or nonlinear map) systems producing chaotic sequences, it is important to answer whether the generated sequences are passing or not standard statistical tests. Fortunately, the cellular automata with ID=101 was already submitted to difficult batteries of such tests by other authors [18], in the context of evolutionary search for CA with very good cryptographic properties.

Notably, the authors mention that only 3 rules: ID=30, ID=86 and ID=101 passed both the FIPS 140-2 standard testing and the battery of 23 strong tests of the Diehard program [23]. But from what was discussed above, only automata networks based on rule ID=101 is *conservative* (giving complex dynamics with no transient times) and *binary synchronizable*. The modifications in the homogenous CA model presented in Section 5.2 were found to have no influence on the results of the statistical tests.

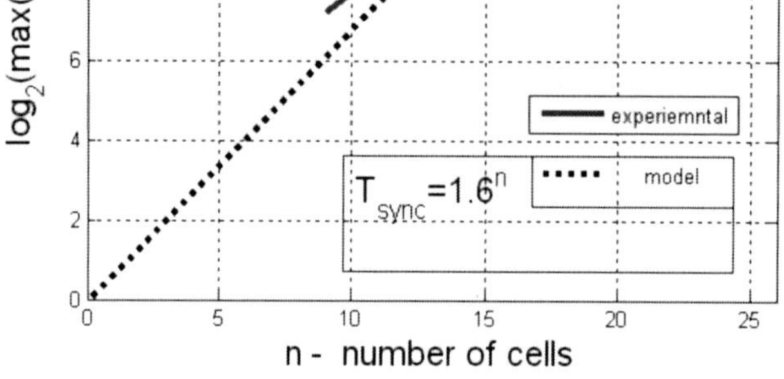


**Fig. 5.4** Experimentally determined average synchronization time and its dependence on the number of cells $n$

## 5.4 FPGA Implementations of Cellular Automata

Usual microcontrollers may be used to implement the cellular automata networks with binary synchronization capability, however maximal speed and efficiency can be achieved when CA models are implemented in a fully parallel fashion. In [20] we provide a convenient methodology to generate VHDL representations using the Algebraic Normal Form conversion. For the case of any elementary CA (with cells defined as 3-inputs Boolean functions) the ANF representation is:

$$y = k_0 \oplus k_1 u_1 \oplus k_2 u_2 \oplus k_3 u_3 \oplus k_4 u_1 u_2 \oplus k_5 u_2 u_3 \oplus k_6 u_3 u_1 \oplus k_7 u_1 u_2 u_3 \quad (5.4)$$

The translation from ordinary (Truth Table) representation to the ANF is done using a procedure in [24]. This procedure, together with other information describing the cellular automata structure (encryption key) is embedded into a specialized software tool which automatically produces the VHDL description of either a single (Tx) cellular automata or of the both systems as required in a transmission chain. Effective implementations were successfully tested on the DE2 University Program board from Altera equipped with Cyclone II EP2C35F672C6 device. The resource allocation on the FPGA device for a particular example (autonomous hybrid CA with rule 101 and 7 cells) is given in Table 5.1.

**Table 5.1** Resource Allocation of a HCA implementation on the Cyclone II - EP2C35F672C6 FPGA device

| | |
|---|---|
| Total logical elements (LE): | 7/33,216(< 1%)<br>Combinational functions: 7<br>Logic registers: 7 |
| Total pins: | 9/475 (2%) |
| Total memory bits: | 0/483,840 (0%) |
| Embedded 9-bit Multipliers: | 0/70 (0%) |

Note the very efficient allocation of one cell per FPGA logic register. The above results confirm that cellular automata with very large $n$ ($n = 33216$ in the case of the chip on the DE2 board) can be easily realized in low cost series FPGA. Compared to other FPGA implementations reported in the literature, ours provides the most compact implementation. The same VHDL description may be used to generate part of specialized sensor chips (e.g. in addition to low power image sensors [25]) using an ASIC design flow. The result would be a fully integrated sensor system with encryption, compression and other capabilities, as discussed in [26]. Such a sensor would perform compressed sensing using a different, more efficient approach than recently proposed compressive sensing [1].

## 5.5 Applications

Since CA models discussed above (sometimes called *chaotic counters*, since they are counting through the maximal length cycle) are similar with any of the previously investigated nonlinear discrete maps using finite computing precision, previously reported applications [27] for such chaotic maps may benefit by replacing them with the proposed CA models. There are three main positive effects:

a) CA models have a higher implementation efficiency than nonlinear maps;

b) CA models will offer maximal length cycles with no transient effects;

c) A simple and efficient mechanism for synchronization is provided. There are however two classes of novel applications particularly suited for the proposed complex sequence generators. They will be briefly discussed next:

### *5.5.1 Compressed Sensing Based on Chaotic Scan*

This class of applications is a compact alternative to the compressive sensing methods [1] and was first proposed in [26]. The simplified model is given in Fig. 5.6, with regards to an image sensor. The idea may be further extended to any other kind of multi-dimensional sensor in order to reduce the number of samples effectively transmitted from the sensor. The method is effective assuming that adjacent elements in the array are highly correlated. Chaotic scan implemented with low implementation complexity chaotic counters perform simultaneously a form of compression as well as ciphering and open the possibility to develop low cost sensors with compressive sensing and radio-transmission facilities. To demonstrate the main idea of the

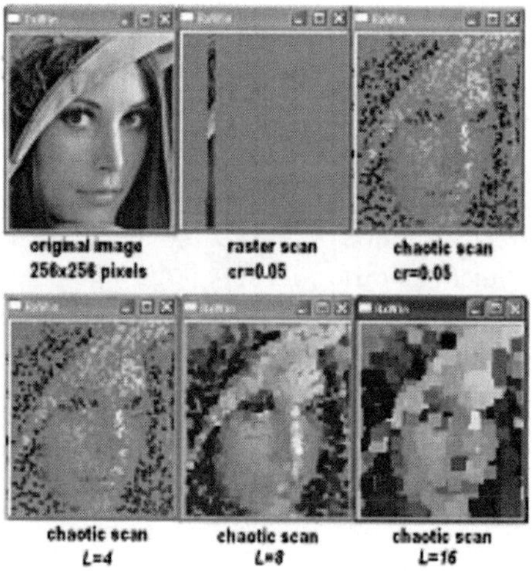


**Fig. 5.5** Upper row: picture coverage with raster scan and chaotic scan after sending 5% of all pixels from the transmitter; Lower row: optimization of the reconstruction error by choosing different sizes L of the pixel neighborhood at receiving point

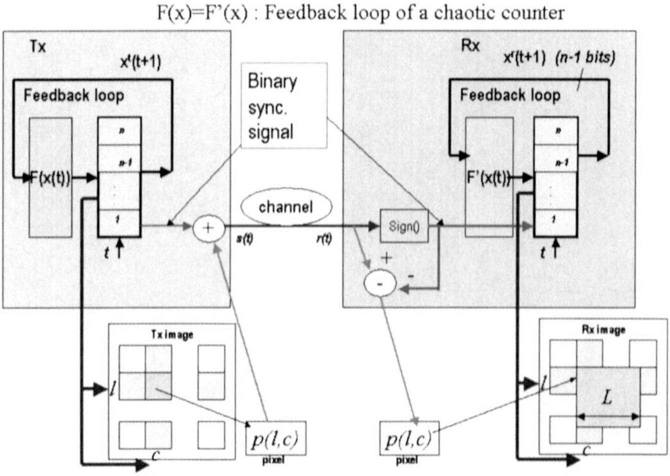


**Fig. 5.6** An alternative model for compressed sensing based on chaotic scan: Pixels in a sensor array are chaotically selected by the chaotic counter. At receiving point a similar chaotic counter, synchronized with the one in Tx, is recomposing the image form the serially received pixel value while estimating its L-sized neighborhood. A small fraction of samples from the Tx image suffices to recover a good quality replica of the original image with a certain acceptable information loss.

method and its effectiveness, for a given image (leftmost one in Fig.5.5) both raster scan (using conventional counters to address the array) and chaotic scan using the hybrid CA rule 101 cellular automaton were performed in order to sample and send a small fraction of only 5% of the pixels. It is clear that by sampling uncorrelated pixels the image content is better retrieved in the case of chaotic scan. On the other hand, controlling the size of the neighborhood filled with the same value of the received pixel allows optimizing the reconstruction error.

Ordinary chaotic maps (i.e. logistic, tent, etc.) cannot be used in such applications since their finite computing precision implementations often produce cycles with only a very small fraction of state vectors (each addressing a pixel in the image sensor array) belonging to the counting cycle. Consequently, only a small fraction of the sensing elements will be addressed, compromising the information aquisition process.

### 5.5.2 *Efficient Generation of Spreading Sequences*

A more detailed treatment of this application is provided in [28]. The principle is exposed in Fig.5.7. An important feature of the chaotic counters described in Section 2 is exploited in this case: The output of each cell can be considered as generating a spreading sequence. Simulations for the case $n = 29$ confirmed that such sequences can be considered orthogonal. When synchronization is needed, the

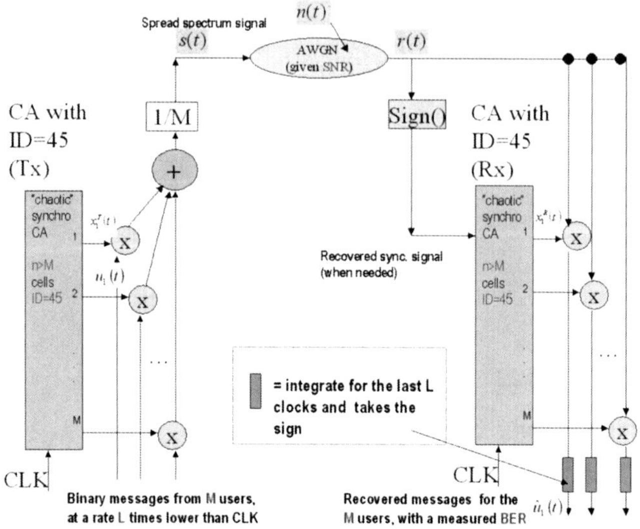


**Fig. 5.7** A model of a low complexity code division multiple access (CDMA) system with spreading sequences generated by each cell of the cellular automata chaotic counter

signal associated with the first user is set to $u_1(t) = M$, while all others are set to 0. When the two counters (in Tx and Rx) are synchronized, the spreading sequences are recovered at the receiving point allowing to decode the specific's user message. In designing such systems, a tradeoff is considered between the number of users $M$ and the spreading factor $L$ (that is, the number of basic clock cycles allocated to each symbol) such that a desired bit error rate BER is achieved. The transmission channel is usually modeled as an additive white Gaussian noise – or AWGN - type. In order to compare the performances of our system, we considered a more complex implementation solution based on logistic maps [29][30]. It turned out that both systems have similar functional performances. It follows that the HCA-based chaotic map with a low complexity implementation may replace more sophisticated systems employing logistic maps without decreasing the functional capabilities. In both cases the ratio $L/M \cong 32$ in order to have bit-error BER=0 when the signal to noise ratio is SNR=10 dB.

## 5.6 Conclusions

This chapter reviews recent results and applications of complex (chaotic) binary sequences generated using non-linear elementary cellular automata. Since these automata were optimized to count most from all $2^n$ possible states, they are also called "chaotic counters". Two architectures were considered, built around cells associated with the Boolean function with 3-inputs ID=101, namely a hybrid model where some properly selected cells have inverted outputs and a "small-worlds" model where some properly selected outputs are swapped. Compared to other nonlinear maps used in discrete time and discrete space (finite computing implementations) cellular automata models presented herein posses several advantages:

a) Binary synchronization property: One single bit (the output of only one of the $n$ cells) suffices to recover the entire $n$-bits state at the receiver through a simple "drive-response" chaos synchronization process.

b) Efficient use of resources: This property makes the proposed automata models highly effective for low power implementations as desired on sensor chips when such functions as compressive sensing and encryption are needed. It is shown that such models have a simple and scalable hardware description language representation producing implementations with allocated resources following a linear law O(n). Such implementations do not use multipliers or other sophisticated operators, as it is the norm in the case of nonlinear maps;

c) The proposed automata models are scalable, making thus possible to build cryptographic systems with high immunity to attacks. As shown, low cost FPGA chips can easily accommodate cellular automata networks with tens of thousands of cells;

d) The proposed CA are *non-linear* unlike most of the previously proposed cellular automata for cryptography that are linear (the most common example is the hybrid CA with rules 90/150). Consequently the CA models proposed herein are in the same category with Non Linear Feedback Shift Registers (NLFSR) that are recently proposed to replace classic Linear Feedback Shift Registers (LFSR) [31] due to their increased immunity to cryptographic attacks. As quoted in [31] $O(2^n)$ observations (consecutive bits in the complex sequence) are needed in the case of nonlinear automata instead of only O(n) in the case of linear ones in order to extract the CA structure and therefore reconstruct the transmitter.

e) As any conservative automata, the models proposed herein are characterized by 0-transient, a property that makes them useful since no supplementary precaution is needed to enter the stationary regime (typical problem with all known nonlinear maps).

f) Since the proposed cellular automata models are optimized to run on a longest possible cycle they act as counters providing the basis for a special kind of compressive sensing [26], with less computational efforts than traditional methods. Further research to compare these methods is in progress. Another application where such cellular automata models provide efficient implementations (as discussed in Section 5) is the generation of quasi-orthogonal binary spreading sequences, as required by the CDMA systems.

Further research is devoted to improve the methods to optimize the encryption key (structure of the automata), a problem which remains difficult to solve when $n > 31$.

## References


1. Baraniuk, R., Cevher, V., Duarte, M., Hedge, C.: Model-based compressive sensing. IEEE Trans. Inform. Theory 56(4), 1982–2001 (2010)
2. Pecora, L.M., Carroll, T.L.: Synchronization in chaotic systems. Phys. Rev. Lett. 821, 821–824 (1990)
3. Kolumban, G., Kennedy, M.P., Chua, L.O.: The role of synchronization in digital communication using chaos– Part II: Chaotic modulation and chaotic synchronization. IEEE Trans. Circuits and Syst. I 1140, 1129–1140 (1998)
4. López-Mancilla, D., Cruz-Hernández, C.: Output synchronization of chaotic systems: Model-matching approach with application to secure communication. Nonlinear Dynamics and Systems Theory 5(2), 141–156 (2005)
5. Alvarez, G., Montoya, F., Romera, M., Pastor, G.: Breaking two secure communication systems based on chaotic masking. IEEE Trans. on Circuits and Systems II: Express Briefs 51, 505–506 (2004)
6. Alvarez, G., Li, S.: Some basic cryptographic requirements for chaos-based cryptosystems. Int. J. Bifurcation Chaos Appl. Sci. Eng. 16, 2129–2151 (2006)
7. Kanso, A., Smaoui, N.: Logistic chaotic maps for binary numbers generations. Chaos, Solitons and Fractals 40, 2557–2568 (2009)
8. Pareek, N.K., Patidar, V., Sud, K.K.: Cryptography using multiple onedimensional chaotic maps. Commun. Nonlinear Sci. Numer. Simul. 10(7), 715–723 (2005)
9. Yi, X.: Hash function based on chaotic tent maps. IEEE Trans. Circuits Syst. II, Exp. Briefs 52(6), 354–357 (2005)
10. Vlad, A., Luca, A., Frunzete, M.: Computational Measurements of the Transient Time and of the Sampling Distance That Enables Statistical Independence in the Logistic Map. In: Gervasi, O., Taniar, D., Murgante, B., Laganà, A., Mun, Y., Gavrilova, M.L. (eds.) ICCSA 2009, Part II. LNCS, vol. 5593, pp. 703–718. Springer, Heidelberg (2009)
11. Takens, F.: Detecting strange attractors in turbulence. In: Rand, D.A., Young, L.-S. (eds.) Dynamical Systems and Turbulence. Lecture Notes in Mathematics, vol. 898, pp. 366–381. Springer (1981)
12. Djemai, M., Barbot, J.P., Boutat, D.: New type of data transmission using a synchronization of chaotic systems. International Journal of Bifurcation and Chaos 15, 207–223 (2005)
13. Dimitriev, A.S., Hasler, M., Kassian, G.A., Khilinsky, A.D.: Chaotic synchronization of 2-D maps via information transmission. In: Proceedings of 2001 International Symposium on Nonlinear Theory and its Applications, vol. 1, pp. 79–82 (2001)
14. Dogaru, R., Chua, L.O., Murgan, A.T.: Secure communication based on binary synchronization of chaos in cellular neural networks. In: Proceedings SCS 1997, Int'l Symposium on Circuits and Systems, Iasi, Romania, pp. 97–100 (1997)
15. Dogaru, R., Dogaru, I., Kim, H.: Binary chaos synchronization in elementary cellular automata. Int. J. Bifurcation Chaos 19(9), 2871–2884 (2009)
16. Chua, L.O.: A Nonlinear Dynamics Perspective of Wolfram's New Kind of Science (Vol I-IV). World Scientific Series on Nonlinear Science, Series A, vol. 57, 68, 76. World Scientic Publishing Company (2006, 2009, 2011)
17. Cho, S.J., Choi, U.S., Kim, H.D., Hwang, Y.H., Kim, J.G., Heo, S.H.: New synthesis of one-dimensional 90/150 linear hybrid group CA. IEEE Transactions on Computer-Aided Design of Integrated Circuits and Systems 26(9), 1720–1724 (2007)
18. Seredynski, F., Bouvry, P., Zomaya, A.Y.: Cellular automata computations and secret key cryptography. Parallel Comput. 30(5-6), 753–766 (2004)

19. Urias, J., Salazar, G., Ugalde, E.: Synchronization of cellular automaton pairs. Chaos 8, 814–818 (1998)
20. Dogaru, I., Dogaru, R.: Algebraic Normal Form for Rapid Prototyping of Elementary Hybrid Cellular Automata in FPGA. In: Proceedings ISEEE 2010, Galati, Romania, pp. 273–276 (September 2010)
21. Dogaru, R.: Hybrid Cellular Automata as Pseudo-Random Number Generators with Binary Synchronization Property. In: Proceedings of the International Symposium on Signals Circuits and Systems (ISSCS 2009), Iasi, Romania, pp. 389–392 (July 2009)
22. Dogaru, R.: HCA101: A chaotic map based on cellular automata with binary synchronization properties. In: Proceedings of The 8th Int'l Conference on Communications, COMM 2010, Bucharest, Romania, June 10-12, pp. 41–44 (2010)
23. Marsaglia, G.: Diehard (2012), `http://stat.fsu.edu/~geo/diehard.html`
24. Ronjom, S., Abdelraheem, M., Danielsen, L.E.: TT and ANF Representations of Boolean functions. In: Online Database of Boolean Functions (2007), `http://www.selmer.uib.no/odbf/help/ttanf.pdf`
25. Fernandez-Berni, J., Carmona-Galan, R., Carranza-Gonzalez, L.: FLIP-Q: A QCIF Resolution Focal-Plane Array for Low-Power Image Processing. IEEE Journal of Solid-State Circuits 46(3), 669–680 (2011)
26. Dogaru, R., Dogaru, I., Kim, H.: Chaotic Scan: A Low Complexity Video Transmission System for Efficiently Sending Relevant Image Features. IEEE Trans. on Circuits and Systems for Video Technology 20(2), 317–321 (2010)
27. Tam, W.M., Lau, F.C.M., Tse, C.K.: Digital Communications With Chaos. Elsevier, Oxford (2007)
28. Dogaru, R., Kim, H., Dogaru, I.: Binary Synchronization in Cellular Automata for Building Compact CDMA Systems. In: Proceedings of the International Symposium on Signals Circuits and Systems (ISSCS 2009), Iasi, Romania, pp. 393–396 (July 2009)
29. Jovic, B., Unsworth, C.P.: Performance comparison of multi-user chaos-based DS-CDMA synchronisation unit within AWGN and Rayleigh fading channel. Electronics Letters 43(18) (August 31, 2007)
30. Jovic, B., Unsworth, C.P.: Chaos-based multi-user time division multiplexing communication system. IET Commun. 1(4), 549–555 (2007)
31. Dubrova, E.: How to speed-up your NLFSR-based stream cipher. In: Proceedings of Design, Automation & Test in Europe Conference & Exhibition (DATE 2009), pp. 878–881 (2009)

# Part II
# Systems' Dynamics Modeling and Simulation with Applications to Real Physical Systems and Phenomena

# Chapter 6
# Fast Switching Behavior in Nonlinear Electronic Circuits: A Geometric Approach

Tina Thiessen, Sören Plönnigs, and Wolfgang Mathis

**Abstract.** In this paper an outline about the geometric concept of nonlinear electronic circuits is given. With this geometric concept the fast switching behavior of circuits, i.e. the jumps in their state space, is illustrated and a jump condition is formulated. Furthermore, the developed geometric approach is adapted to MNA based systems of equations. This new method enables the simulation of such ill-conditioned circuits without regularization and presents an implementation approach for common circuit simulators like SPICE.

## 6.1 Introduction and Motivation

Circuit simulation is a key tool in the design of electronic circuits. Despite the successful development on the construction of robust circuit simulators [1], there are still some open problems, e.g. the simulation of fast switching behavior in nonlinear circuits.

Interesting circuits for our purpose are circuits with fast switching behavior, i.e. circuits with discontinuous changes, which are called "jumps" in state space. Attributes which indicate fast switching behavior are for example topological properties like positive feedback and circuit characteristics like negative differential resistance or port characteristics seen by capacitors and inductors. It is mentionable that many so-called digital circuits belong to these class of circuits, because they are in fact analog circuits that retain information by assuming a certain state. When the information changes, fast transitions may occur. One can show, that those non-regularized circuits contain a "fold" in their manifolds (see Fig. 6.1) and provide examples for the so-called "time-constant problem" of circuit simulation (see [2], [3]).

Tina Thiessen · Sören Plönnigs · Wolfgang Mathis
Institute of Theoretical Electrical Engineering, Leibniz University of Hanover, D-30167, Hanover, Germany
e-mail: {thiessen,ploennigs,mathis}@tet.uni-hannover.de

K. Kyamakya et al. (Eds.): Selected Topics in Nonlinear Dynamics, SCI 459, pp. 99–116.
DOI: 10.1007/ 978-3-642-34560-9_6 

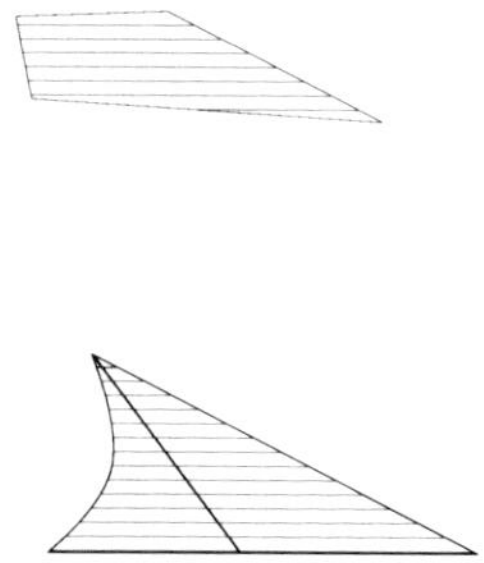

**Fig. 6.1** Folded state space of non regularized network

It can be shown, that the simulation with common circuit simulators of non-regularized networks $\mathcal{N}$ fails, if their state space $\mathcal{S}$ exhibits a fold with respect to a network specific projection direction and if the dynamics is in the vicinity of the fold edge (cf. Fig. 6.1: fold edge is represented by the edge of the shaded area). Such a fold leads to jumps in the embedding space from one point on $\mathcal{S}$ to another, which corresponds to the fast switching in the transient solutions. The points during the jump are no admitted points, because they are outside $\mathcal{S}$. Therefore, we need to introduce an embedding space $\mathcal{E} = \mathbb{R}^k$ and where $\mathcal{S}$ is a subset of $\mathcal{E}$. When the network is $\epsilon$-regularized [4], the jump behavior can be viewed as the limit $\epsilon \to 0$ of the solutions of the singularly perturbed system [5]. The state space $\mathcal{S}_\epsilon$ of the regularized network $\mathcal{N}_\epsilon$ is defined in a different embedding space $\mathcal{E}_\epsilon = \mathbb{R}^{k+c}$ and contains parts of $\mathcal{S}$. However, the fold edge of $\mathcal{S}$ marks - in particular for $\epsilon \to 0$ - the ill-conditioned areas of the dynamics of $\mathcal{N}_\epsilon$. This fold edge corresponds also to the impasse points described in [6], [7]. By adding suitably located $\epsilon$-parasitic inductors L or capacitors C considering Tikhonov's Theorem [8], the network is regularized [9]. Nevertheless, by choosing wrongly located L's and C's, the circuit can be regularized indeed, but the determined transient solutions are inconsistent with respect to $\mathcal{N}$. Another problem are the widely spaced time-constants, which appear due to the fact that the dynamics of a regularized circuit can be divided into a slow and a fast part, leading to the so-called "time-constant problem" of circuit simulation [2]. This difficulty can be circumvented by using stiff solvers (e.g. implicit integration methods like BDF or Gear method) if the reason for the time-constant problem is not related to a jump behavior [3]. In previous publications [10], [11], we have shown the implementation of a geometric concept using geodetic differential equations. In this work we will, for the first time, adapt an alternative geometric concept to systems of equations based on the Modified Nodal Analysis (MNA).

## 6.2 Geometric Approach of Circuits and Fast Switching Behavior

It is known that the state space $\mathcal{S}$ of an electronic circuit can be interpreted as a differential manifold [12]. The branch voltages and currents must satisfy the algebraic nonlinear resistive and Kirchhoff constraints. The Kirchhoffian space $\mathcal{K}$ is defined as the set of all currents and voltages which satisfies Kirchhoff's laws. Moreover the Ohmian space $\mathcal{O}$ is defined as the set of all currents and voltages which satisfies all resistive constitutive relations. Thus the state space of the circuit is defined as the intersection of the Ohmian space $\mathcal{O}$ and the Kirchhoffian space $\mathcal{K}$ by $\mathcal{S} := \mathcal{K} \cap \mathcal{O}$ [13], [14].
The dynamics of a nonlinear dynamic circuit are defined by the set of all solutions of the descriptive differential equations on a sufficient smooth state space $\mathcal{S}$. Therefore the following conditions have to be fulfilled: 1) $\mathcal{S}$ is a smooth manifold. 2) The dynamics can be created on $\mathcal{S}$.

A set possesses the structure of a differentiable (smooth) $m$-dimensional manifold if it is locally equivalent to a $\mathbb{R}^m$. A concrete representation of a manifold can be given by means of a chart (map) that maps a part of $\mathcal{S}$ into $\mathbb{R}^m$. A detailed discussion about differentiable manifolds can be found in the monograph of Guillemin and Pollack [15].
The Kirchhoffian space $\mathcal{K}$ has a vector space structure in linear and nonlinear circuits since Kirchhoff's laws are homogeneous equations. Therefore $\mathcal{K}$ has also the structure of a differentiable manifold. But in general the Ohmian space $\mathcal{O}$ is not a differentiable manifold and even if $\mathcal{O}$ wears the structure of a differentiable manifold it is not obvious that the state space $\mathcal{S}$ wears this structure.
If we consider a circuit by its descriptive equations it means that the intersection of the solution sets of the Kirchhoffian equations and the Ohmian equations is a smooth manifold if these equations are "local" independent. From a geometric point of view this means that the intersection of $\mathcal{K}$ and $\mathcal{O}$ is "transversal" or in a more technical setting: if $\mathcal{K}$ and $\mathcal{O}$ are two submanifolds of $\mathbb{R}^{2n}$ (where $n$ is the number of branches) we call $\mathcal{K}$ and $\mathcal{O}$ transversal, if the following condition is satisfied:

$$x \in \mathcal{K} \cap \mathcal{O} \text{ where } T_x\mathcal{K} \oplus T_x\mathcal{O} = T_x\mathbb{R}^{2n} \text{ or } x \notin \mathcal{K} \cap \mathcal{O} \tag{6.1}$$

Now, we are able to characterize the standard situation in nonlinear dynamical circuits. The state space $\mathcal{S}$ is a smooth manifold if: 1) the Ohmian space $\mathcal{O}$ is a smooth manifold and 2) the state space $\mathcal{S} = \mathcal{K} \cap \mathcal{O}$ is not empty as well as $\mathcal{K}$ and $\mathcal{O}$ are transversal. These properties can be satisfied by applying a suitable remodelling technique with resistive elements [16]. Therefore, this situation is typical or so-called generic and in the following we assume $\mathcal{S}$ to be a smooth manifold.

The second condition requires the construction of a vector field $X$ on the smooth manifold $\mathcal{S}$. We know that based on fundamental physical laws, the relationships between currents and voltages of $\lambda$ capacitors and $\gamma$ inductors are given by means of differential relations. Therefore these differential equations are formulated in $i_L$ and $u_C$ coordinate planes $\mathbb{R}_i^\lambda \oplus \mathbb{R}_u^\gamma$ of the embedding space $\mathbb{R}^{2n}$.

We define a 1-form $\Omega$ and a 2-form $G$ on the space of currents of inductors and voltages of capacitors. Then a projection map $\pi\colon \mathcal{S} \to \mathbb{R}_i^{\lambda} \oplus \mathbb{R}_u^{\gamma}$ is chosen that maps a certain part of $\mathcal{S}$ to the coordinate planes of the inductors and capacitors, respectively. Now we use the map $\pi^*$ to "lift" or "pull-back" $\Omega$ and $G$ on the state space $\mathcal{S}$. This operation is local because there are situations where $\mathcal{S}$ is folded just like in fig. 6.1. In this case, there is more than one part of $\mathcal{S}$ that can be mapped to the same part of the coordinate planes. With respect to the local dynamics of a circuit, the following theorem is fundamental.

If the Ohmian space $\mathcal{O}$ is a smooth manifold, the Kirchhoffian space $\mathcal{K}$ and the Ohmian space $\mathcal{O}$ are transversal and a pullback map $\pi^*$ exists such that a 1-form $\omega := \pi^*\Omega$ and a non-degenerated 2-tensor (bilinear map) $g := \pi^* G$ can be defined on $\mathcal{S}$, then there exists locally a unique vector field $X\colon \mathcal{S} \to T(\mathcal{S})$ which satisfies

$$g(X,Y) = \omega(Y) \tag{6.2}$$

for all smooth vector fields $Y$. With this locally defined vector field $X$ we are able to define the (local) dynamics of a circuit by means of $\dot{\xi} = X \circ \xi$.

### 6.2.1 Singular Points and Jumps

There are several cases where a locally defined vector field $X$ does not exist. If $\mathcal{S}$ is a smooth manifold, then it is essential that $g$ is non-degenerated. The bilinear map $g := \pi^* G$ can be interpreted as an inner product such that the assumed non-degeneracy of $g$ follows from the condition $g(X,Y) = 0$ for all $Y \Leftrightarrow X = 0$. Therefore a degeneracy of $g$ results from defects of $\pi^*$ or $G$. $G$ is degenerated if $L(i)$ or $C(u)$ is zero for some $i$ and $u$, respectively, where these nongeneric cases can be remodelled by parasitic reactances. A defect of $\pi^*$ is related to a dependency of the dynamic variables. With respect to the Kirchhoffian space $\mathcal{K}$ a defect of $\pi^*$ corresponds to loops of capacitors and independent voltage sources or so-called cut-sets of inductors and independent current sources. With respect to the Ohmian space $\mathcal{O}$ a defect of $\pi^*$ is related to a zero of $du_R/di_R$ or $di_R/du_R$ such that above mentioned loops and meshes arise. Also in these cases a remodelling process is available in order to obtain a generic situation of the circuit dynamics. For further details the reader is left to Mathis [16].

These considerations can be discussed in a more concrete manner if circuit topology is included. For this purpose we have to restrict ourself to RLC circuits. Then interconnections of a circuit can be described by oriented graphs and its boundary and coboundary operators or assuming a coordinate system (a chart) by its incidence matrices. If we assume that a proper tree of a graph exists (i.e. a circuit including all capacitor branches and no inductor branches), then no so-called "forced degeneracies" arise. These forced degeneracies are defects of the dynamics related e. g. to meshes of capacitors and cut-sets of inductors.

It is shown by Ichiraku [17] that a point $(i,u)$ of the state space $\mathcal{S}$ is a singular point if and only if the characteristic manifold $\mathcal{O}_R$ and the affine subspace $\mathcal{K}_R$ are not transverse at $(i_R,u_R) := \pi_R(\mathbb{R}^{2n})$ where $\pi_R$ is the natural projection from the embedding space $\mathbb{R}^{2n}$ to the currents and voltages of the resistors. $\mathcal{K}_R$ is the Kirchhoffian space and $\mathcal{O}_R$ is the Ohmian space of the resistive circuit obtained from the given one by open-circuitting all inductor branches and short-circuitting all capacitor branches.

In the following, we exclude forced degeneracies from our discussion (i.e. meshes of capacitors and cut-sets of inductors, as well as $L(i) = 0$ or $C(u) = 0$), which cause the dynamics to be degenerated [18].

## 6.3 Chart Representation of Circuits and Jump Phenomena

By choosing a suitable chart, the circuit equations can be considered as algebro-differential equations (DAEs) [3] in a semi explicit form:

$$\mathbf{C}(\mathbf{x})\dot{\mathbf{x}} = \mathbf{g}(\mathbf{x},\mathbf{y},\mathbf{z}) \qquad \mathbf{g} : \mathbb{R}^k \to \mathbb{R}^n \tag{6.3}$$

$$0 = \mathbf{f}(\mathbf{x},\mathbf{y},\mathbf{z}) \qquad \mathbf{f} : \mathbb{R}^k \to \mathbb{R}^m \tag{6.4}$$

The vector $\mathbf{x} \in \mathbb{R}^n$ corresponds to the capacitor voltages and inductor currents, $\mathbf{z} \in \mathbb{R}^\eta$ is the vector corresponding to independent voltage or current input sources (which can also be a function of time $\mathbf{z}(t)$) and $\mathbf{y} \in \mathbb{R}^m$ is a vector of additional voltages and currents. Here, $\eta$ is the number of independent sources. The matrix $\mathbf{C}(\mathbf{x})$ is related to the dynamical elements and becomes a constant matrix for linear inductances and capacitances. The nonlinear vector field with respect to $\mathbf{x}$, $\mathbf{y}$ and $\mathbf{z}$ is represented by $\mathbf{g}$.

Now, we have to introduce the embedding space $\mathcal{E} = \mathbb{R}^k$ and define $\mathcal{S}$ as a subspace of $\mathcal{E}$. The solution set of the algebraic equations (6.4) represents the state space $\mathcal{S}$ of the circuit, whereas the differential equations (6.3) represent its dynamical behavior. The dimension $k$ of the embedding space can be determined by $k = n + m + \eta$. The state space $\mathcal{S}$ has the dimension $l = n + \eta$ and the codimension is $m = k - l$.

### 6.3.1 *Jumps in State Space*

As mentioned in section 6.2.1, a generic dynamics of a circuit do not exist at points where the projection map $\pi^*$ has singularities. Such singularities arise if the state space is folded, which would result in a jump of the transient solution from one point on $\mathcal{S}$ to another instantaneously. Considering that the energy of capacitors and the charge of inductors is preserved, the voltages across capacitances and currents through inductances have inertia through a jump process and do not change (i.e. the values of $\mathbf{x}$ do not change during the jump). Another restriction is the fixed value of $\mathbf{z}$ during a jump.

Then, with respect to the semi explicit DAE representation, the singular points are defined at points where the local solvability to $\mathbf{y}$ is not guaranteed. These points are specified by the following condition:

$$\det\left(\frac{\partial \mathbf{f}(\mathbf{x},\mathbf{y},\mathbf{z})}{\partial \mathbf{y}}\right) = 0 \text{ where } \mathbf{f}(\mathbf{x},\mathbf{y},\mathbf{z}) = \mathbf{0} \tag{6.5}$$

Therefore we assume eq. (6.5) to be the necessary jump condition (cf. [19] [20], [21]).

The zero set of all points fulfilling the $m+1$ algebraic equations specified by eq. (6.5) is called "jump-set" $\Gamma$ and represents a $l-1$-dimensional subset of $\mathcal{S}$. Of course, the calculation of the solution set of this equation system is difficult. However, we are not interested in all roots of eq. (6.5), but only in the actual chosen jump point during a simulation. Hence, we trace the dynamics on $\mathcal{S}$ (specified by eq. (6.3) and (6.4)) till reaching a stopping point $P_s$. This stopping point is defined as a point, where the step size of the numerical solver reaches a lower boundary (which is related to the machine constant of the simulating computer). In the next step, we search for the "nearest" point on $\Gamma$ (by choosing a suitable norm) and define it as actual jump point $P_j$ (cf. Fig 6.2). Furthermore, we define a straight line $\xi_s$ connecting $P_s$ and $P_j$, which will be used for determining the hit point.

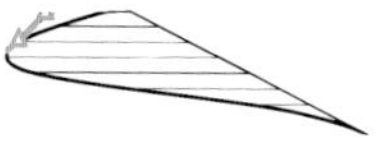

**Fig. 6.2** Concept for jump point calculation; $\Gamma$: green triangles; trajectory: blue circles; $\xi_s$: red line

The sufficient jump condition is first given in a heuristic sense:
A point, which is specified by eq. (6.5) and whose neighborhood includes a Lyapunov stable and an unstable point, is called proper jump point $P_j$. The sufficient jump condition can be verified by calculating the eigenvalues $\lambda_i$ of the characteristic equation

$$\det\left(\frac{\partial \mathbf{f}(\mathbf{x},\mathbf{y},\mathbf{z})}{\partial \mathbf{y}} - \lambda \cdot \mathbf{E}\right) = 0 \tag{6.6}$$

$$\lambda^m + \beta_{m-1} \cdot \lambda^{m-1} + \cdots + \beta_1 \cdot \lambda^1 + \beta_0 = 0\,, \tag{6.7}$$

where $\mathbf{E}$ is the identity matrix (cf. theory of discontinuous oscillators e.g. [21], [22]). If all $\lambda_i$ of eq. (6.7) have negative real parts, the sufficient jump condition is not fulfilled and there are no proper jump points. The difference from the definition in this work to the one in the theory of discontinuous oscillators [21], [22] is, that we interpret $\mathbf{z}$ as variables and therefore extend the approach to all systems with fast switching behavior caused by a fold in their state space manifold and not only to autonomous circuits.

The jump-set $\Gamma$ separate the state space manifold in a stable $\mathcal{S}^-$ and an unstable $\mathcal{S}^+$ part, which are defined by the real part of the eigenvalues

$$\mathcal{S}^- : \quad \Re e\{\lambda_i\} < 0, \quad \forall i \in N, i \leq k \tag{6.8}$$

$$\mathcal{S}^+ : \quad \Re e\{\lambda_i\} > 0, \quad \exists i \in N, i \leq k\,. \tag{6.9}$$

By crossing the jump-set $\Gamma$ between stable and unstable region, there appears either one real positive $\lambda$ or a pair of conjugate complex $\lambda_1, \lambda_2$ with positive real parts. The appearance of more than two $\lambda_i$ with positive real parts is uncommon and will not be analyzed [21]. In the following we will restrict ourself to the case, where only one real positive $\lambda$ appears. Then, for points on $\Gamma$ there is only one eigenvalue equal to zero. As a result the constant term of eq. (6.7) (which corresponds to eq. (6.5) $\beta_0 = \det\left(\partial_{\mathbf{y}}\mathbf{f}(\mathbf{x},\mathbf{y},\mathbf{z})\right)$) is equal to zero. Therefore, the necessary and sufficient jump condition can be tested by the zero crossing of the determinant (6.5).

Because of the fixed values of $\mathbf{x}$ and $\mathbf{z}$ during the jump, the jump takes place in a tangential space of $\mathbb{R}^m$, which corresponds to the coordinate space of $\mathbf{y}$. In the following, the jump space will be denoted by $\mathcal{JS}$. Because we introduced the embedding space, the hit point $P_h$ can be calculated by the intersection of the jump space defined in the jump point and the state space, excluding the jump point ($P_h \in (\mathcal{JS}_{P_j} \cap \mathcal{S}) \setminus \Gamma$). Therewith, it is guaranteed that the values of $\mathbf{x}_j = \mathbf{x}_h$ and $\mathbf{z}_j = \mathbf{z}_h$ before and after the jump are the same. Hence the corresponding "hit-set" is the intersection of the "bundle" of all jump spaces at points of the jump-set and the state space $\mathcal{S}$.

The conditions for fast switching can be summarized as follows:

- The necessary and sufficient jump condition have to be fulfilled: This can be tested by the zero crossing of $\det\left(\partial_{\mathbf{y}}\mathbf{f}(\mathbf{x},\mathbf{y},\mathbf{z})\right)$ for points on $\mathcal{S}$.
- The trajectories hit again the manifold: The intersection of $\mathcal{JS}$ and $\mathcal{S}$ have another solution than the jump point itself, i.e. $(\mathcal{JS}_{P_j} \cap \mathcal{S}) \setminus \Gamma$ is transversal and not empty.

### 6.3.2 Determining the State Space

If one is interested in plotting the state space in a certain area, the following considerations are necessary. Since there are electronic circuits like e.g. the Schmitt Trigger circuit which exhibits a fold respectively the independent input sources $\mathbf{z}$, $\mathcal{S}$ is "near the jump" not unique with respect to $\mathbf{z}$. Thus, for the determination of $\mathcal{S}$, one has to interpret $\mathbf{z}$ as variables and not as a constant or time dependent input value. Furthermore, also the values of $\mathbf{x}$ are fixed during the jump. As a consequence $\mathcal{S}$ is "near the jump" unique with respect to $\mathbf{y}$ and not unique with respect to $\mathbf{x}$ and $\mathbf{z}$ (cf. Fig. 6.3). So, by specifying $l$ components of $\mathbf{y}$, one can determine $\mathcal{S}$. During the determination of $\mathcal{S}$, the determinant criterion can be checked by the local evaluation of eq. (6.5) yielding the jump-set. Nevertheless, there are circuits (e.g. the series connection of resonant tunneling diodes), where the state space is not unique to any coordinate. In this case, the determination of $\mathcal{S}$ is not possible with the method described here. One possibility could be the usage of geodesic coordinates.

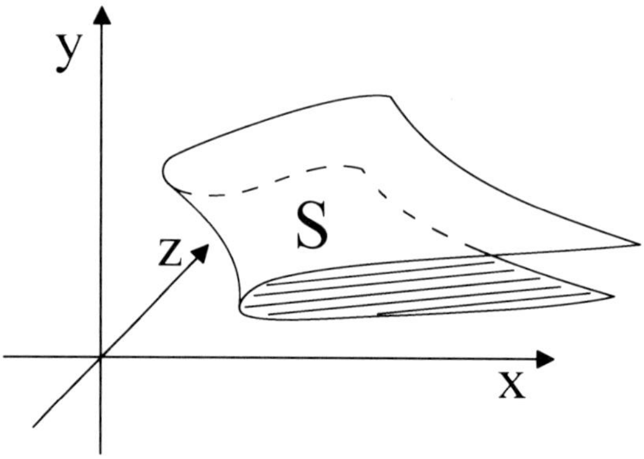


**Fig. 6.3** Uniqueness of the state space with respect to **y**

## 6.3.3 Transient Solution and Hit Point Calculation

For the computation of the transient solution, of course, the time dependency of the input sources $\mathbf{z}(t)$ has to be taken into account. Since the dynamics is not defined at points of the jump-set, it increases very fast nearby the jump-set while tracing it. Therefore we have defined a stopping criterion, which compares the actual step size of the numerical integrator with a lower boundary (see section 6.3.1). When the step size reaches this boundary, the integration is stopped at a point $P_s$. From here, the jump point $P_j$ is calculated as described in section 6.3.1. Then, the corresponding hit-point $P_h$ will be calculated by the intersection of the jump space $\mathcal{JS}$ defined in a point $P_{j'}$ and $\mathcal{S}$ (cf. Fig. 6.4; here $\mathcal{JS}$ is one-dimensional). The point $P_{j'}$ is chosen not to be the actual jump point $P_j$ which lies in $\mathcal{S}$. After calculating the jump point $P_j$, we use the distance $\Delta$ from $P_s$ to $P_j$ on $\zeta_s$ to calculate $P_{j'}$ (see Fig. 6.4). Thereby, we make sure that $P_{j'}$ does not lie in $\mathcal{S}$ and the numerical solver is able to find a unique hit-point $P_{h'}$. If one is interested to get a more exact numerical solution of $P_h$, one can calculate the solution of the intersection of $\mathcal{JS}_{P_j} \cap \mathcal{S}$ with the initial value $P_{h'}$. But, as we will see in section 6.5, the hit point $P_{h'}$ is comparable to the hit point of a regularized circuit. From $P_h$ the dynamics can be traced, till reaching the jump-set again.

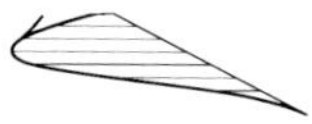

**Fig. 6.4** Concept for hit point calculation

## 6.4 Adaption of the Geometric Approach to MNA Based System of Equations

Common circuit simulators (e.g. SPICE) are based on MNA, which leads in the description of electronic circuits to a high dimensional system of equations. To apply the methods described in section 6.3, several modifications to the classical MNA are necessary [23].

### 6.4.1 *Modification of the System of Equations*

From a given netlist of an electronic circuit, we obtain the system of equations (6.10), which results from applying the MNA stamps [24].

$$\tilde{\mathbf{C}}(\mathbf{n})\dot{\mathbf{n}} + \tilde{\mathbf{G}}\mathbf{n} = \tilde{\mathbf{B}}\mathbf{u}(\mathbf{n},t) \tag{6.10}$$

$\tilde{\mathbf{C}}(\mathbf{n})$ is the matrix related to the dynamical elements which is, for typical circuits, a singular matrix. If there are nonlinear capacitances or inductances, one approach is to separate $\tilde{\mathbf{C}}(\mathbf{n})$ in a linear and a nonlinear part. But, for simplicity, in the following we only consider linear inductances and capacitances, so that $\tilde{\mathbf{C}}$ is a constant matrix. $\tilde{\mathbf{G}}$ is the coefficient matrix related to the non-dynamic elements and $\tilde{\mathbf{B}}$ is the coefficient matrix related to the input sources. The vector $\mathbf{n}$ contains node voltages and currents including at least the currents $i_L$ through inductances. The nonlinear elements are considered as nonlinear dependent current or voltage sources and summarized together with the input sources in the vector $\mathbf{u}(\mathbf{n},t)$. Now, we have to modify the system of equations to apply the methods described earlier.

Since we exclude forced degeneracies from our discussion (i.e. meshes of capacitors and cut-sets of inductors, as well as $L(i) = 0$ or $C(u) = 0$), the rank $r_C$ of $\tilde{\mathbf{C}}$ is equal to the number of capacitances and inductances. Therefore, we add the rows of $\tilde{\mathbf{C}}$, so that there are only $r_C$ non zero rows remaining. Simultaneously, we manipulate the matrices $\tilde{\mathbf{G}}$ and $\tilde{\mathbf{B}}$ in the same manner, yielding:

$$\mathbf{C}^*\dot{\mathbf{n}} + \mathbf{G}^*\mathbf{n} = \mathbf{B}^*\mathbf{u}(\mathbf{n},t)\,. \tag{6.11}$$

By now, the vector $\mathbf{n}$ includes only node voltages and currents. To distinct $\mathbf{n}$ in conserved quantities and non conserved quantities, i.e. in $\mathbf{x}$ and $\mathbf{y}$, we have to insert the capacitor voltages. Therefore, we add algebraic equations which describe the relations between capacitor voltages and the corresponding node voltages (e.g. $U_{C1} = \varphi_{n3} - \varphi_{n7}$) to the system of equations. We summarize all capacitor voltages and inductor currents in the vector $\mathbf{x}$ and all additional node voltages and currents in the vector $\mathbf{y}$. The input vector $\mathbf{u}$ is divided into nonlinear sources $\mathbf{h}(\mathbf{x},\mathbf{y})$, constant bias sources $\mathbf{u}_0$ and independent input sources $\mathbf{z}$ (which can also be a function of time $\mathbf{z}(t)$). The further modifications leads to the system of equations:

$$\mathbf{C}\begin{pmatrix}\dot{\mathbf{x}}\\ \dot{\mathbf{y}}\end{pmatrix}+\mathbf{G}\begin{pmatrix}\mathbf{x}\\ \mathbf{y}\end{pmatrix}=\mathbf{B}\begin{pmatrix}\mathbf{h}(\mathbf{x},\mathbf{y})\\ \mathbf{u}_0\\ \mathbf{z}\end{pmatrix}, \tag{6.12}$$

which can be formulated as semi explicit system of equations:

$$\begin{pmatrix}\mathbf{C}_{11} & 0\\ 0 & 0\end{pmatrix}\begin{pmatrix}\dot{\mathbf{x}}\\ \dot{\mathbf{y}}\end{pmatrix}+\begin{pmatrix}\mathbf{G}_{11} & \mathbf{G}_{12}\\ \mathbf{G}_{21} & \mathbf{G}_{22}\end{pmatrix}\begin{pmatrix}\mathbf{x}\\ \mathbf{y}\end{pmatrix}= \begin{pmatrix}\mathbf{B}_{11} & \mathbf{B}_{12} & \mathbf{B}_{13}\\ \mathbf{B}_{21} & \mathbf{B}_{22} & \mathbf{B}_{23}\end{pmatrix}\begin{pmatrix}\mathbf{h}(\mathbf{x},\mathbf{y})\\ \mathbf{u}_0\\ \mathbf{z}\end{pmatrix}. \tag{6.13}$$

Furthermore, it becomes apparent that the addition of branch voltages is suitable. These additional voltages are stored in the vector **b**, which can be determined by $\mathbf{b}=\mathbf{L}\cdot\mathbf{y}$, where **L** is the matrix relating the node potentials to their branch voltage.

## 6.5 Application on Two Simple Example Circuits

The described concept is illustrated on an emitter coupled multivibrator shown in Fig. 6.5 and on a Schmitt Trigger shown in Fig. 6.13. To analyze the circuits, we use the Ebers-Moll model in forward mode. For the comparison of our non-regularized solution with the solution of the regularized system, the parasitic base-emitter capacitances $C_{reg}$ parallel to the diodes $D_1$ and $D_2$ were added.

### 6.5.1 Emitter Coupled Multivibrator

The design parameters are $R=500\Omega, R_i=100k\Omega, C=33nF, I_S=7fA, V_T=26mV$ and $\alpha_F=0.99$. For the simulation we use a constant bias voltage $U_0=5V$ and a constant bias current $I_1=I_2=0.26mA$.

The resulting vector **u** can be found in Fig. 6.6, the vector **n** in Fig. 6.7 and the matrix dimension in Fig. 6.8. Here the dimension of the embedding space is $k=9$ and because $r_C=1$, we have to split one dynamic equation from the algebraic ones. The state space is one-dimensional ($l=k-m=1$) and the codimension is $m=8$. To display $\mathcal{S}$, we assign the diode voltages to the corresponding node voltages $U_{D1}=\varphi_{n5}-\varphi_{n1}$ and $U_{D2}=\varphi_{n6}-\varphi_{n2}$. In Fig. 6.9 the state space (blue) in the coordinate system $U_{D1}-U_{D2}-U_C$ is shown. The associated jump is represented by the red line (triangles) and, for comparison, the transient solution of the regularized system is shown by the green line (circles).

In Fig. 6.10 the transient solutions of the circuit in relation to different regularization capacitances $\epsilon$ is shown. As one can see, $P_{jump,reg}$ and $P_{hit,reg}$ approaches for $\epsilon\to 0$ the non regularized case. $P_{jump,reg}$ represents the point where the transient solution significantly comes off $\mathcal{S}$ and $P_{hit,reg}$ represents the point, where the transient solution first proceeds in the near of $\mathcal{S}$. Furthermore one can see, that the overshoots of the transient responses after reaching the stable part of $\mathcal{S}$ again is independent from $\epsilon$.

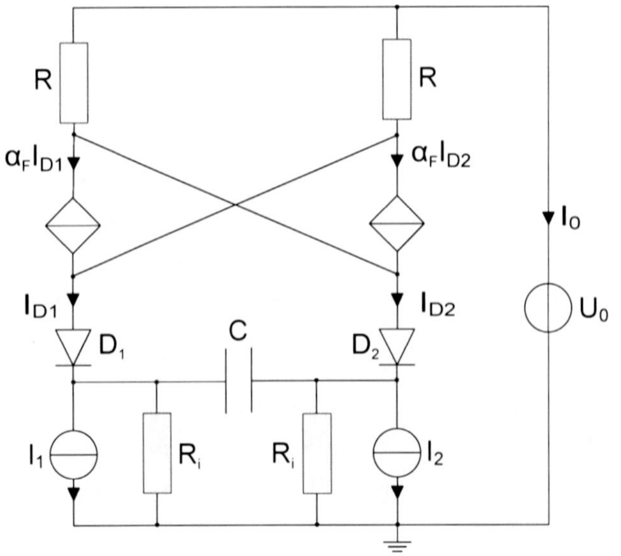


**Fig. 6.5** Emitter coupled multivibrator

$$\mathbf{u} = \begin{bmatrix} I_{D1} \\ I_{D2} \\ I_1 \\ I_2 \\ U_{VS} \end{bmatrix} = \begin{bmatrix} 7 \cdot 10^{-15} \cdot (e^{\frac{\varphi_{n5}-\varphi_{n1}}{26 \cdot 10^{-3}}} - 1) \\ 7 \cdot 10^{-15} \cdot (e^{\frac{\varphi_{n6}-\varphi_{n2}}{26 \cdot 10^{-3}}} - 1) \\ 0.00026 \\ 0.00026 \\ 5 \end{bmatrix}$$

**Fig. 6.6** Vector **u** including input sources and nonlinear dependent sources

$$\mathbf{n} = \begin{bmatrix} \mathbf{x} \\ \mathbf{y} \end{bmatrix} = \begin{bmatrix} U_C \\ \varphi_{n7} \\ \varphi_{n6} \\ \varphi_{n5} \\ \varphi_{n1} \\ \varphi_{n2} \\ I_{VS} \\ I_{D1} \\ I_{D2} \end{bmatrix}$$

**Fig. 6.7** Vector **n** including **x** and **y**

| matrix | rows | columns |
|---|---|---|
| C | 9 | 9 |
| B | 9 | 5 |
| G | 9 | 9 |
| L | 2 | 8 |

**Fig. 6.8** Matrix dimensions

By comparing Fig. 6.11 and 6.12, one can see, that our concept enables a simulation without regularization. Of course, the peaking during the switching process does not exist in our solution, but the main problem, namely the functionality of the circuit, is guaranteed.

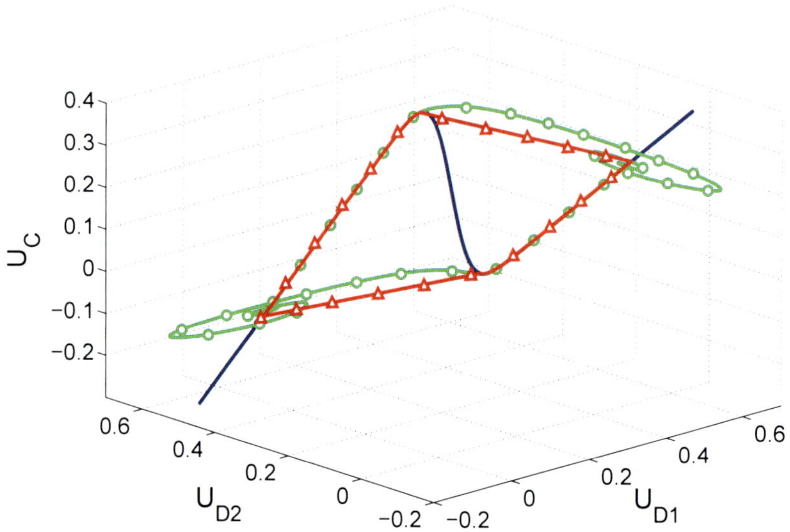


**Fig. 6.9** State space (blue) in $U_{D1} - U_{D2} - U_C$ coordinate system; non-regularized jump solution (red triangles); regularized solution (green circles) $C_{reg} = 500pF$

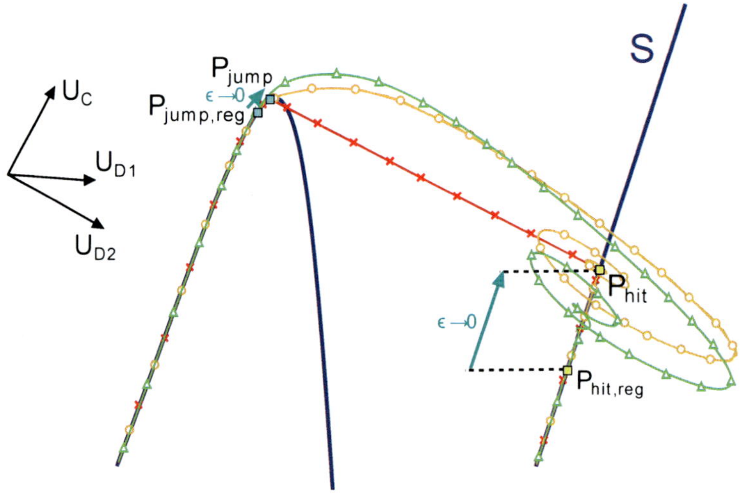


**Fig. 6.10** Transient solution in relation to different regularization capacitances $C_{reg} = \epsilon$ $C_{reg} = 900nF$ (green triangles), $C_{reg} = 10nF$ (yellow circles) and $C_{reg} = 0$ (red crosses)

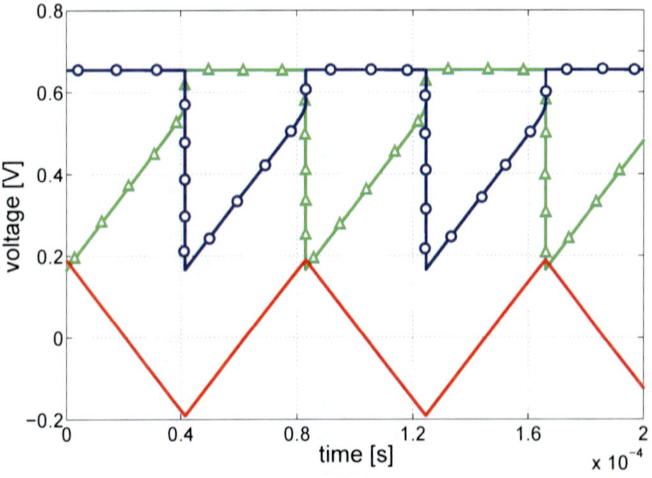


**Fig. 6.11** Transient solution of non regularized multivibrator; $U_{D1}$ (blue circles), $U_{D2}$ (green triangles) and $U_C$ (red)

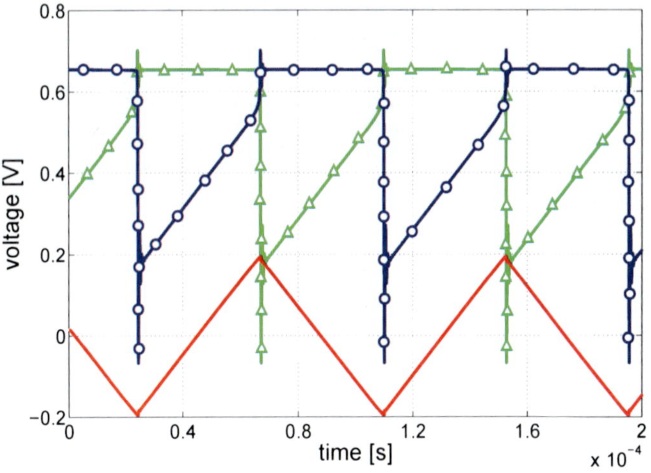


**Fig. 6.12** Transient solution of regularized multivibrator $C_{reg} = 500pF$; $U_{D1}$ (blue circles), $U_{D2}$ (green triangles) and $U_C$ (red)

## *6.5.2 Schmitt Trigger*

The design parameters are $R_{C1} = R_{C2} = 1k\Omega, R_e = 100\Omega, R_1 = 8.2k\Omega, R_2 = 2.7k\Omega, R_V = 4.7k\Omega, I_S = 6.734fA, V_T = 26mV$ and $\alpha_F = 0.99$. For the simulation we use a constant bias voltage $U_0 = 10V$.

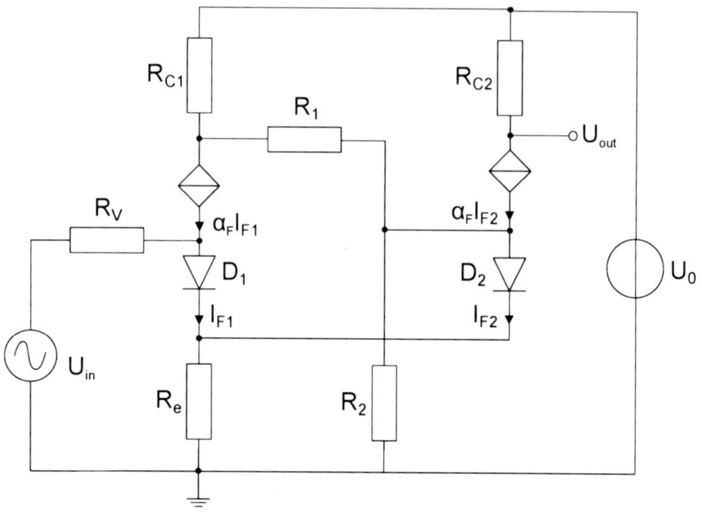


**Fig. 6.13** Schmitt Trigger

$$\mathbf{u} = \begin{bmatrix} I_{D1} \\ I_{D2} \\ U_{V_1} \\ U_{IN} \end{bmatrix} = \begin{bmatrix} 6.734 \cdot 10^{-15} \cdot (e^{\frac{\varphi_{N6}-\varphi_{N7}}{26 \cdot 10^{-3}}} - 1) \\ 6.734 \cdot 10^{-15} \cdot (e^{\frac{\varphi_{N4}-\varphi_{N7}}{26 \cdot 10^{-3}}} - 1) \\ 10 \\ 1.5 + 1.5 \cdot (sin(2 \cdot \pi \cdot 1000 \cdot t)) \end{bmatrix}$$

**Fig. 6.14** Vector **u** including input sources and nonlinear dependent sources

$$\mathbf{n} = \begin{bmatrix} \mathbf{x} \\ \mathbf{y} \end{bmatrix} = \begin{bmatrix} \varphi_{N1} \\ \varphi_{N2} \\ \varphi_{N6} \\ \varphi_{N5} \\ \varphi_{N4} \\ \varphi_{N3} \\ \varphi_{N7} \\ I_{V_1} \\ I_{V_{IN}} \\ I_{D1} \\ I_{D2} \end{bmatrix}$$

**Fig. 6.15** Vector **n** including **x** and **y**

| matrix | rows | columns |
|---|---|---|
| C | 11 | 11 |
| B | 11 | 4 |
| G | 11 | 11 |
| L | 4 | 11 |

**Fig. 6.16** Matrix dimensions

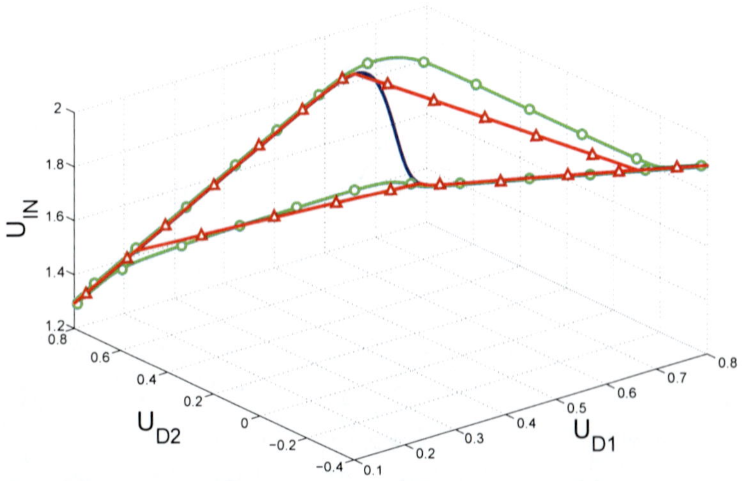


**Fig. 6.17** State space (blue) in $U_{D1} - U_{D2} - U_{IN}$ coordinate system non-regularized jump solution (red triangles); regularized solution (green circles) $C_{reg} = 200pF$

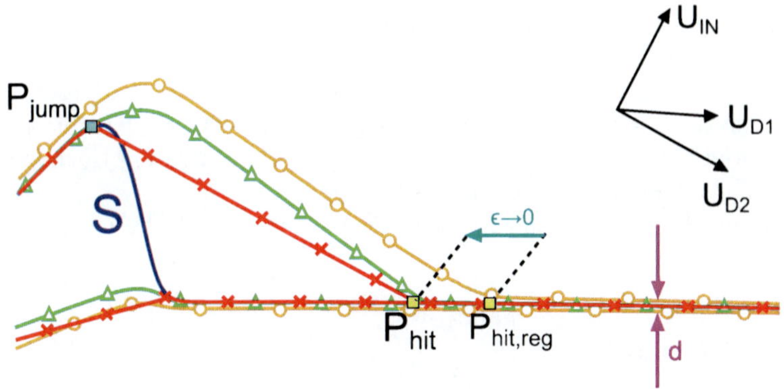


**Fig. 6.18** Transient solution in relation to different regularization capacitances $C_{reg} = \epsilon$ $C_{reg} = 800pF$ (yellow circles), $C_{reg} = 50pF$ (green triangles) and $C_{reg} = 0$ (red crosses)

The resulting vector **u** can be found in Fig. 6.14, the vector **n** in Fig. 6.15 and the matrix dimension in Fig. 6.16. Here the dimension of the embedding space is $k = 12$ and there is one independent input voltage. Therefore state space is one-dimensional ($l = k - m = 1$) and the codimension is $m = 11$. To display $\mathcal{S}$, we assign the diode voltages to the corresponding node voltages $U_{D1} = \varphi_{n6} - \varphi_{n7}$ and $U_{D2} = \varphi_{n4} - \varphi_{n7}$. In Fig. 6.17 the state space (blue) in the coordinate system $U_{D1} - U_{D2} - U_{IN}$ is shown. The associated jump is represented by the red line (triangles) and, for comparison, the transient solution of the regularized system is shown by the green line (circles).

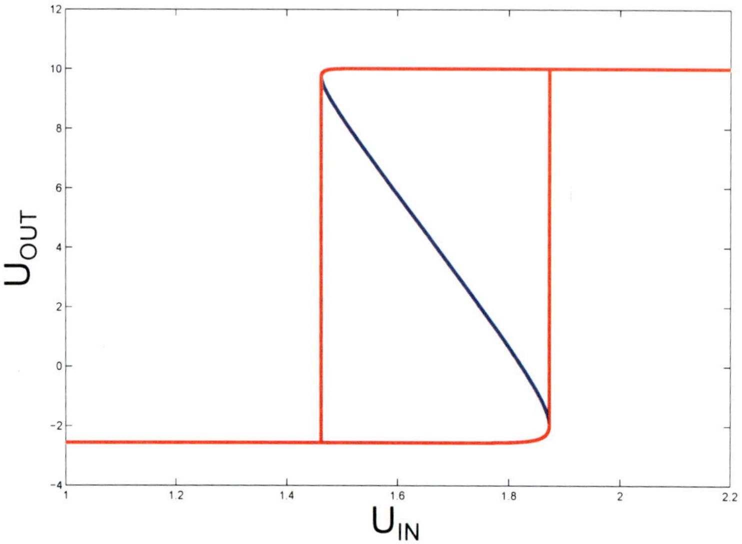


**Fig. 6.19** Transfer characteristic in $U_{IN} - U_{OUT}$ coordinate system state space (blue), transient solution (red)

In Fig. 6.18 the transient solutions of the circuit in relation to different regularization capacitances $\epsilon$ is shown. Here, the regularized solution is far from the state space manifold and approaches for $\epsilon \to 0$ the non regularized case. Also the distance $d$ from transient solution and $\mathcal{S}$ vanishes for $\epsilon \to 0$.

At the end, the known input output characteristic in form of a hysteresis is shown in Fig. 6.19.

## 6.6 Conclusion

In this article we have described why certain circuits with jumps sometimes require adding regularizing capacitors or inductors. We have given a geometric interpretation of jumps in state space and formulated a concrete local criterion to check if a circuit's state space manifold exhibits a fold. With our approach, it will not be necessary to add a regularization to get the transient solutions of a circuit. Furthermore we have shown, for the first time, a complete concept for adapting this geometric approach to a system of equations based on MNA. Therefore, the developed concept can be implemented in an MNA based circuit simulator like SPICE. Finally we have proven the functionality of our concept by numerical results of two BJT circuits. In a further work [25], we have adapted the non MNA based geometric approach to MOS circuits, where the EKV model is used as equivalent circuit diagram. However, the developed theory presented in this paper can be adapted to any electronic circuit containing a fold in their state space.

**Acknowledgements.** The authors would like to thank the German Research Foundation (DFG) for the financial support.

## References

1. IEEE Solid-State Circuits Magazine, 3(2) (2011)
2. Sandberg, I.W., Shichman, H.: Numerical integration of systems of stiff nonlinear differential equations. Bell Syst. Tech. J. 47, 511–527 (1968)
3. Gear, W.: Simultaneous numerical solution of differential-algebraic equations. IEEE Transactions on Circuit Theory CT-18(1), 89–95 (1971)
4. Ihrig, E.: The regularization of nonlinear electrical circuits. Proceedings of the American Mathematical Society 47(1), 179–183 (1975)
5. Sastry, S., Desoer, C.: Jump behavior of circuits and systems. IEEE Transactions on Circuits and Systems 28(12), 1109–1124 (1981)
6. Chua, L.: Dynamic nonlinear networks: State-of-the-art. IEEE Transactions on Circuits and Systems CAS-27(11), 1059–1087 (1980)
7. Venkatasubramanian, V.: Singularity induced bifurcation and the van der pol oscillator. IEEE Transactions on Circuits and Systems I: Fundamental Theory and Applications 41(11), 765–769 (1994)
8. Tikhonov, A.N., Vasil'eva, A.B., Sveshnikov, A.G.: Differential Equations. Springer, Heidelberg (1985)
9. Knorrenschild, M.: Differential/algebraic equations as stiff ordinary differential equations. SIAM J. Numer. Anal. 29, 1694–1715 (1992), http://dx.doi.org/10.1137/0729096
10. Thiessen, T., Gutschke, M., Blanke, P., Mathis, W., Wolter, F.-E.: A numerical approach for nonlinear dynamical circuits with jumps. In: 20th European Conference on Circuit Theory and Design, ECCTD 2011 (2011)
11. Thiessen, T., Gutschke, M., Blanke, P., Mathis, W., Wolter, F.-E.: Differential Geometric Methods for Jump Effects in MOS Systems (presented). In: 2012 IEEE International Symposium on Circuits and Systems, ISCAS 2012 (2012)
12. Smale, S.: On the mathematical foundation of electrical circuit theory. Journ. Differential Geometry 7(1-2), 193–210 (1972)
13. Mathis, W.: Geometric Theory of Nonlinear Dynamical Networks. In: Moreno-Díaz, R., Pichler, F. (eds.) EUROCAST 1991. LNCS, vol. 585, pp. 52–65. Springer, Heidelberg (1992)
14. Thiessen, T., Mathis: Geometric Dynamics of Nonlinear Circuits and Jump Effects. International Journal of Computations & Mathematics in Electrical & Electronic Engineering (Compel 2011) 30(4), 1307–1318 (2011)
15. Guillemin, V., Pollack, A.: Differential Topology. Prentice-Hall, Inc., Englewoods Cliffs (1974)
16. Mathis, W.: Theorie nichtlinearer Netzwerke. Springer, New York (1987)
17. Ichiraku, S.: On singular points of electrical circuits. Yokohama Mathematical Journal 26, 151–156 (1978)
18. Ichiraku, S.: Connecting Electrical Circuits: Transversality and Well-Posedness. Yokohama Mathematical Journal 27, 111–126 (1979)
19. Tchizawa, K.: An Analysis of Nonlinear Systems with Respect to Jump. Yokohama Mathematical Journal 32, 203–214 (1984)
20. Rozov, N.K.: Asymptotic theory of two-dimensional relaxation self-oscillating system. Mathematical Notes 37, 71–77 (1985)

21. Andronov, A., Vitt, A., Khaĭkin, S., Fishwick, W.: Theory of oscillators. Dover books on electronics, electricity, electrical engineering. Dover (1987)
22. Mishchenko, E.F., Rozov, N.K.: Differential Equations with Small Parameters and Relaxation Oscillators. Plenum Press, New York (1980)
23. Thiessen, T., Plönnigs, S., Mathis, W.: A Geometric MNA Based Approach for Circuits with Fast Switching Behavior (presented). In: 2012 IEEE International Symposium on Circuits and Systems, ISCAS 2012 (2012)
24. Vlach, J., Singhal, K.: Computer Methods for Circuit Analysis and Design. Springer, New York (1993)
25. Sarangapani, P., Thiessen, T., Mathis, W.: Differential algebraic equations of mos circuits and jump behavior (accepted). Advances in Radio Science

# Chapter 7
# Dynamics of Liénard Optoelectronic Oscillators

Bruno Romeira, José Figueiredo, Charles N. Ironside, and Julien Javaloyes

**Abstract.** In this chapter we present a comprehensive study on the dynamics of novel nonlinear optoelectronic oscillators (OEO) modeled by Liénard OEO systems. The OEO dynamical systems are based on negative differential resistance resonant tunneling diode oscillators incorporating a photoconductive region and laser diodes. The modeling results are in a good agreement with the wide variety dynamics that has been observed in recent experimental work spanning from self-sustained relaxation oscillations to injection locking and chaotic behaviors in both electrical and optical domains. Potential applications range from generation of periodic and chaotic signals for chaos-based communication schemes to highly stabilized OEOs for microwave-photonic systems.

## 7.1 Introduction

Oscillation is among the simplest of dynamic behaviors to describe mathematically a wide variety of periodic phenomena observed in atmospheric physics, condensed matter, nonlinear optics and electronics, plasma physics, biophysics, biology, etc. [1]. With appropriate perturbation an oscillator system can exhibit highly complex dynamical characteristics ranging from stable, narrow-linewidth oscillation to

Bruno Romeira · José Figueiredo
Center of Electronics Optoelectronics and Telecommunications, Department of Physics, University of the Algarve, 8005-139 Faro, Portugal
e-mail: {bmromeira,jlongras}@ualg.pt

Charles N. Ironside
School of Engineering, University of Glasgow, Glasgow G12 8QQ, United Kingdom
e-mail: charles.ironside@glasgow.ac.uk

Julien Javaloyes
Departament de Fisica, Universitat de les Illes Baleares, C/Valldemossa km. 7'5, E-07122, Palma, Spain
e-mail: julien.javaloyes@uib.es

K. Kyamakya et al. (Eds.): Selected Topics in Nonlinear Dynamics, SCI 459, pp. 117–138.
DOI: 10.1007/ 978-3-642-34560-9_7 

broadband chaos. In recent years, the perturbation schemes that have been more thoroughly studied for optoelectronic and optical communications applications are the optical injection [2, 3], optical feedback [4–6], and optoelectronic feedback [7]. For example, irregular emission behavior of semiconductor lasers with injected coherent light has been intensively investigated to provide bandwidth enhancement, chirp reduction, and noise reduction for optical communication applications. The injection of coherent light into a semiconductor laser is commonly refereed as injection locking [2]. Optical feedback of laser systems has been systematically studied and consists of optical re-injection of a fraction of the light produced by the laser into its active region. In this case the round-trip time of light in the external cavity introduces a delay in the system that is utilized to control and adjust the dynamical behavior [8].

Of increasing interest are optoelectronic oscillator (OEO) dynamical systems subjected to injection and time delayed feedback that can simultaneously generate highly pure signals in both electrical and optical domains [9, 10]. A typical OEO consists of a nonlinear system with a time-delayed feedback loop that can function in the frequency range from tens of GHz down to few kHz. In a OEO system, external elements such as electro-optic modulators, radio-frequency (RF) oscillators, etc., are used to produce nonlinearities, and a laser diode is used only as a light source. Using this configuration, it was demonstrated that delayed-feedback OEOs provide very high-quality microwave-photonics oscillators [11]. Because of the variety of complex dynamical regimes, OEO topologies have been receiving great attention due to the chaos generation and control capabilities that allows increasing the security in optical communication systems [12, 13]. The dynamics of such systems can be easily synchronized and controlled adjusting either the external perturbation or feedback parameters [14–16].

As depicted in Fig. 7.1 a nonlinear time-delay system relies on a nonlinear device, an element to compensate for any losses, and a feedback delay. In this chapter, the investigated OEO system has a simple configuration whose nonlinearities and gain arise from the differential negative conductance (NDC) of a resonant tunneling diode (RTD) [17]. The RTD-based OEO comprises an RTD with a photoconductive region and a laser diode [18, 19], forming an optoelectronic voltage controlled oscillator (OVCO) with both electrical and optical input and output ports [20–22]. As

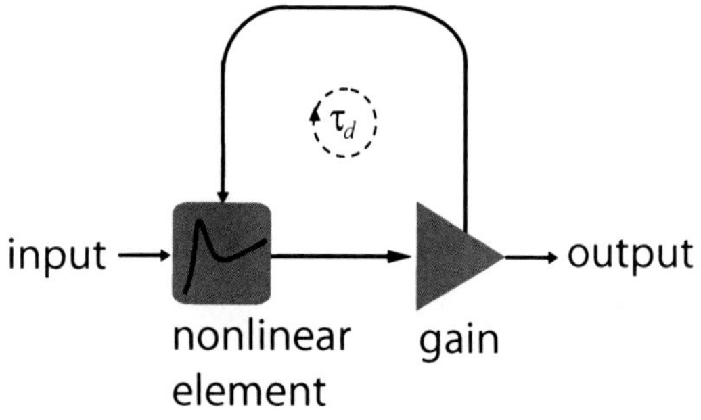


**Fig. 7.1** Schematic of a simple nonlinear time-delay system with feedback, gain, and time delay $\tau_d$.

discussed throughout this chapter, the complex dynamics of RTD-OEOs is modeled by a Liénard OEO system consisting by two sets of coupled differential equations, one describing the electrical properties of the oscillator circuit, which corresponds to a classical Liénard oscillator, and the other modeling the laser diode dynamics using the single mode laser rate equations. The main potential advantages of utilizing the RTD-OEO system instead of optical injection or feedback of optoelectronic/laser systems include the simple implementation, compact solution, tunability of RTD oscillator, and possibility of novel dynamics taking advantage of the nonlinearities of both RTD and LD devices plus delayed feedback.

The chapter is organized as follows. In the next section we present a brief review on RTD-based OEO systems. Then we describe in the detail the generalized forced Liénard OEO model incorporating delayed feedback and discuss in detail its dynamical regimes. Finally, the last section compiles the most relevant concluding remarks and expected developments.

## 7.2 Resonant Tunneling Diode Optoelectronic Oscillators

Resonant tunneling diode OEOs combine the electrical non-linearities of RTD oscillators with photo-detectors and laser diode light sources [17, 18, 21]. An RTD consists of a nano-electronic structure that uses a vertical stacking of epitaxial layers of semiconductor alloys with the active region consisting of a double barrier quantum well (DBQW), in total, about 10 nm thick, that act as a Fabry-Pérot interferometer for the electron wavefunctions. The Fabry-Pérot effect gives rise to a highly nonlinear current-voltage $I(V)$ characteristic with a wide bandwidth NDC region. When DC biased in the NDC region the RTD can act as an electric amplifier and as a relaxation oscillator [17].

Taking advantage of the RTD wide bandwidth nonlinear NDC region, it is possible to implement compact OEO configurations. Figure 7.2 shows a schematic diagram of an experimental RTD-based OEO comprising a RTD monolithic integrated waveguide photo-detector (PD) driving a laser diode, and an optical fiber delay feedback loop. As discussed throughout this chapter the OEO can be operated in a wide variety of operation modes: as an OVCO producing both RF and optical

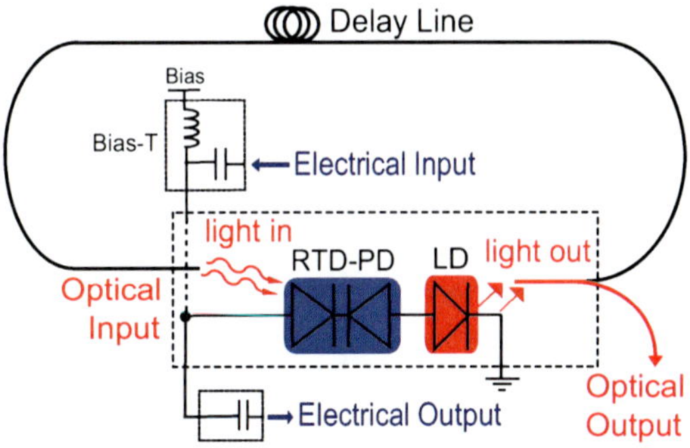


**Fig. 7.2** Schematic diagram of the RTD-laser diode OEO experimental setup with external electrical and optical injection, and time delayed feedback.

modulated signals; as an electrically and optically injection locked OEO; and as a self-synchronized OEO using a time delayed feedback loop configuration.

Next we review in more detail the main features of RTD-OEOs that include the RTD's $I(V)$ characteristic, the photo-detection characteristics, the equivalent circuit model and the laser diode single mode rate equations. The section ends with the formulation of the general Liénard OEO mathematical model derived from the RTD-OEO experimental configuration represented in Fig. 7.2.

### *7.2.1 Resonant Tunneling Diode*

A semiconductor DBQW-RTD consists of a low band-gap semiconductor layer (the quantum well, typically from 5 nm to 10 nm wide) surrounded by two thinner layers of higher band-gap material (barriers, typically from 1.5 nm to 5 nm), both sandwiched between lower band-gap material layers, usually the same as the well material, as shown schematically in Fig. 7.3(a). When both sides are terminated by highly doped semiconductor layers for electrical connection (the emitter and the collector regions/contacts) the structure is called a resonant tunneling diode.

Under applied bias, the DBQW-RTD functions as a Fabry-Pérot filter to charge carrier energy distribution. This is exploited to control the number of carriers that can take part in the conduction through the DBQW resonant levels. The carrier transmission coefficient maxima give rise to a $I(V)$ characteristic with regions of strong nonlinear NDC. In the case of $n$-type RTDs (the case here), the RTD $I(V)$ characteristic depicted in Fig. 7.3(b) can be understood with the help of the lowest conduction band profile shown in Fig 7.3(a).

When the applied bias is small, i.e., $V << V_p$ (peak voltage and local carrier transmission coefficient maximum), the conduction band profile is not much affected, remaining almost flat, see Fig. 7.3(a)(i). As voltage is increased, the energy of the first resonant level is moved downwards to the emitter Fermi level, leading to an almost linear current increase with the voltage, the first positive differential conductance region, till reaching a local maximum $I_p$, ideally, at $V \simeq 2E_{N=1}/q$, when the overlap between the emitter electron Fermi sea energy spectrum and the transmission coefficient around the first resonant level reaches a local maximum, Fig 7.3(a)(ii). A further increase in the applied voltage pulls the first resonant level towards the bottom. This leads to a sharp current decrease, giving rise to the first NDC portion of the device's $I(V)$ characteristic 7.3(b). At a given voltage, known as the valley voltage $V_v$, with $V_v > V_p$, the current reaches a local minimum $I_v$. An additional increase of the bias voltage will further lift up the emitter Fermi level and tunneling through higher resonant levels or through the top regions of the barriers, 7.3(a)(iii), will lead again to a current increase, similar to the behavior of a classical diode $I(V)$ characteristic. The overall N-shaped $I(V)$ characteristic with a NDC region is shown in 7.3(b).

Thus in the NDC region the current density decreases with increasing voltage (increasing electric field across the DBQW), which in general corresponds to an unstable regime. The actual electric response depends, for instance, upon the contact conditions and the attached circuit, which in general contains - even in the absence of external load resistors - unavoidable resistive and reactive components such as lead resistances, lead inductances, package inductances, and packages capacitances. The global $I(V)$ characteristic of an RTD can be calculated from the local $j(\varepsilon)$ relation, where $\varepsilon$ represents the electric field across the DBQW, by integrating the current density $j$ over the cross-section $A$ of the current flow

$$I = \int_A j dx' dy' \tag{7.1}$$

where the electric field $\varepsilon$ over the length $L$ of the sample is obtained by

$$V = \int_0^L \varepsilon dz' \tag{7.2}$$

where $z'$ is the direction of current flow and $x'$ and $y'$ are the perpendicular directions. Unlike the $j(\varepsilon)$ relation, the $I(V)$ characteristic is not only a property of the semiconductor material, but also depends on the geometry, the boundary conditions, and the contacts of the sample. The $I(V)$ relation is said to display NDC when

$$\frac{dI}{dV} < 0. \tag{7.3}$$

For a useful representation of the RTD non-linear $I(V)$ characteristic one have to considerer a wide variety of device structures and materials available, that is, a suitable modeling of the RTD $I(V)$ characteristic has to include, as much as possible, RTD physical parameters such as material properties, layer dimensions, energy levels, dopant concentrations, and the device geometry.

Since a quantum mechanics based model that includes a full description of RTD features is not yet available, several attempts have been made to incorporate the full pronounced nonlinear NDC and RTD high-speed operation characteristics into circuit simulation packages such as SPICE-like CAD tools [23]. In this work we consider the physics-based model proposed by Schulman et al. [24], that consists of a mathematical function which provides a satisfactory fitting of the RTD $I(V)$

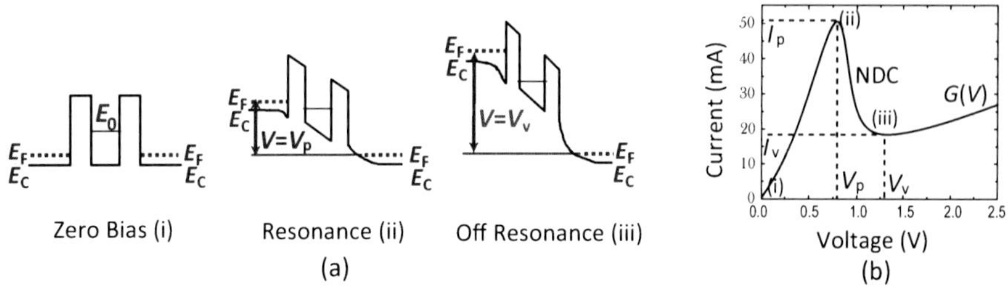


**Fig. 7.3** (a) Lowest conduction band profile under applied voltage. (b) Voltage dependent current source function $G(V)$ of a typical RTD device with 800 $\mu m^2$ area, showing the NDC region.

consisting of InGaAlAs semiconductor compound materials. The expression is given in the form

$$G(V) = A \cdot \ln \left[ \frac{1 + e^{q(B-C+n_1V)/k_BT}}{1 + e^{q(B-C-n_1V)/k_BT}} \right] \cdot \left[ \frac{\pi}{2} + \tan^{-1}\left( \frac{C - n_1V}{D} \right) \right] + H\left(e^{n_2qV/k_BT} - 1\right) \quad (7.4)$$

and contains physical quantities described in detail in [24], which can also be treated as empirical parameters for fitting purposes. In Eq. (7.4), $q$ and $k_B$ are the unit electric charge and the Boltzmann constant, respectively. Figure 7.3(b) shows a typical RTD $I(V)$ characteristic using the fitting parameters $A = 6.42 \times 10^{-3}$ A, $B = 0.0875$ V, $C = 0.1334$ V, $D = 0.013$ V, $H = 4.656 \times 10^{-2}$ A, $n_1 = 0.1502$, and $n_2 = 0.0041$.

The RTD structures described here include also photoconductive layers to operate as a photo-detector which allows RTD oscillators to be controlled by both optical and electrical signals. In a RTD-PD device, the NDC, and hence the current flow, can be controlled by the incident optical power due to the photo-detection characteristics of the structure. By taking advantage of NDC intrinsic gain, RTD-based photo-detectors can exhibit high responsivity and large gain-bandwidth-efficiency [18].

The RTD photo-detector depicted in Fig. 7.2 represents a waveguide photo-detector configuration incorporating a DBQW as reported in [18]. The RTD photo-generated current $I_{ph}(P)$ in response to an optical signal $P(\lambda)$ is given by

$$I_{ph}(P) = \eta_{ph} \frac{q\lambda}{hc} P(\lambda) \quad (7.5)$$

where $\lambda$ is the operation wavelength, $h$ and $c$ are the Planck constant and the speed of light in the vacuum, respectively, and $\eta_{ph}$ is the waveguide photo-detector quantum efficiency given by

$$\eta_{ph} = \kappa(1 - R_{ref})(1 - e^{-\alpha\gamma_{ph}\Lambda}) \quad (7.6)$$

with $\kappa$ being the light coupling factor, $R_{ref}$ is the waveguide facet reflectivity, $\alpha$ is the waveguide core absorption coefficient, $\gamma_{ph}$ is the overlap integral of the electric field and the optical field, and $\Lambda$ is the waveguide photo-detector absorbing active length. For a sinusoidal optical input with average power, $P(\lambda)$, and modulation depth, $m$, the incident power is

$$P(t) = P(\lambda)(1 + m\sin(2\pi f_{in}t)) \quad (7.7)$$

In Table 1 are presented the typical physical parameters used to model the RTD photo-detector characteristics.

### 7.2.2 RTD Photo-Detector Equivalent Electrical Circuit

Using the theory of nonlinear oscillations in terms of differential equations, we have formulated a dynamical model from the analysis of the lumped electrical circuit

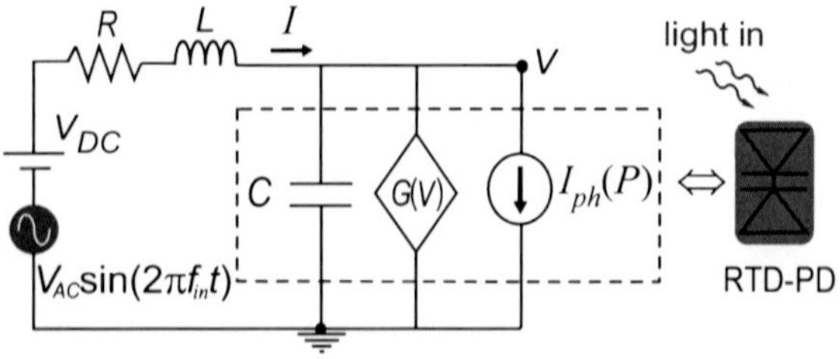


**Fig. 7.4** Schematic of the equivalent electrical circuit of the RTD-PD oscillator subjected to external injection.

shown in Fig. 7.4. The model is used to investigate the dynamics of RTD-PD oscillators perturbed by both electrical and optical modulated signals.

The equivalent circuit of Fig. 7.4 models the forced RTD-PD $RLC$ nonlinear oscillator, with the RTD being represented by its intrinsic capacitance in parallel with a voltage dependent current source $G(V)$, and the photo-detection characteristics being represented by an optical power dependent current source $I_{ph}(P)$. The resistor $R$ and inductor $L$ account for the circuit's resistance and inductance parasitics, respectively. By applying Kirchoff's rules (using Faraday's law) to the circuit of Fig. 7.4, the voltage $V$ across the capacitance $C$ and the current $I$ through the inductor $L$ under external injection are given by the following set of two first-order non-autonomous differential equations

$$\frac{dV}{dt} = \frac{1}{C}\left[I - G(V) - I_{ph}(P)\right] \tag{7.8}$$

$$\frac{dI}{dt} = \frac{1}{L}\left[V_{DC} + V_{AC}\sin(2\pi f_{in}t) - RI - V\right] \tag{7.9}$$

where $V_{DC}$ is the DC bias voltage, $V_{AC}\sin(2\pi f_{in}t)$ is the external electrical periodic force, and $G(V)$ is the mathematical representation of the RTD-PD $I(V)$ characteristic, Fig. 7.3(b). Table 1 summarizes the typical circuit parameters used in the model given by Eqs. (7.8-7.9).

### 7.2.3 *Laser Diode Rate Equations*

In order to correlate the response of the laser diode to its physical parameters we use single-mode rate equations to describe the dynamic behavior of the laser diode [25]. The rate equations relate mathematically the interaction between electrons and photons within the laser cavity and thus describe the nonlinearity in the input-output relationship between the RTD driving current and the optical modulated signal of the laser light output. The rate equations for the photon $S$ and injected carrier $N$ densities in the active region are

**Table 7.1** Description of the typical physical parameters of the electrical circuit and RTD waveguide photo-detector

| Symbol | Quantity | Typical order of magnitude |
|---|---|---|
| $R$ | Resistance | $1\ \Omega$ |
| $L$ | Inductance | $9 \times 10^{-9}$ H |
| $C$ | Capacitance | $5.5 \times 10^{-12}$ F |
| $V_{DC}$ | Bias voltage | 1 V |
| $\lambda$ | Operation wavelength | $1.55\ \mu$m |
| $\kappa$ | Light coupling factor | 0.35 |
| $R_{ref}$ | Waveguide facet reflectivity | 0.3 |
| $\alpha_v$ | Waveguide core absorption coefficient (valley) | 400 $\text{cm}^{-1}$ |
| $\gamma_{ph}$ | Overlap integral of the electric and optical fields | 0.25 |
| $\Lambda$ | Waveguide contact length | $150\ \mu$m |

$$\frac{dN}{dt} = \frac{I_m}{qV_{act}} - \frac{N}{\tau_n} - g_0(N - N_0)(1 - \epsilon_N S)S \tag{7.10}$$

$$\frac{dS}{dt} = \Gamma g_0(N - N_0)(1 - \epsilon_N S)S - \frac{S}{\tau_p} + \Gamma\beta\frac{N}{\tau_n} \tag{7.11}$$

$$\frac{S}{P_f} = \frac{\Gamma\tau_p\lambda_0}{V_{act}\eta_l hc} \tag{7.12}$$

where $I_m$ accounts for the oscillatory current entering in the active region given by Eqs. (7.8-7.9), and the bias current $I_{DC}$. The laser dynamics is modeled employing typical parameters of semiconductor laser diodes. In Table 3 are described the physical parameters of the laser rate equations.

**Table 7.2** Description of the physical parameters of the laser rate equations

| Symbol | Quantity | Typical order of magnitude |
|---|---|---|
| $V_{act}$ | Active region volume | $6.75 \times 10^{-11}\ \text{cm}^3$ |
| $\tau_n$ | Carrier lifetime | $2 \times 10^{-9}$ s |
| $\tau_p$ | Photon lifetime | $1.2 \times 10^{-12}$ s |
| $g_0$ | Gain coefficient | $10^{-6}\ \text{cm}^3\text{s}^{-1}$ |
| $N_0$ | Optical transparency density | $10^{18}\ \text{cm}^{-3}$ |
| $\epsilon_N$ | Nonlinear gain compression | $0.6 \times 10^{-17}\ \text{cm}^3$ |
| $\beta$ | Spontaneous emission | $4 \times 10^{-4}$ |
| $\Gamma$ | Optical confinement factor | 0.44 |
| $\eta_l$ | Differential quantum efficiency per facet | 0.2 |
| $\lambda_0$ | Lasing wavelength | $1.55 \times 10^{-6}$ m |

### 7.2.4 *Forced Liénard OEO System with Time Delayed Feedback*

In what follows, the Liénard OEO system is formulated using dimensionless differential equations from the analysis of RTD-PD oscillator equivalent circuit and laser diode rate equations presented in previous subsections. Figure 7.5 presents a block diagram showing the corresponding mathematical representation of the forced Liénard OEO model with time delayed feedback, where $P(t)$ and $F(t)$ stand for the time-dependent functions that represent the optical and electrical external perturbations, respectively, and $s(t-\tau_d)$ corresponds to the optical time-delayed feedback variable.

The dimensionless equations are obtained from the normalization of RTD-PD differential Eqs. (7.8) and (7.9) and laser diode Eqs. (7.10) and (7.11). In order to normalize Eqs. (7.8) and (7.9), we choose $V_0$ and $I_0$ as scale parameters with physical dimensions of current and voltage ($I_0 = 1$ A and $V_0 = 1$ V), respectively, and rescale $V = xV_0$, $I = yI_0$, $t = \tau\sqrt{LC}$, $V_{DC} = \upsilon_0 V_0$, $V_{AC} = \upsilon V_0$, $f_{in} = \Omega\omega_0$, $\omega_0 = (\sqrt{LC})^{-1}$, and $R = \gamma(V_0/I_0)$. Variables $x$ and $y$ are dimensionless.

Because in real systems noise sources affect the dynamics continuously in an unpredictable manner, we have also included in the model an external Gaussian white noise source, $\chi\xi_y$, to account for the stochastic noise in the OEO, where $\chi$ denote the dimensionless noise strength, and $\xi_y$ represents the Gaussian function.

The dimensionless single mode rate equations are obtained from Eqs. (7.10) and (7.11) making use of the normalized carrier density $n$ and the normalized photon density $s$, rescaling $N = nN_{th}$ and $S = sS_0$, where $S_0 = \Gamma(\tau_p/\tau_n)N_{th}$, with $N_{th} = N_0 + (\Gamma g_0 \tau_p)^{-1}$ being the laser threshold carrier density.

Finally, redefining $\tau$ as $t$ and introducing into the system an optical delayed-feedback $s(t-\tau_d)$, where $\eta$ is the feedback strength and $\tau_d$ is the time-delay with respect to the dimensionless time $t$, we obtain the following dimensionless coupled delay differential equations (DDEs)

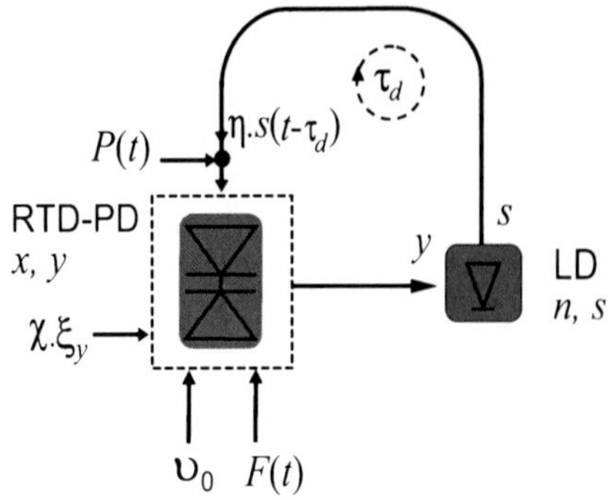


**Fig. 7.5** Block diagram of the forced Liénard OEO dimensionless system with time delayed feedback

$$\frac{dx}{dt} = \frac{1}{\mu}\left[y - g(x) - P(t) - \chi\xi_y - \eta s(t - \tau_d)\right] \tag{7.13}$$

$$\frac{dy}{dt} = \mu\left[v_0 + F(t) - \gamma y - x\right] \tag{7.14}$$

$$\frac{dn}{dt} = \frac{1}{\tau_n'}\left[\frac{i_m}{i_{th}} - n - \frac{n - \delta}{1 - \delta}(1 - \epsilon s)s\right] \tag{7.15}$$

$$\frac{ds}{dt} = \frac{1}{\tau_p'}\left[\frac{n - \delta}{1 - \delta}(1 - \epsilon s)s - s + \beta n\right] \tag{7.16}$$

Equations (7.13)-(7.16) describe the system of equations of the Liénard OEO with time delayed feedback control through the variable $s(t - \tau_d)$, and electrical and optical perturbation through $F(t)$ and $P(t)$, respectively. The Liénard OEO with time-delayed feedback obeys DDEs rather than ordinary differential equations. In DDE systems the state of a dynamic variable at a given time depends on the values of the dynamic variables at both current and previous times.

In Eqs. (7.13) and (7.14), the function $g(x)$ comes from the normalization of $G(V)$ and $\mu = V_0/I_0\sqrt{C/L}$ is dimensionless. Equations (7.15) and (7.16) are the dimensionless rate equations, where $\delta = N_0/N_{th}$ and $\epsilon = \epsilon_N S_0$ are two dimensionless parameters, $i_m$ is the bias current and current modulation given by electrical model, Eqs. (7.13) and (7.14), and $i_{th}$ is the laser diode dimensionless threshold current. The parameters $\tau_n'$ and $\tau_p'$ come from the time rescaling.

The OEO presented here is an example of a Liénard-type system. Such systems were intensely studied during the advent of the radio and vacuum tubes since certain oscillating circuits can be modeled as Liénard systems [26], and are also used in many areas of physics and engineering. Hence, Liénard systems have been the object of intensive analysis by numerous authors (see [27, 28] and the references cited therein). Among the most relevant dynamic behaviors of such systems are the existence of periodic solutions in the form of limit cycles and bifurcations.

A particular example of a Liénard system is the Van der Pol oscillator. Balthasar van der Pol [29] devised the van der Pol oscillator to analyze the nonlinear oscillations in a parallel *RLC* circuit linked to a triode valve as the amplifier, with the anode current in the triode being a nonlinear function of the lumped voltage. The van der Pol equation can describe self-sustained oscillations in the form of limit cycles and injection locking phenomena [30], not only in electronic circuits but also in many other dissipative structures including, among others, chemical reactions, biological and optical systems.

## 7.3 Dynamical Regimes of Liénard OEOs

In what follows, we discuss the dynamics of the Liénard OEO system subjected to external deterministic periodic force and time delayed feedback. The modes of operation and the corresponding dynamical regimes are numerically investigated and mapped in detail.

### 7.3.1 Self-Sustained Oscillations

The Liénard oscillator described by the differential equations (7.13) and (7.14) is a system with at least two dimensions, which means that it is possible to have cyclic or periodic behavior represented by closed loop trajectories in the state space: the limit cycle solution. The motion on a limit cycle in state space represents oscillatory, repeating motion of the Liénard system, which in our case represents the formation of self-sustained oscillations in both electrical and optical domains as observed in the experimental circuit [20, 21].

The analysis of the Liénard oscillator starts with the investigation of limit cycles without external injection and without time delayed feedback by analyzing the location of its fixed points. Fixed points (equilibria) of Eqs. (7.13) and (7.14) occur where $\frac{dx}{dt} = 0$ and $\frac{dy}{dt} = 0$. These points are obtained solving the following equations

$$y = g(x) \tag{7.17}$$

$$y = \frac{1}{\gamma}(v_0 - x) \tag{7.18}$$

Equation (7.17) is the $I(V)$ characteristic of the RTD-PD, and Eq. (7.18) is the load line. Fixed points occur at the points of intersection of these pair of curves. Since Eq. (7.17) describes a curve with two turning points and Eq. (7.18) is a straight line, the general configuration of the fixed points can be deduced. Figure 7.6(a) shows the situation where $\gamma$ is small, that is, when the series resistance, $R$, is sufficiently small ($R < |\frac{dG(V)}{dV}|$). In this situation, the load line is steep and the curves only intersect at a single point as $v_0$ is increased from zero.

At intermediate values of $v_0$ the load line intersects the downward sloping section of $g(x)$. For the situation when the load line is small, the typical form of $g(x)$ with two turning points is sufficient to ensure that all stability transitions are Hopf bifurcations (containing only one fixed point) creating stable periodic orbits. Self-sustained oscillations occur when there is a stable periodic orbit. This is the van der Pol-like situation which results in a stable oscillation if the downward sloping section of $g(x)$ is sufficiently steep.

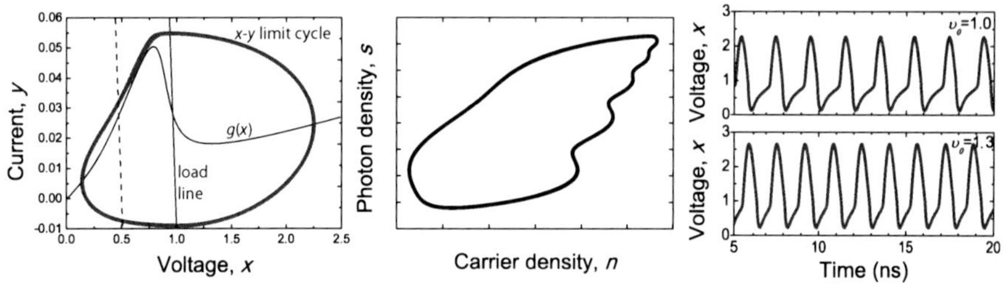


(a) (b) (c)

**Fig. 7.6** (a) $I(V)$ characteristic $g(x)$ function, load lines and $x - y$ limit cycle. (b) $n - s$ limit cycle. (c) Voltage, $x$, relaxation oscillations in time domain.

Figures 7.6(a) and (b) present examples of limit cycles when the load line intersects the downward sloping section of $g(x)$. The analysis of the phase portraits, Figs. 7.6(a) and (b), confirm the existence of a stable limit cycle in the $x$-$y$ phase plane (electrical domain) and $n$-$s$ phase plane (optical domain). Intersections on the other sections of the curve outside the NDC give stable equilibria and no oscillations are present.

The self-sustained oscillations produced are of the type of relaxation oscillations, which are characteristic of negative resistance-type oscillators (e. g. van der Por oscillator). Figure 7.6(c) shows typical relaxation oscillations at 0.498 GHz when $v_0 = 1$ and 0.616 GHz when $v_0 = 1.3$. In the Liénard oscillator the characteristics of relaxation oscillations are determined by parameters $\mu$, $v_0$, and by the voltage dependent current source function $g(x)$. Of particular interest is the frequency dependence of the Liénard oscillator to the change of the $v_0$ parameter. As shown in Fig. 7.6(c), because this change in the parameter will cause the amplitude of the oscillation to change, the Liénard oscillator exhibits a frequency tunability as a function of bias voltage. This regime of operation is called voltage controlled oscillator and has been experimentally verified in previous work [20].

### 7.3.2 *Injection Locking Dynamics*

Frequency locking is a common phenomenon whenever two or more oscillation frequencies (within the same physical system or physically distinct oscillators) interact nonlinearly. Considering the Liénard oscillator, if the oscillator is characterized by a self-sustained oscillation frequency $\Omega_0$ and an external frequency $\Omega$ is perturbing the system, and if the two frequencies are commensurate over some range of control parameter values, that is

$$\frac{\Omega}{\Omega_0} = \frac{p}{q} \tag{7.19}$$

with $p$ and $q$ integers, than we say the frequencies are frequency locked over this parameter range.

In order to study the frequency locking dynamics, first we consider the case of electrical injection of a periodic signal $F(t) = v_e \sin(2\pi\Omega t)$, where $v_e$ is the external amplitude and $\Omega$ the external frequency. The overall dynamics considering both control parameters - external frequency and amplitude - was mapped solving the differential equations (7.13)-(7.16) (in this case $\xi = 0$, $\eta = 0$ and $P(t) = 0$), and plotting the results in the $v_e - \Omega_p$ plane to illustrate the range of frequency ratios over which lockings occur, where $\Omega_p$ is the frequency ratio $\Omega/\Omega_0$ ($\Omega_0$ was set to 0.1109 corresponding to the oscillator natural frequency of 0.498 GHz). In Fig. 7.7 is presented the corresponding locking map, where each color corresponds to a tongue and is related with a periodic region signed with its corresponding number $p/q$ ratio. As $v_e$ increases, the frequency locking regions expand to fill finite intervals along the $\Omega_p$ interval.

Figure 7.7 shows the 1 : 1, 2 : 1, 3 : 1, and 4 : 1, frequency-locking regions. These regions are called Arnold tongues after the Russian mathematician who pioneered the study of frequency-locking [31]. Between each large locking region one can find several types of behavior, including sub-harmonic locking with narrow tongues and aperiodic outputs, namely chaos, which in the graphic of Fig. 7.7 corresponds to the white regions between tongues.

In order to map with more detail the locking regions, the dynamics of the Liénard oscillator was analyzed using $1D-$bifurcation maps. The bifurcation maps were constructed by calculating the time series of a given system variable, such as the voltage or photon density, and plotting the corresponding peak heights as a function of a given circuit control parameter, for example external frequency.

In Fig. 7.8(a) we present the bifurcation map for the photon density, $s$, as a function of frequency ratio $\Omega_p$, in the region of frequencies between $0.002 < \Omega < 0.5$ for a fixed amplitude $v_e = 0.3$. As pictured in Fig. 7.8(a), increasing the external frequency $\Omega$ from close to DC, a stable periodic signal is obtained with period$-p$, followed by a aperiodic region, then a stable period$-p$ region, followed by a aperiodic, and so on. This phenomenon is known as period-adding bifurcation and is controlled by the electrical injection of the Liénard oscillator where the modulated laser output follows the nonlinear dynamics of the electrical Liénard oscillator.

The period-adding sequence presented in the bifurcation map in Fig. 7.8(a) follows the broader locking regions related with the fundamental and harmonic frequencies of the Liénard oscillator. In a close analysis of the bifurcation maps, Fig. 7.8(b), we find a fine structure of locking as a result of the competition between the self-sustained oscillation and the external injected signal. As shown in Fig. 7.8(b), the sub-harmonic frequency locking regions decrease in length (along the $\Omega$ axis) when the denominator in the fraction $p/q$ increases. For example, the sub-harmonic locking region 8 : 3 is shorter than the sub-harmonic locking region 5 : 2. These regions can be easily mapped by constructing the so-called Farey tree that is based in the theory of numbers [32]: if we have two rational fractions $p/q$ and $p'/q'$, the

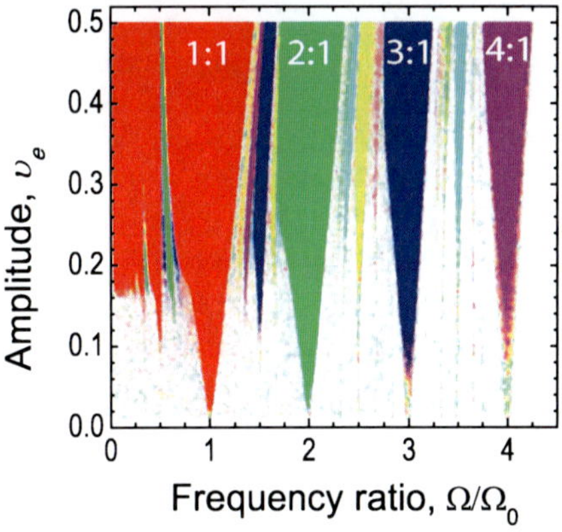


**Fig. 7.7** Arnold tongues map for electrical injection of a periodic signal $F(t)$

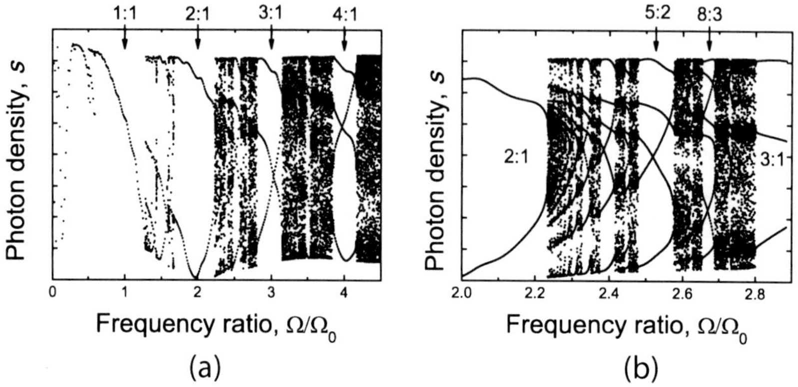


**Fig. 7.8** (a) Bifurcation map for electrical injection of a periodic signal $F(t)$ and fixed amplitude $v_e = 0.3$. (b) Detailed bifurcation map in the region of frequencies $0.22 < \Omega < 0.32$ showing an example of the the Farey tree sequence.

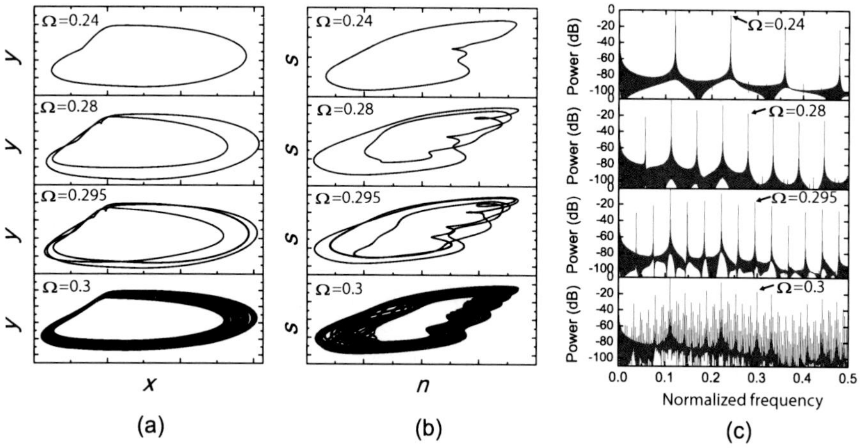


**Fig. 7.9** Electrical (a) and optical (b) phase spaces for the following locking regions 2:1 ($\Omega = 0.24$), 5:2 ($\Omega = 0.28$), 8:3 ($\Omega = 0.295$), and aperiodic region ($\Omega = 0.3$). (c) Corresponding power spectra of the voltage, $x$, output.

rational fraction that lies between and has the smallest denominator is the rational fraction $(p + p')/(q + q')$.

In Figs. 7.9(a) and (b) we show the corresponding trajectories of the locking and aperiodic regions in the electrical and optical domains in the $x - y$ and $n - s$ phase spaces, respectively, represented in the bifurcation map of Fig. 7.8(b). Clear trajectories of limit cycles are observed when the Liénard oscillator is locked at $\Omega = 0.24$, $\Omega = 0.28$, and $\Omega = 0.295$. When the system is not frequency locked and the ratio of frequencies cannot be expressed as a ratio of integers, then the ratio is called irrational and the motion is said to be aperiodic or quasi-periodic because it never exactly repeats itself. An example of this type of output is shown in Figs. 7.9(a) and 7.9(b) at $\Omega = 0.3$. The corresponding Fourier spectrum output characterized by a high harmonic content is shown in Fig. 7.9(c).

We have also studied the dynamics of the oscillator for the case of optical injection locking of a periodic signal $P(t) = \upsilon_{opt} \sin(2\pi\Omega t)$, where $\upsilon_{opt}$ is the external amplitude. In Fig. 7.10(a) we present the Arnold tongues map showing the frequency-locking regions. The tongues are considerable smaller when compared with the electrical injection example discussed previously, and follow the expected experimental results for injection of optical signals with power levels below 10 mW [18]. The tongues can be enlarged increasing the modulation depth parameter or the responsivity parameter of the waveguide photo-detector. In Fig. 7.10(b) we present the bifurcation map for a fixed amplitude $\upsilon_{opt} = 3 \times 10^{-3}$.

The dynamics of the Liénard oscillator subjected to optical injection is similar to the dynamics reported in the electrical injection, which includes the period-adding and the Farey tree sequence. However, we found that for a fixed amplitude the period-adding sequence does not decrease along the $\Omega$ axis but rather increases, as is clearly seen in the map of Fig. 7.10(b) where the harmonic locking regions are larger than the fundamental locking region. This is physically explained by the fact that in this case the nonlinearity of the circuit is externally perturbed in the current component. This behavior can be further studied by adjusting the circuit physical parameters or the circuit configuration in order to follow a similar dynamical sequence presented in the case of the electrical injection.

Both the period-adding and Farey tree sequences found in the optical and electrical injection regimes are good examples of the rich dynamics found in the Liénard oscillator and can be used to construct robust frequency divider optoelectronic oscillators based on the injection-locking dynamics presented here.

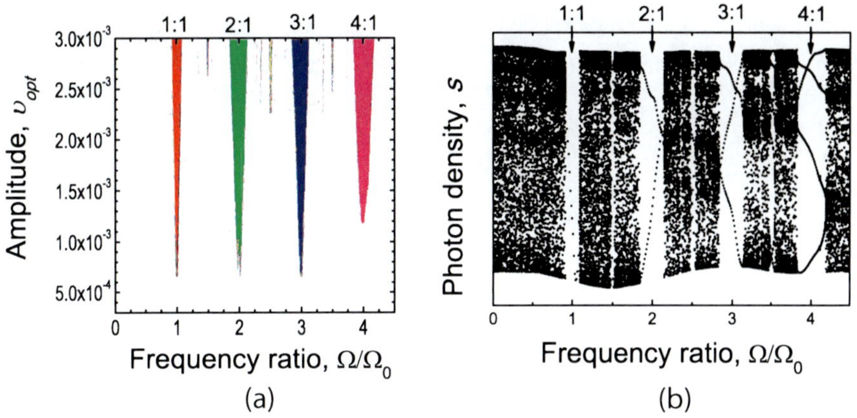


**Fig. 7.10** (a) Arnold tongues map for optical injection of a periodic signal $P(t)$. (b) Bifurcation map for optical injection of a periodic signal $P(t)$ and fixed amplitude $\upsilon_{opt} = 3 \times 10^{-3}$.

### 7.3.3 Quasi-Periodicity and Chaotic Dynamics

Here we analyze the route to chaos found in the Liénard oscillator in the case of electrical injection of a periodic signal $F(t)$. The system follows the quasi-periodic route to chaos scenario and begins with a limit cycle trajectory. As a control parameter is changed a second periodicity appears in the behavior of the system. If the ratio of the period of the second type of motion to the period of the first is not a rational ratio, then we say, as described previously, that the motion is aperiodic or quasi-periodic. Under some circumstances, if the control parameter is changed further, the motion becomes chaotic.

The complex structure observed in the limit cycles of Figs. 7.11(a) and 7.11(b) are examples of electrical and optical chaos generated by the Liénard oscillator. The strongest evidence the system dynamics is chaotic is provided by its Fourier spectrum, Fig. 7.11(c), which is broadband and continuous, a signature of a chaotic signal. The remaining peaks observed in Fig. 7.11(c) correspond to the injected external signal, the free-running signal and to minor vestiges of some quasi-periodicity, which are an indication that the system dynamics evolves into a quasi-periodic route to chaos.

In order to distinguish quasi-periodic orbits from chaotic orbits, the analysis continues further with the determination of the system's Lyapunov exponents. Lyapunov exponents measure the rate of divergence of nearby trajectories, and are a key component of chaotic dynamic studies. A positive Lyapunov exponent is taken as the defining signature of chaos. The Lyapunov exponents also discriminate between the different dynamics: a limit cycle will have one zero exponent, with the other being negative; and a $m-$frequency quasi-periodic orbit will have $m$ zero exponents, with the remaining negative.

Figure 7.11(d) shows the Lyapunov characteristic exponents of the Liénard OEO system. We found one positive exponent, one zero exponent due to the external injection and the remaining exponents are negative. Under these conditions, in the $\Omega$ interval considered, chaotic states occur between the quasi-periodic regions. These chaotic transitions become more evident increasing the amplitude of the driving signal.

Recently, this route to chaos was experimentally validated using electrical injection of an RTD oscillator circuit driving a laser diode [22], confirming many of the details of the model discussed here. The route to chaos is different from several numerical and experimental studies using directly modulated semiconductor lasers including Fabry-Pérot and DFB light sources in sveral optical and optoelectronic configurations [33, 34].

### 7.3.4 Time Delayed Feedback Dynamics

We now analyze the dynamics of the Liénard OEO subjected to time delayed feedback. For purposes of numerical simulation, Eqs. (7.13)-(7.16) were integrated with a standard constant step size Runge-Kutta (RK) method of fourth order [35]. The presence of a delayed contribution in Eq. (7.13) demand a special care.

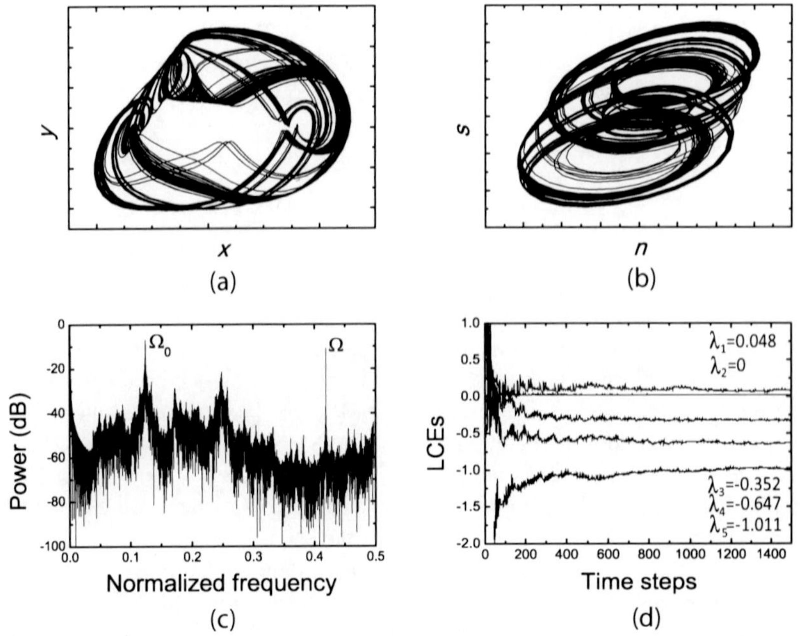


**Fig. 7.11** Electrical and optical chaotic trajectories in the (a) $x - y$, and (b) $n - s$ phase spaces fixing the external amplitude $v_e = 1.505$. (c) Fourier spectrum of the chaotic signal. (d) Lyapunov characteristic exponents.

Indeed, to advance the solution with a step $h$ from $t_n = nh$ to $t_{n+1} = (n+1)h$, the RK algorithm requires evaluating the values of $s(t - \tau_d)$ at intermediate points $t_{mid} = (n + 1/2)h$. However, $s(t_{mid} - \tau_d)$ is not known and must be interpolated from past values, e.g., $(t_{n-1} - \tau_d), s(t_n - \tau_d), s(t_{n+1} - \tau_d)$, etc., with an order consistent with the algorithm of integration. Therefore, in addition of the past values of $s(t)$ we also kept $s'(t)$, that is, a quantity readily available upon time integration which allows building a third order Hermite polynomial between $t_n - \tau_d$ and $t_{n+1} - \tau_d$. By evaluating this interpolant of the delayed term at $t_{mid} - \tau_d$, we ensure an overall fourth order accuracy. At last, the stochastic noise contribution was added after the deterministic step by simply using the Euler method.

We first evaluate the stochastic noise contribution in the free-running oscillator dynamics. Figures 7.12(a) and 7.12(b) show the fundamental voltage $x$ spectra without and with noise contribution, respectively. The introduction of noise into the system smears out the stable limit cycle producing a broader peak in the Fourier domain. It also reduces the signal-to-noise ratio of the free-running oscillation frequency because of the white Gaussian noise contribution.

When the time delayed feedback is included, important dynamical effects are observed that include close-to-carrier noise reduction and the appearance of spurious levels due to the delay contribution. Figures 7.13(a) and 7.13(b) present the voltage $x$ spectrum around the fundamental self-sustained oscillation frequency for a time delay of $\tau_{op} = 0.5\ \mu s$ (corresponding to a normalized time delay of $\tau_d = 2.2 \times 10^3$), fixing the dimensionless noise strength at $\chi = 7 \times 10^{-5}$ and increasing the delayed feedback strength $\eta$. We are assuming the feedback route is an optical time delay $\tau_d$ due to the optical fiber length $L$ given by

$$\tau_{op} = \frac{n_F L}{c} \tag{7.20}$$

with $n_F$ being the optical fiber effective refractive index and $c$ the velocity of light. An optical time delay of $\tau_{op} = 0.5$ corresponds to an optical fiber line around 100 m in length. As shown in Figs. 7.13(a) and 7.13(b), the introduction of time delayed feedback narrows the linewidth of the fundamental oscillation frequency and generates frequency side peaks due to the time-delay $\tau_d$, which corresponds to the free spectral range (FSR) of the OEO given by $FSR = 1/\tau_d$. The presence of these side modes with single-mode-suppression-ratio (SMSR) lower than -40 dBc, Fig. 7.13(b), deteriorates the spectra at offsets around the FSR.

In Fig. 7.14 we investigate the influence of the time delay at a fixed level of feedback $\eta = 5 \times 10^{-4}$. When $\tau_{op} = 0.25\ \mu s$ the carrier frequency peak resembles the broad peak of the free-running oscillation without delayed feedback, Fig. 7.12(b), with some vestiges of side modes at levels close to the noise floor. The influence of the delay is more pronounced in Fig. 7.14(b) at $\tau_{op} = 1\ \mu s$ showing several side modes with a SMSR around -40 dBc. The results provide evidence that phase noise levels can be improved by increasing the fiber length, but there is a compromise between the fiber length and the oscillator stability because increasing the delayed feedback produces high-power side peaks close to the carrier. To overcome this limitation delayed configurations using multiple delayed-feedbacks can be implemented to suppress the spurious frequencies [36]. The numerical simulations presented here follow the experimental dynamics recently reported in [19], where we observed the stability of the OEO can be controlled using an optical fiber delay line.

The large number of side-bands spaced by FSR inversely proportional to the time delay and the feedback strength parameters is an indication that more complex

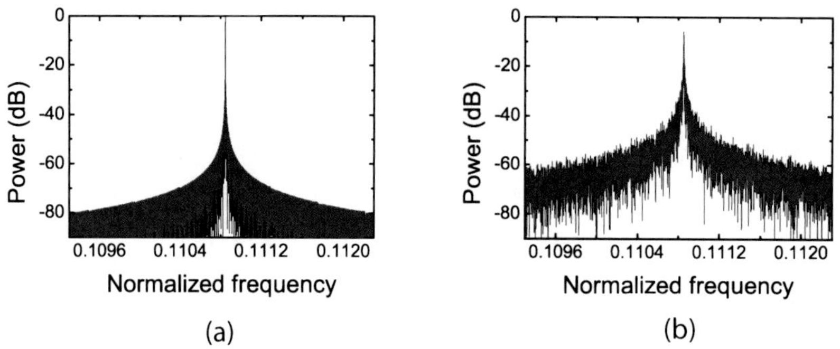


**Fig. 7.12** Simulated spectra of free-running fundamental oscillation (a) without stochastic noise $\chi = 0$, and (b) with noise contribution $\chi = 7 \times 10^{-5}$

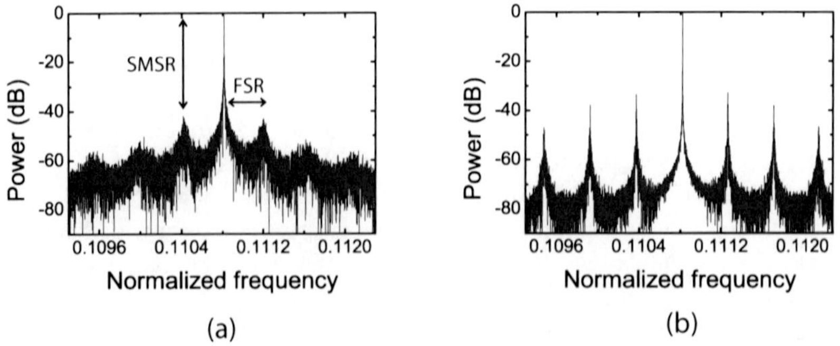


**Fig. 7.13** Simulated spectra of fundamental oscillation with feedback level (a) $\eta = 5 \times 10^{-4}$ and (b) $\eta = 5 \times 10^{-3}$, and fixed time delay of $\tau_{op} = 0.5\ \mu s$

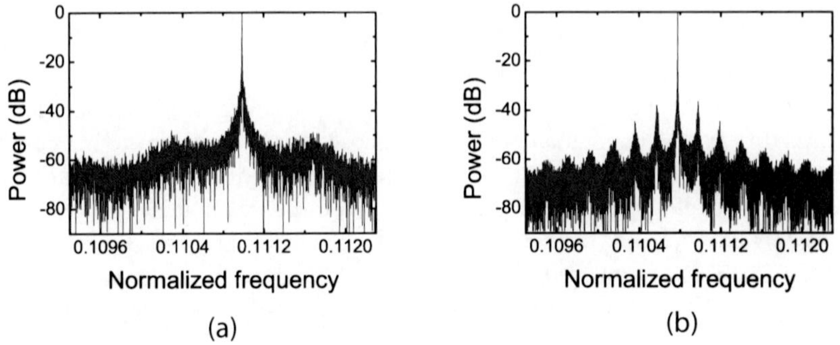


**Fig. 7.14** Simulated spectra of fundamental oscillation with time delay (a) $\tau_{op} = 0.25\ \mu s$ and (b) $\tau_{op} = 1\ \mu s$, and fixed feedback level of $\eta = 5 \times 10^{-4}$

dynamics may occur, which is strongly depended of the feedback level and the length of the external cavity. This can be exploited in innovative applications such as generation of electrical and optical frequency combs [11], or even chaotic transitions, controlled by the delayed feedback parameter. In the results presented here, considering the moderate levels of feedback employed, $\eta << \Delta\frac{I}{I_0}$, where $\Delta I$ is the peak-to-valley current ratio, only stable limit cycles were observed. Currently under study are the instabilities in the Liénard OEO system model associated to the time delay and feedback level.

## 7.4 Conclusion and Future Work

As a conclusion, we have presented a new type of optoelectronic oscillator (OEO) comprising a resonant tunneling diode (RTD) oscillator driving a laser diode. The RTD-OEO is modeled as a Liénard OEO system subject to external perturbations and to time delayed feedback. We have mapped in detail the dynamic regimes spanning from stable and low-noise free-running oscillations with time-delayed feedback, to period-adding bifurcations and Farey-tree sequence dynamics, and quasi-periodicity route to chaos scenarios under external injection.

Considering the variety of dynamical regimes found in general Liénard nonlinear systems, this work suggests the theoretical and experimental study of the synchronization and chaotic dynamics of RTD-based OEOs can find many applications such as in generation of random numbers, novel spread spectrum, ultra wide bandwidth, and optical communication schemes. The main potential advantages of utilizing RTD-based OEO systems instead of using other optoelectronic and laser systems subjected to optical injection or feedback include the simplicity and compactness of our OEO solution, the RTD frequency tunability as a function of the voltage, and the electrical and optical input ports provided by the RTD detector which reduces considerably the number of high-speed electronics required in most of the schemes used to generate chaos.

At the present the RTD-based OEO exploits only RTD's inherent non-linearity, with the laser diode parameters being selected in a way that the laser dynamics does not influence the overall dynamics of RTD-OEO oscillator, e.g., the laser is used only as a light source. In future work novel approaches may be explored in the Liénard OEO system using the laser relaxation oscillation frequency to increase the system's dimensionality and investigate the resultant dynamics. The Liénard OEO system can be explored in further applications not discussed in this chapter, namely chaos-controlled methods using time delayed feedback to stabilize a desired dynamical behavior. This can be used to implement novel microwave-photonic dynamical systems for applications requiring the ability to control the chaotic trajectories.

**Acknowledgements.** Bruno Romeira acknowledges the Fundação para a Ciência e Tecnologia (FCT), Portugal, PhD grant SFRH/BD/43433/2008, and the Fundação Calouste Gulbenkian, Portugal, Programa Estímulo à Criatividade e à Qualidade na Actividade de Investigação, 2009. This work was supported in part by the Fundação para a Ciência e Tecnologia (FCT), Portugal, project grant PTDC/EEA-TEL/100755/2008. Julien Javaloyes acknowledges financial support from the Ramón y Cajal program. The authors would like to thank J. M. Quintana, Universidad de Sevilla, for the fruitful discussions in the RTD numerical simulation.

## References

1. Strogatz, S.H.: Nonlinear Dynamics and Chaos. Addison-Wesley, New York (1994)
2. Simpson, T.B., Liu, J.M., Huang, K.F., Tai, K.: Nonlinear dynamics induced by external optical injection in semiconductor lasers. Quantum and Semiclassical Optics 9, 765–784 (1997)
3. Wieczorek, S., Krauskopf, B., Lenstra, D.: A unifying view of bifurcations in a semiconductor laser subject to optical injection. Optics Communications 172, 1–6 (1999)
4. Lang, R., Kobayashi, K.: External optical feedback effects on semiconductor injection laser properties. IEEE J. Quantum Electron. 16, 347–355 (1980)
5. Dahmani, B., Hollberg, L., Drullinger, R.: Frequency stabilization of semiconductor-lasers by resonant optical feedback. Optics Letters 12, 7876 (1987)

6. Mork, J., Tromborg, B., Mark, J.: Chaos in semiconductor-lasers with optical feedback - theory and experiment. IEEE J. Quantum Electron. 28, 93 (1992)
7. Tang, S., Liu, J.M.: Chaotic pulsing and quasi-periodic route to chaos in a semiconductor laser with delayed opto-electronic feedback. IEEE J. Quantum Electron. 37, 329–336 (2001)
8. Ohtsubo, J.: Semiconductor Lasers: Stability, Instability and Chaos. Springer (2005)
9. Ikeda, K., Kondo, K., Akimoto, O.: Successive higher-harmonic bifurcations in systems with delayed feedback. Phys. Rev. Lett. 49, 1467–1470 (1982)
10. Murphy, T.E., Cohen, A.B., Ravoori, B., Schmitt, K.R.B., Setty, A.V., Sorrentino, F., Williams, C.R.S., Ott, E., Roy, R.: Complex dynamics and synchronization of delayed-feedback nonlinear oscillators. Phil. Trans. R. Soc. A. 368, 343–366 (2010)
11. Yao, X.Y., Maleki, L.: Optoelectronic oscillator for photonic systems. IEEE J. Quantum Electron. 32, 1141–1149 (1996)
12. Callan, K.E., Illing, L., Gao, Z., Gauthier, D.J., Schöll, E.: Broadband chaos generated by an optoelectronic oscillator. Phys. Rev. Lett. 104, 113901 (2010)
13. Argyris, A., Syvridis, D., Larger, L., Annovazzi-Lodi, V., Colet, P., Fisher, I., Garcia-Ojalvo, J., Mirasso, C., Pesquera, L., Shore, K.A.: Chaos-based communications at high bit rates using commercial fibre-optic links. Nature 438, 343–346 (2005)
14. Kouomou, Y.C., Colet, P., Larger, L., Gastaud, N.: Chaotic breathers in delayed electro-optical systems. Phys. Rev. Lett. 95, 203903 (2005)
15. Illing, L., Gauthier, D.J., Roy, R.: Controlling optical chaos, spatiotemporal dynamics, and patterns. Advances in Atomic, Molecular, and Optical Physics 54, 615–697 (2006)
16. Peil, M., Jacquot, M., Chembo, Y.K., Larger, L., Erneux, T.: Routes to chaos and multiple time scale dynamics in broadband bandpass nonlinear delay electro-optic oscillators. Phys. Rev. E 79, 45201 (2009)
17. Figueiredo, J.M.L., Romeira, B., Slight, T.J., Ironside, C.N.: Resonant tunnelling optoelectronic circuits. In: Kim, K.Y. (ed.) Advances in Optical and Photonic Devices (2010), `http://www.intechopen.com/articles/show/title/resonant-tunnelling-optoelectronic-circuits`
18. Romeira, B., Figueiredo, J.M.L., Ironside, C.N., Kelly, A.E., Slight, T.J.: Optical control of a resonant tunneling diode microwave-photonic oscillator. IEEE Photon. Technol. Lett. 22, 1610–1612 (2010)
19. Romeira, B., Seunarine, K., Ironside, C.N., Kelly, A.E., Figueiredo, J.M.L.: Self-Synchronized optoelectronic oscillator based on an RTD photo-detector and a laser diode. IEEE Photon. Technol. Lett. 23, 1148–1150 (2011)
20. Slight, T.J., Romeira, B., Wang, L., Figueiredo, J.M.L., Wasige, E., Ironside, C.N.: A Liénard oscillator resonant tunnelling-laser diode hybrid integrated circuit: model and experiment. IEEE J. Quantum Electron. 44, 1158–1163 (2008)
21. Romeira, B., Figueiredo, J.M.L., Slight, T.J., Wang, L., Wasige, E., Ironside, C.N., Kelly, A.E., Green, R.: Nonlinear dynamics of resonant tunneling optoelectronic circuits for wireless/optical interfaces. IEEE J. Quantum Electron. 45, 1436–1445 (2009)
22. Romeira, B., Figueiredo, J.M.L., Ironside, C.N., Slight, T.J.: Chaotic dynamics in resonant tunneling optoelectronic voltage controlled oscillators. IEEE Photon. Technol. Lett. 21, 1819–1821 (2009)
23. Brown, E.R., McMahon, O.B., Mahoney, L.J., Molvar, K.M.: SPICE model of the resonant-tunneling diode. Electronics Lett. 32, 938–940 (1996)
24. Schulman, J., Santos, H., Chow, D.: Physics-Based RTD current-voltage equation. IEEE Electron Device Lett. 17, 220–222 (1996)
25. Mena, P.V., Kang, S., DeTemple, T.A.: Rate-equation-based laser models with a single solution regime. J. Lightw. Technol. 15, 717–730 (1997)

26. Liénard, A.: Etude des oscillations entretenues. Rev. Gen. Electr. 28, 901–946 (1928)
27. Lins, A., Melo, W., Pugh, C.: On Liénards equation. Lecture Notes in Mathematics, vol. 597. Springer, New York (1977)
28. Hale, J.K.: Theory of functional differential equations. Springer, New York (1977)
29. Van der Pol, B.: A theory of the amplitude of free and forced triode vibrations. Radio Rev. 710, 754–762 (1920)
30. Van der Pol, B., Van der Mark, J.: Frequency demultiplication. Nature 120, 363–364 (1927)
31. Arnold, V.I.: Geometrical methods in the theory of ordinary differential equations. Springer, New York (1983)
32. Zuckerman, H.S., Montgomery, H.L., Niven, I.M., Niven, A.: An Iintroduction to the theory of numbers. John Wiley, New York (1991)
33. Bennett, S., Snowden, C.M., Iezekiel, S.: Nonlinear dynamics in directly modulated multiple-quantum-well laser diodes. IEEE J. Quantum Electron. 33, 2076–2083 (1997)
34. Liu, H.F., Ngai, W.F.: Nonlinear dynamics of a directly modulated 1.55 $\mu$m InGaAsP distributed feedback semiconducor laser. IEEE J. Quantum Electron. 29, 1668–1675 (1993)
35. Press, W.H., Teukolsky, S.A., Vetterling, W.T., Flannery, B.P.: Numerical recipes: the art of scientific computing. Cambridge University Press (2007)
36. Yao, X.S., Maleki, L.: Multiloop optoelectronic oscillator. IEEE J. Quantum Electron. 36, 79–84 (2000)

# Chapter 8
# Application of Coupled Dynamical Systems for Communities Detection in Complex Networks

Nikolai Nefedov

**Abstract.** In this chapter we present a dynamical systems framework and its applications for stable communities detection and missing (or hidden) link predictions utilizing network topology and its dynamics. In particular, we consider the dynamical formulation of modularity extended with a random walk approach, and then generalize it to coupled dynamical systems to detect communities at different hierarchical levels. We introduce attractive and repulsive coupling and study different scenarios for dynamical links updates that allow us to make predictions on a cooperative or a competing behavior of users in the network and analyze connectivity dynamics. The developed methods are tested on benchmark networks and then applied for analysis of real-world mobile datasets to derive a social community structure and to make link predictions/recommendations.

## 8.1 Introduction

The growing spread of smart phones equipped with various sensors makes it possible to record rich-content user data and compliment it with on-line processing. Mobile data processing could help people to enrich their social interactions and improve environmental and personal health awareness. At the same time, mobile sensing data could help service providers to understand better human behavior and its dynamics, identify complex patterns of users' mobility, and to develop various service-centric and user-centric mobile applications and services on-demand. One of the first steps in analysis of rich-content mobile datasets is to find an underlying structure of users' interactions and its dynamics by clustering data according to some similarity measures. Classification and clustering (finding groups of similar elements in data) are well-known problems which arise in many fields of sciences,

---

Nikolai Nefedov
Nokia Research Center, Switzerland
e-mail: nikolai.nefedov@nokia.com
ISI Lab, Swiss Federal Institute of Technology, Zurich (ETHZ)

K. Kyamakya et al. (Eds.): Selected Topics in Nonlinear Dynamics, SCI 459, pp. 139–159.
DOI: 10.1007/ 978-3-642-34560-9_8 

e.g., [2, 9, 22]. In cases when objects are characterized by vectors of attributes, a number of efficient algorithms to find groups of similar objects based on a metric between the attribute vectors are developed. On the other hand, if data are given in the relational format (causality or dependency relations), e.g., as a network consisting of $N$ nodes and $E$ edges representing some relations among the nodes, then the problem of finding similar elements corresponds to detection of communities, i.e., groups of nodes which are interconnected more densely among themselves than with the rest of the network.

The growing interest to the problem of community detection was triggered by the introduction of a new clustering measure called modularity [11, 20]. The modularity maximization is known as the NP-problem and currently a number of different sub-optimal algorithms are proposed, e.g., see [10] and references within. However, most of these methods address static networks partitioning into disjoint communities. On the other hand, in practice communities are dynamic and often overlapping. It is especially visible in social networks, where people are affiliated to different groups, depending on professional activities, family status, hobbies, and etc.

In this chapter we present a framework for communities detection in dynamical graphs and its applications for missing (or hidden) link predictions /recommendations based on the network topology and its dynamics. In particular, we use dynamical formulation of modularity maximization based on a fast greedy search [5, 20] extended with a random walk approach [16] to detect multi-resolution communities beyond and below the resolution provided by max-modularity. We generalize a random walk approach to coupled dynamical systems [3] and then extend it with dynamical links updates to make predictions beyond the given topology. In practice many biological and social systems show a presence of conflicting processes. For example, dynamics of social relations may be described in terms of cooperative and competitive interactions among the participants. In this chapter we introduce attractive and repulsive interactions among the nodes in a graph which allows us to detect and predict cooperative and competitive behavior in evolving social networks. In particular, in case of coupled oscillators the competitive behavior may be modeled as attractive coupling driving oscillators into global synchronization and repulsive coupling forcing system into chaotic/random behavior. Then a dynamical interplay between the given network topology and local interactions drives the connectivity evolution. We evaluated several coupling scenarios using different clustering measures and found that a combination of attractive and initially neutral coupling combined with dynamical links updates provide the the best performance. To deal with overlapping communities we introduce a soft community detection and propose friend-recommendations in social networks, where new link recommendations are made as intra- and inter-clique communities completion and recommendations are prioritized according to topologically-based similarity measures [17] modified to include multiple-communities membership. We show that the proposed prediction rules are in line with the network evolution predicted by coupled dynamical systems. To test the proposed framework we use the benchmark network [23] and then apply developed methods for analysis of multi-layers graphs built from real-world mobile datasets [14].

The chapter is organized as follows. In Section 2 we outline the dynamical formulation of community detection that forms the basis for the rest of the paper. Topology detection using coupled dynamical systems and its extensions to model a network evolution are described in Section 3. Soft community detection for networks with overlapping communities and its applications are briefly outlined in Section 4. Evaluation of the proposed methods in the benchmark network is presented in Section 5. Analysis of some real-world datasets collected during Nokia data collection campaign is presented in Section 6, followed by conclusions in Section 7.

## 8.2 Community Detection

### *8.2.1 Modularity Maximization*

Let's consider the clustering problem for an undirected graph $G = (V,E)$ with $|V| = N$ nodes and $E$ edges. Recently Newman et al [11, 13] introduced a new measure for graph clustering, named a modularity, which is defined as a number connections within a group compared to the expected number of such connections in an equivalent null model (e.g., in an equivalent random graph). In particular, the modularity $Q$ of a partition $\mathcal{P}$ may be written as

$$Q = \frac{1}{2m}\sum_{i,j}\left(A_{ij} - P_{ij}\right)\delta(c_i,c_j)\,, \tag{8.1}$$

where $c_i$ is the $i$-th community; $A_{ij}$ are elements of graph adjacency matrix; $d_i$ is the $i$-th node degree, $d_i = \sum_j A_{ij}$; $m$ is a total number of links, $m = \sum_i d_i/2$; $P_{ij}$ is a probability that nodes $i$ and $j$ in a null model are connected; if a random graph is taken as the null model, then $P_{ij} = d_i d_j/2m$.

By construction $|Q| < 1$ and $Q = 0$ means that the network under study is equivalent to the used null model (an equivalent random graph). Case $Q > 0$ indicates a presence of a community structure, i.e., more links remain within communities than would be expected in an equivalent random graph. Hence, a network partition which maximizes modularity may be used to locate communities. This maximization is NP-hard and many suboptimal algorithms are suggested, e.g., see [10] and references within.

In the following we use the basic greedy search algorithm [20] extended with a random walk approach described below, since it gives a reasonable trade-off between accuracy of community detection and scalability.

### *8.2.2 Communities Detection with Random Walk*

It is well-known that a network topology affects a system dynamics, it allows us to use the system dynamics to identify the underlying topology [3, 4, 16]. First, we review the Laplacian dynamics formalism recently developed in [8, 16].

Let's consider $N$ independent identical Poisson processes defined on every node of a graph $G(V,E), |V| = N$, where random walkers are jumping at a constant rate from each of the nodes. We define $p_n$ as the density of random walkers on node $i$ at step $n$, then its dynamics is given by

$$p_{i,n+1} = \sum_j \frac{A_{ij}}{d_j} p_{j,n} \, . \tag{8.2}$$

The corresponding continuous-time process, described by (8.3),

$$\frac{dp_i}{dt} = \sum_j \frac{A_{ij}}{d_j} p_j - p_i = \sum_j \left( \frac{A_{ij}}{d_j} - \delta_{ij} \right) p_i \tag{8.3}$$

is driven by the random walk operator $\dfrac{A_{ij}}{d_j} - \delta_{ij}$, which in case of discrete time is presented by the random walk matrix $\mathbf{L}_{rw} = \mathbf{D}^{-1}\mathbf{L} = \mathbf{I} - \mathbf{D}^{-1}\mathbf{A}$, where $\mathbf{L} = \mathbf{D} - \mathbf{A}$ is a Laplacian matrix, $\mathbf{A}$ is a non-negative weighted adjacency matrix, $\mathbf{D} = \mathrm{diag}\{d_i\}$, $i = 1,\ldots,N$. For an undirected connected network the stationary solution of (8.2) is given by $p_i^* = d_i/2m$.

Let's now assume that for an undirected network there exist a partition $\mathcal{P}$ with communities $c_k \in \mathcal{P},\ k = 1,\ldots,N_c$. The probability that initially, at $t_0$, a random walker belongs to a community $c_k$ is $\Pr(c_k,t_0) = \sum_{j \in c_k} d_j/2m$. Probability that a random walker, which was initially in $c_k$, will stay in the same community at the next step $t_0 + 1$ is given by

$$\Pr(c_k,t_0,t_0+1) = \sum_{j \in c_k} \sum_{i \in c_k} \left( \frac{A_{ij}}{d_j} \right) \left( \frac{d_j}{2m} \right). \tag{8.4}$$

The assumption that dynamics is ergodic means that the memory of the initial conditions is lost at infinity, hence $\Pr(c_k,t_0,\infty)$ is equal to the probability that two independent walkers are in $c_k$,

$$\Pr(c_k,t_0,\infty) = \left( \sum_{i \in c_k} \frac{d_i}{2m} \right) \left( \sum_{j \in c_k} \frac{d_j}{2m} \right). \tag{8.5}$$

Combining (8.4) and (8.5) we may write

$$\sum_{c_k \in P} \left( \Pr(c_k,t_0,t_0+1) - \Pr(c_k,t_0,\infty) \right) = \frac{1}{2m} \sum_{i,j} \left( A_{ij} - \frac{d_i d_j}{2m} \right) \delta(c_i,c_j) = Q \, . \tag{8.6}$$

In general case, using (8.3), one may define a stability of the partition $\mathcal{P}$ as [8, 16]

$$R_{\mathcal{P}}(t) = \sum_{c_k \in \mathcal{P}} \Pr(c_k, t_0, t_0 + t) - \Pr(c_k, t_0, \infty) \tag{8.7}$$

$$= \sum_{c_k \in \mathcal{P}} \sum_{i,j \in c_k} \left( \left( e^{t(\hat{A}-I)} \right)_{ij} \frac{d_j}{2m} - \frac{d_i d_j}{4m^2} \right), \text{ where } \hat{A}_{ij} = \frac{A_{ij}}{d_j}. \tag{8.8}$$

Then, as the special cases of (8.8) at $t = 1$, we get the expression for modularity (8.6).

Note that $R_{\mathcal{P}}(t)$ is non-increasing function of time: at $t = 0$ we get

$$R_{\mathcal{P}}(0) = 1 - \sum_{c_k \in \mathcal{P}} \sum_{i,j \in c_k} \frac{d_i d_j}{4m^2} \tag{8.9}$$

and $\max_{\mathcal{P}} R(0)$ is reached when each node is assigned to its own community. Note that (8.9) corresponds to collision entropy or Rényi entropy of order 2.

On the other hand, in the limit $t \to \infty$, the maximum of $R_{\mathcal{P}}(t)$ is achieved with Fiedler spectral decomposition into 2 communities. In other words, time here may be seen as a resolution parameter: with time $t$ increasing, the $\max_{\mathcal{P}} R(t)$ results in a sequence of hierarchical partitions $\{\mathcal{P}_t\}$ with the decreasing numbers of communities. Furthermore, as shown in [8], we may define a time-varying modularity $Q(t)$ by linear terms in time expansion for $R(t)$ at $t \approx 0$,

$$R(t) \approx (1 - t) \cdot R(0) + t \cdot Q = Q(t)\,, \tag{8.10}$$

which after substitution (8.6) and (8.9) gives

$$Q(t) = (1 - t) + \sum_{c_k \in \mathcal{P}} \sum_{i,j \in c_k} \left( \frac{A_{ij}}{2m} t - \frac{d_i d_j}{4m^2} \right). \tag{8.11}$$

In the following we apply time-dependent modularity maximization (8.11) using the greedy search to find hierarchical structures in networks beyond modularity maximization $Q_{max}$ in (8.1). This approach is useful in cases where maximization of (8.1) results in a very fragmental structure with a large number of communities. Also it allows us to evaluate the stability of communities at different resolution levels. However, since the adjacency matrix **A** is not time dependent, the time-varying modularity (8.11) can not be used to make predictions beyond the given topology.

## 8.3 Topology Detection Using Coupled Dynamical Systems

### *8.3.1 Laplacian Formulation of Network Dynamics*

Let's consider an undirected weighted graph $G = \{V, E\}$ with $N$ nodes and $E$ edges, where each node represents a local dynamical system and edges correspond to local coupling. Dynamics of $N$ locally coupled dynamical systems on the graph $G$ is described by

$$\dot{x}_i(t) = q_i(x_i(t)) + k_c \sum_{j=1}^{N} A_{ij}\psi\left(x_j(t) - x_i(t)\right) , \tag{8.12}$$

where $q_i(x_i)$ describes a local dynamics of state $x_i$; $A_{ij}$ is a coupling strength between nodes $i$ and $j$; $\psi(\cdot)$ is a coupling function; $k_c$ is a global coupling gain.

In case of weakly phase-coupled oscillators the dynamics of local states is described by Kuramoto model [1, 15]

$$\dot{\theta}_i(t) = \omega_i + k_c \sum_{j=1}^{N} A_{ij} \sin\left[\theta_j(t) - \theta_i(t)\right] . \tag{8.13}$$

Linear approximation of coupling function $\sin(\theta) \simeq \theta$ in (8.13) results in the consensus model [21]

$$\dot{\theta}_i(t) = k_c \sum_{j=1}^{N} A_{ij}\left[\theta_j(t) - \theta_i(t)\right] , \tag{8.14}$$

which for a connectivity graph $G$ may be written as

$$\dot{\Theta}(t) = -\, k_c \mathbf{L}\, \Theta(t) , \tag{8.15}$$

where $\mathbf{L} = \mathbf{A} - \mathbf{D}$ is the Laplacian matrix of $G$. The solution of (8.15) in the form of normal modes $\omega_i(t)$ may be written as

$$\omega_i(t) = k_c \sum_{j=1}^{N} V_{ij}\theta_j(t) = k_c\, \omega_i(t_0) \mathrm{e}^{-\lambda_i t} , \tag{8.16}$$

where $\lambda_1,\ldots,\lambda_N$ are eigenvalues and $\mathbf{V}$ is the matrix of eigenvectors of $\mathbf{L}$. Note that (8.16) describes a convergence speed to a consensus for each nodes. Let's order these equations according to the descending order of their eigenvalues. Then it is easy to see that nodes are approaching the consensus in a hierarchical way, revealing at the same time a hierarchy of communities in the given network $G$. Note that (8.15) has the same form as (8.3), with the difference that the random walk process (8.3) is based on $\mathbf{L}_{rw} = \mathbf{D}^{-1}\,\mathbf{L}$. It allows us to consider random-walk-based communities detection in the previous section as a special case of coupled oscillators synchronization.

Similarly to (8.15), we may derive the Laplacian presentation for locally coupled oscillators (8.13). In particular, the connectivity of a graph may be described by the graph incidence $(N \times E)$ matrix $\mathbf{B}$: $\{\mathbf{B}\}_{ij} = 1$ (or $-1$) if nodes $j$ and $i$ are connected, otherwise $\{\mathbf{B}\}_{ij} = 0$. In case of weighted graphs we use the weighted Laplacian defined as

$$\mathbf{L}_{\mathrm{A}} \triangleq \mathbf{B}\mathbf{D}_{\mathrm{A}}\mathbf{B}^T . \tag{8.17}$$

Based on (8.17) we can rewrite (8.13) as in [19]

$$\dot{\Theta}(t) = \Omega - k_c \mathbf{B}\mathbf{D}_{\mathrm{A}} \sin\left(\mathbf{B}^T\Theta(t)\right), \tag{8.18}$$

where vectors and matrices are defined as follows: $\Theta(t) \triangleq [\theta_1(t),\cdots,\theta_N(t)]^T$; $\Omega \triangleq [\omega_1,\cdots,\omega_N]^T$; $\mathbf{D}_\mathrm{A} \triangleq \mathrm{diag}\{a_1,\ldots,a_E\}$, $a_1,\ldots,a_E$ are weights $A_{ij}$ indexed from 1 to $E$. In the following we use (8.18) to describe different coupling scenarios.

## 8.3.2 Dynamical Structures with Different Coupling Scenarios

Synchronization in static networks is well-established topic and has being under studies for several decades. Recently the interest is moving to networks with changing topology, where topology evolves as a result of dynamical network interactions, thus creating an interplay between structure and dynamics [6]. Below we address this problem by including time-dependent topology into Kuramoto model.

Usually in Kuramoto model a global order parameter $k_c$ is used to characterize the level of synchronization among oscillators. To take into account local effects we consider a local order parameter $r_{ij}$ measuring correlations between instant phases of oscillators,

$$r_{ij}(t) = \langle cos\left[\theta_j(t) - \theta_i(t)\right]\rangle, \tag{8.19}$$

where the average $\langle\cdot\rangle$ is taken over initial random phases $\theta_i(t=0)$. Following [3,4] we may define a dynamical connectivity matrix $\mathbf{C}_t(\eta)$, where two nodes $i$ and $j$ are connected at time $t$ if their local phase correlation is above a given threshold $\eta$,

$$C_t(\eta)_{ij} = 1 \quad \text{if} \quad r_{ij}(t) > \eta$$

$$C_t(\eta)_{ij} = 0 \quad \text{if} \quad r_{ij}(t) < \eta\,. \tag{8.20}$$

We select communities resolution level (time $t$) using a random walk as in Section 2. Next, by changing the threshold $\eta$, we obtain a set of connectivity matrices $\mathbf{C}_t(\eta)$ which reveal dynamical topological structures for different correlation levels. Since the local correlations $r_{ij}(t)$ are continuous and monotonic functions in time, we may also fix $\eta$ and express dynamical connectivity matrix (8.20) in the form $\mathbf{C}_\eta(t)$ to present the evolution of connectivity in time for a fixed correlation threshold $\eta$. Using this approach we consider below several scenarios of networks evolution with dynamically changing coupling.

### 8.3.2.1 Attractive Coupling with Dynamical Updates

As the first step, let's introduce dynamics into static attractive coupling (8.13). Using the dynamical connectivity matrix (8.20) we may write

$$\dot{\theta}_i(t) = \omega_i + k_c \sum_{j=1}^{N} F_{ij}^{(\eta)}(t) \sin\left[\theta_j(t) - \theta_i(t)\right], \tag{8.21}$$

where matrix $\mathbf{F}^{(\eta)}(t)$ describes dynamical attractive coupling, $F_{ij}^{(\eta)}(t) = A_{ij}C_{\eta}(t)_{ij} \geq 0$. Then, similar to (8.18), the attractive coupling with a dynamical update may be described as

$$\dot{\Theta}(t) = \Omega(t) - k_c \mathbf{B}(t)\mathbf{D}_{\mathrm{F}}(t) \sin\left(\mathbf{B}(t)^T \Theta(t)\right), \tag{8.22}$$

where initial conditions are defined by $A_{ij}$; $\mathbf{D}_{\mathrm{F}}(t)$ is formed from $\mathbf{D}_{\mathrm{A}}$ with elements $\{a_k\}$ scaled according to $\mathbf{C}_{\eta}(t)$.

#### 8.3.2.2 Combination of Attractive and Repulsive Coupling with Dynamical Links Update

Many biological and social systems show a presence of a competition between conflicting processes. In case of coupled oscillators it may be modeled as the attractive coupling (driving oscillators into the global synchronization) combined with the repulsive coupling (forcing system into a chaotic/random behavior). In the following we call this type of coupling as *AR* scenario.
To allow positive and negative interactions we use instant correlation matrix $\mathbf{R}(t) = \mathbf{R}^+(t) + \mathbf{R}^-(t)$, and separate attractive and repulsive parts [19]

$$\begin{aligned} \dot{\theta}_i(t) =& \omega_i + k_c^+ \sum_{j=1}^{N} r_{ij}^+(t)\ A_{ij} \sin\left[\theta_j(t) - \theta_i(t)\right] - \\ & k_c^- \sum_{j=1}^{N} |r_{ij}^-(t)|\ A_{ij} \sin\left[\theta_j(t) - \theta_i(t)\right], \end{aligned} \tag{8.23}$$

where superscripts denote positive and negative correlations [1].
Note that in this case the total number of links in the network does not change, at a given time instant each link performs either attractive or repulsive function.

To obtain the Laplacian presentation we define a dynamical connectivity matrix $\mathbf{F}(t)$ as element-by-element matrix product

$$\mathbf{F}(t) = \mathbf{R}(t) \circ \mathbf{A} = \mathbf{F}^+(t) + \mathbf{F}^-(t), \tag{8.24}$$

and present dynamic Laplacian as the following

$$\mathbf{L}_{\mathrm{F}}(t) = \mathbf{B}(t)(\mathbf{D}_{\mathrm{F}^+}(t) + \mathbf{D}_{\mathrm{F}^-}(t))\mathbf{B}^T(t). \tag{8.25}$$

It allows us to write

$$\begin{aligned} \dot{\theta}_i(t) =& \omega_n + k_c^+ \sum_{m=1}^{N} F_{ij}^+(t) \sin\left[\theta_j(t) - \theta_i(t)\right] - \\ & k_c^- \sum_{m=1}^{N} F_{ij}^-(t) \sin\left[\theta_j(t) - \theta_i(t)\right], \end{aligned} \tag{8.26}$$

[1] For presentation clarity we omit here the correlation threshold $\eta$.

or in matrix form

$$\dot{\Theta}(t) = \Omega - k_c^+ \mathbf{B}(t)\mathbf{D}_{\mathrm{F}^+}(t) \sin\left(\mathbf{B}^T(t)\Theta(t)\right) + k_c^- \mathbf{B}(t)\mathbf{D}_{\mathrm{F}^-}(t) \sin\left(\mathbf{B}^T(t)\Theta(t)\right). \tag{8.27}$$

#### 8.3.2.3 Combination of Attractive and Initially Neutral Coupling with Dynamical Links Update

Negative correlations (resulting in repulsive coupling) are typically assigned between nodes which are not initially connected. However, in many cases this scenario is not realistic. For example, in social networks, the absence of communications between people does not necessary indicate conflicting (negative) relations, but often has a neutral meaning. To take this observation into account we modified second term in (8.23) such that it sets neutral initial conditions to unconnected nodes in adjacency matrix **A**. This case is referred below as *AN* scenario. In particular, system dynamics with links update (8.24) and initially neutral coupling is described by

$$\dot{\theta}_i(t) = \omega_i + k_c^+ \sum_{j=1}^{N} F_{ij}^+(t) \sin\left[\theta_j(t) - \theta_i(t)\right] + k_c^- \sum_{j=1}^{N} F_{ij}^-(t) \cos\left[\theta_j(t) - \theta_i(t)\right], \tag{8.28}$$

or in the matrix form

$$\dot{\Theta}(t) = \Omega - k_c^+ \mathbf{B}(t)\mathbf{D}_{\mathrm{F}^+}(t) \sin\left(\mathbf{B}^T(t)\Theta(t)\right) - k_c^- \mathbf{B}(t)\mathbf{D}_{\mathrm{F}^-}(t) \cos\left(\mathbf{B}^T(t)\Theta(t)\right) \tag{8.29}$$

Then a dynamical interplay between the given network topology and local interactions drives the connectivity evolution. Note that according to (8.29), if negative correlations take place between selected nodes at a certain time instant, the probability that negative relations would appear at the next time instant in attractive-neutral scenarios is lower than in the *AR* scenario. In social relations it may be interpreted as a forgetting factor for occasional negative events, it facilitates different forms of co-operation, starting from communities formation (clustering) up to global consensus (full synchronization).

#### 8.3.2.4 Dynamical Balance between Attractive and Repulsive Coupling with Dynamical Links Update

As it was mentioned, the interplay between coupled dynamics and topology in general results in a complicated evolution of network topology. For example, dynamical link updates running at a given connected graph may generate a wide range of topologies ranging from fully connected graphs and dynamical clustering to random networks. On the other hand, it would be interesting to find an allocation of attractive and repulsive coupling among nodes such that it maintains (possibly with small fluctuations) a balance between attractive and repulsive interactions and keeps the given topology stable. A detailed analysis of this problem is to be addressed elsewhere. In this chapter we only briefly outline a special case relevant to social network analysis.

In particular, given social interactions as a graph, we want to find an allocation of attractive and repulsive relations in the dynamical network which maintains its stability. In general this combinatorial problem is NP-hard and does not have unique solution. However, by using network evolution with dynamical links update we can find approximate solutions. The suggested approach is to find values $k_c^-$ and $k_c^+$ in (8.29) such that after some transition time the dynamical connectivity matrices describing attractive $\mathbf{F}^-(t)$ and repulsive coupling $\mathbf{F}^+(t)$ are becoming time-invariant (with possible small fluctuations). In the following we call this scenario as attractive-neutral stable (*ANS*) coupling. It allows us to derive underlying pattern of attractive and negative interactions in the social networks.
Another approach to the find positive and negative relations is to consider dynamics of connectivity matrices, then select edges with repulsive interactions and estimate its duration. Obviously, the longer repulsive dynamics exists between selected edges, the stronger are negative interactions between users.

In the following we use coupled system dynamics approach to predict networks' evolution and to make missing links predictions and recommendations. Furthermore, the suggested approach allows us also to predict repulsive relations in the network based on the network topology and links dynamics.

## 8.4 Overlapping Communities

### *8.4.1 Multi-membership*

In social networks people belong to several overlapping communities depending on their families, occupations, hobbies, etc. As the result, users (presented by nodes in a graph) may have different levels of membership in different communities. This fact motivated us to consider multi-community membership as edge-weights to different communities. As an example, we can measure a membership $g_j(k)$ of node $k$ in $j$-th community as a number of links (or its weight for a weighted graph) between the $k$-th node and other nodes within the same community, $g_j(k) = \sum_{i \in c_j} w_{ki}$. Then, for each node $k$ we assign a vector $\mathbf{g}(k) = [g_1(k), g_2(k), \ldots, g_{N_c}(k)]$, $k \in \{1, \ldots, N\}$ which presents the node membership (or participation) in all detected communities $\{c_1, \ldots, c_{N_c}\}$. In the following we refer $\mathbf{g}(k)$ as a soft community decision for the $k$-th node.

To illustrate the approach, overlapping communities derived from benchmark karate club social network [23] are depicted at Fig.8.1. Modularity maximization here reveals 4 communities shown by different colors. However, the multi-communities membership results in overlapping communities illustrated by overlapping ovals (Fig.8.1).

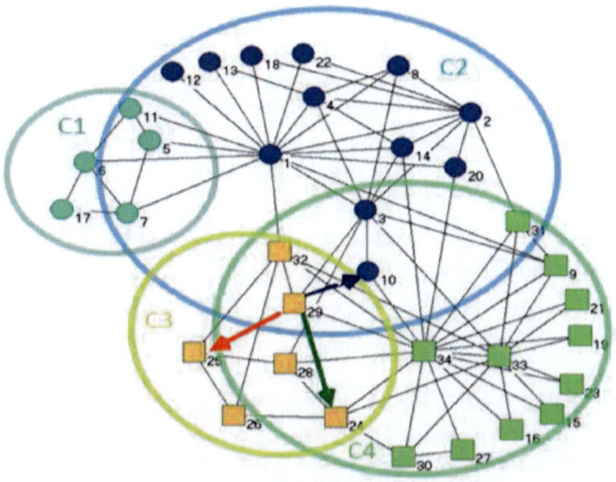


**Fig. 8.1** Overlapping communities in karate club social network: colors indicate communities. Intra- and inter-community link recommendations are shown by arrows

## 8.4.2 *Application of Soft Community Detection for Recommendation Systems*

In online social networks a recommendation of new social links may be seen as an attractive service. Recently Facebook and LinkedIn introduced a service "People You May Know", which recommends new connections using the friend-of-friend (FoF) approach. However, in large networks the FoF approach may create a long and often not relevant list of recommendations, which is difficult (and also computationally expensive, in particular in mobile solutions) to navigate. Furthermore, in mobile social networks (e.g., Nokia portal Ovi Store) these kinds of recommendations are even more complicated because users' affiliations to different groups (and even its number) are not known. Hence, before making recommendations, communities are to be detected first.

### 8.4.2.1 Recommendations as Communities Completion

Based on soft communities detection we suggest to make the FoF recommendations as follows [18]:

(i) detect communities, e.g., by using one of the methods described above;
(ii) calculate membership $g_j(k)$ in all relevant communities for each node $k$;
(iii) make new recommendations as communities completion following rules below;
(iv) use multiple-membership to prioritize recommendations.

To make new link recommendations in (iii) we suggest the following rules:

- each new link creates at least one new clique (the FoF concept);
- complete cliques within the same community (intra-cliques) using the FoF concept;

- if there is no FoF links, then complete cliques towards to the fully-connected own intra-community;
- complete inter-cliques (where nodes belong to different communities);
- prioritize intra-clique and inter-clique links completion according to some measure based on multi-membership.

To assign priorities we introduce several similarity measures outlined below. We will show in next sections that these rules are well in line with link predictions made by coupled dynamical systems described in Section 3.

#### 8.4.2.2 Modified Topology-Based Predictors

Let'a define sets of neighbors of node $k$, which are inside and outside of community $c_i$ as $\Gamma_i(k) = \{\Gamma(k) \in c_i\}$ and $\Gamma_{\setminus i}(k) = \{\Gamma(k) \notin c_i\}$, respectively. This allows us to introduce a set of similarity measures by modifying topology-based base-line predictors listed in [17] to take into account the multiple-membership in overlapping communities. As an example, for the intra-clique completion we may associate a quality of missing link prediction (or recommendation) between nodes $k$ and $n$ within $c_i$ community by modifying the base-line predictor scores as follows [18]:
- Preferential attachment: $S^{(i,i)}_{\mathsf{PA}}(k,n) = |\Gamma_i(k)| \cdot |\Gamma_i(n)|$;
- Jaccards score: $S^{(i,i)}_{\mathsf{JC}}(k,n) = |\Gamma_i(k) \cap \Gamma_i(n)| / |\Gamma_i(k) \cup \Gamma_i(n)|$;
- Adamic/Adar score: $S^{(i,i)}_{\mathsf{AA}}(k,n) = \sum_{z \in \Gamma_i(k) \cap \Gamma_i(n)} (\log|\Gamma(z)|)^{-1}$;
- Katz score (intra-community):

$$S^{(i,i)}_{\mathsf{KC}}(k,n) = \sum_{l=1}^{\infty} \beta^l |\mathsf{path}(k,n)^{(l)}| = \left\{(\mathbf{I} - \beta\mathbf{A}^{(i)})^{-1} - \mathbf{I}\right\}_{(k,n)},$$

where $|\mathsf{path}_i(k,n)^{(l)}|$ is number of all paths of length-$l$ from $k$ to $n$ within $c_i$; $\mathbf{I}$ is the identity matrix, $\mathbf{A}^{(i)}$ is the (weighted) adjacency matrix of community $c_i$, $\beta$ is a dumping parameter, $0 < \beta < 1$, such that $\sum_{ij} \beta A_{ij} < 1$.
The measures above consider communities as disjoint sets and may be used as the 1st order approximation for link predictions in overlapping communities. To take into account both intra- and inter-community links we use multi-community membership for nodes, $g_i(k)$. In general, for nodes $k \in c_i$ and $n \in c_j$, the inter-community relations may be asymmetric, $g_j(k) \neq g_i(n)$. In the case of undirected graphs we may use averaging and modify the base-line predictors $S(k,n)$ as

$$S^{(i,j)}(k,n) = \frac{g_j(k) + g_i(n)}{2m} S(k,n). \tag{8.30}$$

For example, modified Katz score which take into account multi-communities membership is defined as

$$S^{(i,j)}_{\mathsf{KC}}(k,n) = \frac{g_j(k) + g_i(n)}{2m} \left\{(\mathbf{I} - \beta\mathbf{A}^{(C_{n,k})})^{-1} - \mathbf{I}\right\}_{(k,n)}, \tag{8.31}$$

where $k \in c_i, n \in c_j$; $\mathbf{A}^{(C_{n,k})}$ is an adjacency matrix formed by all communities relevant to nodes $n$ and $k$.

In the next section we compare Katz predictor (8.31) with dynamics-based predictors.

## 8.5 Methods Testing in Benchmark Networks

### *8.5.1 Zachary Karate Club: Communities and Its Dynamics*

To test algorithms we use karate club social network [23] as the benchmark network. A number of communities at different resolution levels is presented at Fig.8.2. As one can see, *max*-modularity partition and *max*-stable partition are not necessary the same, the most stable partition in case of karate club network is the partition into 2 communities.

Comparison of coupling scenarios *AR* and *AN* are presented at Fig.8.3 - Fig.8.5. Pair-wise correlations between oscillators at $t = 1$ for coupling scenarios *AR* (on the left) and *AN* (on the right) are presented at Fig.8.3. Coupling scenario *AN* reveals clearly communities structure, while in case of *AR* the negative coupling dominates over the attractive coupling and forces the system into a chaotic behavior. Dynamic connectivity matrices reordered by communities for the attractive-neural coupling at $t = 1$ (on the left) and $t = 10$ (on the right) are depicted at Fig.8.4. As one can see from Fig.8.4 and Fig.8.5 in case of *AN* the number of connections with the attractive coupling is growing in time, while the repulsive connections are decreasing.

For a given initial graph $G(V,E)$, prediction process increases total number of edges $E_p$ in the network. To compare different prediction schemes we introduce prediction depth $k_p = E_p/E$, $1 < k_p < N^2/2E$. In case of dynamical systems $k_p$ corresponds to a time instant $t_p$ when total number of edges reaches $E_p$. Upper part at Fig.8.6 depicts the adjacency matrix for Zachary karate club (red circles) and links predicted by dynamics (blue dots) at time instants corresponding to $k_p = \{1.58, 2.44, 5.44\}$. As expected, the dynamical links prediction tends to make more connections within the established communities first, followed by merging communities and creating the higher hierarchical level partitions.

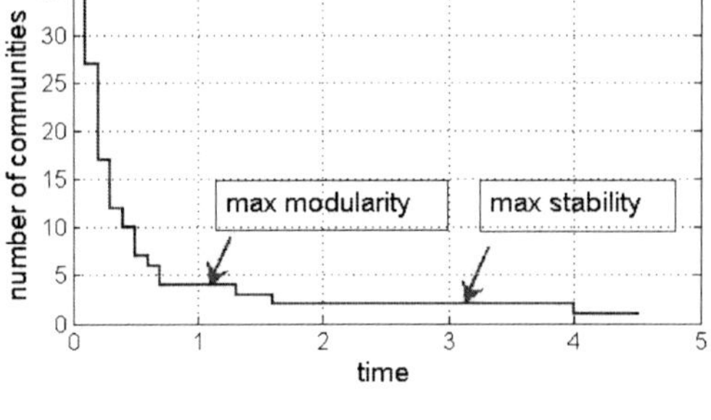


**Fig. 8.2** Karate club network: stability of communities at different resolution levels

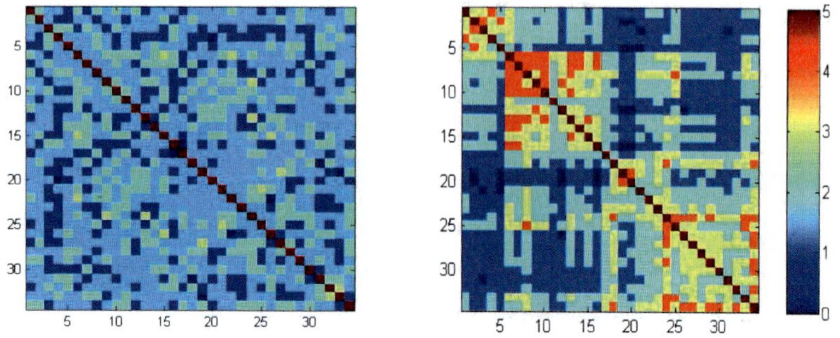


**Fig. 8.3** Karate club: pair-wise correlations (scaled by 5) between oscillators at $t = 1$. Re-ordered by communities. Coupling scenarios: $AR$ ($k_c^+ = k_c^-$) on the left; $AN$ ($k_c^+ = k_c^-$) on the right

This observed behavior looks similar to the recommendations made by community completion introduced in Section 4. For a comparison, the lower part of Fig.8.6 presents predictions made by topology-based Katz predictor (8.31) based on soft community detection at the same prediction depths $k_p$.

## 8.5.2 *Comparison of Different Predictions Schemes*

To compare topologies produced by different predictors we use the dynamical systems framework described in Section 3. Let us consider a network of $N$ identical particles connected by elastic strings according to adjacency matrix $\mathbf{A}$ and described by motion equations

$$\ddot{x}_i + \sum_{j=1}^{N-1} A_{ij}(x_i - x_j) = 0 \tag{8.32}$$

where $x_i$ is the coordinate of $i$-the particle. Vibrational frequencies $\omega_a$ of this network are defined by eigenvalues $\gamma_a = -\omega_a^2$ of Laplacian $\mathbf{L}_A$ of matrix $\mathbf{A}$. Laplacian spectrum of a graph is often called as the vibrational spectrum [7]. We measure difference between two graphs using Laplacian spectrum. In particular, we present spectral density $\rho(\omega)$ for a graph as a sum of narrow Lorentz distributions [12],

$$\rho(\omega) = K \sum_{a=1}^{N-1} \frac{\gamma}{(\omega - \omega_a)^2 + \gamma^2} \tag{8.33}$$

where $\gamma$ is the width of the Lorentz distributions, $K$ is a normalization coefficient such that $\int \rho(\omega) d\omega = 1$. Using spectral densities (8.33), distance $d(G_1, G_2)$ between two graphs $G_1$ and $G_2$ may be defined as

$$d(G_1, G_2) = \int_0^\infty [\rho_1(\omega) - \rho_2(\omega)]^2 d\omega. \tag{8.34}$$

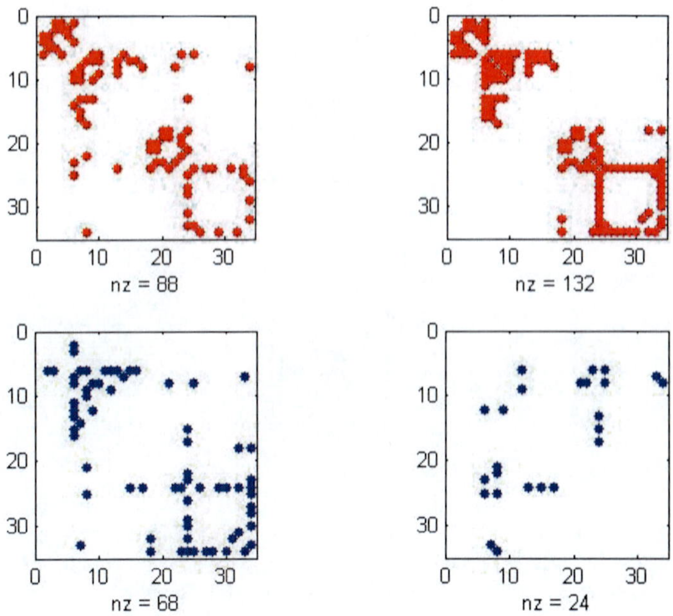


**Fig. 8.4** Karate club: dynamic connectivity matrices for attractive (shown on the top in red color) and repulsive (shown at the bottom in blue color) coupling at $t = 1$ (left) and $t = 10$ (right); coupling scenario *AN*; nodes are reordered by communities

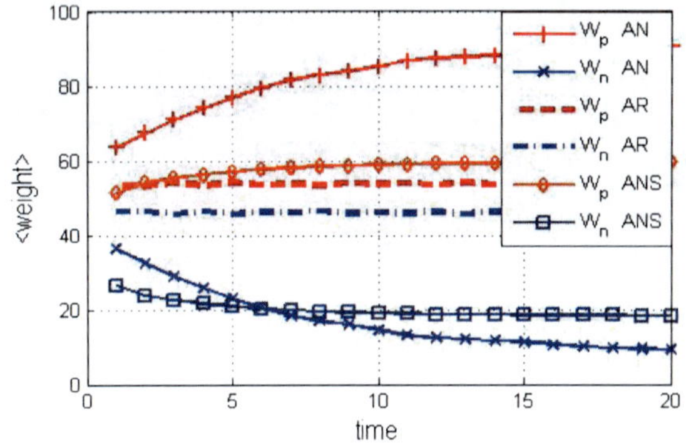


**Fig. 8.5** Karate club: evolution of averaged attractive and repulsive weights for coupling scenarios *AR* ($k_c^+ = k_c^-$), *AN* ($k_c^+ = k_c^-$) and *ANS* ($k_c^+ = 10\, k_c^-$)

As an illustration, vibration spectra and distance between graphs for considered predictors are depicted at Fig. 8.7 and Fig.8.8, respectively. Even though both predictors are approaching the fully connected graph, the network evolution for each

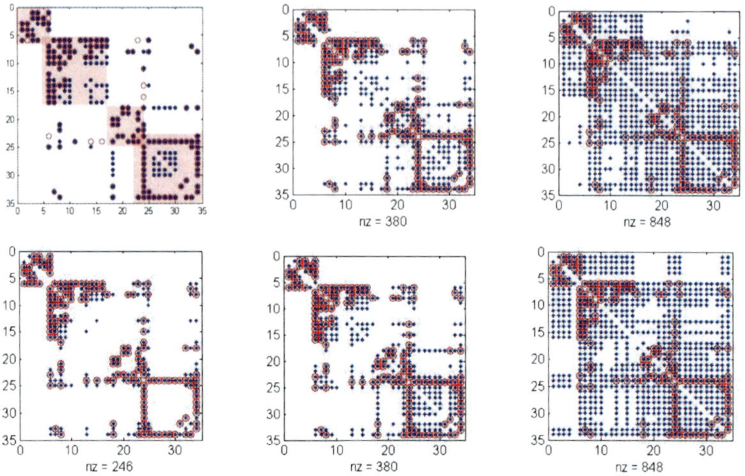

**Fig. 8.6** Karate club: adjacency matrix is shown by red circles, detected communities by pink squares, predicted links are shown by blue dots. The upper part (a)-(c): predictions made by dynamical systems at different time scales. The bottom part (d)-(f): recommendations made by the modified Katz predictor at different values of $\beta$ while keeping the same number of edges $E_p$ as in the dynamical predictor.

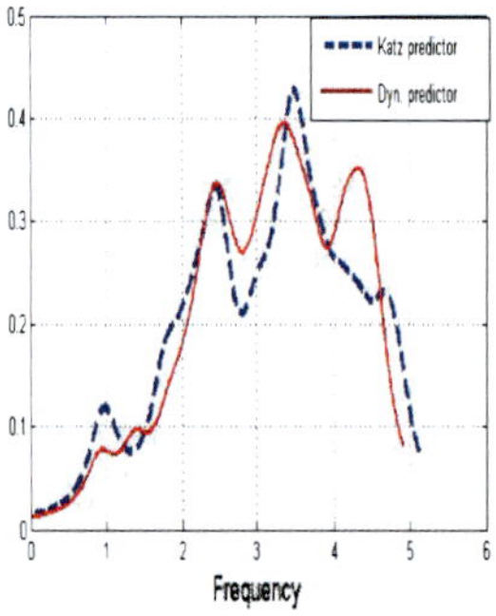


**Fig. 8.7** Karate club: vibration spectra of networks formed by Katz predictor (dashed blue) and dynamical predictor (red line); prediction depth $k_p = 2.4$, $\gamma = 0.05$

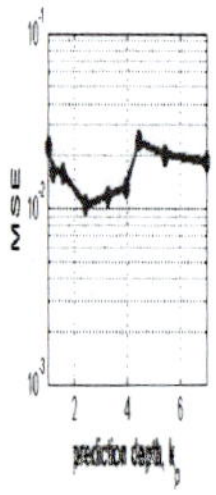


**Fig. 8.8** Karate club: mean square error between vibration spectra as function of prediction depth $k_p$, where total number of edges is $E_p = k_p E$ and $E$ is number of edges in the original networks

of predictors may follow different trajectories. In particular, at small values of prediction depth, $k_p < 3$, the network evolution is similar for both cases. However, at intermediate values of $k_p$ the evolution trajectories may follow different paths (cf. Fig.8.8), which in turn results in different predictions (Fig.8.6, middle part).

### 8.5.3 Detection of Negative Relations

In this section we illustrate applications of *ANS* scenario to predict repulsive relations based on network topology and links dynamics. Following *ANS* approach our task here is to find dynamical relations among nodes which make the given topology stable by balancing total weights of attractive and repulsive interactions. For karate network we found that this balance may be achieved if $k_c^+ = 10\, k_c^-$ in (8.29). Simulation results for *ANS* scenario presented at Fig.8.5 show that after some transition period the time-evolution of weights (lines marked by squares and rhombi) becomes practically time-invariant. Also for all simulated initial phase realizations we did not observe changes in connectivity matrices (cf. Fig.8.4) provided that $t > 10$.

Recall that allocation of attractive and repulsive relations to preserve the given topology is not unique. One of possible solutions for karate club network under *ANS* scenario is shown at Fig.8.9. Here we present normalized weights (intensity) for attractive and repulsive relations (left and right parts, respectively). Nodes are reordered according to communities. As one can see, negative relations (light-blue dots at Fig.8.9, right) more frequently appear at the communities' boarders, revealing overlapping communities structure. Furthermore, this approach allows to spot negative relations within communities under assumption that topology is stable. At the same time the intensity of positive relations are not the same for different nodes (Fig.8.9, left). Note that the allocation at Fig.8.9 is only one of possible solutions. Fig.8.10 (left part) presents another solution example where nodes are ordered in natural order to verify relations using Fig.8.1. Averaging several solutions allows us

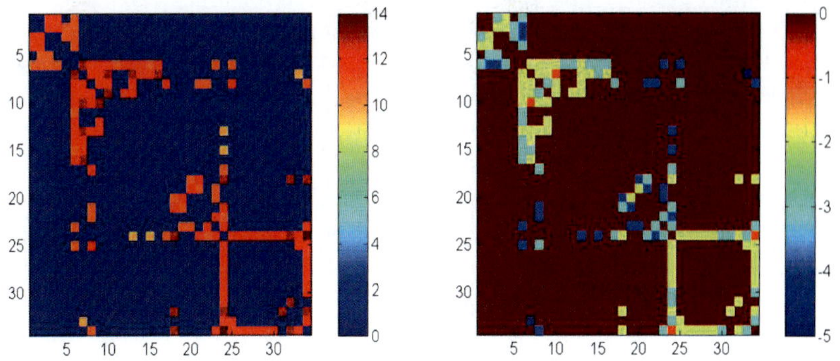


**Fig. 8.9** Karate club: Intensity of attractive and repealing (negative) relations; nodes are ordered according to communities

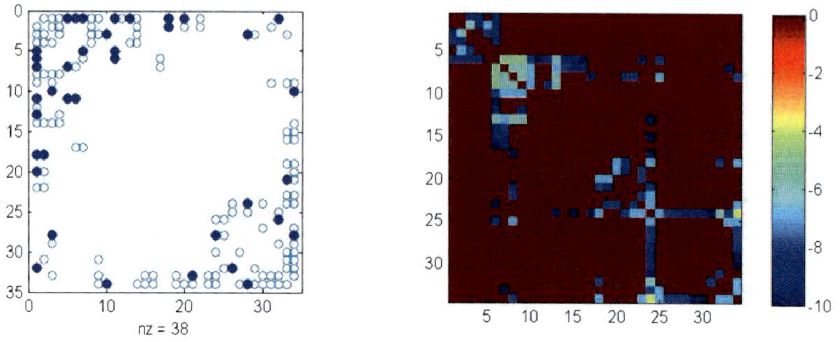


**Fig. 8.10** Karate club: allocation of repulsive relations to maintain the stability of karate network. One solution example, natural ordering of nodes (left); intensity of negative relations by averaging over 10 solutions, nodes are reordered according to communities (right)

more reliably find negative relations within communities which are more difficult to detect than nodes with multi-community membership.

## 8.6 Applications for Mobile Networks Data

To analyze mobile users behavior and underlying social structure Nokia Research Center/Lausanne organized mobile data collection campaign at EPFL university campus [14]. Rich-content datasets (including data from mobile sensors, call-logs, bluetooth (BT) and WLAN proximity, GPS coordinates, information on mobile and applications usage and etc) are collected from about 200 participants for the period from June 2009 till October 2010. Besides the collected data, several surveys before and after the campaign have been conducted to profile participants and to form a basis for the ground truth. Due to lack of space in this section we only briefly outline applications of dynamical systems presented in previous sections for analysis of social affinity graphs constructed from call-logs and BT users proximity.

Fig.8.11 shows voice-call and SMS connections of 134 participants socially connected within the data collection campaign. To find communities we used modularity maximization with the greedy search algorithm [20] which identifies 14 communities after the 3d iteration (Fig.8.11). However, similar to karate network, we found that max modularity partition is not the most stable one. Fig.8.12 presents stability of communities at different hierarchical levels detected by the random walk for the network shown at Fig.8.11. As one can see, the max-modularity partition with 14 communities is highly unstable and hardly could be used for reliable predictions; the stable partitions appear at the higher hierarchical levels starting from 8 communities. We rely on this fact to build the ground truth references for clustering evaluation.

As discussed above, one of applications of coupled systems dynamics may be seen in new links predictions/recommendations. To illustrate the approach we consider coupled oscillators interconnected according to call-logs network at Fig.8.11.

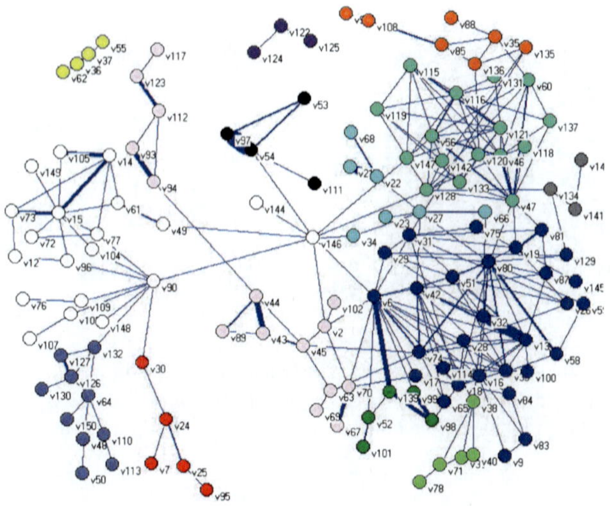

**Fig. 8.11** Nokia mobile data collection campaign: voice-calls and SMS network recorded for 134 users; *max*-modularity is reached at 14 communities (color coded)

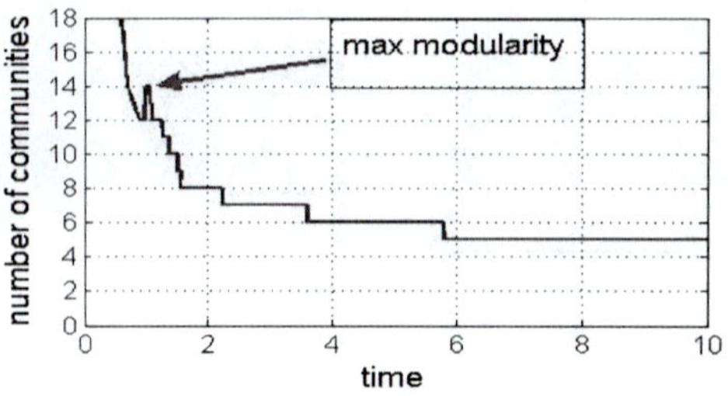


**Fig. 8.12** Nokia call-log network: stability of communities at different resolution levels

Scaled pair-wise correlations between oscillators at $k_p = 2.4$ clearly reveal the community structure shown at Fig.8.13 (left). Right part of Fig.8.13 depicts the call-log adjacency matrix (shown by circles) and connections predicted by coupled oscillators dynamics (blue dots) at time corresponding to one of stable hierarchical levels, $k_p = 2.4$. This method allows us to predict/recommend connections between people and study their connectivity evolution.

Detection of overlapping communities and spotting of competing (repulsive) and cooperative (attractive) social relations in combined call-log and BT networks according to *ANS* coupling strategy are illustrated at Fig.8.14. In particular, light-blue dots at center of Fig.8.14 indicate a separation of two large communities confirmed by questionnaires.

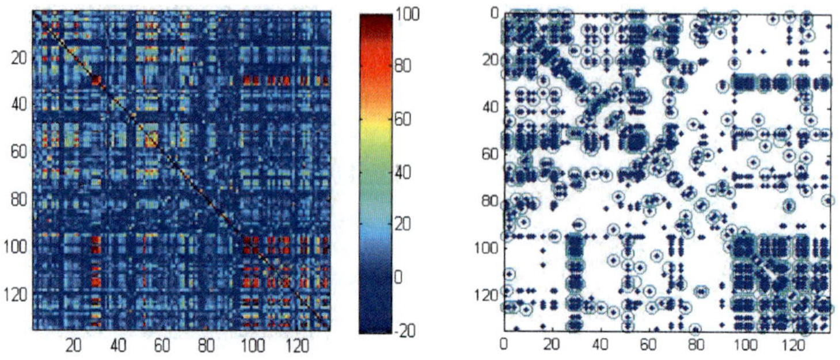

**Fig. 8.13** Nokia call-log network: scaled pair-wise correlations between oscillators at time corresponding to prediction depth $k_p = 2.4$, scenario *AN* (left); links predicted by dynamics at $k_p = 2.4$ are shown by dark-blue dots, adjacency matrix is shown by circles (right)

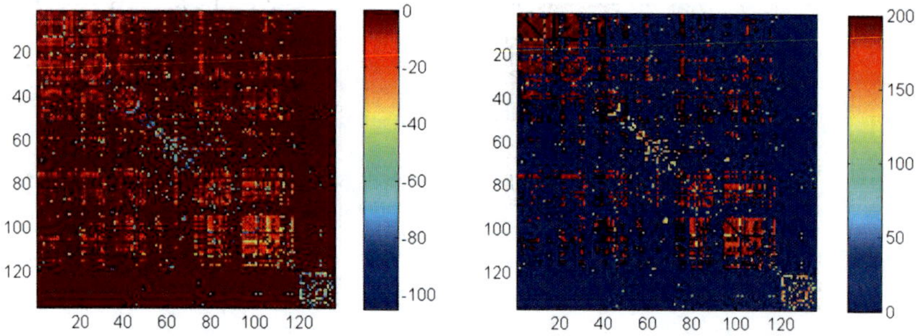

**Fig. 8.14** Nokia data: combined call-log and BT networks: Intensity of repulsive (left) and attractive (right) relations, average over 100 realizations

## 8.7 Conclusions

In this chapter we propose coupled dynamical systems framework and its applications for stable communities detection and links predictions/ recommendations utilizing network topology. The method is based on the dynamical formulation of modularity using a random walk and then is extended to coupled dynamical systems to detect communities at different hierarchical levels. We introduce attractive and repulsive coupling and study different scenarios for dynamical link updates that allow us to make predictions on a cooperative or a competing behavior of users in the network and analyze connectivity dynamics.

The developed methods are first tested on benchmark networks and then applied for analysis of datasets recorded during Nokia mobile-data collection campaign to derive social community structures and to make link predictions/recommendations.

## References

1. Acebrón, J., Bonilla, L., Pérez-Vicente, C., et al.: The Kuramoto model: A simple paradigm for synchronization phenomena. Reviews of Modern Physics 77(1), 137–185 (2005)
2. Albert, R., Barabási, A.L.: Statistical mechanics of complex networks. Reviews of Modern Physics 74, 47–97 (2002)
3. Arenas, A., Díaz-Guilera, A., Pérez-Vicente, C.: Synchronization reveals topological scales in complex networks. Physical Review Letters 96, 114102 (2006)
4. Arenas, A., Diaz-Guilera, A., Kurths, J., et al.: Synchronization in complex networks. Physics Reports 469, 93–153 (2008)
5. Blondel, V., Guillaume, J.L., Lambiotte, R., et al.: Fast unfolding of communites in large networks. Journal of Statistical Mechanics: Theory and Experiment 1742-5468(10), P10008+12 (2008)
6. Boccaletti, S., Latora, M.Y., et al.: Complex networks: Structure and dynamics. Physics Reports 424(4-5), 175–308 (2006)
7. Chung, F.R.K.: Spectral Graph Theory, CMBS Lectures Notes 92. Amer. Math. Society (1997)
8. Evans, T.S., Lambiotte, R.: Line Graphs, Link Partitions and Overlapping Communities. Physical Review E 80, 016105 (2009)
9. Flake, G., Lawrence, S., Giles, C.: Self-organization and identification of Web communities. IEEE Computer 35, 66–71 (2002)
10. Fortunato, S.: Community detection in graphs. Physics Reports 486, 75–174 (2011)
11. Girvan, M., Newman, M.E.J.: Community structure in social and biological networks. Proc. Natl. Acad. Sci. USA 99, 7821–7826 (2002)
12. Ipsen, M., Mikhailov, A.: Evolutionary reconstruction of networks. Physical Review E 66, 046109 (2002)
13. Newman, M.E.J., Girvan, M.: Finding and evaluating community structure in networks. Physical Review E 69, 026113 (2004)
14. Kiukkonen, N., Blom, J., Dousse, O., et al.: Towards Rich Mobile Phone Datasets: Lausanne Data Collection Campaign. In: Proc. ACM Int. Conf. Pervasive Services, Berlin (2010)
15. Kuramoto, Y.: Lectuer Notes in Physics, vol. 30. Springer NY (1975)
16. Lambiotte, R., Delvenne, J.C., Barahona, M.: Laplacian Dynamics and Multiscale Modular Structure in Networks. ArXiv:0812.1770v3 (2009)
17. Liben-Nowel, D., Kleinberg, J.: The Link Prediction Problem for Social Networks. ACM Int. Conf. on Information and Knowledge Management (2003)
18. Nefedov, N.: Multiple-Membership Communities Detection in Mobile Networks. In: Proc. ACM Int. Conf. on Web Intelligence, Mining and Semantics (WIMS 2011), Norway (2011)
19. Nefedov, N.: Applications of System Dynamics for Communities Detection in Complex Networks. In: IEEE Int. Conf. on Nonlinear Dynamics and Sync (INDS 2011), Austria (2011)
20. Newman, M.E.J.: Fast algorithm for detecting community structure in networks. Physical Review, E 69, 066133 (2004)
21. Olfati-Saber, R., et al.: Consensus and Cooperation in Networked Multi-Agent Systems. IEEE Proceedings 95(1), 215–233 (2007)
22. Wasserman, S., Faust, K.: Social Network Analysis. Cambridge University Press, Cambridge (1994)
23. Zachary, W.: An information flow model for conflict and fission in small groups. Journal of Anthropological Research 33, 452–473 (1977)

# Chapter 9
# Infinite Networks of Hubs, Spirals, and Zig-Zag Patterns in Self-sustained Oscillations of a Tunnel Diode and of an Erbium-doped Fiber-ring Laser

Ricardo E. Francke, Thorsten Pöschel, and Jason A.C. Gallas

**Abstract.** A remarkably regular organization of spirals converging to a focal point in control parameter space was recently predicted and then observed in a nonlinear circuit containing two diodes. Such spiral organizations are relatively hard to observe experimentally because they usually emerge very compressed. Here we show that a circuit with a tunnel diode displays not one but two large spiral cascades. We show such cascades to exist over wide parameter ranges and, therefore, we expect them to be easier to observe experimentally.

## 9.1 Introduction

Numerical simulations have recently uncovered a number of surprising and unexpected regularities in the control parameter space of certain dissipative flows. Such regularities were observed in systems as diverse as electrical circuits containing either piecewise-linear or smooth nonlinearities, in certain lasers, in chemical oscillators and in several other paradigmatic flows covering a large spectrum of practical applications [1]– [21]. More specifically, a wide-ranging regular organization of

Ricardo E. Francke
Instituto de Física da UFRGS, 91501-970 Porto Alegre, Brazil
e-mail: ricardo.francke@ufrgs.br

Thorsten Pöschel · Jason A.C. Gallas
Institute for Multiscale Simulation, Friedrich-Alexander Universität, D-91052 Erlangen, Germany
e-mail: Thorsten.Poeschel@cbi.uni-erlangen.de

Jason A.C. Gallas
Departamento de Física, Universidade Federal da Paraíba, 58051-970 João Pessoa, Brazil
Instituto de Física da UFRGS, 91501-970 Porto Alegre, Brazil
e-mail: jgallas@if.ufrgs.br

K. Kyamakya et al. (Eds.): Selected Topics in Nonlinear Dynamics, SCI 459, pp. 161–177.
DOI: 10.1007/ 978-3-642-34560-9_9 

spirals was anticipated numerically to exist in the control parameter space of simple electronic circuit. This organization consists of a doubly infinite hierarchy of spirals converging to focal centers called "periodicity hubs" [2, 3]. Such hubs are very interesting accumulation points of a doubly infinite sequence of spirals: an infinite family of spirals characterized by periodic oscillations which is intercalated with an infinite family of spirals characterized by chaotic oscillations. Every periodic spiral has a characteristic waveform which evolves continuously along the spiral with a period that grows without any bound, diverging at the focal point. Loosely, hubs work like crowded bus stations with busses represent spirals: when arriving at such "station" following an *ingoing spiral* one is presented with a doubly-infinite choice for changing to an outgoing "bus", i.e. to an *outgoing spiral*. This is so because there is an infinite choice of periodic as well as an infinite choice of chaotic patterns to choose from at the focal point. The selection may be simply accomplished by suitable selection of parameters. Examples of such hubs and spirals may be seen in Fig. 9.2 below. That the predicted spiral organization indeed exists in real systems was confirmed experimentally very recently at the ETH in Zürich [22] using a slight variation of the original circuit where they were numerically anticipated [2, 3]. Periodicity hubs were shown to be not isolated points but, instead, to emerge forming infinite hierarchical *networks of points* responsible for the organization of all stable periodic and chaotic phases [8].

Of particular interest for applications is that periodicity hubs are robust against parameter changes and imply a wide-range of predictable regularity in control parameter space. This is important because knowledge of the details of the regular organization of physical parameters allows one to select suitable numerical values to tailor the operation of circuits, lasers, and all sorts of nonlinear oscillators. By constructing detailed phase diagrams, i.e. by constructing detailed *stability charts* displaying the precise location in parameter space of the dynamical phases, one obtains a powerful instrument to perform accurate parameter changes allowing one to indeed *control* the system, not merely *to perturb* it without having a minimal ability of predicting in which new dynamic state the system will land after parameters are changed. Of course, parameter charts also allow one to perform big changes of control parameters, not just infinitesimally small changes.

So far, the spiral organization around periodicity hubs was observed in electronic circuits containing piecewise-linear elements [2–4, 22]. This type of circuits have two features that complicate experimental measurements. First, spirals usually emerge strongly distorted as, for example, in the paradigmatic circuit of Chua [3, 4]. Second, although it is known that spirals arise in infinite hierarchical networks [8], so far only a single isolated spiral has been detected experimentally [22].

The main reason complicating the observation of spiral networks is that the parameter regions containing them become significantly compressed making it hard to record them, particularly in noisy systems. An additional reason is that to observe networks one first needs to locate adequate two-parameter cuts in an usually high-dimensional control parameter space. This last task (parameter tuning) may

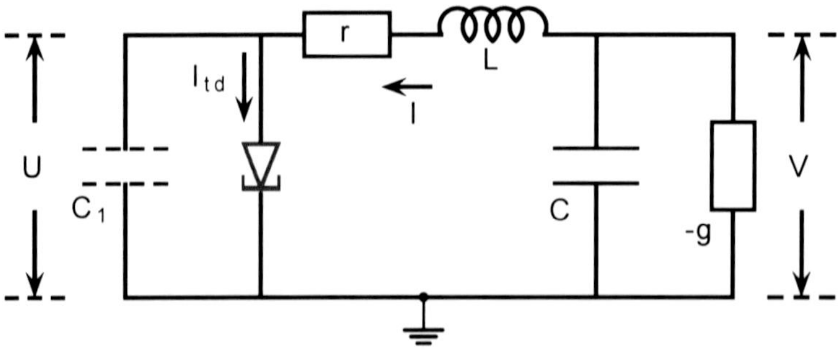


**Fig. 9.1** Schematic representation of the tunnel diode circuit leading to Eqs. (9.13)–(9.15). The voltage applied to the diode is denoted by $U$.

be rather difficult to perform experimentally. In this case computer simulations are of great help in locating suitable regions to search experimentally for hubs. On interesting additional byproduct is that computations may reveal shortcomings of the theoretical description of the electronic components (diodes, etc) in the sense that discrepancies between computations and measurements may emerge.

Here, our aim is to describe a simple autonomous electronic circuit, shown in Fig. 9.1, which we found to display clear and easily accessible sequences of spirals in its parameter space, as illustrated in Fig. 9.2 below. Apart from standard capacitances, inductances and resistances, the circuit contains two active elements, namely a linear negative conductance $-g$ and a tunnel diode.

Chaotic oscillations in diodes were studied quite early in pioneering works by Pikovsky and Rabinovich [23–25] and other authors, e.g. [26]– [31], in several configurations, autonomous or not. Tunnel diodes were found to display very rich dynamical scenarios when their control parameters are varied [24, 26]. Although the chaotic dynamics of circuits with tunnel diodes seems nowadays to have simply felt in oblivion, we wish to point out that they contain an unsuspected richness of dynamics to offer both for convenient experimental exploration as well as to help developing novel theoretical tools to deal with new complex phenomena being discovered like, e.g. periodicity hubs, which are yet far from understood.

## 9.2 The Flow Defined by a Simple Circuit with a Tunnel Diode

In this Section we derive the equations governing the self-excited oscillator illustrated in Fig. 9.1, containing a tunnel diode. At the end of the Section we comment an approximation in the original expressions in the literature [24].

From Fig. 9.1, where $I$ denotes the current through the inductance, $U$ the voltage across $C_1$, and $V$ the voltage across $C$, using Kirchhoff's laws we get:

$$V - U = rI + \frac{dI}{dt}, \tag{9.1}$$

$$-I = -gV + C\frac{dV}{dt}, \tag{9.2}$$

$$I = F(U) + C_1\frac{dU}{dt}. \tag{9.3}$$

With the help of an auxiliary variable $W \equiv V - rI$ these equations become

$$\frac{dI}{dt} = \frac{W - U}{L}, \tag{9.4}$$

$$\frac{dW}{dt} = -I\frac{1-gr}{C} + \frac{gL - rC}{LC}W + \frac{r}{L}U, \tag{9.5}$$

$$\frac{dU}{dt} = \frac{I - F(U)}{C_1}. \tag{9.6}$$

Handy adimensional equations can be obtained by introducing the following changes of variable

$$\tau = \sqrt{\frac{1-gr}{LC}}\,t \equiv \omega t, \quad I = (x+1)I_0, \quad U = (z+1)U_0, \quad y = \frac{W - U_0}{\omega L I_0}. \tag{9.7}$$

In addition, we need to replace $F(U)$ by its transformed $f(z)$ in the variable $z$, obtaining then:

$$\frac{dx}{d\tau} = y - \frac{U_0}{\omega I_0 L}z, \tag{9.8}$$

$$\frac{dy}{d\tau} = -x + \frac{gL - rC}{\omega LC}y + \frac{rU_0}{\omega^2 L^2 I_0}z + \Big(-1 + \frac{gU_0}{\omega^2 I_0 LC}\Big), \tag{9.9}$$

$$\frac{dz}{d\tau} = \frac{I_0}{\omega C_1 U_0}\big(x - f(z)\big). \tag{9.10}$$

Now, by introducing the following abbreviations

$$\delta = \frac{U_0}{\omega I_0 L}, \qquad 2\gamma = \frac{gL - rC}{\omega LC}, \qquad \alpha = \frac{rU_0}{\omega^2 L^2 I_0}, \tag{9.11}$$

$$\beta = -1 + \frac{gU_0}{\omega^2 I_0 LC} = \alpha - 1 + 2\gamma\delta, \qquad \mu = \frac{\omega C_1 U_0}{I_0}, \tag{9.12}$$

the equations can be written in a much simpler form, namely,

$$\frac{dx}{d\tau} = y - \delta z, \tag{9.13}$$

$$\frac{dy}{d\tau} = -x + 2\gamma y + \alpha z + \beta, \tag{9.14}$$

$$\mu\frac{dz}{d\tau} = x - f(z). \tag{9.15}$$

These equations coincide with those of Pikovsky and Rabinovich [23–25]. However, we obtain them using $\omega^2 = (1 - gr)/(LC)$ (see Eq. (9.7)) instead of the approximation $\omega^2 = 1/(LC)$ used by them. Both expressions agree when $gr \ll 1$.

Equations (9.13)–(9.15) are used below to study the dynamics of the tunnel diode. In Eq. (9.15), the nonlinear function $f(z)$ represents the characteristic function of the tunnel diode which, for simplicity, we assume to be a cubic function: $f(z) \equiv z^3 - z$.

Before proceeding we mention that Eqs. (9.13)–(9.15) were investigated theoretically in 1989 by Carcasses and Mira [32]. Using a Poincaré surface of section, these authors associated a two-dimensional diffeomorphism $T$ to the differential equations and then considered the *qualitative* bifurcation structure of $T$ in the $\mu \times \beta$ parameter plane. Here, however, we consider the quantitative bifurcation structure observed in the $\gamma \times \delta$ parameter plane as generated directly by Eqs. (9.13)–(9.15), not by an approximate Poincaré proxy.

## 9.3 The Slow-Fast Dynamics of the Circuit with a Tunnel Diode

Slow-fast systems (also known as singularly perturbed or systems with multiple time scales) are ubiquitous systems in physics, engineering, and biology in which two or more processes take place on different time scales [33, 34]. They are vector fields of the generic form

$$\mu \dot{x} = f(x,y,\mu), \tag{9.16}$$

$$\dot{y} = g(x,y,\mu), \tag{9.17}$$

where $\mu$ is a small parameter.

In this context, the flow defined by Eqs. (9.13)–(9.15) is particularly interesting because the parameter $\mu$ in front of the derivative $\dot{z}$ may be conveniently tuned to induce dynamical effects happening at different time scales. When $\mu$ is small, motions in the phase space can be divided into *slow motions*, corresponding to trajectories on the surface $x = f(z)$, and *fast motions*, corresponding to the straight lines $x =$ constant and $y =$ constant. As described by Rabinovich [24], the system has three states of equilibrium for a broad interval of values of the parameters $\alpha, \beta, \gamma$, and $\delta$, one state located at the origin, and the remaining pair located symmetrically on the surface of slow motions. All three states are unstable. If the untwisting of the paths near the unstable foci, say $A$ and $A'$, is not too fast, the mapping point cannot leave the region containing all three states of equilibrium: the mapping point moves outward away from the point $A$ along the spiral and, having reached the line $x = \pm 1$ along which the surface of slow motion bends over, it enters the neighborhood of the symmetrically located state $A'$. It then follows the paths leaving this point thus returning to the neighborhood of $A$, repeating the sequence again and again.

An important property of flows like the one above is that two trajectories lying arbitrarily close to one another near the boundary at which they break off from the slow-motion surface, may behave completely differently. Those lying inside the path tangential to $|x| = 1$ remain on the slow-motion surface and complete one additional

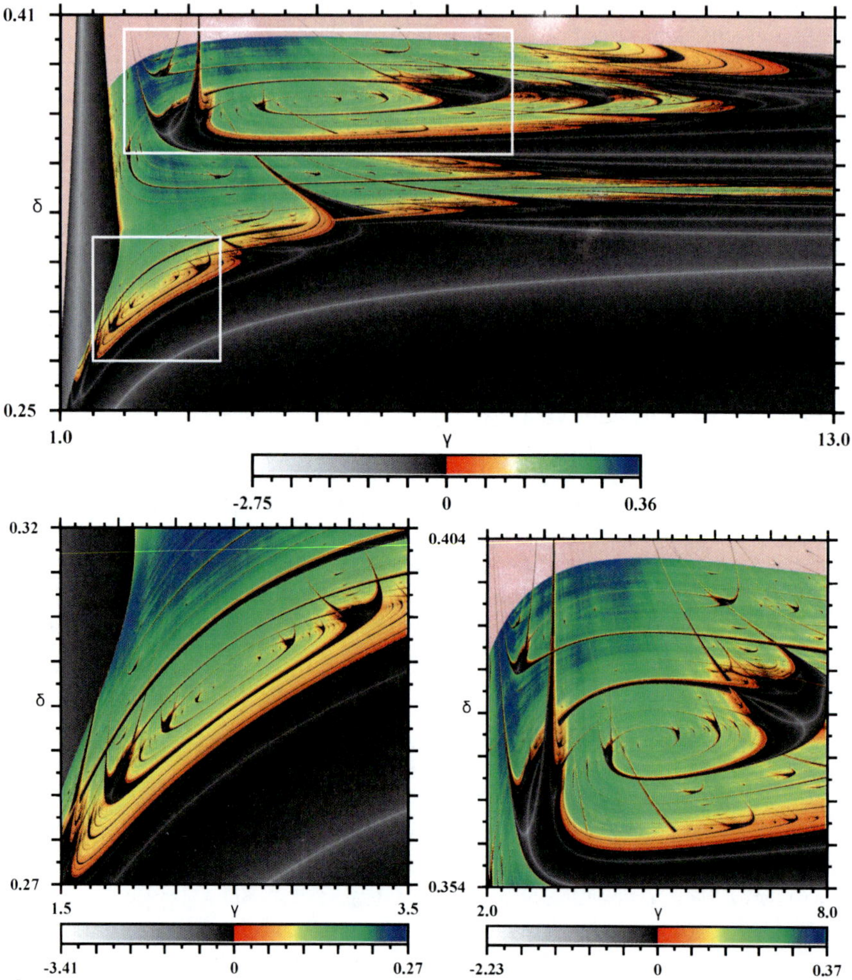


**Fig. 9.2** (Color online) Top panel: Global view of the control parameter space of the tunnel diode circuit, Eqs. (9.13)–(9.15), with boxes indicating the location of two periodicity hubs and spirals of large influence. Bottom panels: magnifications of the white boxes in the upper panel. Pink denotes divergent solutions. Here $\alpha = -0.013$, $\beta = 0$, $\mu = 0.1$. Each individual panel displays $2400 \times 2400 = 5.76 \times 10^6$ Lyapunov exponents.

turn around the equilibrium point. However, trajectories that are arbitrarily close to it but located outside this tangential path, fall downward (or rise upward) and enter the neighborhood of the symmetric state of equilibrium. Thus, as pointed out by Rabinovich [24], the future of these trajectories depends on fine details of their past.

Apart from $f(z)$, the flow defined by Eqs. (9.13)–(9.15) involves only linear terms, facilitating the theoretical analysis. Mathematically, the equations representing the circuit with a tunnel diode look quite similar to the ones governing the dynamics of the simple piecewise-linear resistive circuit were periodicity hubs were originally discovered [2, 3, 8]. Note that numerical simulations do not depend on the restriction of $\mu$ being a small parameter.

## 9.4 Phase Diagrams

This Section presents several high-resolution Lyapunov phase diagrams discriminating the nature (chaotic or periodic) of the dynamical behavior observed in the $\gamma \times \delta$ control parameter plane.

Lyapunov phase diagrams are generated by solving numerically the equations of motion (here with a standard fixed-step fourth-order Runge-Kutta integrator) and using the solutions obtained to compute all Lyapunov exponents for the system and plotting the largest nonzero exponent. As it is well-known, Lyapunov exponents are convenient numerical indicators used to discriminate the dynamical nature of the asymptotic oscillations observed in dynamical systems, i.e. they allow one to discriminate between periodic oscillations (which lead to negative exponents) and chaos (positive exponents).

Figure 9.2 shows Lyapunov phase diagrams summarizing what happens over a wide portion of the $\gamma \times \delta$ control space of the tunnel diode, discriminating periodic from chaotic phases. As indicated by the color scales, colors represent positive Lyapunov exponents, i.e. regions where chaos is prevalent. In contrast, periodic phases are represented using darker shadings. Note that the color scales representing negative and positive exponents vary independently from each other on both sides of zero, i.e. the variation is not uniform from the negative minimum to the positive maximum of the scales. Further, the color table of each enlargement is renormalized according to the minimum and maximum exponents so that colors may vary as one magnifies specific regions of the parameter space.

In Fig. 9.2 it is possible to recognize something that is very desirable for experiments: the existence of two large-size groups of nested spirals accumulating into distinct focal points, where the periodicity hubs are located. Converging to each focal point one sees two groups of intertwined spirals, defined by periodic and by chaotic oscillations. Both groups seem to contain an infinite number of spirals. As parameters approach the focal point, the waveforms of the periodic oscillations evolve continuously and their periodicity grows without bound. Note that sequences of *shrimps* [35–37] occur along the periodicity spiral arranged in consecutive pairs at each half-turn. Many other interesting parameter domains worth investigating may be also recognized in Fig. 9.2. For details see Ref. [8].

In Fig. 9.2 and in other phase diagrams here we vary $\gamma$ and $\delta$ over experimentally accessible ranges. As it is clear from the definitions in Eqs. (9.11)–(9.12), specific values of $\gamma$ and $\delta$ may be conveniently achieved in more than one way by suitably selecting numerical values for the several reactances in the circuit. Thus, all circuit elements are equally important, not just the tunnel diode.

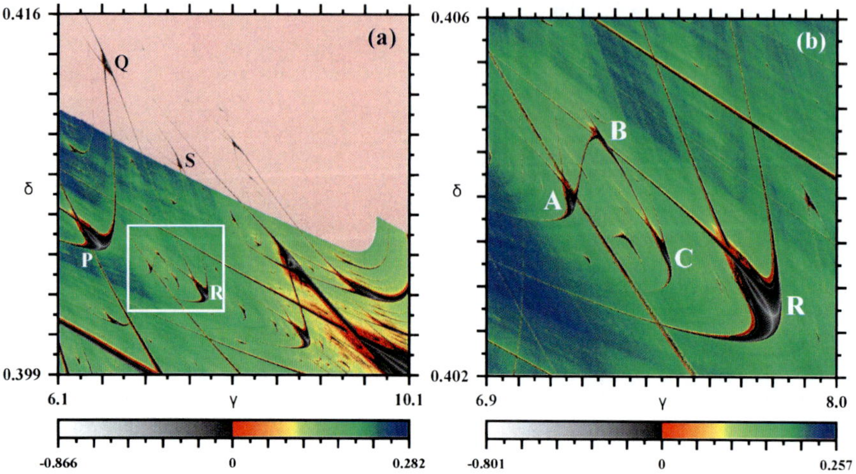


**Fig. 9.3** Examples of $V$-connections in the control space of the tunnel diode. (a) A zig-zag pattern $PQRS$ formed by "gluing" $V$-connections together. The zig-zag continues beyond $S$ but the additional shrimps are too small to be seen in the scale of the figure. Pink denotes parameter regions leading mainly to unbounded solutions (divergence); shrimps $Q$ and $S$ are embedded in it. (b) The $V$-connection $ABC$ in the white box in (a). One of the legs of $R$ allows passing between $R$ and $B$ via continuous parameter changes. A complex network of periodicity domains interconnects these shrimps. Here $\alpha = -0.33, \beta = 0, \mu = 0.1$.

Figure 9.3 shows a curious and abundant type of interconnection among distinct clusters of periodicity ("shrimps" [35]), computed here for $\alpha = -0.33, \beta = 0, \mu = 0.1$. Each panel of Fig. 9.3 shows $1200 \times 1200$ Lyapunov exponents, the same resolution used in Figs. 9.5 and 9.6 below. Figure 9.3b displays an upside-down "$V$-connection" or "$V$-bridge", as indicated by the letters $ABC$. This type of connection can be seen in Fig. 3 of a recent paper by Celestino et al. [15], who used a discrete map to study the properties of the unbiased current in the ratchet transport of particles. The shrimps in Fig. 9.3b are identical to those that combine to form the infinite chain that composes the continuous spirals in Fig. 9.2. Shrimps were originally described forming regular sequences of parallel clusters of periodicity, apparently disconnected from each other [36]. Here, however, the clusters of stability $A$ and $B$ are clearly interconnected by $B$, forming a structure that resembles an upside-down $V$. The periodicity clusters $A$, $B$, and $C$ are contained in the white box in Fig. 9.3a which contains many such connections forming zig-zag sequences in parameter space.

As mentioned in the introduction, knowledge of the existence of such parameter paths interconnecting distinct clusters of periodicity may be obviously used as a simple and powerful technique to control the system, i.e. to efficiently implement with a single operation macroscopic parameter changes leading to desirable changes in the behavior of the system in a *predictable way*, allowing one to precisely select which change to implement. In sharp contrast to "control techniques" which rely

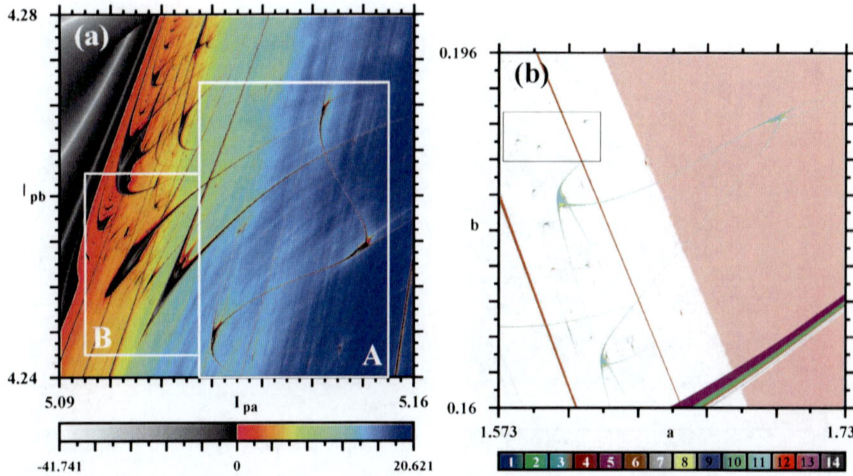


**Fig. 9.4** (Color online) $V$-connections observed in other systems. (a) in an erbium-doped fiber-ring laser (inside box A), and (b) in the discrete-time Hénon map of Eqs. (9.18)–(9.19). The two wide periodicity regions inside box B are high-order structures studied in detail for the Hénon map in Ref. [42]. The black box in (b) contains several additional $V$-connections which are too small to be seen in the scale of this figure [43]. Each panel displays results for $2400 \times 2400$ parameter points. Pink denotes parameter regions leading mostly to divergence.

on infinitesimal changes and are totally unable to target any specific final orbit, knowledge of parameter charts allows one to perform parameter changes of any arbitrary size and may move to any nearby *stable* orbit either with a single parameter jump or with sequences of controlled parameter changes, if so desired.

Figure 9.4 shows that $V$-connection provinding bridges among periodicity clusters are not difficult to find in other flows and even in the discrete-time models, i.e. in maps. For example, inside box A of Fig. 9.4a one sees a clear $V$-interconnection to be present in the control space of an erbium-doped dual-ring fiber laser [38–40]. Several others interconnections like this one exist over wide range of parameters [41]. The equations of motion and parameters adopted for this laser are given in the Appendix. Noteworthy in box B of this figure are the cuspidal island and the large island near it. Such structures appear profusely in parameter space. They have not been studied so far, although some results are known [42]. After the shrimps, the cuspidal island and the large island near it are the structures observed more frequently in the parameter space of flows and maps.

Figure 9.4b illustrates a $V$-connection for the paradigmatic discrete-time Hénon map defined here as follows [36, 37]:

$$x_{t+1} = a - x_t^2 + by_t, \tag{9.18}$$

$$y_{t+1} = x_t. \tag{9.19}$$

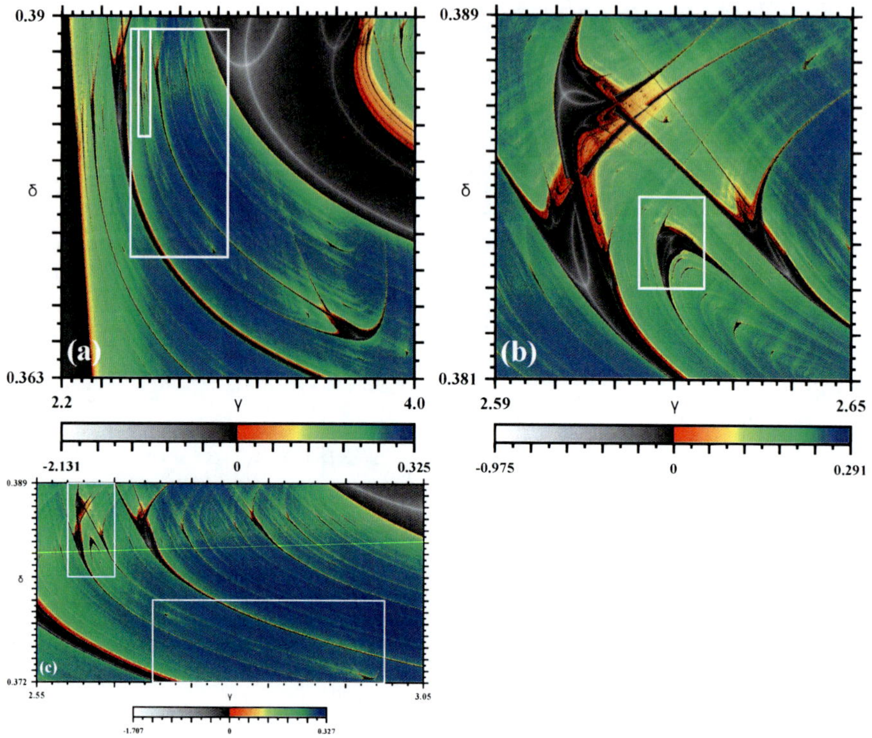


**Fig. 9.5** Successive enlargements illustrating a continuous spiral "arising" from a $V$-connection in the control space of the tunnel diode. (a) Global phase diagram, with the pair of boxes indicating the regions magnified in the other two panels. (b) The $V$-connection part of a spiral. The white box is magnified in Fig. 9.6. (c) Magnification of the largest box in (a), showing the $V$-connection (left box) and the spiral (inside the large white rectangle on the right). Several other analogous spirals and hubs exist although most of them are restricted to rather small parameter windows. Here $\alpha = -0.33, \beta = 0, \mu = 0.1$.

The Hénon map displays a profusion of $V$-connections, in addition to several other connections with complex forms that are quite difficult to classify systematically. The number of interconnections of all sorts is so great that one has the impression that in the end, all clusters of periodicity might in fact compose just a vast single network of connected domains fixed by the equations of motion. A more detailed investigation of the parameter space of the Hénon map is presented elsewhere [43].

Figure 9.5 illustrates a situation where, instead of the zig-zag patterns seen in Fig. 9.3, the $V$-connection gives origin to an infinite sequence of shrimps coiling up to form a continuous spiral. It seems appropriate to recall that a proper and encompassing mathematical description of spiral organizations in parameter space is still to be done. The only scenario that is presently reasonably well-understood is one associated with a theorem by L. Shilnikov [9, 10].

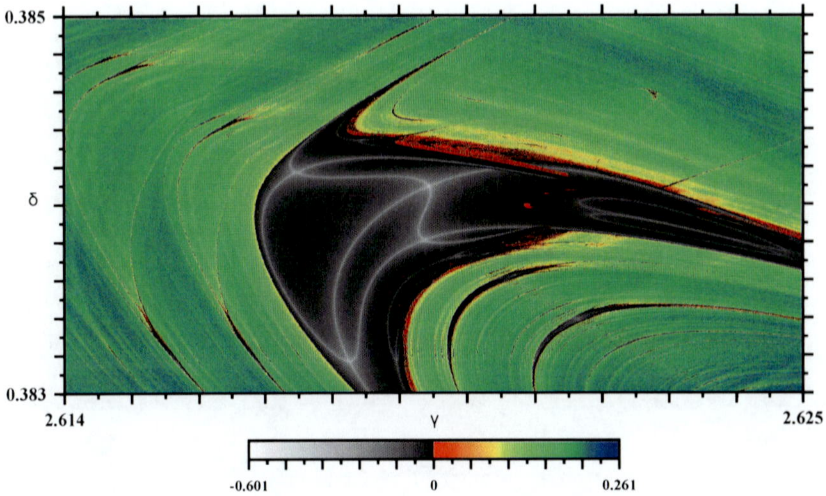


**Fig. 9.6** Magnification of the white box in Fig. 9.5b showing a complex periodicity cluster resembling a shrimp but containing a rather intricate network of loci that resemble "super-stable" loci, a concept properly defined for one-dimensional multi-parameter maps. Here $\alpha = -0.33, \beta = 0, \mu = 0.1$.

However, considerably richer scenarios are possible in higher-dimensional slow-fast systems, particularly when period-doubling cascades follow a Hopf bifurcation and subsequent canard explosion, producing alternations of periodic and chaotic oscillations. As the amplitude of the chaotic attractors grows one observes a spiking regime consisting of large pulses separated by irregular time intervals in which the system displays small-amplitude chaotic oscillations. This scenario, reminiscent of Shilnikov's homoclinic chaos despite the fact that no homoclinic connections are involved, has been observed very recently in ground-breaking experimental studies of a semiconductor laser with optoelectronic feedback by Al-Naimee et al. [44] and in the equations governing a light emitting diode (LED) subjected to the same feedback [45]. Such experiments provide new insight, showing that key concept of excitability needs to be extended beyond that familiar to fixed points, into the realm of higher-dimensional attractors of maps and flows as anticipated theoretically [46]. They equally show that slow-fast systems are relatively poorly understand and need to be investigated in more detail. Interestingly, lasers and circuits with LEDs open now the possibility for probing experimentally such elusive and unexplored phenomena. For details, see Refs. [7, 45].

Figure 9.6 displays a structure that looks like a shrimp but contains a much more intricate arrangement of parameters as represented by the white curves inside the wide periodicity cluster. Such curves look very much like "superstable loci". However, as pointed before [2], superstable loci are only defined for one-dimensional maps, where they mark trajectories passing through at least one "critical point" of the map, i.e. a point where the derivative of the map is zero [47]. Although Fig. 9.6

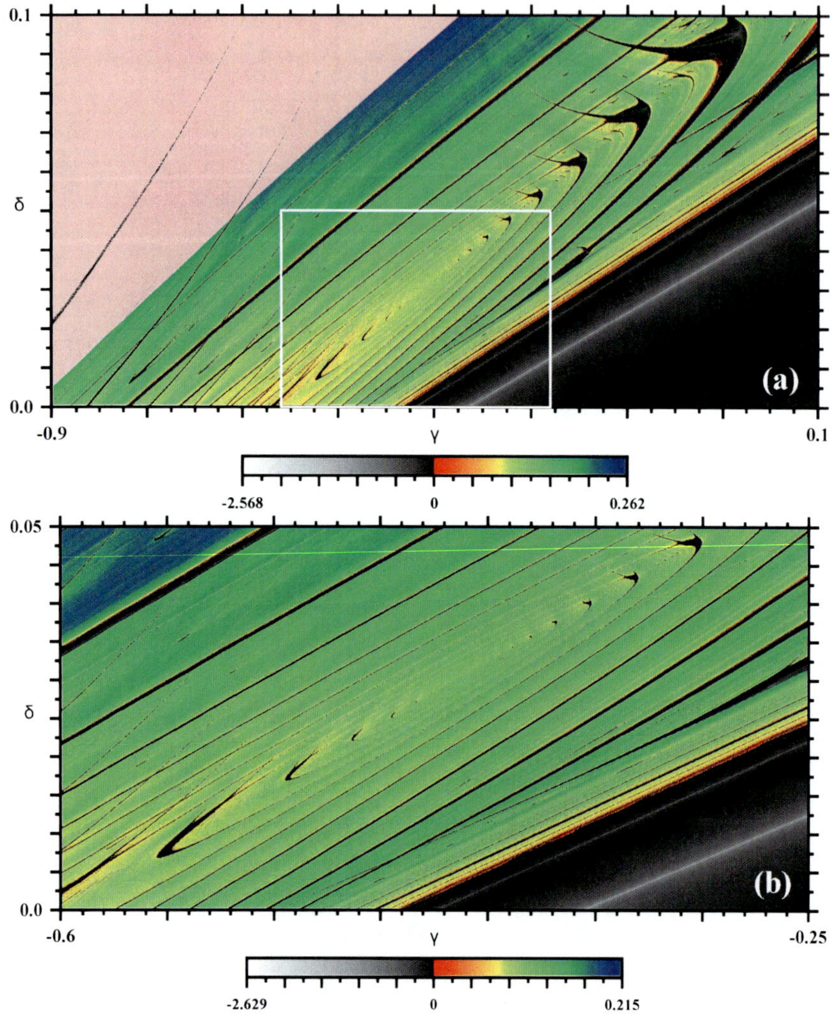


**Fig. 9.7** Illustration of a spiral structure extending over a very wide parameter range of the tunnel diode. Similar spiral arrangements exist over wide ranges for many other choices of parameters. This structure of the parameter space looks quite similar to the one found in the control space of Chua's circuit [3, 4, 8]. Each panel displays $1600 \times 1600 = 2.56 \times 10^6$ Lyapunov exponents. Here $\alpha = -0.33, \beta = 0, \mu = 0.1$.

displays a phase diagram for a flow (not a map), for lack of a proper definition and a better name we loosely refer to the white curves as being "superstable loci". Two important points may be recognized from Fig. 9.6: first, the existence of rather complex periodicity clusters not yet considered theoretically and, second, the necessity of adequately generalizing some known concepts in order to also deal with pressing situations that emerge abundantly when considering periodicity clusters of flows.

We remark that even though stability diagrams for flows display rather interesting networks of the aforementioned "superstable loci", there is still no theoretical prescription which would allow one to predict their existence and compute them for flows. In fact, even a proper name is still to be invented for them.

Figure 9.7 displays spiral structures which extend over very wide parameter ranges and that, we believe, should be relatively easy to observe in experiments. Of course, experimental resolution sets a limit on the number of turns of the spiral that can be observed. Important here is that the regular distribution of the successive shrimps gives an indication that a spiral has been spotted. For instance, it should not be difficult to unveil the regular organization simply by plotting bifurcation diagrams passing through the diagonal line containing the main body of the shrimps. The spiral organization seen in Fig. 9.7 looks very similar to spirals reported for Chua's circuit when operating both with piecewise-linear or cubic nonlinearities, as might be seen from Fig. 4 of Ref. [4] or from Fig. 5 of Ref. [8]. How could one objectively quantify the isomorphism among these systems?

## 9.5 Conclusions and Outlook

This work presented several high-resolution Lyapunov phase diagrams showing that a simple circuit containing a tunnel diode displays a pair of large continuous spiral networks with rich intertwined structures extending over a wide region in control parameter space. Near them one finds an infinite sequence of smaller spirals, as described in Refs. [8, 9]. The large pair of spiral networks makes tunnel diodes quite interesting testground to probe experimentally intricate and elusive dynamical properties described recently in the literature. We also described the abundance of a class of shrimp arrangement, certain $V$-connections [15], which we have shown to be capable of forming quite long zig-zag paths and networks in parameter space. Analogous features were also observed in the control space of an erbium-doped dual-ring fiber laser and of the much simpler Hénon map, known to represent well the dynamics of loss-modulated $CO_2$ lasers [1]. Spirals and zig-zag patterns offer an interesting way to move in a *controlled and systematic way* between families of stable solutions, quite distinct from the nowadays so popular method of randomly perturbing trajectories without having any control of the final state to be reached after application of the perturbation.

Parameter spirals of periodicity (and of chaoticity) emerge from and accumulate at periodicity hubs: mathematically, such hubs are associated in phase-space with very small regions of quite strong curvature, sometimes (but not necessarily) involving homoclinic bifurcation curves of a common saddle-focus equilibrium. These homoclinic bifurcation curves are arranged in fractal-like sheaves in the parameter plane [9]. The specific organization of hub networks *depends strongly on the interaction between the homoclinic orbits and the global structure of the underlying attractor* [9]. A challenging problem now is to describe what is causing the complex organization of periodicity clusters in phase diagrams of flows not involving homoclinic orbits. Note that presently there is no mathematical framework to predict and describe the genesis of hubs and the associated spirals in more general scenarios where the celebrated theorem of Shilnikov does not apply [7, 9].

The present work also shows that Lyapunov phase diagrams are quite valuable exploratory tools for practical applications allowing one to understand global features of complex attractors. We believe that the use of Lyapunov phase diagrams can significantly augment and speed-up the understanding of physical models. Lyapunov phase diagrams focus exclusively on *stable solutions*, i.e. on features that are directly measurable experimentally. Lyapunov phase diagrams reveal the occurrence of many global bifurcations without recourse to more specialized and demanding numerical techniques. They are therefore a very powerful way to begin the analysis of nonlinear systems and can also be applied to laboratory experiments which, of course, only detect stable structures. As described elsewhere in detail [7], note that *there is absolutely no need to compute Lyapunov exponents from experimentally measured data*. For experimental data it is enough to simply construct "binary" black-and-white phase diagrams discriminating between two states: presence or absence of periodicity. A complementary tool of great utility in analyzing dynamical systems is the direct study of the *periodicity and the number of extrema of the oscillations* as parameters are tuned [48, 49] (without resource to secondary and somewhat artificial quantities derived from the period like, e.g. when artificially introducing *pairs* of frequencies in phenomena where such pairs are not naturally present or not quite justifiable [49, 50]).

We hope the findings reported here to motivate their experimental investigation. From a theoretical point of view, at present it is totally unclear where to expect hubs and spirals to be found in flows. It is equally unclear which type of flows should be expected to contain hubs and spirals, particularly in high-dimensional systems. Thus, the only way to learn about them is through detailed numerical simulations and experiments. A related open question is how to optimize the search for the "most convenient" sections of the high-dimensional surface in control parameter space so as to better expose the intricacies and the structure of phase diagrams. In other words, to find an efficient way of quickly asserting the impact of changing *all control parameters*. From an experimental pointo of view, an interesting challenge is to investigate how realistic the simple cubic function used here is to describe the dynamics of real-life tunnel diodes. Obviously, high-resolution phase diagrams have the power of revealing eventual shortcomings of the mathematical formulation of models of natural phenomena. Phase diagrams can show where models need to be improved to better reproduce experimental measurements.

**Acknowledgements.** We are indebted to Arkady Pikovsky for helpful email exchanges concerning certain approximations that they used to obtain their equations. This work was supported by the Deutsche Forschungsgemeinschaft through the Cluster of Excellence *Engineering of Advanced Materials*. JACG was supported by CNPq, Brazil, and by the US Air Force Office of Scientific Research, Grant FA9550-07-1-0102. All computations were done in the CESUP-UFRGS clusters. A preliminary version of these results was presented at the *Workshop on Nonlinear Physics and Applications*, NOLPA, in João Pessoa, Brazil, September 5-9, 2011.

## References


1. Bonatto, C., Garreau, J.C., Gallas, J.A.C.: Phys. Rev. Lett. 95, 143905 (2005)
2. Bonatto, C., Gallas, J.A.C.: Phys. Rev. Lett. 101, 054101 (2008); Phil. Trans. Royal Soc. London, Series A 366, 505 (2008)
3. Ramírez-Ávila, G.M., Gallas, J.A.C.: Revista Boliviana de Física 14, 1–9 (2008)
4. Ramírez-Ávila, G.M., Gallas, J.A.C.: Phys. Lett. A 375, 143 (2010)
5. Freire, J.G., Field, R.J., Gallas, J.A.C.: J. Chem. Phys. 131, 044105 (2009)
6. Kovanis, V., Gavrielides, A., Gallas, J.A.C.: Eur. Phys. J. D 58, 181 (2010)
7. Freire, J.G., Gallas, J.A.C.: Phys. Rev. E 82, 037202 (2010)
8. Gallas, J.A.C.: Int. J. Bif. Chaos 20, 197 (2010)
9. Vitolo, R., Glendinning, P., Gallas, J.A.C.: Phys. Rev. E 84, 016216 (2011)
10. Barrio, R., Blesa, F., Serrano, S., Shilnikov, A.: Phys. Rev. E 84, 035201(R) (2011)
11. Bragard, J., Pleiner, H., Suarez, O.J., Vargas, P., Gallas, J.A.C., Laroze, D.: Phys. Rev. E 84, 037202 (2011)
12. Castro, V., Monti, M., Pardo, W.B., Walkenstein, J.A., Rosa Jr., E.: Int. J. Bif. Chaos 17, 956 (2007)
13. Zou, Y., Thiel, M., Romano, M.V., Kurths, J., Bi, Q.: Int. J. Bif. Chaos 16, 3567 (2006)
14. Albuquerque, H.A., Rubinger, R.M., Rech, P.C.: Phys. Lett. A 372, 4793 (2008)
15. Celestino, A., Manchein, C., Albuquerque, H.A., Beims, M.W.: Phys. Rev. Lett. 106, 234101 (2011)
16. Oliveira, D.F.M., Robnik, M., Leonel, E.D.: Chaos 21, 043122 (2011)
17. Oliveira, D.F.M., Leonel, E.D.: New J. Phys. 13, 123012 (2011)
18. Stegemann, C., Albuquerque, H.A., Rubinger, R.M., Rech, P.C.: Chaos 21, 033105 (2011)
19. Stegemann, C., Albuquerque, H.A., Rech, P.C.: Chaos 20, 023103 (2010)
20. Viana, E.V., Rubinger, R.M., Albuquerque, H.A., de Oliveira, A.G., Ribeiro, G.M.: Chaos 20, 023110 (2010)
21. Cardoso, J.C.D., Albuquerque, H.A., Rubinger, R.M.: Phys. Lett. A 373, 2050 (2009)
22. Stoop, R., Benner, P., Uwate, Y.: Phys. Rev. Lett. 105, 074102 (2010)
23. Pikovsky, A.S., Rabinovich, M.: Sov. Phys. Dokl. 23, 183 (1978); Dokl. Akad. Nauk SSSR 239, 301–304 (1978)
24. Rabinovich, M.: Sov. Phys. Usp. 21, 443–469 (1978); Usp. Fiz. Nauk 125, 123–168 (1978)
25. Pikovsky, A.S., Rabinovich, M.: Physica D 2, 8 (1981)
26. Gollub, J.P., Brunner, T.O., Daly, B.G.: Science 200, 48 (1978)
27. Gollub, J.P., Romer, E.J., Socolar, J.E.: J. Stat. Phys. 23, 321 (1980)
28. Linsay, P.S.: Phys. Rev. Lett. 47, 1349 (1981)
29. Testa, J., Perez, J., Jeffries, C.: Phys. Rev. Lett. 48, 714 (1982)
30. Octavio, M., DaCosta, A., Aponte, J.: Phys. Rev. A 34, 1512 (1986)
31. Su, Z., Rollins, R.W., Hunt, E.R.: Phys. Rev. A 40, 2698 (1990)
32. Carcasses, J.P., Mira, C.: In: Mira, C., Netzer, N., Simo, C., Targonski, G. (eds.) Proc. Int. Conf. on Iteration Theory: ECIT 1989, Batschuns, World Scientific, Singapore (1991)
33. Jones, C.K.R.T., Khibnik, A.I.: Multiple-Time-Scale Dynamical Systems, Mathematics and its Applications, vol. 122. Springer, NY (2000)
34. Grasman, J.: Asymptotic methods for relaxation oscillations and applications, Applied Mathematical Sciences, vol. 63. Springer, NY (1987)

35. "Shrimps" refer to wide-reaching structures in parameter space formed by a regular set of adjacent windows centered around a main pair of usually intersecting 'superstable' parabolic arcs (see discussion of Fig. 9.6). Thus, a shrimp is a doubly-infinite *mosaic of periodicity domains* composed by an innermost *main* domain plus all the adjacent periodicity domains arising from two symmetrically located period-doubling cascades together with their corresponding domains of chaos [36]. Shrimps should not be confused with their innermost main domain of periodicity or with superstable loci. For details see Refs. [36, 37]
36. Gallas, J.A.C.: Phys. Rev. Lett. 70, 2714 (1993); Physica A 202, 196(1994); Appl. Phys. B 60, S203 (1995), special supplement issue: Festschrift Herbert Walther; Hunt, B.R., Gallas, J.A.C., Grebogi, C., Yorke, J.A., Koçak,H.: Physica D 129, 35(1999)
37. Lorenz, E.N.: Physica D 237, 1689 (2008)
38. Luo, L., Tee, T.J., Chu, P.L.: J. Opt. Soc. Am. B 15, 972 (1998)
39. Senlin, Y.: Chaos 17, 013106 (2007)
40. Zhang, S., Shen, K.: Chin. Phys. 12, 149 (2003)
41. Pöschel, T., Gallas, J.A.C.: The distribution of self-pulsing and chaos in control space of an erbium-doped fiber-ring laser, preprint
42. Endler, A., Gallas, J.A.C.: Comptes Rendus Mathem (Paris) 342, 681 (2006)
43. Gallas, J.A.C.: Shrimps and the eigenvalue structure of the Hénon map, preprint
44. Al-Naimee, K., Marino, F., Ciszak, M., Abdalah, S.F., Meucci, R., Arecchi, F.T.: Eur. Phys. J. D 58, 187 (2010); New J. Phys. 11, 073022 (2009)
45. Marino, F., Ciszak, M., Abdalah, S.F., Al-Naimee, K., Meucci, R., Arecchi, F.T.: Phys. Rev. E 84, 047201 (2011)
46. Marino, F., Marin, F., Balle, S., Piro, O.: Phys. Rev. Lett. 98, 074104 (2007)
47. For a quite early and very nice review of the role of critical points as originally used by Schröder, Fatou and Julia, see H. Cremer, Jahresber. Deutsche Math. Ver. 33,185 (1924)
48. Freire, J.G., Gallas, J.A.C.: Phys. Lett. A 375, 1097 (2011)
49. Freire, J.G., Gallas, J.A.C.: Phys. Chem. Chem. Phys. 13, 12191 (2011)
50. Freire, J.G., Pöschel, T., Gallas, J.A.C.: Stern-Brocot tree: A unifying organization of oscillations for a broad class of phenomena, submitted for publication

## Appendix: The Erbium-Doped Dual-Ring Fiber Laser

This Appendix collects the equations for the continuous-time model of the erbium-doped dual-ring fiber laser.

We follow Luo et al. [38] and consider the erbium-doped dual-ring fiber laser with the lasing fields in the two rings frequency locked through a coupler $c_0$ with phase change of $\pi/2$ from one ring to the other. In this case, the equations for the fundamental system are [38–40]:

$$\frac{dE_a}{dt} = -k_a(E_a + c_0 E_b) + g_a E_a D_a, \tag{9.20}$$

$$\frac{dE_b}{dt} = -k_b(E_b - c_0 E_a) + g_b E_b D_b, \tag{9.21}$$

$$\frac{dD_a}{dt} = -(1 + I_{pa} + E_a^2)D_a + I_{pa} - 1, \tag{9.22}$$

$$\frac{dD_b}{dt} = -(1 + I_{pb} + E_b^2)D_b + I_{pb} - 1, \tag{9.23}$$

where $E_a$ and $E_b$ are the lasing fields and $D_a$ and $D_b$ are the population inversion in rings $a$ and $b$, respectively. The parameters $k_a, k_b, g_a, g_b$ represent the decay rate and the gain coefficient of the lasing fields $a$ and $b$, as indicated. $I_{pa}$ and $I_{pb}$ represent pump intensity in the respective fiber rings. Note that this laser model contains cubic nonlinearities, similarly to the one present in the tunnel diode model.

For the model above, an interesting paper by Zhang and Shen [40] reported hyperchaotic dynamics, in particular for the following set parameters

$$k_a = k_b = 1000, \quad c_0 = 0.2, \quad g_a = 10500, \quad g_b = 4700.$$

These are the parameter values adopted here to compute the phase diagram in Fig. 9.4. However, we emphasize that the laser phase diagram is not at all sensitive to these specific values in the sense that similarly looking diagrams are obtained for a wide range of parameter choices in addition to the above ones [41].

# Chapter 10
# Study of Dynamics of Atmospheric Pollution and Its Association with Environmental Parameters

Siwek Krzysztof, Osowski Stanislaw, and Swiderski Bartosz

**Abstract.** The chapter is devoted to the study of the dynamic processes of creation of atmospheric pollution (particulate matter, $SO_2$, $NO_2$ and ozone) and its association with the environmental atmospheric parameters, like the temperature, wing, humidity, insolation, etc. We analyse the problem of nonlinearity of these processes, their chaocity as well as the interrelationships between the concentration of the particular pollutant and the environmental variables. The results of these studies are applied in building the prediction models of the concentration of the particular pollutants.

## 10.1 Introduction

The analysis and forecasting of the air quality parameters are important topics of atmospheric and environmental research today due to the health impact caused by air pollution [1,4]. The main pollutants of the air are the particulate matters (PM), nitrogen dioxide ($NO_2$), sulphur dioxide ($SO_2$) and ozone. Forecasting their concentrations represents a difficult task due to the complexity of the physical and chemical processes involved. On the other side the relevant information is contained in the dynamic structures of the measured atmospheric variables [5,11]. Data mining of these structures may increase our understanding of the processes and apply this knowledge in the prediction procedure. In this chapter we will study different aspects of pollution dynamics changing in time scale of hours, trying to catch the

Siwek Krzysztof
Warsaw University of Technology, Warsaw, Poland
e-mail: ksiwek@iem.pw.edu.pl

Osowski Stanislaw
Warsaw University of Technology & Military University of Technology, Warsaw, Poland
e-mail: sto@iem.pw.edu.pl

Swiderski Bartosz
University of Life Sciences, Warsaw, Poland

K. Kyamakya et al. (Eds.): Selected Topics in Nonlinear Dynamics, SCI 459, pp. 179–190.
DOI: 10.1007/ 978-3-642-34560-9_10 

most important interrelations between the level of pollution and other environmental parameters like temperature, wind and humidity. The most important is the problem of linearity or nonlinearity of these processes. We will investigate it by applying the statistical methods to time series measurement results. In checking the linearity of the process we apply the Hinich test [10]. We have checked also the chaotic character of the process applying the short term largest Lyapunov exponent as the measure of chaocity [2,3]. From the prediction point of view the most important relations are between the pollution and the environmental variables. We study these interrelations considering the correlation and other measures of dependencies (for example the collinearity) among the variables. On basis of these results we have built the nonlinear prediction model by using the Support Vector Machine and compared its performance with the linear ARX model.

## 10.2 Analysis of the Pollution Time Series

The presented study will be based on the measurements made by the meteorological stations placed in southern part of Warsaw (Ursynow). This quarter of Warsaw is a typical suburb deprived of the industry. The source of air pollution is mainly the traffic and private heating of the houses, especially important in the winter. The analyzed results of measurement refer to hourly registration of pollutants. Different pollution types are analyzed: $PM_{10}$ (the PM of the diameter up to 10 $\mu m$) $SO_2$, $NO_2$ and ozone. Fig. 1 presents the typical time series of these pollutants, measured within the last two years.

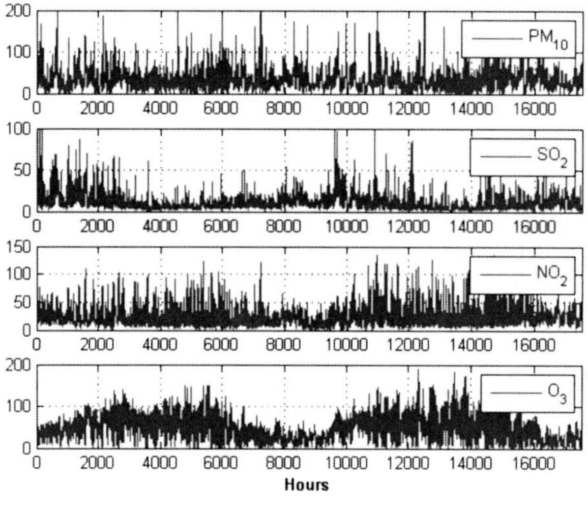


**Fig. 10.1** The succeeding hourly values of $PM_{10}, SO_2, NO_2$ and ozone concentrations measured in Warsaw within the last two years (the data start from January 1st)

We can observe the significant changes of the distribution patterns of these pollutants. There are visible abrupt jumps of pollution concentrations, occurring most often in $PM_{10}$, $SO_2$ and $NO_2$. In assessing the degree of prediction difficulty it is important to know some statistical characteristics of these time series. First we have calculated the mean value and standard deviation for each of them. Very important factor is the ratio between the standard deviation and the mean. The higher this ratio the more difficult is the prediction task. In Table I we present the mean values and standard deviations as well as their ratios for the measured pollution data presented in Fig. 10.1. The ratios std/mean assume high values, and differ for each pollution type. The most difficult case is presented by the data of $PM_{10}$ and $SO_2$, for which these ratios assume the highest values.

**Table 10.1** The mean values and standard deviations of different pollutants types registered in the last two years of measurements

| | Mean $[\mu g/m^3]$ | Std $[\mu g/m^3]$ | Std/mean |
|---|---|---|---|
| $PM_{10}$ | 30.26 | 22.77 | 0.75 |
| $SO_2$ | 8.74 | 7.49 | 0.86 |
| $NO_2$ | 23.69 | 15.87 | 0.67 |
| Ozone | 46.64 | 31.35 | 0.66 |

The fundamental issue in time series analysis is the study of linearity or nonlinearity of the process. The basic idea behind the nonlinearity test is that if the third order cumulant of the process is zero, then the bispectrum and its bicoherence are also zero [7]. If the bispectrum is nonzero then the process is non-Gaussian (potentially nonlinear). In the case of a non-Gaussian and linear process, the bicoherence is a nonzero constant. In practice the so-called "probability of false alarm" (PFA), that is the probability that we will be wrong in assuming that the data have a nonzero bispectrum, has been implemented to test the Gaussianity. If this probability is small, reject the assumption of zero bispectrum and reject also the assumption of the Gaussianity of the process. In the case of non-Gaussian process, the linearity test, checking whether the squared bicoherence is constant for all frequencies f1 and f2 reveals the eventual linearity of the process. In practice the bicoherence is usually not flat. In testing for linearity or nonlinearity of the non-Gaussian processes we may rely on the comparison of the so called empirical and theoretical sample interquartile ranges [10]. If their values are comparable the process is linear. In the case of high differences the process is regarded as nonlinear. In checking the nonlinearity we have applied the function *glstat.m* of Matlab [10]. After applying it on the investigated pollution data we have got in all cases PFA=0, which means that the assumption of Gaussianity should be rejected. In checking linearity or nonlinearity of the process we compared in this test the estimated ($R_e$) and theoretical ($R_t$) values of the interquartile ranges. In each case we have got relatively large values of the ratio $R_e/R_t$. In the case of $PM_{10}$ this ratio was equal 1.92, for $SO_2$ – 2.35, for $NO_2$ – 2.1 and for ozone 3.52. It means that the process of pollution creation for each pollutant is weakly nonlinear. The highest value of $R_e/R_t$ has been observed for ozone. Observation of the time series of each pollutant suggests the chaotic nature of the process. To prove the chaocity of the observed time series we have applied the Lyapunov exponent measure [2].The Lyapunov exponents have been proven to be the most useful dynamic diagnostic measure for the chaotic sys-tems. Lyapunov exponents define the average exponential rate of divergence or convergence of the nearby orbits in the phase space. It may be described in the form $d(t) = d_0 e^{Lt}$, where $L$ means the Lyapunov exponent and $d_0$ the initial distance of two nearby orbits. The magnitude of the exponent reflects the time scale on which the system dynamics become unpredictable [2]. Any system containing at least one positive Lyapunov exponent is defined to be chaotic and the nearby points, will diverge to any arbitrary separation. Two trajectories adjacent to each other at the beginning evolve exponentially in time and their distances change significantly with time (small distance $\Delta xij$ at the beginning and large distance $\Delta \mathbf{x}_{ij}(\Delta t)$) after time increase $\Delta t$. For chaotic time series the function $\Delta \mathbf{x}_{ij}(\Delta t)$) is dependent on the starting point $\mathbf{x}_0$, $\Delta \mathbf{x}_{ij}(\Delta t) = \Delta \mathbf{x}_{ij}(\mathbf{x}_0, \Delta t)$ and will behave erratically. The estimation $L$ of the short-term largest Lyapunov exponent $STL_{max}$ was done according to the formula [2,3]

$$L = \frac{1}{N\Delta t} \sum_{i=1}^{N} \log_2 \frac{|\Delta \mathbf{x}_{ij}(\Delta t)|}{|\Delta \mathbf{x}_{ij}(0)|} \tag{10.1}$$

where $\Delta \mathbf{x}_{ij}(0) = \mathbf{x}(t_i) - \mathbf{x}(t_j)$ is the displacement vector at $t_i$, that is the perturbation of the fiducial orbit observed at time $t_j$ with respect to $t_i$, while $\Delta \mathbf{x}_{ij}(\Delta t) = \mathbf{x}(t_i + \Delta t) - \mathbf{x}(t_j + \Delta t)$ is the same vector after time $\Delta t$. The vector $\mathbf{x}(t_i)$ is the point in the fiducial trajectory for $t = t_i$ and $\mathbf{x}(t_j)$ is properly chosen vector adjacent to $\mathbf{x}(t_i)$ in the phase space [3]. The time increase $\Delta t$ is the evolution time, that is the time which is allowed for $\Delta \mathbf{x}_{ij}$ to evolve in the phase space. For time given in hours the value of $L$ is in bits/hour. $N$ is the number of local $STL'_{max}s$ that will be estimated within the period $T$ of the data segment, where $T = N\Delta t + (p-1)\tau$. For the exponent $L$ to be a good estimate of $STL_{max}$ the candidate vector $\mathbf{x}(t_j)$ should be chosen in such a way that the previously evolved displacement vector $\Delta \mathbf{x}_{i-1,j}(\Delta t)$ is almost parallel to the candidate displacement vector $\Delta \mathbf{x}_{ij}(0)$. Moreover $\Delta \mathbf{x}_{ij}(0)$ should be small in magnitude to avoid computer overflow in the evolution within chaotic region. The choice of the parameters of evolution, such as the embedding dimension $p$, the evolution time $\Delta t$, the lag time $\tau$, the parameters used for selection of $\mathbf{x}(t_j)$ and the length of data segment $T$ are adjusted according to [3]. In studying chaotic behavior of the investigated time series we have applied the following parameters: period $T$ equal 1 year (8760 hours), shift of the calcula-tion window equal one quarter of the year (2190 hours), $\tau$=19 corresponding to the first minimum of the autocorrelation function, embedding dimension of the phase space equal $p$=10, evolution time $\Delta t = 86$. From this we got $dt = \Delta t/T = 0.0098$. All simulations have been performed using the available data from the time period 2004-2011 with the help of Matlab [6]. We have applied Iasemidis approach [3] to the estimation of the largest Lyapunov exponent. The following additional parameters used in the

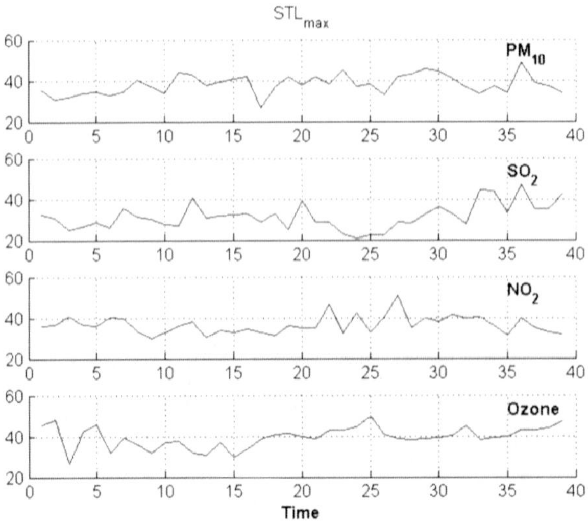


**Fig. 10.2** The time changes of the estimation of the largest Lyapunov exponents for the investigated pollutants. The horizontal axis represents time and its units are given in months.

algorithm were taken according to the hints from the paper [3]: $B$=0.05; $C$=0.1, $IDIST_1$=19, $IDIST_2$= $IDIST_3$=163. The results in the form of changes of the largest Lyapunov exponent with time for each pollutant are presented in Fig. 10.2. The results prove that all time series under investigation are highly chaotic and the average Lyapunov exponents for all pollutants take similar values.

The interesting (from the prediction point of view) is the correlation between the actual pollution level and its previous values. To study this relations we have plotted the autocorrelation function for different delays. The corresponding plots for each pollutant are presented in Fig. 10.3. It is evident that the correlation of pollution levels between distant hours is rather weak (except the first 24 hours). More-over, we can notice the significant differences of the shapes of autocorrelation curves corresponding to different pollutants (for example between $PM_{10}$ and ozone).

Another interesting question is the cross correlation existing among concentrations of different pollutants. High correlation (close to one) means that on the basis of measurement of one type pollutant we can predict the level of the other one.

**Table 10.2** The cross correlation coefficients between 4 types of pollutants

| | $PM_{10}$ | $SO_2$ | $NO_2$ | Ozone |
|---|---|---|---|---|
| $PM_{10}$ | 1 | 0.4467 | 0.5509 | -0.2868 |
| $SO_2$ | 0.4467 | 1 | 0.2688 | -0.2138 |
| $NO_2$ | 0.5509 | 0.2688 | 1 | -0.5592 |
| Ozone | -0.2868 | -0.2138 | -0.5592 | 1 |

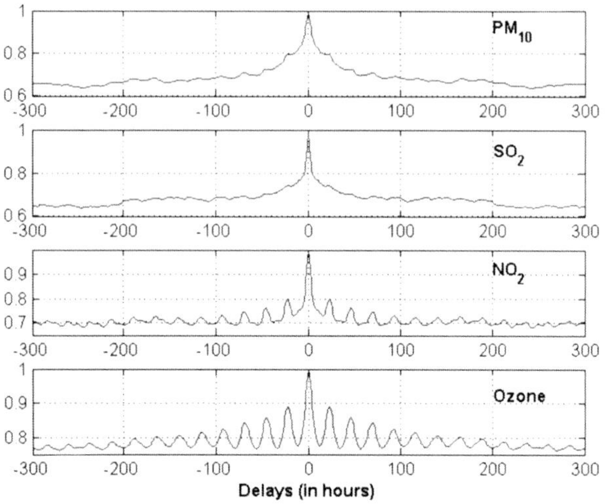


**Fig. 10.3** The autocorrelation functions of the concentration of different pollutants versus the delays measured in hours

Table 10.2 depicts the values of the cross correlation coefficients between 4 investigated pollutants. The results show that there is a weak correlation existing among different pollutants. The values of all coefficients are changing in the range (-0.5592 ÷ 0.5509).

We have analyzed also the mean values of pollution levels corresponding to different seasons of the year. Table 10.3 depicts these results for all investigated pollutants. We can see quite significant differences among them. Different pollutants show peak values corresponding to different seasons of the year. These results convince us to consider the season of the year as one of the input attribute taken into account at the prediction process.

**Table 10.3** The mean values of four pollutants for different seasons of the year

| | $PM_{10}$ | $SO_2$ | $NO_2$ | Ozone |
|---|---|---|---|---|
| Winter | 41.04 | 16.16 | 26.01 | 34.22 |
| Spring | 35.67 | 10.89 | 28.15 | 63.39 |
| Summer | 28.92 | 5.77 | 20.63 | 62.91 |
| Autumn | 37.65 | 8.97 | 24.83 | 35.95 |

The interesting question is the correlations among different seasonal pollution levels. The calculations of these correlations using Matlab [6] have shown very little linear dependence for all investigated pollutants. Table 10.4 shows the exemplary results for $PM_{10}$. Similar levels of correlations have been obtained for other pollutants.

**Table 10.4** The mean values of four pollutants for different seasons of the year

| | Winter | Spring | Summer | Autumn |
|---|---|---|---|---|
| Winter | 1 | 0.0362 | 0.0146 | 0.0335 |
| Spring | 0.0362 | 1 | 0.1406 | -0.0001 |
| Summer | 0.0146 | 0.1406 | 1 | 0.0746 |
| Autumn | 0.0335 | -0.0001 | 0.0746 | 1 |

All values of the cross correlations are far below one (close to zero level). It confirms once again our conjecture, that season of the year should be included as one of the input attribute in the predicting system.

## 10.3 The Relations between the Pollution and the Environmental Parameters

It is well known that the main factors influencing the pollution level are the temperature, strength and direction of wind, humidity and insolation [1,4]. We study the interrelation between the pollution level and these environmental variables, considering the correlation and other measures of dependencies (for example the collinearity) among these variables.

Fig. 10.3 presents the exemplary relations between the concentration of the $SO_2$ and the values of four environmental parameters: the wind speed, temperature, insolation and humidity. We can see complex distribution of points corresponding to different hours of the days. There are no clear functional relationships, indicating

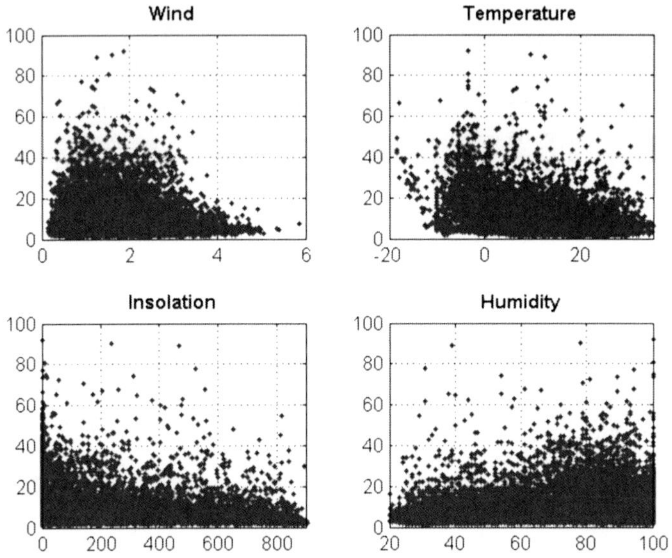

the need for more advanced tools of prediction. Similar relations are observed for the other types of pollution.

These distributions of points are the evidence of the lack of correlation between the pollution level and the environmental parameters. Table V depicts this observation in a numerical form showing the appropriate correlation coefficients. In most cases they are very small. However, it does not mean that there are no relationships between them. We should take into account that the correlation coefficients are only the evidence of the linear dependence existing among the investigated variables. They say nothing about the nonlinear relations between variables. Therefore in the pollution prediction systems these parameters are treated as important exogenous variables that are input to the predicting system.

**Table 10.5** The cross correlation coefficients between four types of pollutants and the considered environmental variables

| | Wind | Temperatur | Humidity | Insolation |
|---|---|---|---|---|
| $PM_{10}$ | -0.3260 | -0.2018 | 0.0818 | -0.1310 |
| $SO_2$ | -0.0224 | -0.3584 | 0.0961 | -0.0478 |
| $NO_2$ | -0.4395 | -0.2423 | 0.2094 | -0.3380 |
| Ozone | 0.3239 | 0.6080 | -0.7723 | 0.6019 |

However, the other fact is that in the predicting systems we should avoid the multi-collinear input features. To check the multi-collinearity among the meteorological parameters we have performed different tests using Matlab. Assuming the tested vector as follows: **x**=$[temp, wind_x, wind_y, hum, insol]$, where *temp* means temperature, $wind_x$ and $wind_y$ – the wind speed in $x$ and $y$ directions, *hum* – the humidity and *insol* – the insolation, we have applied the F-test, testing the joint hypothesis that all coefficients $\alpha_i$ of linear equation

$$pollution = \alpha_0 + \alpha_1 temp + \alpha_2 wind_x + \alpha_3 wind_y + \alpha_4 hum + \varepsilon \qquad (10.2)$$

are all equal zero, where $\varepsilon$ is iid $\sim N(0,1)$. We have tested the null hypothesis:

$$H_0 : \alpha_1 = \alpha_2 = \alpha_3 = \alpha_4 = 0 \qquad (10.3)$$

against the alternative one $H_1$:

$$H_1 \alpha_1 \neq 0 \vee \alpha_2 \neq 0 \vee \alpha_3 \neq 0 \vee \alpha_4 \neq 0 \qquad (10.4)$$

For all pollutants we have got the results $Fstat > 40$ and $p\_value < 0.001$ voting against the null hypothesis. We have applied the next test, variance inflation factor (VIF) quantifying the severity of multi collinearity in an ordinary least squares regression analysis. It measures how much the variance of an estimated regression coefficients is in-creased because of collinearity. The results of this analysis performed in Matlab [11] are depicted in Table 10.6.

**Table 10.6** The results of VIF test

| Exogenous variable | VIF value |
|---|---|
| Temperature | 1.70 |
| $Wind_x$ | 1.02 |
| $Wind_y$ | 1.01 |
| Humidity | 2.02 |
| Insolation | 1.75 |

None of the atmospheric variables for all pollutants was characterized by the VIF value above 5. This is the evidence of the lack of multi-collinearity among them [11]. The performed tests suggest that for all pollutants there is no evident collinearity between the exogenous variables. It means that all exogenous variables are important in the prediction process. The results of these studies allow us to propose the mathematical model of the pollution, which can be used for prediction of the mean concentration of the particular pollutant for the next day on the basis of the actual measurement of the environmental variables and the past history. The results of the previous analysis suggest that model should be rather nonlinear. In this research we have applied the Support Vector Machine in regression mode (SVR) of Gaussian kernel [8]. For comparison we have also used the classical linear autoregressive model with exogenous variables (ARX).

## 10.4 Comparison of the Linear and Nonlinear Prediction Models

The results of the previous analysis have been taken into account at building the particular model of pollutant, enabling to predict the generally unknown, next day mean value of pollution [9]. As the input variables we have used the past val-ues of pollutant and the values of the atmospheric parameters predicted by the meteorologists for the next day. The results of previous experiments have shown that the known pollution history can be limited to only one past day. The general supervised model of the pollution forecast for $d$th day has been assumed in the following mathematical form [9]

$$\hat{P}(d) = f(\mathbf{w}, wind_x, wind_y, temp, hum, insol, r, s, P(d-1)) \tag{10.5}$$

In this expression $\mathbf{w}$ represents the vector of adjusted parameters of the model, $wind_x$, $wind_y$ – the wind speed in $x$ and $y$ directions, $temp$ – temperature, $hum$ – humidity, $insol$ – the insolation level, $r$ – type of the day and $s$ – the season of the year. The symbol $\hat{P}(d)$ represents the predicted pollutant concentration in the air and the $P(d-1)$ written without hat – the known exact value of the pollution of the previous day. All of them are delivered as an input information to the particular predictor. To provide the appropriate representation of the wind, we have applied its two components $wind_x$ and $wind_y$ in the rectangular coordination system. Additionally we take into account the type of the day under prediction (binary representation of weekend or working day: 1 for work day and 0 for weekend) and the season of the year (binary representation of four seasons: 11 – winter, 10 – spring, 01 – summer and 00 - autumn). To provide similar impact of all input variables, the data samples should be normalized. The normalization may take different forms, from which the simplest one (applied in this work) is to divide the real value by the mean of the data base, corresponding to the years taking part in experiments.

The particular form of the applied predictor depends on its structure and way of learning. In this work we have investigated the nonlinear neural model based on Support Vector Machine with Gaussian kernel [8], working in regression mode. To make the comparison with linear predicting system we have applied the linear ARX model [11].

SVR in its principle minimizes the weights of the network, while keeping the output signals as close as possible to the their destination values with the predefined tolerance limit $\varepsilon$ [8]. The regularization constant C is applied for weighting between the values of weights and the prediction error on the learning data, while $\varepsilon$ is the tolerance value in learning process (the prediction error below this tolerance is neglected). The most important advantage of SVR over other solutions is the fact that its learning algorithm is based on the quadratic programming with linear constraints of one, well defined global minimum. Moreover SVRs are known from very good generalization properties. The details of learning SVR network may be found in the excellent book [8].

To assess the obtained results in a most objective way we have applied different measures of the prediction quality. Each of these measures determine the quality of prediction from different point of view. Five measures have been used in experiments.

- The mean absolute error (MAE)

$$MAE = \frac{1}{p}(\sum_{i=1}^{p} |t_i - y_i|) \tag{10.6}$$

- The mean absolute percentage error (MAPE)

$$MAPE = \frac{1}{p}(\sum_{i=1}^{p} \frac{|t_i - y_i|}{d_i}) * 100\% \tag{10.7}$$

- The root mean squared error (RMSE)

$$RMSE = \sqrt{\frac{1}{p}\sum_{i=1}^{p} |t_i - y_i|^2} \tag{10.8}$$

- The correlation coefficient (R) of the observed and predicted data

$$R = \frac{R_{yt}}{std(y)std(t)} \tag{10.9}$$

- The index of agreement (IA)

$$IA = 1 - \frac{\sum_{i=1}^{p}(t_i - y_i)^2}{\sum_{i=1}^{p}(|t_i - \bar{t}| + |y_i - \bar{t}|)^2} \tag{10.10}$$

In these relations $p$ is the number of data points, $y_i = \hat{P}_i$ is the predicted value, $t_i$ – the really observed value, $\bar{t}$ the average of really observed data, $R_{yt}$ is the co-variance value between the really observed and predicted data points of pollutant concentration, and *std* denotes standard deviation of the appropriate variable. To get the most objective assessment of the proposed prediction system we have applied 10-fold cross validation. This means repeating 10 times the sessions of learning and testing the system using the learning and testing data organized in different way. In each session we have generated randomly the set of learning (800 out of available 1460 samples) and testing (the remaining 660 samples) data. The final results of learning and testing are the mean of all trials. In presenting the results we limit ourselves only to the testing data, not taking part in learning, since these data are representative for future operation of the system in the most objective way. The linear ARX model of prognosis was implemented in Matlab [6] by using its built in learning algorithm. The ARX adaptation procedure has shown that the best results have been obtained at the rank of denominator $N_a = 4$ and numerator $N_b = 1$ of the ARX model. The SVR predictor was adapted at regularization constant $C$=100, the tolerance limit $\varepsilon = 0.01$ and the width of the Gaussian kernel $\sigma = 0.8$. All other parameters of this model (the number of support vectors and the positions of theirs centers) have been automatically adjusted by the learning algorithm.

**Table 10.7** The statistical results (mean±std) of SVR and ARX predictors in 10 cross validation experiments on the testing data for $PM_{10}$, $SO_2$, $NO_2$ and ozone

| | $PM_{10}$ SVR | $PM_{10}$ ARX | $SO_2$ SVR | $SO_2$ ARX | $NO_2$ SVR | $NO_2$ ARC | Ozone SVR | Ozone ARX |
|---|---|---|---|---|---|---|---|---|
| MAE [$\mu g/m^3$] | 10.5±0.78 | 11.2±0.99 | 9.74±0.82 | 10.4±1.12 | 11.5±0.76 | 11.6±0.77 | 6.2±0.39 | 6.5±0.45 |
| RMSE [$\mu g/m^3$] | 15.0±3.71 | 15.5±3.84 | 14.7±3.74 | 14.8±3.91 | 17.9±3.78 | 17.5±3.69 | 7.8±1.87 | 8.1±1.97 |
| MAPE [%] | 28.4±1.39 | 32.4±3.27 | 28.5±2.95 | 33.2±3.45 | 32.7±2.58 | 35.1±2.60 | 23.7±1.29 | 24.6±1.67 |
| R | 0.77±0.09 | 0.74±0.11 | 0.78±0.11 | 0.77±0.09 | 0.70±0.12 | 0.69±0.78 | 0.84±0.11 | 0.83±0.11 |
| IA | 0.83±0.07 | 0.84±0.11 | 0.85±0.09 | 0.86±0.12 | 0.68±0.07 | 0.70±0.06 | 0.91±9.08 | 0.90±0.11 |

The results depicted in Table 10.7 present the mean values and standard deviation (the numbers behind ± sign) of the appropriate quality measures. It is evident that the nonlinear model of prediction (SVR) performs much bet-ter than the linear one. All quality measures related to it are much better. This con-firms the nonlinear character of the predicted time series. The highest accuracy was obtained for ozone, the pollutant of the smallest std/mean ratio and of the greatest balance of the succeeding hourly values of concentration.

## 10.5 Conclusions

The results presented in this work have shown that the mechanism of pollution creation, irrespective of the type of pollutant, is very complex and belongs to the nonlinear, chaotic process not easy in modeling. To obtain the highest quality of prediction results we should apply the nonlinear model of the process, better taking into account the complex relations between the concentration of pollutant and the basic atmospheric parameters influencing the mechanisms of creation and spreading the pollution. The numerical experiments confirmed the superiority of the nonlinear SVR model applying Gaussian kernel over the classical ARX linear model.

Further work will be concentrated on the development of the more accurate prediction method. In our opinion the main source of improvement will be in consideration of more factors taken into account as input attributes in the prediction process. We should consider additional potential prognostic features such as longer history of pollution development, peak pollution values and their long term influence on the mechanism of pollution creation, etc. Additional tools of feature selection, well associated with the prediction problem will be developed.

**Acknowledgements.** This research activity was financed from the fund intended for science development as the R& D project within the years 2010-2012.

## References

1. Grivas, G., Chaloulakou, A.: Artificial neural network models for predictions of PM10 hourly concentrations in greater area of Athens. Atmospheric Environment 40, 1216–1229 (2006)
2. Holzfuss, J., Lauterborn, W.: Lyapunov exponents from a time series of an acoustic chaos. Physical Review A 39, 2146–2152 (1989)
3. Iasemidis, L.D., Principe, J.C., Sackellares, J.C.: Measurement and quantification of spatio-temporal dynamics of human epileptic seizures. In: Akay, M. (ed.) Nonlinear Signal Processing in Medicine, pp. 1–27. IEEE Press, New York (1999)
4. Kukkonen, J., Partanen, L., Karpinen, A., Ruuskanen, J., Junninen, H., Kolehmainen, M., Niska, H., Dorling, S., Chatterton, T., Foxall, R., Cawley, G.: Extensive evaluation of neural networks models for the prediction of NO2 and PM10 concentrations, compared with a deterministic modeling system and measurements in central Helsinki. Atmospheric Environment 37, 4539–4550 (2003)
5. Li, S.T., Shue, L.Y.: Data mining to aid policy making in air pollution management. Expert Systems with Applications 27, 331–340 (2004)
6. Misiti, M., Oppenheim, G., Poggi, J.M., Misiti, Y.: User manual of Matlab, MathWorks, Natick, USA (2010)
7. Nikias, L., Petropulu, A.P.: Higher-order spectral analysis a nonlinear signal processing framework, Englewood Cliffs, NJ (1993)
8. Schlkopf, B., Smola, A.: Learning with kernels. MIT Press, Cambridge (2002)
9. Siwek, K., Osowski, S., Widerski, B.: Study of PM10 Pollution Using Signal Analysis Methods, pp. 228–233. ISTET, Klagenfurt (2011)
10. Swami, A., Mendel, J., Nikias, C.L.: HOSA toolbox of Matlab 5.1, MathWorks, Natick, USA (1993)
11. Wilks, D.S.: Statistical Methods in the Atmospheric Sciences. Academic Press, San Diego (1995)

# Chapter 11
# System Dynamics Modeling of Intelligent Transportation Systems
# Human and Social Requirements for the Construction of Dynamic Hypotheses

Oana Mitrea

**Abstract.** The current article structures the knowledge and requirements for the (system dynamics) modeling of distributed actions in the intelligent transportation systems from the perspective of the sociology of technology.

## 11.1 Introduction

Challenging the discrete view of separate events and decisions system dynamics comes with a novel purpose of assembling a formal structure "that can reproduce by itself, without exonegenous explanations the essential characteristics of a dynamic problem" [1]. Its key conceptual tools are: the endogenous perspective; the feed-back thinking; the stocks and flows; the feed-backs, and the dominance of non-linearity (id). The importance of system dynamics modeling and simulation has recently increased in the field of traffic engineering and traffic planning [2-3]. As Raux (2003) emphasizes, in practice the complex non-linearity has often been neglected in the existing models due to the "heaviness of calculations and the involved complexity" [4]. However, this neglect should not cast shadows on its chances and promises. Particularly the fact that changeable system structures can lead to unexpected behavior through a complex chain of causes and effects has encouraged extensive preoccupation with system dynamics modeling in the field of the intelligent traffic [5]. The central concept of this approach is to model how all the objects in a system interact with each other through "feedback loops, where a change in one variable affects other variables over time, which in turn affects the original variable, and so on" [5]. The key advantages of this approach are the better control of the increasingly high complexity of the modern systems and the faster and easier sensitivity analyses, through the quick implementation of tests on the structure of the models [4]. System dynamics models have been constructed particularly at

Oana Mitrea
University of Klagenfurt
e-mail: oana.mitrea@uni-klu.ac.at

K. Kyamakya et al. (Eds.): Selected Topics in Nonlinear Dynamics, SCI 459, pp. 191–206.
DOI: 10.1007/ 978-3-642-34560-9_11 

the macroscopic level of the traffic flow to explore the interaction of transportation, economy, and urban planning [3-4] or to evaluate the effect of different transport policies on emissions of pollutants from European countries [6]. Also microscopic system-dynamics based models for traffic simulation can be mentioned [5] [7-9]. Since system dynamics is much involved with complexity and adaptability, we believe that there is still room for interdisciplinary ideas flowing into the "engineering" perspective. The current article aims to examine whether the results of socio-logical research about the distributed agency in hybrid systems (systems in which humans and collective actors interact with intelligent technology) can be useful for tracing the basic formal structure of the system (in accordance with the framework provided by Forrester (1969) (see below) and for the production of dynamic hypotheses[10]:

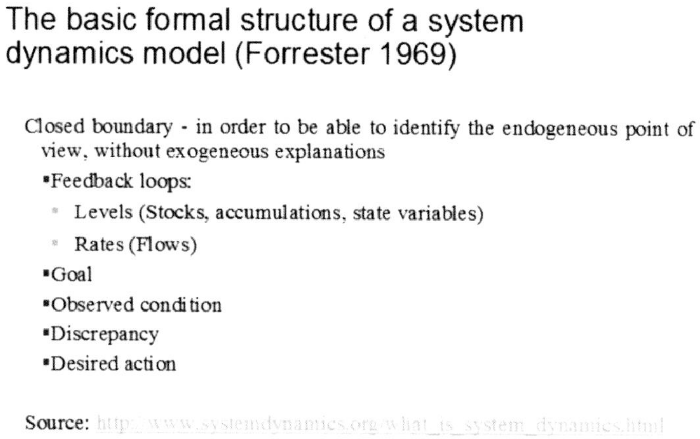


**Fig. 11.1** The formal structure of the system dynamics models after Forrester (1969)

## 11.2 Interaction and Interactivity in Intelligent Transportation Systems

Intelligent Transportation Systems are defined by the Intelligent Transportation Systems Society as "a category of systems and concepts utilizing synergistic technologies and systems engineering concepts to develop and improve transportation systems of all kinds"[1] . The most recent efforts have been done in the areas of driver assistence (ADAS) and cooperative traffic systems with the purpose of increasing safety, efficiency, comfort, and the ecological sustainability of driving.

In general the "intelligence" of the ITS is mainly related to the way that the data received from sensors are processed and turned into meaningful information for various "on-demand" situations. The following components are thereby involved [11]:

[1] http://www.ewh.ieee.org/tc/its/

- the active in- and -out vehicle environment sensing, respectively road infrastructure sensing which will provide users with improved safety, spontaneity, efficiency and comfort
- intelligent spaces defined as "environments that continuously monitor what's happening in them, communicate with their inhabitants and neighbourhoods, make related decisions, and act on these decisions" [11] (Wang et al. 2006, p. 68 )
- agent-based control – distributed operational agents letting networked transportation systems operate on a management-on-demand or service-on-demand basis (Wang et al. 2006, p. 69 )

Over time, modeling and simulation of human behavior have become increasingly important for the development of innovative ITS solutions and their evaluation [12]. According to Möbus et al, common aims of driver models are: "to predict and generate driver behavior emitted by individual drivers sometimes in interaction with assistance systems; to identify situations or maneuvers and classify behavior (e.g. anomalous vs. normal) of ego driver; provide a robust and valid mapping from human sensory data to human control actions; be learnt from time series of raw data or empirical probability distributions with statistical sound (machine-learning) procedures with only a few non-testable ad hoc or axiomatic assumptions; should be able to learn new patterns of behavior without forgetting already learnt skills (stability-plasticity dilemma)" [13]. The user-centered approaches to driver modeling have mainly focused on the development of interface design for in-vehicle navigation systems that are adapted to driver behavior in driving contexts. In general, the driver behavior has been modeled in order to understand the mechanism of driving performance in road traffic environments; to estimate a driver's intention prior to various maneuvers; to access the benefit of in-vehicle information systems; and to predict driver behavior from previous and current observations [14]. A great deal of such studies employed probabilistic models (including Markov dynamic models and Bayesian network model) in which the random driving operations observed commonly in a real environment were regarded as stochastic events [13-14]. Their aim has been mainly a (psychologically valid) representation of the traffic agent (e.g. driver behavior).

One cannot deny the importance of driver's psychology for a correct specification of a driver model. It is critical to know which internal characteristics influence the driver in a given situation. However, the specification of the majority of the properties of human internal capability (such as: "information reception, perception and processing; neuromuscular dynamics with threshold, time delay and limitations; preview, anticipation; adaptation/learning; planning capacities (path and speed); driving experience; risk behavior; concentration, tiredness/ stress and emotion" [12]) does not satisfactorely surprend the way that the new "intelligent" technologies and humans interact.

The situation analyzed by the classical HCI (Human Computer Interaction) - persons sitting in front of computers - point towards a (quite) predictible, and quite perceptible static setting. However, the interactivity of humans with advanced mobile technologies is more complex, dynamic; and involves more senses. A newer branch of Human Computer Interaction (HCI) strongly concentrates now on the more complex matters of interactivity between humans and the smart ubiquitous computing technologies that imply multisensory interactivity with everywhere interfaces and the pro-active environment intelligence [15-23].

In the last time the ITS (Intelligent Transportation Systems) have started to gain in relevance for the sociology of technology, mainly because they illustrate how advanced technologies can "act" in cooperation with human and social systems to achieve the global goals of safety, efficiency and sustainability.

**At the microscopic level** any external observer can confirm nowadays that the users of driver assistence systems are less and less the only ones acting autonomeously, although they still take decisions and perform a variety of tasks during driving. The agents in on-board units stay behind them and mix in human decisions and actions. The intelligent cooperation of technical units is achieved if their behavior, although based on algorithms, can be described as situated cooperativeness or "artificial" interaction rather than determined operation [24]. In this way, advanced technologies cannot be represented as passive objects, reduced at neutral instrumentality anymore. On the contrary, they entangle with humans and make room for a hybrid system – consisting of human and nonhuman decision makers [25]. The sociology of **hybrid systems** is highly preoccupated by the issues of automomy and control and could provide some interesting findings for the construction of microscopic system dynamics models in ITS [25-26].

In hybrid systems humans experience the agents as communicative counterparts, which reactions may deviate from human expectations and develop in a contingent manner [24]. This type of interactivity between humans and intelligent agents is characterized by complexity, contingence, and symbolic mediated communication [24]. While it is true that agents "stay" in place of humans in various areas, the agent representation of human wishes, decisions, motoric reactions is usually more then the name suggests: a "substitution", "acting in place of another". It is rather a re-construction of humans' behaviors and decisions from a given perpective, in accordance with an underlying theoretical model (see the variety of driver models). There is no exaggeration to tell that in some cases the cooperation of agents in advanced technologies leads to unexpected or at least for the involved humans, difficult to foresee effects [2] . At the same time, the human that has already interacted with the agents is not the same anymore. He/She has reveived in-advance information on ways not possible without the mediation of advanced technologies; has been able to react faster to the events on the road; sometimes has just simply complied with recommendations without questioning their backgrounds (as considered too complicated for further enquiring). The "oriented human" [27] after interaction with

[2] This perspective is consistent with Intelligent Transportation Systems as complex adaptative systems: Complex systems comprised of multiple components (agents) interacting via nonlinear feedback processes. Interesting here is that: "Their macroscopic properties, emerging from the collective interaction, contrast with the properties of the individual components. In particular, in complex adaptive systems, agents are able to adapt to the environment and co-evolve with the system as a whole." http://www.sg.ethz.ch/teaching/cas_2010

advanced technologies does not think and act in the same way as the human that has handled with instrumental, non-intelligent ICTs so far.

After these considerations, first questions relevant for the construction of dynamic hypotheses for system dynamics model at the microscopic level:

- How do human and non-human elements interact and co-evolve? [28].
- How do some consequences of this interaction (like for instance short-time adaptativity and reactive-passivity [25]) propagate through the system?

If we take three examples: the decentralized intersection control, the hierarchical (policemen and traffic lights controlled) and the cooperative intersaction control (involving manifold communication between humans, smart agents embedded in vehicles, and street infrastructure), interesting comparisons can be performed. The effective functioning of a totally decentralized intersection (without advanced technologies) is based on the learned and internalized norm-conforming behaviour. This is also the case for the classical hierarchical traffic control by policeman or by traffic lights. As Weyer (2009) emphasizes, the strategical driver bevahior is still possible (based on expectations and experience) in this situation. If the green light blinks, one can asess his/her own possibilities to pass the intersection in time [25]. However, the situation is different in the cooperative intersection control. Its declared aim is to enhance traffic safety and flow through through intelligent applications such as: journal plans, messaging, collision avoidance, feed-back and sensory inputs [29]. In this demand-oriented intersection the behavioral control rules might be clearly (and manageable) formulated, nevertheless the human perception of complexity and contingency of technology arises from the multitude of possible system states (and possibly from the different implementation strategies derogating from human expectations [29]. The degrees of freedom can no longer move within the normatively expectable behavior space. Manifold (obscure) independent activity is attributed to an intelligent artifact. This may lead, according to Weyer, to the decrease of the self-perception about human autonomy and to a passive-reactive behaviour in driving [25]. In this context of interactivity, further research questions with potential of generating dynamic hypotheses are:

- Which decisions and actions (tactical, strategical, and operational) can be described in the intelligent driving? (capturing of state variables, their levels, flows)
- How are the sequences of actions distributed? On what instances?
- How do humans assess inputs, real-time recommendations from the intelligent traffic system? How do they respond in real-time to them?What happens in time with the humans' capacity to plan and anticipate their actions (predictive, strategic thinking) under the real-time condition?
- To what extent are voluntary decisions, human autonomy, and their ability to recognize situations in the cooperative traffic preserved? How do humans react in time to system failure and false alarms, distraction, cognitive overload, unintended consequences of actions?
- How do perceptions about the ease of use the usefullness of applications (acceptance) evolve in time?

- Which collective actors are involved? What are their perspectives and objectives?
- How can/do the agents adapt to the changes occurred to humans in time?

At the macroscopic level the pragmatical sociology of technology extends the view about the "intelligent driving" to the the level of the heterogeneous constellation of humans, collective actors, technical agents, infrastructures, and signs [24], [30-31]. According to this view, technology in action can be understood and modeled in interaction with technical elements (called "intra-action" Rammert [24] ); and in interactivity with human and social systems [24]. The second type of interactivity is related to the embedding of the human-vehicle-environment system into a broader socio-technical constellation (comprising of institutional actors, guiding principles, visions about mobility systems, social norms and values) [24]. The recent advances in the sociology of technology stress the fact that the performance of "intelligent" mobility systems represent a result of a complex interface coordination among the activities of human drivers, car technology, environment, social, and political actors. Due to the distributiveness of actions to technological, human, social and political systems in hybrid constellations, it is possible that the performance of the whole system decreases, even if the driving behaviour, car control or telecommunications become smarter. The performance of the intelligent mobility system is dependent on how the hybrid driver-vehicle-environment system is embedded in the socio-technical configurations of the transport system. This include a variety of actors: city and traffic planners, producers of alternative mobility technologies, telecom providers, the structure of their products, etc., who decide whether with individual vehicles or not, with how much freedom, in what combination with other means and under what guidance and regulation system is mobility performed [24] (Rammert 2007, p. 129). As a consequence the system dynamics model needs to adopt a holistic perpective, including all relevant actors of a given socio-technical constellation. This shall integrate the microscopic time-space perspectives of the individuals,

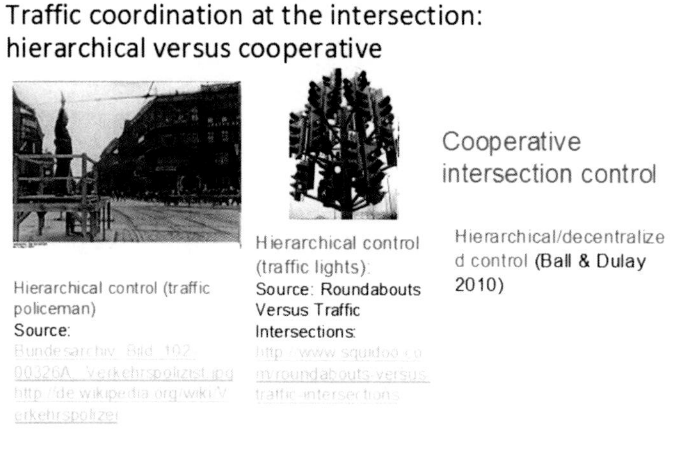


**Fig. 11.2** Hierarchical versus cooperative intersection control

the interactivity with the intelligent agents, and the macroscopic views of the involved actors in different spatial and temporal contexts. According to these considerations, the tracing of boundaries for the macroscopic system dynamics modeling and simulations (and evaluations) can be performed at the major interface between the system driver-vehicle-environment and the socio-technical constellation of the collective actors.

## 11.3 Particularization: Human and Social Requirements for the System Dynamics Modeling of Cooperative Traffic Scenarios

Cooperative traffic systems are defined as: "systems in which a vehicle communicates wirelessly with another vehicle (Vehicle-to-vehicle communication, or V2V) or with the road infrastructure (vehicle-to-infrastructure - or V2I communications, and infrastructure-to-vehicle communication or I2V" [32]. The majority of the cooperative traffic services are aimed at the foresighted driving and an early detection of hazards (for instance through ESP, ACC, operative platooning, PReVent, WILL-WARN (Wireless Local Danger Warning), INTERSAFE, etc.). Examples of already implemented cooperative traffic services are: incident management, road/weather condition warning, roadwork information, lane utilisation information, in-vehicle variable speed limit information, traffic congestion warning. Services to be introduced later address navigation and driver/management support in a more general sense: ISA (Intelligent Speed Adaptation) with infrastructure link, international

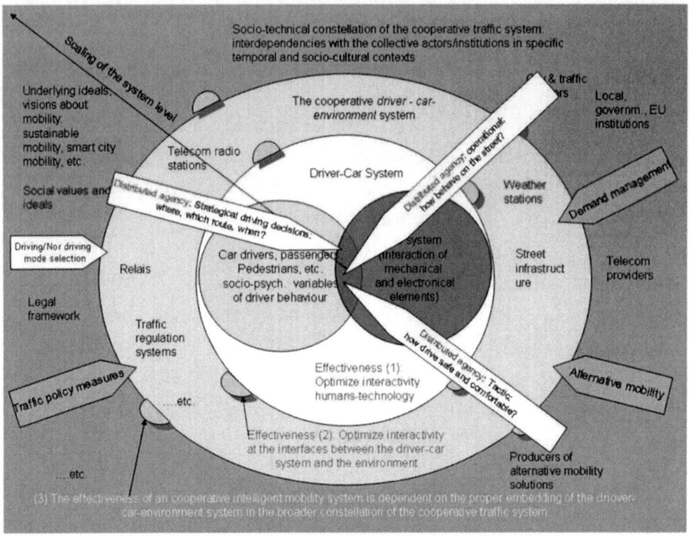


**Fig. 11.3** The anatomy of a cooperative system, adapted after the analysis of the intelligent mobility system (Rammert 2007) [24]

service handover, road charging to influence demand, route navigation - estimated journey time, route navigation - recommended next link, route navigation - map information update, floating car data (FCD)"[3].

A microscopic view:

*"I'm approaching the X crossroad, driving home after work in a distant city. It is raining. I would prefer to travel by train, but unfortunately the schedule does not suit me. I am tired, but anxious to see my family and to spend a nice evening together. I'm entering the outskirts of my city. Unfortunately, the warning system of my car informs me that 200 meters ahead, after the intersection, the road suddenly blocked due to an accident. Between 16:00 - 17:00 this spot should be a nightmare, a perfect image of chaos, but the reality is dazzling: nothing happens. Every car seems to quickly adapting to the spontaneous changes and to fit in the newly created niches: some change the lane and turn to the left, some take the next right turn, some wait patiently. I'm waiting for my turn. After a fast communication with other cars and infrastructure, my intelligent car recommends me the optimal maneuvers for the fastest alternative route. I'm accepting them, driving safely and being just in time in front of my house. Now, let's suppose I don't have THIS car, stuffed with sensors and devices which can send and receive information to/from other cars and infrastructure. It is still possible that nothing happens. Partially because I am an experienced driver, executing the right maneuvers, partially because the other "intelligent" cars can perceive me as a moving potential danger and can avoid me. However, so I put so significant strain on the effective functioning of the system: its safety and flow. The accumulation of such small events can lead to delays and coordination problems. If I want to be really bad, I can simply ignore warnings and recommendations from my intelligent car, or act against them – because I still have the overhand about what to do with the information received (blended decision making). I eventually comply with the rules of the "intelligent cooperative traffic" because I don't want endanger other traffic members."*

This story depicts some key elements of the cooperative driving; among them the foresighted driving, the early detection of hazards. From the microscopic perspective the focus is placed on the continuous experience of the modeling unit (car-driver system) with space, time, exchanged information, and driving coordination. The Finnish Strategic Research Agenda "Cooperative Traffic ICT" however stresses the necessity to follow a red line through all this complexity – while defining the cooperative traffic system in accordance to its purpose [33]: "We have to understand the transport system as a stratified entity of decision making where both the quality and quantity of traveling and transporting are influenced by societal and individuals' values and extending to a single maneuver at the wheel. The cooperative aspect is embedded in the capability of different actors either passively or actively to acquire data and share it with other parts and players of the traffic system."[33]

[3] Described in in WP3000: COOPERS services and value chains, concerning operator/ user behaviour, Integration of services in the co-operative system, 2010, http://www.coopers-ip.eu/fileadmin/results/deliverables/D3600-3700_COOPERS_SERVICES_AND_VALUE_CHAINS_VO4_APPROVED.pdf

If one conceives the cooperative driving as a continuous "cooperative" action distributed on human and technical instances, the interest lays with the exact characterization of the actions chain for each individual unit and with the rules that favor the passage from one action to another. Our expectation is that this "passage" is influenced, among others, by the socio-technical distribution of the cooperative agency (by what/who/where decides about next action sequence). Further we deal with the embedding of the driving actions in the broader configuration of the cooperative traffic system involving policy driven system goals: safety, fuel economy, environmental protection.

The process of model building at both microscopic and macroscopic levels includes several steps ranging from qualitative description to quantification. The understanding of the real situation is crucial for the next formal description of the process: key variables and their linear and non-linear interactions. It is about finding out how humans and the cooperative driving agents embedded in cars and infrastructure can cope with the daily driving needs in real-time in the context of the interaction with a variety of other systems (alternative mobility, communication, and urban infrastructure, social and political systems). For the description of actions chains in the hybrid constellation of the cooperative traffic we employ and complete the frame of reference advanced by the Cooperative Traffic Agenda (2009), which distinguishes among four levels of approach: (1) societal goals, levels of individuals, 2) strategic behavior, 3) tactical behavior, 4) operational behavior [33].

1. At the broader of the socio-technical constellation of the cooperative traffic system, relevant are basic decisions about driving / not driving a car, the amount of driving, and choice of travel means. The action of driving is highly influenced by the dynamic of social and individual values embedded in the objectives of transport policies (visions about the sustainable mobility, pricing, alternative travel options, measures for environmental sustainability), and by the interaction with the systems of alternative mobility.
2. Once the decision to drive an intelligent car on the cooperative traffic system taken, there is now room for strategic decisions about how to plan the route in space and time. The drivers do not accomplish this alone, nor do the warning and recommender smart agents. Instead, a joint strategy emerges from action distribution on drivers, passengers, other involved traffic members, collective actors AND travel information agents, traffic management assistance on the route, infrastructures, etc. This wider socio-technical constellation exerts an important influence on the type of information required through the embedding of mobility visions, social values in the design of travel information services and in traveler's expectations.
3. At the tactical level are relevant decisions about how to travel safer, faster, and more comfortable. Various traffic management, assistance, services, traffic info, location-based- services are here involved. It should be also mentioned that mobility visions and social values remain embedded in services design and traveler's expectations.
4. At the operational level are important decisions about how to behave on the road, how to control the car. Many assistive applications are designed for instance to

support a more "sustainable" behavior of the driver (safe driving, eco-driving). Agency distribution is dominated by the smart agents with contingent and even intentional properties. Also here are social visions (for instance about sustainable mobility) embedded in their construction. Human agency capacity may decline to short-time adaptability, with unexpected future consequences.

A forthcoming research project aims at modeling the distributed actions in cooperative traffic systems from the microscopic perspective. A scenario of "between-vehicle cooperation" will be selected. For a detailed comprehension of the actions and interactions in cooperative driving it is necessary to follow a progressive exploration strategy (see Rammert's approach [24]. The first set of research questions regards the description of actions in system: "where/what/who acts". Specific interactions and forms of interactivity between humans and agents within this frame are further explored. The last questions relate the hybrid actions to the broader context of the "hybrid constellation" of the cooperative mobility.

### *11.3.1 Description of the Pro-active and Co-operative Agency [24]:*

This set of questions aims at describing the action chains in detail, formally characterizing the distributiveness of actions and decisions on humans and agents:

- How does cooperative driving work? What are the involved instances?
- What decisions (tactical, strategical, and operational) are taken at the different scales in different contexts?
- What actions occur (communication, information access, monitoring, generating solutions, selecting solutions, implementing cooperative solutions, denying solutions, etc.?) How can their degree of automation be described (from manual control to full automation [25])?
- Who/what performs the actions in detail?
- How are the actions distributed? On what instances?
- What label reaches the action (1-causal, 2- contingent, 3-intentional) for the specific instances?
- Which order feature the action chains in various cooperative scenarios for particular individuals/ technical systems?
- What are the unintended consequences of the sequencies of actions?
- What conflicts can be identified? How do they evolve in time?

### *11.3.2 The level of Interpersonal Interaction, Intra-activity (Interaction among Technical Agents) and Interactivity with Human and Social Systems [24]*

These questions in-depth explore the interaction human-humans, among technical elements and the interactivity between humans technical components and agents

- How do humans cooperate for the accomplishment of mobility goals in dynamic mobility systems?
- What type of interaction occurs among technical agents in such systems?
- How do humans assess inputs, real-time recommendations from the cooperative traffic system? How do they respond in real-time to them?
- How human decisions and actions are delegated in real-time to the agents?
- What happens with the humans' capacity to plan and anticipate their actions (predictive, strategic thinking) under the real-time condition?
- To what extent are voluntary decisions, human autonomy, and their ability to recognize situations in the urban traffic preserved?
- How do humans react to system failure and false alarms, distraction, cognitive overload?

### *11.3.3 The "Hybrid Constellations" of Pro-active and Cooperative Agency [24]*

The questions here are about the contextual frames the should be taken into account to link the cooperative action chains to the elements of the heterogeneous constellation. Important is also to learn how the social behavior of humans in the intelligent traffic is shaped through their real-time interaction and coordination of mobility.

In particular:

- Which visions about social cooperation, systems of mobility, alternative mobility are involved, and are dynamically interacting with the cooperative traffic system?
- Which collective actors are involved with particular actions? What are their roles: for coordination, control?
- Which general rules apply (street, city regulation on road sections)?
- Which contextual elements from the broader constellation (like for instance weather, telecommunication systems) have influence on particular driving actions? What influence?

These research questions will be answered through the direct experiencing of the source system (participant observation) and by the extraction of knowledge and perceptions from users, experts, and stake-holders by means of the COMPRAM methodology and Constellation Analysis. Constellation Analysis will provide a structured visualization of interdisciplinary points of view to the complex interplay of actors and contexts of action in different times and contexts. The method has been mainly designed as a tool for interdisciplinary collaboration at the Center for Technology and Society at the TU Berlin. Its main advantage is that traffic situations can be analyzed from different perspectives as respect to their diversity and heterogeneity. COMPRAM stands for the Complex Problem Handling Method, a framework specially developed to handle complex societal problems DeTombe [34]. Although the method is developed for the usage at the macro-perspective; we believe it will be particularly useful for the generation of dynamic hypotheses about the system behavior and for the construction of the conceptual system dynamics model.

The research design includes the following steps:

1. The generation of system knowledge (from awareness to complete description). A systematical and detailed knowledge about decisions and actions during cooperative driving in general and in various "between-vehicle cooperation" scenarios will be obtained from literature surveys, participant observation, secondary analyses of test beds data, past small-scale experiments of technological interventions/evaluations. In-depth interviews. The results will be examined by expert teams (involved in the COMPRAM approach).
2. The construction of the simulation model for the selected "between-vehicle cooperation" scenario: In this phase, the obtained system knowledge will be converted into a simulation model that takes into account both qualitative and quantitative aspects. The output of the model will allow the calibration according to the selected scenario, simulation, and prediction of effects. An array of cooperative traffic settings is considered for the modeling and simulation in the current project. They are selected (in cooperation with involved stake-holders) in such a way that makes possible the evidentiation of the eventual conflict potential and problems.

## 11.4 Implications for the Modeling of the User Acceptance

One of the most important non-technical factors for the prediction of the market penetration of such innovations is user acceptance [35]. This is defined as "the degree to which individual users will use a given system when usage is voluntary or discretionary" [35]. Among the many theoretical models of user acceptance, one of the most widely adopted in the technology literature is the Unified Theory of Acceptance and Use of Technology (UTAUT), developed by Venkatesh, Morris, David and David 2003 [36].

The challenge for the system dynamics modeling is to integrate the microscopic model of the cooperative driving and co-operative traffic findings into another model explaining and predicting user acceptance of the intelligent transportation sys-tem. The link between the user acceptance and the hybridization of actions could represent an interesting subject for system dynamics modeling. Recent studies about autonomy and control in hybrid systems have indicated that the consequence of the hybridization of actions in hybrid systems (system in which intelligent agents interact with human and social systems) represents for humans a successive transfer from their strategic to the adaptative agency [25]. While the capacity for strategic agency implies instrumental rationality, the possibility to anticipate the consequences of actions, regularity and predictability, the adaptative agency of humans translates into rapid reactions to situation-adjusted solutions with emergent character generated by the IT systems, which cannot be previously predicted in all the details [25]. Although Weyer considers aviation the most adequate prototype of a new work world which is more and more shaped by "autonomous" technology, the cooperative traffic system may face similar promises and problems [25-26]. According to Weyer, it is important to identify the conflict potential emerging from the relation between the transparence degree of technological agency and the driver adaptability and autonomy. Recent sociological studies of technology emphasize the fact that the distribution of activities among humans and machines in hybrid socio-technical constellations of driving should be gradually balanced in system design, leaving room for human self-initiative; own responsibility; control of personal data; intervention capacity; and the human decision about the real usefulness of applications [24].

## 11.5 Conclusion

At the microscopic level, the dynamic hypotheses about "intelligent" driving actions can be formulated by including driving sequences in loops of information feedback and circular causality. Based on the identification of levels for variables describing the actions and their inflows and outflows, the final intention is to build a diagram capturing the stock-and-flow/causal feedback structure. This diagram will be finalized in the research in progress. In the article we concentrate on the potential of the theory of the distributiveness of actions in heterogeneous constellations (sociology of technology [24] [30-31]) to enhance the formulation of dynamic hypotheses for system dynamics modeling of ITS.

At the macroscopic level we highlight the importance of closing the system boundaries in order to achieve endogeneous explanations. The performance of the Intelligent Transportation Systems is dependent not only on the interactivity between the vehicle-human systems and the environment, but also by the impact of socio-political and other contextual factors from a broader socio-technical constellation. Valide and complete information about the system structure can be obtained only by extending the perspective to the level of the socio-technical constellation of the system [24], or in any case, by specifying the limits of the interpretation.

As the present paper shows, the design of intelligent technologies relies more and more on the comprehension of broader (non-technical) mechanisms. At the same time, there is the possibility to use simulation models to develop sociological models of societies and organizations and to reflect about the emergence of non-intended effects in the mobile information and communication society.

**Acknowledgements.** The author thanks Prof. Kyandoghere Kyamakya for the captivating discussions, the kind contributions to the clarification of the research theme and methodology, and for my opportunity to get acquainted with the system dynamics perspective

## References

1. Richardson, G.P.: System Dynamics. In: Gass, S., Harris, C. (eds.) Encyclopedia of Operations Research and Information Science. Kluwer Academic Publishers (1999/2011)
2. Kyamakya, K., Bouchachi, A., Chedjou, J.C.: Recent Advances of Computational Intelligence. In: Nonlinear Dynamics and Synchronization, p. 250. Atlantis Press, Paris (November 30, 2009); (Atlantis Computational Intelligence Systems)

3. Abbas, K.A., Bell, M.G.H.: System dynamics applicability to transportation modeling. Transportation Research A 28A(5), 373–400 (1994)
4. Raux, C.: A system dynamics model for the urban travel system. Paper Presented at the European Transport Conference, Strasbourg, October 8-10 (2003)
5. Xue, W., Hudson, C.: System Dynamics Traffic Flow Simulation (2004), http://www.systemdynamics.org/conferences/2004/SDS_2004/PAPERS/161HUDSO.pdf
6. EEA, European Environmental Agency, TREMOVE: an EU-wide transport model (2011), http://ec.europa.eu/environment/air/pollutants/models/tremove.html
7. Punzo, V., Ciuffo, B.: How Parameters of Microscopic Traffic Flow Models Relate to Traffic Dynamics in Simulation. Transportation Research Record 2124(-1), 249–256 (2009), http://trb.metapress.com/openurl.asp?genre=article&id=doi:10.3141/2124-25
8. Farhi, N., Goursat, M., Quadrat, J.P.: About Dynamical Systems Appearing in the Microscopic Traffic Modeling. Arxiv preprint arXiv09114672: 38 (2009), http://arxiv.org/abs/0911.4672
9. Mehmood, A., Saccomanno, F., Hellinga, B.: Application of System Dynamics in Car-Following Models. Journal of Transportation Engineering 129(6), 625 (2003), http://link.aip.org/link/JTPEDI/v129/i6/p625/s1&Agg=doi
10. Forrester, J.: Urban Dynamics. M.I.T. Press, Cambridge (1969)
11. Wang, F., Zeng, D., Yang, L.: Smart Cars on Smart Roads: An IEEE Intelligent Transportation Systems Society Updat. IEEE Pervasive Computing, 68–69 (October-December 2006)
12. Rudigier, M., Horn, M.: Concepts for Modeling Drivers of Vehicles Using Control Theory. In: Düh, J., Hufnagl, H., Juritsch, E., Pfliegl, R., Schimany, H.-K., Schönegger, H., et al. (eds.) Data and Mobility. AISC, vol. 81, pp. 27–38. Springer, Heidelberg (2010)
13. Möbus, C., Eilers, M., Garbe, H., Zilinski, M.: Probabilistic and Empirical Grounded Modeling of Agents in (Partial) Cooperative Traffic Scenarios. In: Duffy, V.G. (ed.) Digital Human Modeling, HCII 2009. LNCS, vol. 5620, pp. 423–432. Springer, Heidelberg (2009)
14. Sato, T., Akamatsu, M.: Modeling and prediction of driver preparations for making a right turn based on vehicle velocity and traffic conditions while approaching an intersection. Transportation Research Part F: Traffic Psychology and Behaviour 11(4), 242–258 (2008)
15. Mariani, M.: COMUNICAR: Designing multimodal interaction for advanced in-vehicle interfaces. In: de Waard, D., Brookhuis, K.A., Moraal, J., Toffetti, A. (eds.) Human Factors in Transportation, Communication, Health, and the Workplace Maastricht, pp. 113–120. Shaker, The Netherlands (2002)
16. Erzin, E., Yemez, Y., Tekalp, A.M., Ercil, A., Erdogan, H., Abut, H.: Multimodal Person Recognition for Human-Vehicle Interaction. IEEE MM 13(2), 18–31 (2006), http://ieeexplore.ieee.org/lpdocs/epic03/wrapper.htm?arnumber=1621030
17. Friedewald, M.: Ubiquitous Computing: Ein neues Konzept der Mensch-Computer-Interaktion und seine Folgen. In: Hellige, H.D. (ed.) Das Mensch-Computer-Interface: Geschichte, Gegenwart und Zukunft, pp. 259–280. Transcript Verlag, Bielefeld (2008)

18. Liu, H., Li, W., Li, J.: Harmonious Human-Computer Interaction in Pervasive Environment. 2008 Third International Conference on Pervasive Computing and Applications, pp. 776–780. IEEE (2008), http://ieeexplore.ieee.org/xpl/freeabs_all.jsp?arnumber=4783714
19. Grifoni, P.: Multimodal Human Computer Interaction and Pervasive Services. Engineering (2009), http://services.igi-global.com/resolvedoi/resolve.aspx?doi=10.4018/978-1-60566-386-9
20. Jaimes, A., Sebe, N.: Multimodal human computer interaction: A survey. Computer Vision and Image Understanding 108(1-2), 116-134 (2007), http://linkinghub.elsevier.com/retrieve/pii/S1077314206002335
21. Fischer, P., Nurnberger, A.: Adaptive and multimodal interaction in the vehicle. In: Systems Man and Cybernetics, SMC 2008, pp. 1512–1516 (2008), http://ieeexplore.ieee.org/stamp/stamp.jsp?arnumber=04811500
22. Vilimek, R., Hempel, T., Otto, B.: Multimodal Interfaces for In-Vehicle Applications. In: Jacko, J.A. (ed.) HCI 2007. LNCS, vol. 4552, pp. 216–224. Springer, Heidelberg (2007)
23. Mller, C., Friedland, C.: Multimodal interfaces for automotive applications (MIAA). In: IUI 2009, Proceedings of the 13th International Conference on Intelligent User Interfaces, vol. 493 (2008), http://portal.acm.org/citation.cfm?doid=1502650.1502732
24. Rammert, W.: Technik Handeln Wissen: zu einer pragmatischer Technik- und Sozialtheorie. VS Verlag Sozialwissenschaften, Wiesbaden (2007)
25. Weyer, J.: Die Kooperation menschlicher Akteure und nicht-menschlicher Agenten. Ansatzpunkte einer Soziologie hybrider Systeme. In: Berger, W., Getzinger, G. (eds.) Das Ttigsein der Dinge. Beitrge zur Handlungstrgerschaft von Technik, pp. 61–92. Profil, Mnchen und Wien (2009)
26. Weyer, J.: Autonomie und Kontrolle. Arbeit in hybriden Systemen am Beispiel der Luftfahrt. Technikfolgenabschtzung - Theorie und Praxis 16(2), 35–42 (2007)
27. Kyamakya, K.: DOM - Der Orientierte Mensch: Einige Anstze zu Ortung und Navigation. Shaker Verlag (2003)
28. Rammert, W., Schulz-Schaeffer, I. (eds.): Knnen Maschinen handeln? Soziologische Beitrge zum Verhltnis von Mensch und Technik. Campus, Frankfurt (2002)
29. Ball, R., Dulay, N.: Enhancing Traffic Intersection Control with Intelligent Objects (2010), http://www.webofthings.org/urban-iot/2010/pdfs/Ball.pdf
30. Rammert, W.: Where the action is: Distributed agency between humans, machines, and programs, The Technical University Technology Studies Working Papers, TUTS-WP-4-2008 (2008), http://www.ts.tu-berlin.de/fileadmin/fg226/TUTS/TUTS_WP_4_2008.pdf (February 2, May 23, 2011)
31. Schulz-Schaeffer, I.: Die Frage nach der Handlungstrgerschaft von Technik und das Konzept gradualisierten Handelns. In: Berger, W., Getzinger, G. (eds.) Das Ttigsein der Dinge. Beitrge zur Handlungstrgerschaft von Technik, pp. 37–59. Profil Verlag, Mnchen (2009)
32. *** Kooperative Mobilitt in der Stadt, http://www.polis-online.org/fileadmin/hot_topic/CVIS/POL_CVIS_Handbook_DE_03_WEB.pdf

33. Laitinen, J., Hakala, H.: Cooperative Traffic ICT, Strategic Research Agenda (2008), `http://its-finland.fi/SRACooperativeTrafficICTv11.pdf`
34. DeTombe, D. J. Compram, a Method for Handling Complex Societal Problems. DeTombe, D.J. (Guest Editor) Feature Issue: Complex Societal Problems, European Journal of Operation Research; Slowinski, D.J.,Teghem, R., Wallenius, J. (Eds). vol. 128(2), pp. 266–282. Elsevier, North-Holland (January 16, 2001)
35. Bankosegger, D.: COOPERS: Driver Acceptance Assessment of Cooperative Services. Results from the Field Test in Austria. In: Düh, J., Hufnagl, H., Juritsch, E., Pfliegl, R., Schimany, H.-K., Schönegger, H., et al. (eds.) Data and Mobility. AISC, vol. 81, pp. 39–47. Springer, Heidelberg (2010)
36. Venkatesh, V., Morris, M.G., Davis, G.B., Davis, F.D.: User acceptance of information technology: toward a unified view. MIS Quarterly 27(3), 425–478 (2003)

# Part III
# Electromagnetics Theory, Modeling and Simulation of Real Physical Electromagnetic Prototypes

# Chapter 12
# Electromagnetics, Systems Theory, Fluid Dynamics, and Some Fundamentals in Physics

Alfred Fettweis

**Abstract.** Based on strict validity of Maxwell's equations in vacuum, concepts of *field velocity* **v**, *rest field*, *basal* electromagnetic (EM) field etc. are defined. Their properties lead to *flow equations* that have the same structure as those of fluid dynamics and thus in fact describe an *EM fluid* whose flow velocity is equal to v. For a *photon* a model is presented whose inner structure consists of such an EM fluid and exhibits all known photon properties. The corresponding model of an *electron* is sufficiently complete to analyse its dynamic properties, which turn out to be as required by classical relativity. The (time-independent) *Schrödinger equation* is obtained within the frame of the present theory.

## 12.1 Introduction

This paper goes well beyond [1], on which it is crucially based. Essential results already published are recapitulated if needed, but for most major new results outlines of proofs are sketched. As in [1], strict validity of *Maxwell's equations* in vacuum, even down to the smallest dimensions, is assumed, including the relativistic transformation rules of these equations.

The concepts of *field velocity* and *rest field* of an electromagnetic (EM) field are introduced and precisely defined. The known equations involving field momentum and stress tensor as well as the equation relating Poynting vector and energy are then found to be equivalent to new equations, say *flow equations*, that are of the same type as the corresponding equations of fluid dynamics and thus lend themselves to a consistent mechanistic interpretation. We speak about this as an *observation* at the *primary* or *basic level*. At this basic level, the EM field thus behaves like an *electromagnetic fluid (EM fluid)* that is moving under the influence of the surface

Alfred Fettweis
Ruhr-Universität Bochum, Germany
e-mail: fettweis@nt.rub.de

K. Kyamakya et al. (Eds.): Selected Topics in Nonlinear Dynamics, SCI 459, pp. 209–234.
DOI: 10.1007/ 978-3-642-34560-9_12 

and volume forces acting in it. In particular, changes of energy densities are caused by to two entirely distinct effects: *convection* and the *work done* by these forces.

On the other hand, one can combine these two mechanisms into a single effect and characterize the overall energy migration simply by means of an effective *energy velocity*. We refer to this as an *observation* at the *secondary level*. This energy velocity reaches twice the field velocity at the low end and becomes equal to it at the speed of light. Whether at the primary or the secondary level, the flow equations clearly show that an EM field inherently has inertia and thus mass.

At the secondary level, thus when using the energy velocity, the original flow equations admit a form that exhibits relevant properties, at least in as far as the expressions for mass and energy are concerned, that are in agreement with classical *relativistic dynamics* [2-4]. At the primary level, however, the results are only compatible with those of an alternative relativistic dynamics that was first mentioned in [5] and had then gradually been refined in a sequence of papers, the last one of which being [7] (although some of the claims mentioned in these papers cannot be upheld in view of newer results such as those discussed in [1] and the present paper). Even at the secondary level, this alternative theory still plays a decisive role and may therefore not simply be replaced by the classical theory.

The fields are assumed throughout to be *autonomous* (self-sustaining), and frequently, more specifically, to be *basal*. For basal fields the flow equations become particularly elegant. Elementary particles, at least EM particles such as electrons, positrons, and photons, are found to be nowhere point objects but condensed fields. They have an inner structure whose fine details can be observed at the two levels already mentioned. At the *tertiary level* of observation these details are ignored and only the movement of a particle as a whole is considered; it is found to follow the laws of classical relativistic dynamics.

For a *photon* model the resulting equations can be solved analytically. This leads to a long list of features that comprises all photon properties known to this author. Puzzling properties such as the *wave-particle duality* and the identity of *Planck's constant* $\hbar$ , defined as the proportionality constant between energy and frequency of a photon, with the photon spin and the (double value of the) electron spin are fully explained. For an *electron* (positron) important steps have been achieved towards finding a complete representation. Due to the results already available, a variety of properties have been verified.

The results available for an electron have made it possible to examine its *behaviour in* an *external field*. This way, the dynamic behaviour of an electron is found to follow exactly the laws of classical relativity. Hence, *relativistic dynamics* and thus, in the limit, *Newtonian dynamics* are proved to be direct consequences of Maxwell's equations. In particular, inertia and thus inertial as well as gravitational mass have turned out to be inherent properties of an EM field, not the result of some extraneous influence. At least for photons and electrons (positrons), the existence of an Higgs particle as source for mass has no relevance.

The last section concerns the application to quantum physics. In particular, the celebrated (time-independent) *Schrödinger equation* is derived within the frame of the present theory. The *Schrödinger function* is found to stand for any of the field components of an *EM co-field* associated with the electron EM field spread out inside of the atom in which it is incorporated. The precise form of the co-field depends on the contingencies of the process by which the electron has been incorporated into the atom. Consequently, the actual randomness is *not* due to *intrinsic* (inherent to the nature of the objects), but to *extrinsic probability*, i.e. probability caused by complexity, like in thermodynamics.

## 12.2 Electromagentic Field in Vacuum: Maxwell's Equations and Related Results

Let RF be the *reference frame* under consideration. A *point* P in RF is characterized by its *position* coordinates $x, y, z$, or compactly by

$$\mathbf{r} = (x, y, z)^T,$$

and by its *time* coordinate $t$, altogether thus by its four *coordinates* $x, y, z, t$. In this text we exclusively consider *electromagnetic (EM) fields* in vacuum. Such fields can be described in RF by the *field variables*,

$$\mathbf{E} = (E_x, E_y, \quad E_z)^T, \quad \mathbf{H} = (H_x, H_y, H_z)^T, \quad \mathbf{i} = (i_x, i_y, o_z)^T \quad and \quad q,$$

and by the two constants $\varepsilon_0$ and $\mu_0$. While **E**, **H**, H, $\varepsilon_0$, and $\mu_0$ represent quantities in standard notation (although we systematically make use of **H** instead of the frequently preferred **B**), we are designating the *charge density* by $q$, not by $\rho$ as is commonly done. This allows us to represent consistently the density, whether per unit area or per unit volume, of any relevant quantity by a meaningful small letter, and the corresponding full quantity for, say, a particle by the respective capital letter. We also make systematic use of the transposition operator $T$ when handling vectors and matrices; this has several advantages in our context. We do of course assume all coordinate systems to be right-handed.

The EM field itself is described by what is commonly called *Maxwell's equations*, i.e., by

$$\varepsilon_0 \frac{\partial \mathbf{E}}{\partial t} + \mathbf{i} = \nabla \times \mathbf{H}, \tag{12.1a}$$

$$\mu_0 \frac{\partial \mathbf{H}}{\partial t} + \mathbf{i} = -\nabla \times \mathbf{E}, \tag{12.1b}$$

$$\mathbf{q} = \varepsilon_0 \nabla^T \mathbf{E}, \tag{12.2a}$$

$$\nabla^T \mathbf{H} = 0, \tag{12.2b}$$

where

$$\nabla = \left( \frac{\partial}{\partial x}, \frac{\partial}{\partial y}, \frac{\partial}{\partial z} \right)^2.$$

From the point of view adopted throughout in this paper, we assume the field to be *autonomous (self-sustaining)*. In other words, we assume $q$ and $\mathbf{i}$ to be not given sources, but properties of the field that can be determined from $\mathbf{E}$ and $\mathbf{H}$ by (12.1a) and (12.2a). While (12.1) and (12.2) are standard in the literature, it is apparently usually overlooked that these equations admit an equivalent *state-space form*, which is more advantageous to use in many situations. For this form, the set (12.1) and (12.2) is to be replaced by

$$\varepsilon_0 \frac{\partial \mathbf{E}}{\partial t} + \mathbf{i} = \nabla \times \mathbf{H}, \quad \mu_0 \frac{\partial \mathbf{H}}{\partial t} = -\nabla \times \mathbf{E}, \quad \frac{\partial q}{\partial t} = -\nabla^T \mathbf{i} \tag{12.3}$$

with both

$$q = \varepsilon_0 \nabla^T \mathbf{E} \quad and \quad \nabla^T \mathbf{H} = 0 \quad for \quad t = t_0, \tag{12.4}$$

where $t_0$ is some fixed time, for instance the *initial time*.

From Maxwell's equations, thus from (12.1) and (12.2) or from (12.3) and (12.4), two further useful equations can be derived, which, in our notation, can be written as follows:

$$\frac{\partial \mathbf{j}}{\partial t} + (\nabla^T \mathbf{T}_c)^T + \mathbf{f}_c = \mathbf{0}, \tag{12.5a}$$

$$\frac{\partial \omega}{\partial t} + \nabla^T \mathbf{S} + \mathbf{i}^{\mathbf{T}} \mathbf{E} = 0, \tag{12.5b}$$

To these we add the original equations defining $q$ and $i$, thus

$$q = \varepsilon_0 \nabla^T \mathbf{E}, \tag{12.6a}$$

$$\mathbf{i} = \nabla \times \mathbf{H} - \varepsilon_0 \frac{\partial \mathbf{E}}{\partial t}, , \tag{12.6b}$$

In (12.5),

$$\mathbf{j} = \frac{1}{c^2} \mathbf{S} = \frac{1}{c^2} \mathbf{E} \times \mathbf{H} \tag{12.7}$$

is the classically known *momentum density* of the field,

$$\mathbf{f}_c = q\mathbf{E} + \mu_0 \mathbf{i} \times \mathbf{H} \tag{12.8}$$

the *classical Lorentz force density*, and $T_c$, i.e.,

$$\mathbf{T}_c = w\mathbf{1} - \varepsilon_0 \mathbf{E}\mathbf{E} - \mu_0 \mathbf{H}\mathbf{H}, \tag{12.9}$$

a classical stress tensor equal to (the negative of) what is known as *Maxwell's stress tensor*. Furthermore, the *field energy density* $\omega$ in RF and the speed of light, $c$, are defined by

$$w = \frac{1}{2}(\varepsilon_0 E^2 + \mu_0 H^2), \quad E^2 = \mathbf{E}^T \mathbf{E}, \quad H^2 = \mathbf{H}^T \mathbf{H}, \quad c = \frac{1}{\sqrt{\varepsilon_0 \mu_0}} \tag{12.10}$$

and *the Poynting vector* **S** by

$$\mathbf{S} = (S_x, S_y, S_z)^T = \mathbf{E} \times \mathbf{H}. \tag{12.11}$$

While (12.5b) is commonly encountered, (12.5a) as well as (12.9) are less well known. Note that the systematic use of the transposition operator has made it possible to present these equations in a very compact form.

The definitions we have adopted for $E$ and $H$ (cf. (12.10)) do not specify the signs of these quantities. For fully fixing them we could assume $E = |\mathbf{E}|$ and $H = |\mathbf{H}|$, thus $E \geq 0$ and $H \geq 0$ , but in order to preserve our freedom for later purposes we will not impose that unnecessary restriction. A corresponding remark holds for most other vector quantities throughout this paper. This unconventional way of proceeding allows us to express many of the results to be derived in a simpler way than otherwise feasible.

Based on (12.5), it is standard to interpret **j** as a momentum density, $\mathbf{T}_c$ as a measure for surface forces (forces acting per unit surface), and $\mathbf{f}_c$ as a volume force density (force acting per unit volume). A fully satisfactory justification for this standard interpretation of $\mathbf{T}_{c'}$, however, cannot be found in the literature and, as will become clear, is not feasible. The equations (12.5) should indeed be critically compared to the corresponding conservation equations of fluid dynamics, as will be done hereafter.

## 12.3 Fluid Dynamics

The fluid dynamics equations to which (12.5) should be compared are the conservation equations (essentially the Navier-Stokes equations)

$$\begin{aligned} &\frac{\partial \mathbf{j}}{\partial t} + (\nabla^T(\mathbf{v}\mathbf{j}^T))^T + (\nabla^T\mathbf{T})^T + \mathbf{f}_g = \mathbf{0}, \\ &\frac{\partial w}{\partial t} + \nabla^T(w\mathbf{v}) + \nabla^T(\mathbf{T}\mathbf{v}) + \mathbf{v}^{\mathbf{T}}\mathbf{f}_g = 0 \end{aligned} \tag{12.12}$$

which concern the rate of change of the momentum density **j**=m**v** and the energy density $w$ of the fluid. In the first expression (12.12), the term $(\nabla^T(\mathbf{v}\mathbf{j}^T))^T$ is due to the fact that the momentum density **j**, which is proportional to the velocity **v**, is in turn subject to *convection* and is thus travelling itself with velocity **v**. Furthermore, the volume and the surface force densities are represented, respectively, by the vector $\mathbf{f}_g$ (usually due to gravitation) and the matrix (tensor) **T**, which is determined by pressure and the viscosity forces. The corresponding terms in the second expression (12.12) describe the work done by these forces, while $\nabla^T(w\mathbf{v})$ is due to the convection of the energy density.

A proper analogy between (12.5) and (12.12), clearly, is not possible. On the one hand, the term $(\nabla^T(\mathbf{v}\mathbf{j}^T))^T$ is missing in (12.5a). In order to justify the interpretation of **j** one cannot therefore, for instance, simply integrate (12.5a) over an arbitrary volume $V$ of finite extent and delimited by a surface $F$ and then apply

Gauss' theorem. It is therefore customary to argue by considering the limit when $F$ goes to infinity and to assume that the field on $F$ vanishes sufficiently fast so that the total flux of **j** across $F$ goes to zero. In (12.5b), on the other hand, the term in **S** assumes a role similar to that of the convection term (second term) in (12.12), but while the work done by the volume forces is taken into account by the third term, there is no corresponding term involving the surface forces, or else, (12.5b) suggests that the electromagnetic surface forces cannot do work.

## 12.4 Field Velocity, Rest Field, and Energy Velocity

The dilemmas just explained are obviously due to the absence in (12.1) to (12.4) of a velocity like in (12.12). It is crucial therefore that we associate with an EM field a further local property that we call the *field velocity*. We represent it by the symbol **v**, its associated *normalized* velocity by **fi**, and we have,

$$\mathbf{v} = (v_x, v_y, v_y)^T, \quad \beta = (\beta_x, \beta_y, \beta_z)^T = \frac{\mathbf{v}}{c}, \tag{12.13}$$

where $c$ is the speed of light, as always in this paper. Let us briefly recall how these new concepts are introduced.

Since the Poynting vector obviously is intimately related to a flow phenomenon, it appears logical to require **v** to be parallel to **S** and to have $\mathbf{v} = \mathbf{0}$ for $\mathbf{S} = \mathbf{0}$. If at a point P in the given reference frame RF we actually have $\mathbf{v} = \mathbf{0}$ , the field will be said to be there *at rest*. For equally obvious reasons we also impose the following requirement: Let RF' be a reference frame moving with constant velocity $\mathbf{v_0}$ with respect to RF, P' a point in RF', and P the corresponding point in RF. If the field is at rest at P' we require its velocity **v** at P to be equal to the one RF' has with respect to RF, thus to have $\mathbf{v} = \mathbf{v_0}$ . In common terms, this amounts, for instance, to stating that a passenger at rest in an airplane has with respect to ground the same velocity as the airplane itself.

Applying these principles and using concepts of relativity theory, in particular the relativistic transformation rules for electromagnetic fields, one finds [1],

$$\frac{\beta}{1+\beta^2} = \frac{\mathbf{S}}{2cw}, \quad \beta^2 = \beta^2\beta = |\beta|^2 = \frac{v^2}{c^2}, \quad v^2 = \mathbf{v}^T\mathbf{v}. \tag{12.14}$$

According to these conditions, **v** is nowhere necessarily positive (cf. the above discussion concerning $E$ and $H$). There always exist two solutions for **v**, one with $|\beta| \geq 1$ and one with $|\beta| \leq 1$ , but only the latter is physically acceptable. We may thus always assume

$$0 \leq |\beta| \leq 1, \quad 0 \leq |v| \leq c.$$

The limit $\beta = \pm 1$, thus $v = \pm c$ is reached if and only if the field is locally planar, thus such that **E** and **H** are orthogonal and carry both the same energy density. On the other hand, $v = 0$ everywhere if the field is either electrostatic or magnetostatic. We clearly have (cf. (12.11) and (12.14)),

$$\mathbf{v}^T\mathbf{E} = \mathbf{v}^T\mathbf{H} = 0, \quad \beta^T\mathbf{E} = \beta^T\mathbf{H} = 0. \tag{12.15}$$

With the field at P is associated a *rest field* characterized by the rest field quantities $\mathbf{E_0}$, $\mathbf{H_0}$, $\mathbf{i_0}$, and $q_0$, with $\mathbf{S}_0 = \mathbf{E}_0 \times \mathbf{H}_0 = \mathbf{0}$, thus such that

$$\mathbf{E}_0 = \gamma E_0, \quad \mathbf{H}_0 = \gamma H_0, \quad \gamma^T\gamma = 1, \quad \gamma = (\gamma_x, \gamma_y, \gamma_z)^T,$$

$\gamma$ being a unit vector. They are related to the original quantities by

$$\mathbf{E}_0 = \frac{1}{\alpha}(\mathbf{E} + \mu_0\mathbf{v} \times \mathbf{H}, \quad \mathbf{H}_0 = \frac{1}{\alpha}(\mathbf{H} - \varepsilon_0\mathbf{v} \times \mathbf{E}), \tag{12.16a}$$

$$\mathbf{E} = \frac{1}{\alpha}(\mathbf{E}_0 - \mu_0\mathbf{v} \times \mathbf{H}_0, \quad \mathbf{H} = \frac{1}{\alpha}(\mathbf{H}_0 + \varepsilon_0\mathbf{v} \times \mathbf{E}_0), \tag{12.16b}$$

$$\mathbf{i}_0 = \mathbf{i} - \frac{1}{\alpha}\left(q - \frac{\mathbf{v}^T\mathbf{i}}{(1+\alpha)c^2}\right)\mathbf{v} = \mathbf{i} - \frac{1}{\alpha}\left(cq - \frac{\beta^T\mathbf{i}}{1+\alpha}\right)\beta, \tag{12.17a}$$

$$q_0 = \frac{1}{\alpha}\left(q - \frac{1}{c^2}\mathbf{v}^T\mathbf{i}\right) = \frac{1}{\alpha}\left(q - \frac{1}{c}\beta^T\mathbf{i}\right), \tag{12.17b}$$

where $\alpha = \sqrt{1-\beta^2}$, and we obtain

$$\mathbf{v}^T\mathbf{E}_0 = 0, \quad \mathbf{v}^T\mathbf{H}_0 = 0. \tag{12.18}$$

$$\varepsilon_0 E^2 - \mu_0 H^2 = \varepsilon_0 E_0^2 - \mu_0 H_0^2. \quad \mathbf{E}^T\mathbf{H} = \mathbf{E}_0^T\mathbf{H}_0 = E_0 H_0. \tag{12.19}$$

For the resulting *rest energy density* $w_0$ we have,

$$\alpha^2 w = (1-\beta^2)w = (1+\beta^2)w_0, \quad w_0 = \frac{1}{2}\left(\varepsilon_0 E_0^2 + \mu_0 H_0^2\right), \tag{12.20}$$

and from (12.19),

$$\begin{aligned} w_0^2 &= \frac{1}{4}(\varepsilon_0 E_0^2 - \mu_0 H_0^2) + \frac{1}{c^2}(E_0 H_0)^2 \\ &= \frac{1}{4}(\varepsilon_0 E^2 - \mu_0 H^2) + \frac{1}{c^2}(\mathbf{E}^T\mathbf{H})^2 = w^2 - \frac{1}{c^2}|\mathbf{S}|^2, \end{aligned} \tag{12.21}$$

It is thus Lorentz invariant, and the same can be shown to hold for $E_0$ and $H_0$ provided the signs of these two scalars are properly chosen within the freedom available as discussed in the paragraph following (12.11). Furthermore, $0 \leq w_0 \leq w$, with $w_0 = 0$ if an only if $v = \pm c$, which obviously is reminiscent of a basic photon property.

The difference $w_k$ defined by

$$w_k = w - w_0 = \frac{2\beta^2}{1+\beta^2}w \tag{12.22}$$

is the *kinetic energy density* of the field, the second expression following from the first by making use of (12.20).

In view of the interpretation usually given to $\mathbf{S}$ and $w$, a (classical) *energy velocity* $\mathbf{v}_c$ may be defined by

$$\mathbf{S} = \mathbf{v}_c w, \quad \textit{thus with} \quad \mathbf{v}_c = \frac{2}{1+\beta^2}\mathbf{v}, \quad |\mathbf{v}_c| \geq |\mathbf{v}|, \tag{12.23}$$

equality holding only for $\mathbf{v} = 0$ and $|\mathbf{v}| = c$. For small values of $|\beta|$ we have $\mathbf{v}_c = 2\mathbf{v}$. The ratio $|\mathbf{v}_c|/|\mathbf{v}|$ decreases monotonically for increasing values of $|\mathbf{v}|$. Defining

$$\beta_c = \frac{1}{c}\mathbf{v}_c, \quad \beta_c^T\beta_c = \beta_c^2, \quad \alpha_c^2 = 1 - \beta_c^2,$$

we derive

$$\beta_c = \frac{2}{1+\beta^2}\beta, \quad \alpha_c = \frac{\alpha^2}{1+\beta^2} = \frac{1-\beta^2}{1+\beta^2}, \quad \frac{2\mathbf{v}}{\alpha^2} = \frac{\mathbf{v}_c}{\alpha_c}, \quad w = \frac{w_0}{\alpha_c}. \tag{12.24}$$

## 12.5 The Flow Equations

### 12.5.1 General Form of the Flow Equations

Making use of the relevant results in Sections 2 and 4 we can split $T_c$ according to

$$\mathbf{T}_c = \mathbf{T}_0 + \mathbf{v}\mathbf{j}^T,$$

where

$$\mathbf{T}_0 = \mathbf{T}_0^T := w_0\mathbf{U}, \quad \mathbf{U} = \mathbf{U}^T = (1 - 2\gamma\gamma^T). \tag{12.25}$$

This way, we can rewrite (12.5), $\mathbf{f}_c$ being still given by (12.8), in the form

$$\begin{aligned} &\frac{\partial \mathbf{j}}{\partial t} + (\nabla^T(\mathbf{v}\mathbf{j}^T))^T + (\nabla^T\mathbf{T}_0)^T + \mathbf{f}_0 = \mathbf{0}, \\ &\frac{\partial w}{\partial t} + \nabla^T(w\mathbf{v}) + \nabla^T(\mathbf{T}_0\mathbf{v}) + \mathbf{i}^\mathrm{T}\mathbf{E} = 0. \end{aligned} \tag{12.26}$$

To these we may add

$$\frac{\partial q}{\partial t} + \nabla^T\mathbf{i} = 0. \tag{12.27}$$

We call (12.26) and (12.27) the *flow equations* of the EM field.

The *stress tensor* $\mathbf{T}_0$ in (12.26) clearly differs from Maxwell's stress tensor. As follows by comparing (12.25) with (12.9), $\mathbf{T}_0$ is equal to the expression one obtains by replacing everywhere in $\mathbf{T}_c$ the actual field by the rest field. It comprises only one dyadic product instead of two. The matrix $\mathbf{U}$ in (12.25) is orthogonal and is equal to what is known in numerical mathematics as a *Householder matrix* [8]. Clearly,

$\mathbf{T}_0$ may be claimed to be simpler than $\mathbf{T}_{c'}$ , but the latter reduces to the former for $\mathbf{v} = \mathbf{0}$.

Clearly, the equations in (12.26), which are strictly equivalent to the original ones in (12.5), have the same general structure as the fluid dynamics equations (12.12). Thus, the terms describing convection of the momentum density $\mathbf{j}$, the velocity $\mathbf{v}$, and the energy density $w$ are present precisely in the way needed. The same is true for the stress tensor $\mathbf{T}_0$ that describes the surface forces and the work done by these forces. Hence, we may interpret (12.26) as describing an *electromagnetic fluid*, or short, an *EM fluid*. The velocity of that fluid can be identified with the flow velocity $\mathbf{v}$.

Accordingly, a mass density for obtaining $\mathbf{j}$ should be introduced; we call it the *internal mass density*, $m_i$. In view of (12.25) we may write

$$\mathbf{j} = m_i \mathbf{v}, \quad m_i = \frac{2w_0}{c^2 \alpha^2}, \quad m_{i0} = m_i\big|_{\mathbf{v}=\mathbf{0}} = \frac{2w_0}{c^2}, \quad w_0 = \frac{1}{2} m_{i0} c^2. \tag{12.28}$$

This reveals a surprising result: The relationships between $m_i$ and $m_{i0}$ as well as those between $m_{i0}$ and $w_0$ apparently violate classical relativity but are in perfect agreement with the alternative theory presented in [5-7]. Since classical relativity is known to be perfectly confirmed by many experiments, this obviously requires some closer examination. This will be done hereafter primarily in the context of an important subclass of autonomous fields that is of prime concern to us.

### 12.5.2 Flow Equations of a Basal Electromagnetic Field

The above equations simplify if the EM field is *basal*, i.e. if in the adopted reference frame RF we have everywhere $\mathbf{i}_0 = \mathbf{0}$ or, equivalently, $\mathbf{i} = q\mathbf{v}$. The existence of such an RF is a plausible assumption to make for an individual autonomous field. It amounts indeed to requiring that in RF there cannot exist a current without charges travelling with some non-vanishing velocity. In particular, (12.8) becomes

$$\mathbf{f}_c = q(\mathbf{E} + \mu_0 \mathbf{v} \times \mathbf{H}), \quad \mathbf{f}_0 := q_0 \mathbf{E}_0, \tag{12.29}$$

and we derive,

$$\mathbf{v}^T \mathbf{f}_c = \mathbf{v}^T \mathbf{f}_0 = 0, \quad \mathbf{i}^T \mathbf{E} = 0. \tag{12.30}$$

The flow equations now become

$$\frac{\partial \mathbf{j}}{\partial t} + (\nabla^T (\mathbf{v}\mathbf{j}^T))^T + (\nabla^T \mathbf{T}_0)^T + \mathbf{f}_0 = \mathbf{0}, \tag{12.31}$$

$$\frac{\partial w}{\partial t} + \nabla^T (w\mathbf{v}) + \nabla^T (\mathbf{T}_0 \mathbf{v}) = 0, \quad \frac{\partial q}{\partial t} + \nabla^T (q\mathbf{v}) = 0. \tag{12.32}$$

These results are remarkable, as will become more evident in the course of our further analysis. Note that the contributions by the surface and the volume forces in (12.31) and (12.32) depend only on the rest field. On the other hand, the volume forces do not contribute to the energy balance, contrary to the surface forces. This expresses that in a basal field energy can be transmitted only by convection (term $\nabla^T(w\mathbf{v})$) and by work done by the surface forces (term $\nabla^T(\mathbf{T}_0\mathbf{v})$). It is in perfect agreement with Maxwell's claim that actions at a distance are unphysical, or else, that any effect upon a remote location can only be exerted by means of propagation, say by a phenomenon in which properties are handed on, so to speak, from one elementary cell to the next via the surface separating them.

In terms of the state-space form of Maxwell's equations a basal field is described by the partial differential equations

$$\varepsilon_0 \frac{\partial \mathbf{E}}{\partial t} = -q\mathbf{v} + \nabla \times \mathbf{H}, \quad \mu_0 \frac{\partial \mathbf{H}}{\partial t} = -\nabla \times \mathbf{E}, \quad \frac{\partial q}{\partial t} = -\nabla^T(q\mathbf{v}), \tag{12.33}$$

where $\mathbf{v}$ has to be expressed in terms of $\mathbf{E}$ and $\mathbf{H}$ by means of (12.11), (12.13), and (12.14). They comprise seven individual equations and the same number of dependent variables, i.e., the six components of $\mathbf{E}$ and $\mathbf{H}$ and the scalar $q$. This underlines the adequacy of the concept of a basal field. The set (12.32) is in general non-linear, as is needed for achieving stable field configurations.

Let us now return to the relativistic conflict mentioned above. Clearly, all equations (12.29) to (12.33) only involve the field velocity $\mathbf{v}$. In what in mechanics is called a **deformable medium**, however, one must distinguish between various types of propagation velocities. In an elastic medium, for instance, there may occur shear, pressure, and surface waves etc., all with their own, distinct velocity. Familiar distinct phenomena in a fluid are its flow velocity and the speed of sound. More specifically, energy is migrating due not only to convection but also to work done by the forces. As an example, sound energy propagates at a remarkably high speed even if the fluid is essentially at rest. We must therefore expect something similar to occur in an EM fluid.

Accordingly, let us combine in the first equation (12.32) the effects due to convection and to work done. Using $\mathbf{T}_0\mathbf{v} = w_0\mathbf{v}$ we can write (cf. (12.22) and (12.24)),

$$w\mathbf{v} + \mathbf{T}_0\mathbf{v} = (w + w_0)\mathbf{v} = w\mathbf{v}_c.$$

Hence, the changes $w$ undergoes due to the two original effects can be combined into a single energy transport phenomenon that takes place with an equivalent *energy velocity* $\mathbf{v}_c$ equal to what we have called the classical or effective energy velocity. Although this result could have been obtained quite directly, the present way of deriving it is highly instructive. Altogether, (12.32) can now be replaced by

$$\frac{\partial w}{\partial t} + \nabla^T(w\mathbf{v}_c) = 0. \quad \frac{\partial q}{\partial t} + \nabla^T(q\mathbf{v}) = 0. \tag{12.34}$$

The interest in using $\mathbf{v}_c$ is further enhanced by considering a *mass density*, $m$, that we define by using $\mathbf{v}_c$ instead of $\mathbf{v}$. We then obtain from (12.24) and (12.28),

$$\mathbf{j} = \frac{1}{c^2}\mathbf{S} = m_i\mathbf{v} = m\mathbf{v}_c, \quad m = \frac{m_0}{\alpha_c}, \quad m_0 = \frac{1}{2}m_{i0} = \frac{w_0}{c^2}, \tag{12.35}$$

where **j** is equal to the momentum density used before. Clearly, the relationships between $m$ and $m_0$ as well as between $m_0$ and $w_0$ are precisely as to be expected from classical relativity theory, and we have $m_{i0} = 2m_0$, explains the appearance of the factor $\frac{1}{2}$ in some of the results of the alternative theory. Nevertheless, it would not be permitted to build the analogy with fluid dynamics by interpreting $\mathbf{v}_c$ as the velocity of the EM fluid. Indeed, as (12.31) to (12.33) clearly show, the convection of **j**, $w$, and $q$ as well as the work done by the surface forces definitely occur with velocity **v**, not $\mathbf{v}_c$. Obviously, if the field is autonomous but not basal, quite similar, although less complete conclusions can be drawn.

As a consequence of these results the full meaning of the concept of an EM fluid becomes visible. For fully understanding the behaviour we have to distinguish between three levels of observation. At the *first* or *basic level*, the relevant form of the equations has the same structure as those of a fluid in classical mechanics. Nevertheless, a proper relativistic interpretation is possible only by appealing to the alternative theory mentioned before [5-7]. At the *second level*, we do not exclusively observe the fluid flow as such but, more specifically, also the movement of its energy. We are then faced with a split situation: While the energy flow follows classical relativity, the convective flow of momentum and charge is feasible only within the alternative theory. Beyond this, the evidence of the EM fluid suggests that an EM particle, although extremely small, is in fact not point-like. Its inner structure may be pictured as an EM fluid that is likely to be rotating around an axis and is held together by its own inner forces. At the *third level of observation*, the particle is examined when it is moving as a whole, thus without paying attention to what is precisely taking place in its inside. In any case, since the particle cannot move separately from its energy, it is bound to behave according to the laws of classical relativity. This is confirmed by results to be presented later.

According to the discussion just given we may also distinguish between an *internal* and an *external behaviour of the EM fluid*, the basic level of observation corresponding to the internal behaviour, and the secondary level to the external behaviour. More specifically, we may consider the flow velocity **v** to be in fact the *internal velocity* of the EM fluid, and $\mathbf{v_c}$ to be its external velocity. This also justifies calling, as we have done, $m_i$ the internal mass, since its use is indeed closely associated with **v**.

### 12.5.3 Field Rotating around an Axis

We consider an EM fluid that is rotating around an axis. For this we adopt standard spherical coordinates $r, \theta, \phi$ and assume the axis of rotation to be coincident with the $\theta-$axis. The only non-zero components of **v**, $\mathbf{v}_{c'}$, and **j** are those in the $\phi-$direction. We designate them by $v$, $v_c$, and $j$, respectively, and we then obtain from (12.28) and (12.35)

$$j = mv_c = m_i v. \tag{12.36}$$

Comparing this with (12.22) we can write,

$$w_k = \omega l = vj = mvv_c, \quad v = \omega R, \quad j = mv_c,$$
$$l = jR, \quad k = \frac{2\pi}{\lambda} = \frac{\omega}{v} = \frac{1}{R} \tag{12.37}$$

where $\omega$ is the *angular frequency*, $R$ the distance from the axis, $k$ the *wave number*, $\lambda$ the *wavelength*, and $l$ the *angular momentum density*, all at the point P under consideration.

Let then $V$ be the relevant volume occupied by the EM fluid, $W_k$ its *total internal kinetic energy*, $L$ its *total angular momentum*, $W$ its *total energy*, and $M$ its *total mass*. We have,

$$W_k = \int_V w_k dV, \quad L = \int_V l dV, \quad W = c^2 M = \int_V w dV, \quad J = \int_V j dV, \tag{12.38}$$

where we have added a quantity, $J$, that plays a role similar to an "equivalent total momentum". Using (12.37) and (12.38), we first consecutively define a *nominal angular frequency* $\bar{\omega}$, a *nominal field velocity* $\bar{v}$, a *nominal lateral radius* $\bar{R}$, a *nominal energy velocity* $\bar{v}_c$, and a *nominal wave number* $\bar{k}$ by

$$W_k = \bar{\omega} L = \bar{v} J, \quad L = \bar{R} J, \quad J = \bar{v}_c M = \bar{k} L, \tag{12.39}$$

from which we derive

$$\bar{v} = \bar{\omega}\bar{R}, \quad W_k = \bar{v}\bar{v}_c M, \quad \bar{k} = \frac{2\pi}{\bar{\lambda}} = \frac{\bar{\omega}}{\bar{v}} = \frac{1}{\bar{R}}, \quad \bar{R} = \frac{L}{\bar{v}_c M}. \tag{12.40}$$

If, for instance, $\omega$ is constant we have $\bar{\omega} = \omega$, and if the range covered is narrow, as will usually be the case at least for $\omega$, $\bar{\omega}$ will essentially be equal to the value at the centre of that range.

Consider then again a reference frame RF' that moves with constant velocity $\mathbf{v}_0$ with respect to the original reference frame RF. Let $\mathbf{E}_l$ and $\mathbf{H}_l$ be the longitudinal components of $\mathbf{E}$ and $\mathbf{H}$, i.e., their components in the direction of $\mathbf{v}_0$, and let $\mathbf{E'}_l$ and $\mathbf{H'}_l$ be the corresponding components in RF'. As can be shown [1] we have

$$\lim_{|\mathbf{v}_0| \to c} \mathbf{E'}_l = \lim_{|\mathbf{v}_0| \to c} \mathbf{H'}_l = \mathbf{0}. \tag{12.41}$$

Assume next that the EM fluid in RF is rotating around an axis parallel (i.e., in our terminology, either co-parallel or anti-parallel) to $\mathbf{v}_0$ and thus possesses in RF and therefore in RF' an angular momentum parallel to $\mathbf{v}_0$. The presence of such a momentum implies that both $\mathbf{E}_l$ and $\mathbf{H}_l$ are non-vanishing. On the other hand, in the limit $|\mathbf{v}_0| = c$ the field is seen in RF' as travelling with velocity equal to the speed of light. Hence, we conclude from (12.41) that if a field is travelling at the speed of light, an angular momentum parallel to the direction of propagation can be determined at most indirectly, but not by observing the field itself. (See also the discussion given in the last paragraph of Section 8.1.)

## 12.6 A Photon Model

We consider an autonomous EM field that is travelling in the $x$-direction with field velocity $\mathbf{v} = (c,0,0)^T = \mathbf{v}_c$, in which case $\alpha = 0$ and $w_0 = 0$ (cf. Section 4). A detailed analysis [1] reveals the existence of a solution for which

$$E_x = H_x = 0, \quad \sqrt{\varepsilon_0}E_y = \sqrt{\mu_0}H_z, \quad \sqrt{\varepsilon_0}E_z = -\sqrt{\mu_0}H_y,$$

$$E_y = \frac{1}{2\pi\varepsilon_0}\int_{-\infty}^{\infty}\int_{-\infty}^{\infty}\frac{(y-\hat{y})q(\tau,\hat{y},\hat{z})}{\hat{d}^2}d\hat{y}d\hat{z},$$

$$E_z = \frac{1}{2\pi\varepsilon_0}\int_{-\infty}^{\infty}\int_{-\infty}^{\infty}\frac{(z-\hat{z})q(\tau,\hat{y},\hat{z})}{\hat{d}^2}d\hat{y}d\hat{z}, \quad \tau = t - \frac{x}{t}.$$

In there, $\tau$ is defined as given, and $\hat{d}$ is the distance between the points $(y,z)$ and $(\hat{y},\hat{z})$ for $x$ and $t$ held fixed, thus for $\tau$ constant. This field is inherently basal. One crucial point in deriving the model is the assumption that it does not consist of a signal comprising just a single frequency, but instead possesses a spectrum of essentially finite extent, thus with a bandwidth that is nowhere zero although very small compared to its *nominal* (centre) *frequency*. According to standard theory of communication signals, a signal with zero bandwidth, thus a pure sinusoid, is indeed only a mathematical idealization and cannot have any actual physical meaning. Another crucial point concerns the charge density. The model comprises domains with positive and others with negative density that balance each other.

Assuming appropriate symmetry conditions, the model has been confirmed to possess the following photon properties:

1. It propagates strictly along a *straight line* in a single direction, thus without any sideward scattering.
2. Its velocity is equal to the *speed of light, c*.
3. The field is *transversal*.
4. The model exhibits an effective circular or linear *polarization*. In other words, the field is, on the average, circularly or linearly polarized, although the instantaneous field orientation is slightly oscillating around the ideal average position.
5. It has *zero rest energy* and *zero rest mass*.
6. Its *total energy* (cf. (12.38)) is *proportional* to the *nominal frequency* $\Omega$. More precisely, the ratio $\tilde{\hbar} = W/\Omega$ is independent of $\Omega$. Since it is reminiscent of $\hbar$, we have designated it (temporarily) by $\tilde{\hbar}$.
7. Its *momentum* is equal to *W/c*.
8. Let RF be the reference frame for which the model is derived and let us observe it in a reference frame RF' that is moving with respect to RF with constant velocity $\mathbf{v}_0$ parallel to the direction of propagation of the model in RF and thus in RF'. The field observed in RF' is then of exactly same type as the one in RF. It thus represents again a photon, although with $\Omega$ replaced by

$$\Omega' = \Omega\sqrt{\frac{1-\beta_0}{1+\beta 0}}.$$

This is identical to what is known as *longitudinal relativistic Doppler effect* [2]. It implies that for any photon the specific value of $\Omega$ that is encountered is simply a question of the reference frame in which the field is observed.

9. The *local charge density* inside of the model is nowhere zero. More precisely, there are interleaved domains with positive and negative charge densities, respectively, but the *total charge*, thus the charge seen from a distance, is zero.
10. The *total magnetic moment* is zero.
11. While the charges are *oscillating*, the *total positive* charge as well as the *total negative* charge remain individually *constant.*
12. Items 9 to 11 suggest that the photon does indeed consists of an electron and a positron that form an interleaved pair. This explains immediately the *pair production* of an electron and a positron as well as the *annihilation* of such a pair.
13. In view of item 8, there exists only a single type of photon, or else, all photons are the same if we observe each one in its own *basic reference frame* $RF_b$, i.e. the one in which $W_b = 2W_e$ where $W_b$ is the value of $W$ in $RF_b$, and $W_e$ is the energy of an electron at rest. According to item 6, we have $W_b = \tilde{\hbar}\Omega_b$, where $\Omega_b$ is the nominal angular frequency of the photon in $RF_b$. Assuming circular polarisation, we may identify $\Omega_b$ with the angular frequency with which the EM fluids of the electron and the positron are rotating.
14. Since $w_0 = 0$, we also have $W_0 = 0$ for the total rest energy of the model, and hence $W = W_k$, where $W_k$ is the *total kinetic energy*, thus $W_b = W_{kb}, W_{kb}$ being the total kinetic energy in $RF_b$. Identifying $\bar{\omega}$ in (12.39) with the present $\Omega_b$ we have $W_b = \Omega_b L_b$. In there, $L_b$ is the angular momentum and is thus necessarily equal to $2L_e$, where $L_e$ is the angular momentum of as well an electron as a positron. Identifying thus $L_e$ with the *electron spin* we may write $L_e = \hbar/2$, thus $W_b = \Omega_b\hbar$ and hence (cf. item 13), $\tilde{\hbar} = \hbar$. The photon energy is solely due to the energy of the hidden rotational movement of the EM fluid (cf. last paragraph of Section 5.3).
15. For the *gravitational redshift* the same value is obtained as that known from general relativity.
16. Assume again reference frames RF and RF' as above but with $\mathbf{v}_0$ not parallel but *perpendicular* to the direction of propagation. The field in RF' is then again that of a photon but with shifts in frequency and propagation direction as known from the theory of the *lateral relativistic Doppler effect* [2].
17. The theory according to item 16 also reveals that the mass of a photon cannot fully be characterized by a single scalar, but requires two parameters, an *axial mass* $M = W/c^2$ that is equal to the classical relativistic mass, and a *radial mass* equal to $2M$. This leads immediately to the *deflection of a photon* by a star identical to that predicted by general relativity.
18. The spread in *position* $x$ and *momentum* $J$ of the photon model satisfies the classical *uncertainty relation*, and this exactly in the form $\Delta \times \Delta J \geq \hbar/2$.
19. The field is concentrated in a small volume, thus *localized*. It may therefore be said to be like a *particle*.

20. The model behaves like a general *modulated signal*, although with suppressed *carrier*, the carrier frequency corresponding to the *nominal frequency* $\Omega$. The photon thus behaves like a *wave*.
21. In a dispersive medium the *carrier* travels with phase velocity and the *energy* with *group velocity*. This confirms the wave-like behaviour.
22. In view of items 19 to 21, the photon model is *simultaneously particle and wave*, without any ambiguity or apparent contradiction. It thus offers a natural explanation for the *wave-particle duality*.

This list covers all photon properties known to the author. In that sense, it is claimed to be likely quite complete.

## 12.7 Towards a Model of an Electron

### *12.7.1 Purely Electromagnetic Approach*

Consider a basal EM field that is rotating around an axis and assume it to have circular symmetry about its axis and appropriate symmetry with respect to an equatorial plane. We also assume the rotation to be steady, i.e. independent of time, adopt standard spherical coordinates $r$,$\theta$, and $\varphi$, and express all results in terms of appropriately normalized, dimensionless quantities $\hat{E}_r, \hat{E}_\theta, \hat{H}_r, \hat{H}_\theta, \hat{q}, \hat{r}, \hat{w}$. For the three relevant partial differential equations we then obtain [1],

$$\hat{r}\frac{\partial \hat{E}_\theta}{\partial \hat{r}} + \hat{E}_\theta - \frac{\partial \hat{E}_r}{\partial \theta} = 0, \quad \hat{r}\frac{\partial \hat{H}_r}{\partial \hat{r}} + 2\hat{H}_r + \frac{\partial \hat{H}_\theta}{\partial \theta} + \hat{H}_\theta \cot\theta = 0, \tag{12.42}$$

$$\beta(\hat{r}\frac{\partial \hat{E}_r}{\partial \hat{r}} + \hat{E}_r + \frac{\partial \hat{E}_\theta}{\partial \theta} + \hat{E}_\theta \cot\theta) = \hat{r}\frac{\partial \hat{H}_\theta}{\partial \hat{r}} + \hat{H}_\theta - \frac{\partial \hat{H}_r}{\partial \theta}, \tag{12.43}$$

where $\beta$ is given by

$$\frac{2\beta\hat{w}}{1+\beta^2} = \hat{E}_r\hat{H}_\theta - \hat{E}_\theta\hat{H}_r, \quad \hat{w}\frac{1}{2}(\hat{E}_r^2 + \hat{E}_\theta^2 + \hat{H}_r^2 + \hat{H}_\theta^2), \tag{12.44}$$

and the normalized charge density $\hat{q}$ can be determined by means of

$$\hat{r}\hat{q} = \hat{r}\frac{\partial \hat{E}_r}{\partial \hat{r}} + \hat{E}_r + \frac{\partial \hat{E}_\theta}{\partial \theta} + \hat{E}_\theta \cot\theta.$$

These equations remain satisfied if we multiply $E_r, E_\theta, H_r, H_\theta$, and $q$ and thus the corresponding normalized quantities by a same arbitrary constant, say $K$, while leaving $v$ unchanged. They all involve only dimensionless quantities and are of purely mathematical nature in the sense that they are free of any physical parameter. Hence, if a solution exists it will itself be independent of any physical parameter. The set formed by (12.42) to (12.44), clearly, is nonlinear, which is one of the reasons why it is far more difficult to handle the electron than the photon.

The *total charge* $Q$ (normalized: $\hat{Q}$) and the *total angular momentum* $L$ (normalized: $\hat{L}$), are given by

$$\hat{Q} = \int_{\hat{V}} \hat{q} dV', \quad \hat{L} = \int_{\hat{V}} (\hat{E}_r \hat{H}_\theta - \hat{E}_\theta \hat{H}_r) \hat{r} \sin\theta d\hat{V},$$

where $\hat{V}$ is the normalized relevant volume. Hence, the quantity $\tilde{F}$ defined by

$$\tilde{F} = \frac{Q^2}{|L|}\sqrt{\frac{\mu_0}{\varepsilon_0}} = \frac{\hat{Q}^2}{|\hat{L}|}$$

is a pure number. It should be compared to Sommerfeld's *fine-structure constant, F*, which is the dimensionless quantity defined by

$$F = \frac{Q_e^2}{2h\varepsilon_0 c} = \frac{Q_e^2}{2h} Z = \frac{Q_e^2}{8\pi L_e} Z, \quad Z = \sqrt{\frac{\mu_0}{\varepsilon_0}},$$

where $Qe$ is the *electron charge, Z* the *impedance of free space*, $h = 2\pi\hbar$ the *Planck constant*, and $L_e = \hbar/2$ the *electron spin* [9]. The value of $F$ is known to be quite close to 1/137. If we then set $L = L_e$, as in Section 6, item 14, and $Q = Q_e$ we find $f = \tilde{F}/8\pi$. A characteristic radius $R_e$ that is helpful for finding a rough estimate of the electron size is given by

$$R_e = \sqrt{R_c R_B} = \frac{\hbar}{c M_e} = 3.8616 \times 10^{-13}, \quad with \quad \frac{R_c}{R_e} = \frac{R_e}{R_B} = F, \qquad (12.45)$$

where $R_c$ and $R_B$ are the *classical electron radius* and the *Bohr radius, F* again the fine-structure constant, and $M_e$ the electron mass.

### 12.7.2 Incompleteness of the Original Formulation

In principle, the arbitrariness of the constant $K$ mentioned in Section 7.1 allows us to make $Q$ become equal to $Q_e$. It is definitely excluded, however, that $Q_e$ can fully be determined from the equations in Section 7.1. The only physical parameters involved are indeed $\varepsilon_0$ and $\mu\_0$, and as a simple dimensional analysis shows, no combination of these can produce the dimension of a charge. In addition, there exist strong arguments why the above-mentioned equations cannot have a real (in the mathematical sense), thus a physically acceptable solution. An additional, necessarily attractive force must be present that becomes strong in the innermost part of the electron, thus in that part where the charge density approaches zero but where the magnetic field, which reaches its maximum at the centre, is large.

The obvious logical source for such an attractive force is *gravitation*. However, it cannot simply be gravitation in form of the classical *Newtonian gravitational field* since even $\varepsilon_0$, $\mu_0$, and the *gravitational constant* $\Gamma$ still can not produce a charge. Nevertheless, since the forces due to gravitation are in most regions smaller by many orders of magnitude than those due to the EM field this will not measurably affect the determination of the fine structure constant discussed in Section 7.1.

Consequently, the problem of accurately modelling the electron remains unsolved. It definitely requires a generalization of Maxwell's equations. It appears impossible to do this by starting from their original form, say as given in (12.1) to (12.4). A more realistic approach would be to make use of the flow equations. Thus, since *inertial mass* implies *heavy mass*, the presence of $w$, thus of $m = w/c^2$, implies the presence of a gravitational field $\mathbf{G}$. In terms of Newtonian gravity theory we then have

$$\nabla^T \mathbf{G} = 4\pi\Gamma m. \tag{12.46}$$

From this we could conclude that there should exist an associated field of $\mathbf{G}$, say $\tilde{\mathbf{G}}$, such that

$$\frac{1}{4\pi\Gamma}\frac{\partial \mathbf{G}}{\partial t} + \mathbf{j} - \nabla \times \tilde{\mathbf{G}} = 0. \tag{12.47}$$

Indeed, in view of (12.46) and the first equations in (12.34) and (12.35), the vector formed by the first two terms in (12.47) has a vanishing divergence and can therefore differ from $\mathbf{0}$ only by the curl of some other vector.

Furthermore, $\mathbf{f}_0$ in (12.31) would have to be replaced by a more general force density $\mathbf{f}$ that also comprises the term $-m\mathbf{G}$ as well as a contribution due to $\tilde{\mathbf{G}}$. Since the volume forces may not contribute to the work that is done, we would then still have to require $\mathbf{j}^T\mathbf{f} = 0$. It is more likely, however that a proper theory of the electron can be developed only within the framework of *general relativity*. This remains a challenging task, which is well beyond the scope of this paper. Such a complete theory, however, would not affect the symmetry properties that have been mentioned in Section 7.1 and that will play a role in the coming Section 8.

A dimensional analysis might be helpful in order to gain some rough inside about the order of magnitude of the distances at which additional gravitational effects, whether due to $\tilde{\mathbf{G}}$ or to spatial curvature, might become effective. The parameter $R_g$ defined by

$$R_g = KQ_e\sqrt{\frac{Z\Gamma}{c^3}},$$

where $K$ is a dimensionless constant and $Z = \sqrt{\mu 0/\varepsilon_0}$ is again the impedance of free space, can indeed be shown to have the dimension of a length. If $K$ can be determined by some analytic process, it is unlikely that it differs from 1 by many orders of magnitude. Setting thus $K = 1$ we find

$$R_g = 4.89 \times 10^{-36} \quad \textit{meter}.$$

## 12.8 Travelling Particles

### *12.8.1 Electron-Like Particle Observed in Different Reference Frames*

We consider an electron model as discussed in Section 7 and assume it to be given in a reference frame RF. The rest field vectors $\mathbf{E}_0$ and $\mathbf{H}_0$ may be decomposed according to

$$\mathbf{E}_0 = \mathbf{E}_{0ax} + \mathbf{E}_{0ra} \quad and \quad \mathbf{H}_0 = \mathbf{H}_{0ax} + \mathbf{H}_{0ra}$$

into an *axial* and a *radial* component, thus into a component *parallel* to the axis of rotation and one *perpendicular* to it. A corresponding decomposition then holds for the rest energy density $w_0$ and thus for the total rest energy $W_0$, for which we may thus write

$$W_0 = \int_V w_0 dV = W_{0ax} + W_{0ra},$$

$V$ being again the relevant volume of the field.

Let RF' be a reference frame moving with respect to RF with constant velocity $\mathbf{v}_0 = (v_0, 0, 0)^T$. Let $W'$ be the total energy of the field in RF', $\mathbf{v}'_c = (v'_{cx}, v'_{cy}, v'_{cz})^T$ the energy velocity in RF', $W'_x$ the total energy flowing per unit time in the $x$' - direction of RF'. Making use of the symmetry properties of the field and, assuming the axis of rotation to be parallel to $\mathbf{v}_0$, one finds,

$$W'_x + v_0 W' = \alpha_0 v_0 (W_{oax} - W_{ora}), \tag{12.48}$$

where $\alpha_0^2 = 1 - \beta_0^2, \quad \beta_0 = v_0/c$ and

$$W' = \int_{V'} w' dV', \quad W'_x = \int_{V'} v'_{cx} w' dV'.$$

On the other hand, seen from RF', that same energy flow is given by $-v_0 W'$. Hence, we conclude from (12.48),

$$W_{0ax} = W_{0ra} = \frac{1}{2} W_0. \tag{12.49}$$

A field such as the one we are dealing with is modelling a particle, say Pa. Quantities that refer to Pa as a whole will be characterized by a subscript "$p$". Although at any fixed position $x$, $y$, $z$ the field described in Section 7 is independent of $t$ there is of course internal rotational movement. For an outside observer this internal movement may be ignored, and in that sense Pa will then be said to be at rest. Correspondingly, we designate its total energy by $W_{p0}$ instead of $W$.

Assume next that Pa is at rest in RF' and thus seen in RF as travelling as a whole with velocity $-\mathbf{v}_0$, with $\mathbf{v}_0$ as mentioned. We may interpret this by saying that Pa is travelling with a (*particle*) *velocity* $\mathbf{v}_p = -\mathbf{v}_0$. Let

$$\mathbf{J}_p = \int_V \mathbf{j} dV \quad and \quad W_p = \int_V w dV \tag{12.50}$$

be the resulting total *momentum* and total *energy*, respectively, of Pa. These quantities are related by (cf. (7.9) in [1])

$$W_p - \mathbf{v}_p^T \mathbf{J}_p = \alpha_p W_{p0}, \quad \alpha_p^2 = 1 - \beta_p^2, \quad \beta_p^2 = \frac{\mathbf{v}_p^T \mathbf{v}_p}{c^2}. \tag{12.51}$$

A further general relation (cf. (7.17) and (7.28) in [1]) simplifies to

$$\alpha_p W_p - W_{p0} = \beta_p^2 (W_{0ra} - W_{0ax}) \tag{12.52}$$

if $\mathbf{v}_p$ is parallel to the axis of rotation, as will be assumed hereafter. Together with (12.49) we thus obtain,

$$\begin{aligned} W_p &= \frac{W_{p0}}{\alpha_p} = c^2 M_p, \quad \mathbf{J}_p = \frac{W_p}{c^2}\mathbf{v}_p = M_p \mathbf{v}_p, \\ W_{p0} &= c^2 M_{p0}, \quad M_p = \frac{M_{p0}}{\alpha_p}, \end{aligned} \tag{12.53}$$

which are exactly as required by *classical relativity*.

Another result of interest concerns $L$ and $L'$, i.e., the angular momenta of Pa in RF and RF', respectively, with $\mathbf{v_0}$ again parallel to the axis of rotation, as above. Adopting again standard spherical coordinates, we derive from (12.6) in [1] for the component, i.e. the only non-vanishing component of the momentum density $\mathbf{j}$,

$$j'_\phi = \frac{1}{\alpha_0} j_\phi, \tag{12.54}$$

Let $R$ and $R'$ be the distances from the axis in RF and RF', respectively. Integrating (12.54) over the relevant volume $V'$ and observing that $R'=R$ we find,

$$L' = \int_{V'} j'_\phi R' dV' = \frac{1}{\alpha_0} \int_{V'} j_\phi R dV' = \int_V j_\phi R dV = L, \tag{12.55}$$

thus $L' = L$ (a result that is of interest for instance in the context of Section 5.3, last paragraph, and Section 6, item 14,).

In order to justify (12.55), recall first that the integrals comprising $dV' = dx'dy'dz'$ have to be evaluated for $t'$ constant. Hence, the Lorentz transformation has to be used in the form

$$x' = -v_0 t' + ff_0 x, \quad y' = y, \quad z' = z,$$

which yields $dV' = \alpha 0 dV$, where $dV = dxdydz$. In the last integral in (12.55), $j_\phi$ has to be interpreted as $j_\phi(x,y,z,t)$, where, in principle, we have to set $t = ff_0 t' + v_0 x/c^2$. But since the field in RF is independent of $t$, the choice of this variable is irrelevant and that last integral is therefore indeed equal to $L$. Note that proofs of the earlier results mentioned in the present section make use of similar arguments.

### 12.8.2 Dynamic Equations of an Electron-Like Model

Let EF be the EM field of, say, the particle Pa. We assume Pa to be an electron and look at the effect an *external electromagnetic field* EFe has upon it. In principle, EFe is characterized by $\mathbf{E}_e$,$\mathbf{H}_e$, $\mathbf{i}_e$, $q_e$, but "external" as used here implies that the sources of EFe are sufficiently far removed from the relevant volume $V$ of Pa. For simplicity we restrict ourselves to an electrostatic EFe. Inside of $V$ we may therefore assume

$$\frac{\partial \mathbf{E}_e}{\partial t} = \mathbf{0}, \quad \nabla \times \mathbf{E}_e = \mathbf{0}, \quad q_e = \nabla^T \mathbf{E}_e = 0, \quad \mathbf{H}_e = \mathbf{i}_e = \mathbf{0}.$$

Adding such a field to EF does not affect its equations (cf. (12.1) to (12.4)). Two conclusions can be drawn from this: Firstly, if Pa is in a basic reference frame, thus in a reference frame where it is at rest, EF may be assumed to be as discussed in Section 7.1. Secondly, in order to examine the effect EFe exerts upon Pa some generalization of Maxwell's theory is needed; or better, of the flow equations. These must however be used in their original form (12.26) and (12.27), while (12.31) and (12.32) hold only if Pa is at rest in the given reference frame.

We first look at $\mathbf{f}_c$ (cf. (12.8)). As it contains the term $q\mathbf{E}$, an additional term $\mathbf{f}_e = q\mathbf{E}_e$ must be added, although with opposite sign since it corresponds to a force acting upon EF, not one effected by EF. Furthermore, the term $\mathbf{i}^T\mathbf{E}$ in the second equation (12.26) should be complemented correspondingly by a term $\mathbf{i}^T\mathbf{E}_e$. Hence, (12.26) should be replaced by

$$\frac{\partial \mathbf{j}}{\partial t} + (\nabla^T(\mathbf{v}\mathbf{j}^T))^T + (\nabla^T \mathbf{T}_0)^T + \mathbf{f}_c = \mathbf{f}_e, \quad \mathbf{f}_e = q\mathbf{E}_e, \tag{12.56}$$

$$\frac{\partial w}{\partial t} + \nabla^T(\mathbf{v}w) + \nabla^T(\mathbf{T}_0\mathbf{v}) + \mathbf{i}^T\mathbf{E} = \mathbf{i}^\mathbf{T}\mathbf{E}_e. \tag{12.57}$$

While $\mathbf{E}_e$ may depend on $t$, we assume it to be independent of $r$ and, at least for the time being, to be permanently parallel to the initial direction of the axis of rotation of Pa. Due to symmetry, this axis then remains permanently parallel to $\mathbf{E}_e$. If we then integrate (12.56) and (12.57) over the relevant volume $V$ and take into account that $w$ and $w_0$ are vanishing at least with the forth power of the distance, we obtain after carefully taking into account the symmetry properties of the various terms involved [1],

$$\frac{\partial \mathbf{J}_p}{\partial t} = \mathbf{F}_e, \quad \mathbf{F}_e = Q\mathbf{E}_e, \quad \frac{\partial W_p}{\partial t} = \mathbf{v}_p^T\mathbf{F}_e, \quad W_p = M_p c^2, \tag{12.58}$$

where Q is the total *charge* of Pa, $\mathbf{v}_p$ the particle velocity, and where $\mathbf{J}_p$ and $W_p$ are as defined by (12.50). At any fixed time instant we may choose $\mathbf{v}_0 = -\mathbf{v}_p$ and this way associate with Pa another reference frame in which Pa is at rest. Hence, the results (12.53), which have been obtained for a similar scenario, are also valid for the present quantities $\mathbf{J}_p$, $W_p$, $M_p$, and $\mathbf{v}_p$.

The results discussed so far in Sections 8.1 and 8.2 concern an *axial* movement of the field, thus a movement parallel to its axis of rotation. Consider next a *radial* movement of an electron, thus a movement perpendicular to its axis, as has already been mentioned for a photon in Section 6, items 16 and 17. We then find the same need for characterizing the inertia, thus the mass, not simply by a single scalar, but by an *axial mass* $M_{ax}$ and a *radial mass* $M_{ra}$. These are found to be given by

$$M_{ra} = \frac{3}{2} M_{ax}, \quad M_{ax} = M_p,$$

$M_p$ being as above.

Similar results are found for the kinetic energies

$$W_{pkax} = W_{pax} - W_{p0} \quad and \quad W_{pkra} = W_{pra} - W_{p0}, \qquad (12.59)$$

that are associated, respectively, with an axial and a radial movement. In (12.59), $W_{pax}$ and $W_{pra}$ are the total energies for a movement in the axial and the radial direction, respectively. Assuming the same velocity $v_p$ in both cases we find,

$$\frac{W_{pkra}}{W_{pkax}} = \frac{3 + \alpha p}{2}, \quad \alpha_p = \sqrt{1 - \beta_p^2}, \quad \beta_p = \frac{v_p}{c}.$$

Hence, for a given velocity the radial kinetic energy is between $2(v_p = 0)$ and $3/2(v_p \pm c)$ times as large as the axial one. The energy needed for bringing an electron to any given velocity is therefore minimum for a movement in axial direction. The only stable position for the axis of rotation of an electron to occupy in a trajectory is therefore to be parallel to its velocity. Consequently, as soon as an electron (which is not a rigid body, thus not simply a classical gyroscope!) is subjected to a force it will immediately reshuffle its internal field in such a way that its axis becomes parallel to the local tangent, thus such that the assumptions made for arriving at (12.58) are locally fulfilled. The momentum to be used in (12.58) is therefore effectively given by

$$\mathbf{J}_p = M_{ax}\mathbf{v}_p = M_p\mathbf{v}_p,$$

and the angular momentum, thus the electron spin, is oriented like $\mathbf{v}_p$. In particular, the movement of an electron in an external electric field is governed by the laws of classical relativity, also if the trajectory is arbitrarily curved, provided the external field is always practically homogeneous within the tiny relevant volume the particle occupies. This remains obviously true if forces due to, say, a magnetic or a gravitational external field are acting, and it may therefore be assumed to hold also for particles of nature other than an electron.

## 12.9 Quantum Mechanics

### *12.9.1 Problems with the Conventional Approach*

The movement of an electron Pa in a shell of an atom also occurs, in a sense, under the influence of an external electrostatic field, i.e., the one created by the charge of the nucleus. Due to the strong forces involved one must expect, however, that the field of which Pa consists will be even more spread out. The results of Section 5.3, which have been obtained without assuming circular symmetry, remain applicable.

Where appropriate, the symbols for nominal values in Section 5.3 will hereafter be replaced by capital letters that refer to the particle nature of the field. Some of the quantities will have values that differ from those for an isolated electron. Due to conservation laws, $L_p$ and $M_{p0}$ will of course remain unchanged. Gravitational effects as alluded to in Section 7.2 are totally irrelevant.

In the conventional elementary approach to wave mechanics one uses expressions that involve, in particular, a unique frequency, $\Omega$, a unique wave number, $K$, the particle mass $M_p$, the particle velocity $v_p$, and the kinetic energy $W_{pk}$. Using the de Broglie relation $J_p = M_p v_p = \hbar K$, one finds [9],

$$K^2 = 2\frac{M_p}{\hbar^2} W_{pk}, \quad \textit{where} \quad W_{pk} = \frac{1}{2} M_p v_p^2. \tag{12.60}$$

In view of ample experimental evidence the first one of these expressions is definitely correct as long as $v_p$ is non-relativistic, in which case $W_{pk}$ is indeed given by the second expression. By analogy with the photon, one then equates $W_{pk} = \hbar\Omega$, finds $v_p = d\Omega/dK$, and interprets that result by claiming the particle velocity to be equal to the group velocity of the associated wave. In the light of a detailed analysis of the widely misinterpreted concepts of group velocity and group delay [10-12] and the thorough discussion in Appendix E3 of [1], such an interpretation is untenable.

From the point of view adopted in this text, however, $\Omega$ is simply a nominal frequency, say the appropriately defined centre of a non-vanishing frequency band, and $K$ the correspondingly defined nominal wave number. As follows from the discussion in the last paragraph of Section 5.2 (cf. (12.39)), the general law relating $W_{pk}$ and $\Omega$ is **not** $W_{pk} = \hbar\Omega$, but

$$W_{pk} = L_p \Omega, \tag{12.61}$$

with $L_p = \hbar$ for a photon and, for an electron,

$$L_p = \hbar/2 \quad \textit{and} \quad \textit{thus} \quad W_{pk} = \hbar\Omega/2. \tag{12.62}$$

This way, the velocity of Pa is found to be indeed given by $v_p = \Omega/K$, and no conflict arises with a correct use of the group-velocity concept. This agrees with a multidimensional (sufficient in practice: two-dimensional) Fourier analysis of a uniformly travelling particle that is definitely not point-like but consists of a distributed field and thus gives naturally rise to ranges of wavelength and frequency with non-vanishing width.

### 12.9.2 Schrödinger Equation

Define $K_p$,$\Lambda$, and $\Lambda_p$ by

$$K_p = \frac{2\pi}{\Lambda} = \frac{M_p v_p}{2Lp} = \frac{1}{2\bar{R}} \tag{12.63a}$$

$$\Lambda_p = \frac{\Lambda}{2} = 2\pi\bar{R} \tag{12.63b}$$

where the last equality in (12.63a) follows by making use of (12.39), while $M_p$,$L_p$,$v_p$ etc. now take over the roles assumed in Section 5.3 by $M$,$L$,$\bar{v}_c$ etc. But $2\pi\bar{R}$ was the length of the circumference along which the energy, thus also $M$, was travelling periodically with nominal velocity $\bar{v}$ . We may therefore conclude that the particle is "nominally" travelling with constant velocity $v_p$ along a circular path and that the wavelength associated with that periodic movement is equal to $\Lambda_p$ . In a periodic movement, however, the wavelength of the underlying field is double that of its energy and is thus given by $\Lambda = 2\Lambda_p$ . This points to $K_p$ being indeed the wave number of a true electromagnetic phenomenon. Altogether, the results we have obtained suggest the following interpretation:

The spread-out EM field of an electron in an atomic shell will, in practice, never assume its perfect shape. The details of this are unpredictable and depend on the specifics of the process by which the electron has originally been incorporated into the atom. Hence, the ideal field, say the EM *main field*, will be accompanied by an EM *co-field* that is equal to the difference between the actual field and the main field. The co-field has no sources of its own and is tied to its main field. In the relevant domain it is described by a standard wave equation

$$\frac{\partial^2 \psi}{c^2 \partial t^2} - \Delta\psi = 0, \quad \Delta = \frac{\partial^2}{\partial x^2} + \frac{\partial^2}{\partial y^2} + \frac{\partial^2}{\partial z^2} \tag{12.64}$$

where $\psi$ stands for any of the altogether six components of the electric and the magnetic field.

If we solve (12.64) in a standard way for $\psi = \Psi e^{j\Omega t}$ , where $\Psi$ is independent of $t$ and where $i$ designates in the present context the imaginary unit, we obtain the time-independent (Helmholtz) equation

$$\Delta\Psi + K_p^2\Psi = 0, \quad K_p = \frac{\Omega}{c} = \frac{2\pi}{\Lambda}, \quad \Omega = cK_p, \tag{12.65}$$

where $K_p$ and $\Lambda$ are as in (12.63) and where $\Omega$ is assumed to have been chosen according to the last expression in (12.65). This result holds as long as $\Omega$ is independent of $t$, but $\Omega$ and $K_p$ may very well be functions of $\mathbf{r}$, in particular thus of $r$.

If the velocities are non-relativistic, $K_p$ becomes identical to $K$. Taking into account $2L_p = \hbar$, the Helmholtz equation in (12.65) can then be written in the standard form of the time-independent *Schrödinger equation*, $\frac{\hbar^2}{2M_{p0}}\Delta\Psi + W_{pk}\Psi = 0$, while the corresponding time dependent equation is in fact (12.64). If, however, the non-relativistic approximation is not permitted, we may still use (12.65), but for $M_p$ and its relation to $M_{p0}$ and $W_{pk}$ we must now use $M_p = M_{p0}/\alpha_p$, $W_{pk} = W_p - W_{p0} = \left(\frac{1}{\alpha_p} - 1\right) c^2 M_{p0}$, whence we conclude, $K_p = \frac{M_p v_p}{\hbar} = \frac{M_{p0} v_p}{\alpha_p \hbar}, \frac{1-\alpha_p}{\alpha_p} = \frac{W_{pk}}{\alpha_p M_p} = \frac{W_{pk}}{M_{p0}}, \alpha_p = \sqrt{1-\beta_p^2}, \beta_p = \frac{v_p}{c}$. These expressions allow us to determine first $v_p$ in terms of $W_{pk}$ and $Mp0'$, and then $\bar{R}$ by means of $\bar{R} = R_e/2\bar{\beta_p}$ (cf. (12.40) and (12.45)). Note that all these results have been obtained without making use of the de

Broglie relation. The universal validity of this relation in its original form is indeed doubtful since one obtains from (12.39) not only (12.61), but also $J_p = K_p L_p$ , not $J_p = K_p \hbar$ .

The unpredictability of $\Psi$ recalls its probabilistic interpretation in standard quantum physics. According to that classical interpretation, however, the randomness is of *intrinsic* nature and thus belongs to the inherent constitution of quantum objects. In our interpretation, however, it is of *extrinsic* nature. In other words, it is brought about from the outside and is simply due to the great complexity of the actual phenomena. This puts it into the same category as, for instance, properties such as the use of probability theory for dealing with pressure and temperature in the kinetic theory of gases.

## 12.10 Conclusion

The paper centres on electromagnetic (EM) fields in vacuum that are autonomous (self-sustaining). Maxwell's equations and their relativistic transformation rules have been shown to imply the existence of flow equations that are structured exactly like the dynamic equations of a conventional fluid and thus effectively describe an *EM fluid*. That fluid possesses inertia, hence mass, and is therefore also subject to gravitational properties. Photons and electrons have an inner structure that consists of such a fluid. For a photon the partial differential equations (PDEs) to be satisfied can be solved explicitly, and the resulting model possesses an impressive list of properties that seems to comprise all those that are known to hold.

For an electron (or a positron), the PDEs describing the model are nonlinear, and the role of the gravitational field in its innermost part is not yet completely resolved. Nevertheless, highly important results have been obtained: Firstly, a particle based on this model behaves in an external field exactly as required by the relativistic dynamic equations of an electron. It therefore possesses the complete amount of mass that is amply confirmed by experiment, and this without any need to rely on a still speculative Higgs particle. Secondly, introducing *nominal values* for velocity, energy, mass etc. the (time-independent) Schrödinger equation is derived, thus as a direct consequence of Maxwell' equations and their relativistic transformation rules.

Consequently, photons and electrons are nowhere point-like. They possess an inner structure that has been determined by appealing to long known theories. There is no contradiction whatsoever between their behaviour as particles and as waves. The core of the problem is the same as for a distributed mass in classical mechanics, for example an elastic body or a fluid. For these, an analysis of their detailed behaviour requires to handle the relevant partial differential equations, but for examining some highly important aspects it is sufficient to consider only the movement of just one point, the centre of gravity, thus to proceed as if we were dealing with a point mass. The coordinates of the centre of gravity, its velocity etc. are then true nominal values in the above sense, and we are again faced with a true dualism: On the one hand, behaviour as if we were dealing with a point-like particle, on the other, phenomena that are characteristic of distributed systems, including typical wave-like properties.

It would obviously be wrong to see in this something like a wave-particle duality, thus a source of contradiction that is unexplainable in terms of classical mechanics. There is no reason why this should be different for an EM fluid.

According to the theory, presented in this paper, the Schrödinger function stands for any of the six components of a co-field, which is the difference field between the actual and the ideal one. The co-field itself depends on the contingencies of the process by which the electron is incorporated into the atom. Consequently, its precise behaviour is in practice unpredictable and can therefore be assessed only by probabilistic methods. This, however, is an issue of probability due to complexity, say of *extrinsic probability*, thus of probability imposed by conditions exterior to the object. This contradicts the standard Copenhagen interpretation, which assumes *intrinsic probability*, thus probability inherent to the object itself.

Some important further consequences should be mentioned. According to the theory here presented, while EM fields are not composed of point-like particles, as is widely believed, EM particles are, in a sense, composed of fields. Furthermore, there is no reason why EM fields should only exist in condensed form, thus in form of EM particles. On the contrary, it is most likely that the entire cosmos is permeated by an EM field, as thinned out as it may widely be. Since this omnipresent *vagabonding field* is subject to gravitation it will at least provide some contribution to the mysterious dark matter, and its density will be higher inside of galaxies. Except for its own contribution to gravity the vagabonding field is undetectable since only photons, cosmic (particle) rays, electron showers etc. can be registered on earth.

It is likely that similar principles are also of relevance for other types of particles, either elementary or composite ones, possibly with other types of fields involved. This implies that the specific field configuration that forms a particle must be stable, at least to the degree required for achieving the respective lifetime. Stable configurations in turn are only feasible if the fields are altogether non-linear, but, as has been shown, this is indeed the case for autonomous EM fields. Individual stable configurations can exist only as isolated entities. Such an individual *quantum state* can turn into another one only if some threshold is exceeded, either due to some outside influence (collision, absorption of a photon, merger etc.) or simply spontaneously. The latter phenomenon can easily be explained because due to the myriad of permanent external influences the actual field configuration will in reality be constantly fluctuating around the ideal one and will thus from time to time exceed a relevant threshold (spontaneous photon emission, radioactivity, tunnelling etc.).

To the best of the author's knowledge, all results obtained so far are in perfect agreement with confirmed experimental evidence.

## References

1. Fettweis, A.: Electrical communications, fluid dynamics, and some fundamental issues in physics, Nordrhein-Westfälische Akademie der Wissenschaften und der Künste, publication IW 31. Ferdinand Schöningh, Paderborn (2010)
2. Einstein, A.: Zur Elektrodynamik bewegter Körper. Annalen der Physik 17, 891–921 (1905)

3. Born, M.: Die Relativitätstheorie Einsteins, 5th edn. Springer, Germany (1969); Also: Einstein's Theory of Relativity. Dover, New York (1965)
4. Bergmann, P.G.: Introduction to the Theory of Relativity. Prentice-Hall, Englewood Cliffs (1942); Dover, New York (1976) (enlarged and corrected)
5. A. Fettweis, The wave digital method and some of its relativistic implications. IEEE Transactions on Circuits and Systems-I: Fundamental Theory and Applications, vol. CAS-49(6), 862-868 & 49(10), 1521 (2002)
6. Fettweis, A.: Losslessness in nonlinear Kirchhoff circuits and in relativity theory. Problems of Nonlinear Analysis in Engineering Systems 9(3(19)), 141–163 (2003) (Issue in memory of Ilya Prigogine)
7. Fettweis, A.: Kirchhoff circuits, relativity theory, and beyond. In: European Conference on Circuit Theory and Design (ECCTD), Cork, Ireland, August 29-September 2 (2005); Also (improved and corrected): Proceedings of the International Conference on Signals and Electronic Systems (ICESES 2006), Lodz, Poland, September 17-20, pp. 19–43 (2006)
8. Hamming, R.W.: Numerical Methods for Scientists and Engineers, 2nd edn. McGraw-Hill, New York (1973); Dover, Mineola, New York (1986)
9. Krane, K.: Modern Physics. Wiley, New York (1996)
10. Fettweis, A.: On the significance of group delay in communication engineering. Archiv für Elektronik und Übertragungstechnik 31(9), 342–348 (1977)
11. Fettweis, A.: Elemente nachrichtentechnischer Systeme, vol. 2. Teubner, Stuttgart (1996)
12. Fettweis, A.: Can signals truly be faster than light? Signal Processing 83(8), 1583–1596 (2003)

# Chapter 13
# Fundamentals of Electrodynamics Essential Overview of EM Theory

Branko Mišković

**Abstract.** Instead of the strict causal exposition, this is an original cross-section through EM theory, as its brief overview. All the equations are transparently presented in the pairs or tables, thus pointing up their symmetries and relations, with enough verbal announcement of their derivations. The characteristic problems are treated in the form: thesis – antithesis – synthesis. A few apparent antinomies of EM theory are thus presented and resolved, surpassing some, more or less obsolete, principal views. The text is of educational character, on the lower university level, also accessible to the wider public of non-expert readers.

## 13.1 Introduction

Usual scientific texts represent the formal sequences of mathematical procedures and logical conclusions. Though convenient in deductive expositions, this method is not effective in inductive elaboration, demanding some generalizations, with active imagination and intuition, above the formal procedures. Electric circuits and EM fields are the pair of the causal and imaginative thoughts. The circuits are in fact line cross-sections through surrounding fields. Concerning the fields, our text insists on visual images of the processes and exhaustive comparison of the quantities and their relations, from various angles of view.

Hard inductive development of a scientific theory demands a sequence of formal concepts to be introduced. Their formalistic application sometimes follow into obvious contradictions, as the theses and antitheses. The syntheses demand critical revision of the concepts and even of some principal views. The original aspect of this text considers some antinomies of EM theory. Instead of the classical or relativistic approaches, the valid elements of both are affirmed and supplemented by some new ideas resolving the former problems. The application of the two distinct theories to the same physical reality cannot be justified.

Branko Mišković
Independent
e-mail: aham.brami@gmail.com

K. Kyamakya et al. (Eds.): Selected Topics in Nonlinear Dynamics, SCI 459, pp. 235–254.
DOI: 10.1007/ 978-3-642-34560-9_13 

The ready EM theory may be compared with a three-storied house, built of EM *carriers*, *fields* and *potentials*. The *central laws* form its fundament, *differential equations* – the walls, and *algebraic* – the floors. Due to the incomplete basis, the walls had been founded on the intuition and experience. The attempt of the cover installation has generated SRT. With orientation to the walls, all the floors, including the fundament, are here elaborated. The obtained results enable the consistent exposition of all the aspects of the complete theory. Their interpretation, however, demands radical revision of some traditional principal views.

The classical and modern concepts of elementary particles, their fields and interactions, are here surpassed. Instead of the source or carrier of its fields, a particle is reduced to the field center. Being the partial causes of the summary macro-effects, the elementary fields are understood as the rigid structures, stable orientated in space. In the zero sum of two opposite fields, the moving component is producing the kinetic effects, independently of the opposite, unmoving field. The fields interact directly, at each point of space separately. Even Maxwell's equations do not take into account any time of the force transfer.

The sequence of the new views and results is the basis for systematic elaboration and methodical exposition of EM theory. Before the known differential equations, algebraic ones are consistently affirmed. The complementarities of the two formal approaches and their results are understood. All this demands some arrangement of the terminology and notation. The obsolete terms are substituted by more adequate ones. The majority of traditional symbols is kept, with one systematic unification, based on Maxwell's convention: *discrete quantities* are denoted by lower case letters, and *physical fields* by respective capitals.

## 13.2 Basic Concepts

With respect to the vague essence of EM phenomena, respective theory is founded inductively, starting from the sensory effects of these phenomena. In the aim of convenient interpretation of the physical processes, some formal concepts are introduced hypothetically. The central of them are electric *poles* or *charges (q)*, as the material agents in EM interactions. The former term is more convenient for *discrete*, and the latter for *distributed* quantities. The explanation of the attractive and repulsive forces relies on the two types of electric poles, *positive* and *negative* ones. Two equipolar charges mutually repel, and opposite ones attract each other. This is the only way for their comparison and distinction.

Apart from (radial) *static* forces, only dependent on mutual *distance* of interacting charges, the additional EM forces also depend on motion. Two charges moving in parallel, as the convectional currents (qv), interact by (transverse) *kinetic* forces, dependent on the *speed product* and superimposed to the static ones. Finally, the alternating speed of a pole affects all the present such poles, including this pole itself, by the (axial) *dynamic* forces, dependent on *acceleration*. Acting on all the present poles, the static and dynamic forces are covered by the same concept of collinear *electric field*. The kinetic forces, transverse to the speeds of moving poles, are expressed by the embracing *magnetic field*.

Two opposite electric poles, somehow connected to each other on a mutual distance $(r = r_p - r_n)$, form respective $dipole(p)$. The rotation of such a dipole around one of its poles forms the circular current and magnetic $moment(m)$. The new concepts are defined by the following equations:

$$p = qr, \qquad m = qr \times v \tag{13.1}$$

The magnetic moment is perpendicular to the plane of rotation. Unlike the electric dipole, composed of the two separable poles, magnetic moment cannot be split into apparent poles. In spite of this distinction, the parallelism between electric and magnetic phenomena, encouraged by the formal symmetries of their relations, has been exploited in the former development of EM theory. On the basis of behavior of the two defined objects, the two EM fields, electric (E) and magnetic (B), had initially been introduced:

$$t_e = p \times E, \qquad t_m = m \times B \tag{13.2}$$

$$\delta f_e = (p \cdot \nabla)E, \qquad \delta f_m = (m \cdot \nabla)B \tag{13.3}$$

## 13.3 Static & Kinetic Interactions

With respect to the separate electric poles, as the constituent parts of the dipoles, EM theory started by their treatment. The image of the central electric field (Fig. 13.1.) is thus obtained. The field lines determine the directions of respective forces at each point, acting on other such poles and dipoles. The density of the lines points to the local field intensity and global non-homogeneity. The full field flux through a concentric sphere does not depend on the radius. As the Gaussian theorem, this is generalized to any closed surface embracing a given charge. The inverse square function, known as the *electrostatic central law*, had been confirmed independently, by the strong Coulomb's experimental procedure.

The application of the equation (13.3a) to the central field, in the surroundings of a separate pole, gives the force decreasing by third power of the distance. There can be shown that the force between two dipoles decreases by the forth power. In general, the interaction of two multi-poles decreases by the power equal to the sum of all their poles. Therefore, the interactions of statistically neutral bodies practically annul, but are manifest in the very close contacts. In fact, all these interactions are based on the mentioned electrostatic Coulomb's law.

Though the attempts of parallel treatment of the fictional magnetic poles and their central fields made a role in the former theory, they have been mainly abandoned. The excessive insisting on formal symmetries may conceal or fully miss the essence of a natural phenomenon. Instead of the magnetic poles, electric currents and their (kinetic) interactions are observed. Circular magnetic fields are noticed around current carrying conductors. In accord to the axial symmetry, such a field decreases by the first power of distance. In the case of the magnetic moment of a contour current, it has the form of a thoroidal vortex (Fig. 13.2.).

This field is axially directed only in the plane of the contour: in the course of the moment m – inside, and opposite – outside the contour. At the vertical hyperboloid cutting the plane through the contour, this field is parallel to the contour plane. Just in this region, it performs the main interactions with other similar contours, attracting parallel, and repelling anti-parallel currents.

The contour *current*, magnetic field and respective kinetic *force* represent the trihedral vectors. As the consequence of the initial field introduction, in accord to its action upon the apparent dipoles, the formulation of the general *kinetic* law has been hindered. Instead, its *incomplete case* (13.4b), valid at common motion (v = V) at least, is here presented in parallel with the *static law* (13.4a):

$$f_s = nc^2 r_0, \qquad f_k = -n(V \cdot v) r_0 \tag{13.4}$$

$$n = \frac{\mu q_1 q_2}{4\pi r^2}, \qquad c^2 = \frac{1}{\varepsilon\mu} \tag{13.5}$$

Due to easier formal manipulation and comparison of the two laws, the substitutive factor n is defined by (13.5a). The known Maxwell's relation (13.5b) links the speed of light propagation and the two EM constants. With respect to the similar relation

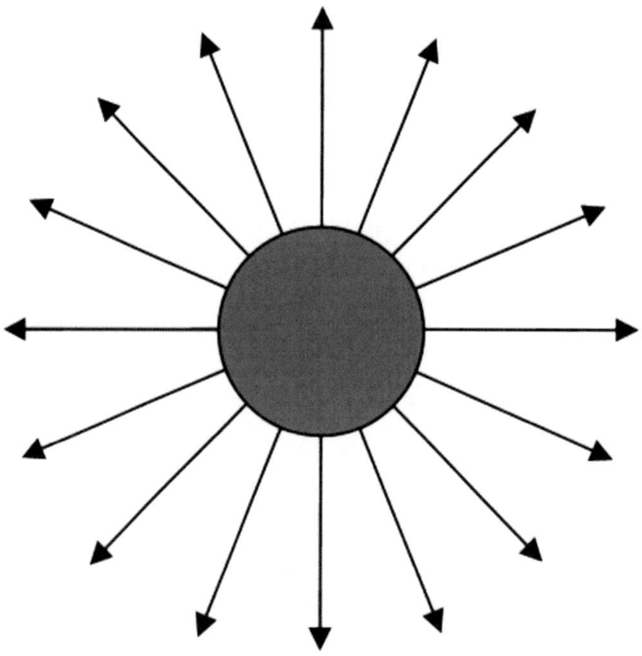

**Fig. 13.1** Central field

in fluid-mechanics, the former constant ($\varepsilon$) expresses *compressibility or elasticity*, and the latter ($\mu$) *mass density* of the medium. This analogy points to aero-dynamic interpretation of EM quantities and their relations. Both forces (13.4) are in the central form, as the inverse square functions. The direct comparison points to their two distinctions: the product of two – instead of the square of the constant speed, with opposite signs of the two forces. As if that the (general) static law might be a special case of the (incomplete) kinetic one, at the speed: $v = V = ic$. The common speed of all the particles points to their continual motion along the temporal dimension, manifest by the known cosmic expansion. The imaginary unit ($i$) expresses the force of attraction. These facts will be physically clearer later, at gradual development of this exposition.

## 13.4 Dynamic Interaction

Apart from the two above interactions, static and kinetic ones, dependent on the *distance* and *motion* of the charges, let us now introduce the *dynamic* EM forces, dependent also on *acceleration*. As the preceding act in this aim, the integration of the two forces (13.4), via the factor $n$ – from $r$ up to $\infty$ – gives the new factor: $m = nr$. Instead of the forces – in the *usual laws* (13.4), their integrals give respective potential energies, expressed by the two *alternative laws*:

$$w_s = mc^2, \qquad w_k = -mV \cdot v \tag{13.6}$$

$$m_{1,2} = \frac{\mu q_1 q_2}{4\pi r_{1,2}}, \qquad m = \frac{\mu q^2}{4\pi r} \tag{13.7}$$

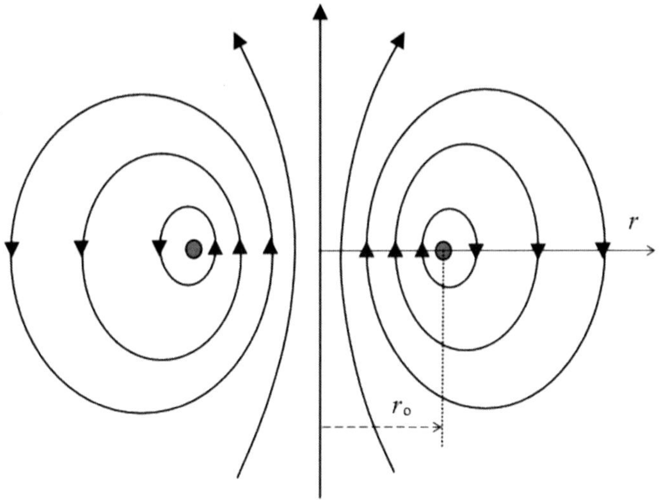


**Fig. 13.2** Thoroidal vortex

With respect to Einstein's equation (13.6a), the factor $m$ is nothing else than mass, *mutual* (13.7a) and *proper* (13.7b) ones. These two masses are in fact elementary factors of *induction* and *self-induction*, respectively, known from electric circuits. The symbol $r$ – in the proper mass (13.7b), represents the particle radius, as the distance of the surface charge from its own center. The simplest (Lorentz') particle model, as the elastic sphere, with evenly distributed surface charge, is understood. Not only that this model is very convenient in calculation, but gives acceptable results. It will be finally fit into the full particle model, in 4D space.

As the measure of its variation, time derivative of the kinetic energy (13.6b), partially – per the product $mV$, with the speed $v$ – as the parameter, gives the power $(v \cdot f)$ of the energy transfer, performed by the dynamic forces. This gives the *force action law*, expressing the dynamic forces of accelerated charges. The mass of a neutral body – as the multi-pole, may be explained by the inertia of the elementary charged particles in its structure. With respect to mutual cancellation of the distant opposite fields, these masses are reduced to the nearer fields, close to the particles. This is the reason of the mass-defect. In absence of some other basis of the inertial phenomena, all points to EM nature of all the matter. However, in current physics mass and charge are treated as the two separate concepts.

The central forces (13.4) act in reverse upon their carrier – as the object of its own fields, proportionally to the force difference distributed about the surface. With respect to the same speed, the incomplete kinetic law (13.4b) is sufficient. The difference equals to the static force scaled by the factor: $g_2 = 1 - v_2/c_2$. Starting from the unit – at rest, this factor approaches zero at the speed $v = c$. The force is ever positive, tending to expand the particle. The particle persistence points to another force in the opposition. This may be some constant pressure of a hypothetical omnipresent *quantum fluid*. The force balance is enabled by compression of the particle in accord to the function: $r = r_0 g$. Therefore, with respect to (13.7b), the mass tends into infinity by inverse function of the same factor: $m = m_0/g$.

Well, lesser particles are of the greater masses. This means that proton radius is lesser than that of en electron, in the inverse ratio of their masses. This contradicts to experience where the greater bodies of the same texture are also more massive. At elementary particles, however, the masses are located in their electric fields. The field domain increases with speed on account of the particle volume, just in the region of the strongest fields. The particle inside may be fully fluidic, without hard material parts. The current physics, however, expects to explain the material essence by possible or fictional structure of the particles.

The factor $g$, and the both parameters of the particle dependent on it, above the speed $c$ are of imaginary values. This points to the upper limiting speed of massive particles, and also of the structures composed of them. The generalization of this restriction to all the speeds in nature, postulated for the sake of SRT, has neither a real basis nor justification. As if, our view is confirmed by the newest experimental result, where a particle slightly exceeds the strict speed $c$. However, the technical conditions of this registration have not been taken into account. The strong fields change the conditions even of light propagation. Comparable measurement of this speed in the same conditions should be also performed.

Fig. 13.3. presents the *classical kinetic energy* (dashed) and Lorentz' *mass function*. The two functions are close at small speeds, but at greater ones separate from each other. The former function keeps the finite values, but the latter one strives into infinity approaching the boundary speed c. Apart from the quantitative, there is the interpretative distinction between the two functions. The classical one concerns the energy dependent on speed, at constant mass. In the latter one, however, the values of mass and energy grow with increasing speed. Mutually proportional, these two quantities are equaled in the *natural units* (NU).

The comparison of the equations (13.5b) and (13.6a) points to the physical essences of these two quantities. Mass plays the role of the *medium density* ($\mu$), and energy – of *elasticity modulus* ($1/\varepsilon$). Their quantitative ratio as if depends on the system of quantities applied. Namely, in SI a small mass accords to the vast energy, but in NU ($c = 1$) energy in vacuum equals to mass, and in material media is lesser from it. In EM waves mass concerns the full wave energy, in both medium layers, but the energy is restricted to vacuum, as the unique layer for transfer of the traveling wave energy, at the standard speed of propagation ($c$).

Unlike the proper mass (13.7b), dependent on the charge radius, and thus – on its speed, the mutual mass (13.7a) depends on the distance of two interacting particles, irrespective of their motion. With respect to the mutual sense of the gravitational

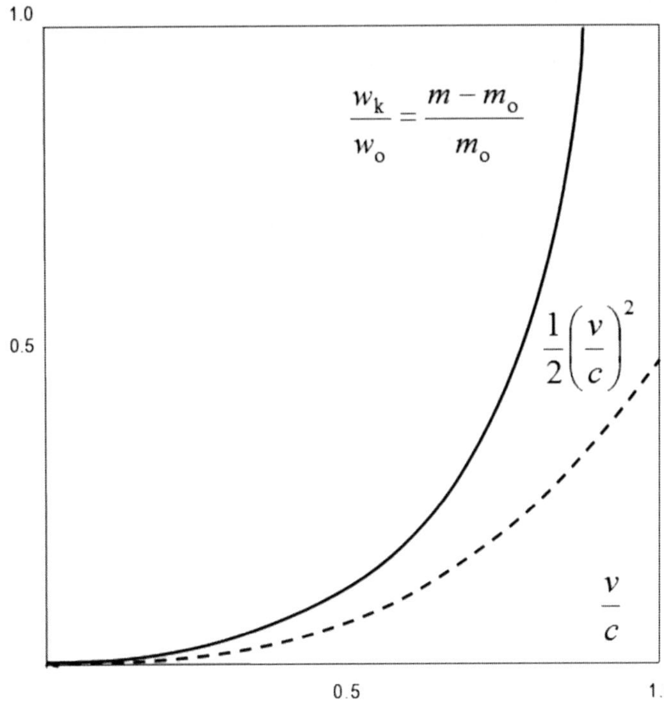


**Fig. 13.3** Mass and energy functions

mass, similar invariance may be also expected from this quantity. This indication calls in question its equivalence with the inertial mass, postulated for the sake of GRT. This theory itself is thus called in question. It, however, is not related with EM theory, and does not deserve our significant attention.

Linear momentum in force action law may be expressed as the product of three factors ($mvv_0$), with its time derivative consisting of three terms. The two former terms form the *inertial*, and latter one – *centrifugal* forces:

$$f_i = \frac{-m\partial_t v}{g^2}, \qquad f_c = \frac{mv^2}{r} \tag{13.8}$$

They are the functions of the variable mass. In relation to resting mass ($m_0 = mg$), the forces are scaled by the factors $g^3$ or $g$. The inertial force keeps the inherited speed, and centrifugal one supports the strait direction.

## 13.5 Central Distributions

The two EM fields were introduced empirically, in accord to their actions upon respective dipoles (13.2) & (13.3). They have been theoretically applied later, to interactions of electric poles and currents (13.4). The action upon an object charge, as the field definition (13.10a), resolves the static law (13.4a) into this definition and distribution of the central field of a carrying charge (13.9a). The equations (a) thus express the *central distribution* and *action* of electric fields:

$$E = \frac{q_1 r_0}{4\pi\varepsilon r^2}, \qquad f_c = \frac{\mu q_1 V \times r_0}{4\pi r^2} \tag{13.9}$$

$$f_e = q_2 E, \qquad f_m = q_2 v \times B \tag{13.10}$$

Such resolution of the incomplete kinetic law (13.4b) would not be correct. The two partial laws (b) were introduced inductively, before the complete law itself. The field of a convectional current ($qV$), in the form (13.9b), is known as Ampere's theorem. At interaction of two parallel line currents, the force is perpendicular to this field and the two currents. In reverse to the resolution of the static law (a), the elimination of the magnetic field from (b) should give the kinetic law. However, apart from the radial force (13.4b), an additional axial component, not satisfying action-reaction symmetry, is thus obtained. As the torque on a moving dipole, this force was not confirmed by Trouton-Noble experiment.

Alike above definitions of the central fields, from respective forces, the resolution of the two alternative laws (13.6), as the energies, gives the distributions and actions of the two central potentials, *static* (a) and *kinetic* (b):

$$\Phi = \frac{q_1}{4\pi\varepsilon r}, \qquad A = \frac{\mu q_1 V}{4\pi r} \tag{13.11}$$

$$w_s = q_2\Phi, \qquad w_k = -q_2 v \cdot A \tag{13.12}$$

The validity of these equations is confirmed by their equivalence with (13.9) or (13.10). Due to the scalar product, the equation (13.12b) is relevant only in the longitudinal direction, for the parallel components of the two speeds.

## 13.6 Algebraic Relations

Algebraic equations relating the two EM fields or their objects are introduced inductively. The actions of moving fields are described by the *convective* (13.13), and reactions upon dissimilar objects – by *relative* relations (13.14):

$$B = \varepsilon\mu V \times E, \qquad E = B \times U \tag{13.13}$$

$$E_{eq} = v \times B, \qquad B_{eq} = \varepsilon\mu E \times u \tag{13.14}$$

Here $u$ is the speed of a magnet, and $U$ – of magnetic field. *One moving field produces the other* (13.13). The central fields (13.9) mutually compared give the relation (13.13a), and the substitution of the forces (13.10) – the *equivalent* electric field (13.14a). The *inverse convective* relation (13.13b) has been established empirically, with the *inverse relative* relation (13.14b) introduced as the set completion. With respect to (13.9b)& (13.10b), magnetic field – embracing the causing current – acts by transverse kinetic forces on the object currents, including free or bound electricity in moving bodies. For these reasons, all above relations are in the form of the vector products of a field with its own speed (13.13), or with that of dissimilar object (13.14). The convective and relative relations in fact describe the two phases of the same process: production of dissimilar fields (13.13) and reactions of respective external fields (13.14). The former equations observe EM interactions from the moving fields, and latter ones – from moving objects. The two crosswise relations are usually used: the convective (13.13b) – for moving magnetic field, and the relative (13.14a) – for electricity moving through such a field. Similarly, the potentials (13.11) compared and the energies (13.12) equaled, give the two *alternative* relations (a), *convective* and *relative* ones:

$$A = \varepsilon\mu\Phi V, \qquad J = QV \tag{13.15}$$

$$\Phi_{eq} = -v \cdot A, \qquad Q_{eq} = -\varepsilon\mu v \cdot J \tag{13.16}$$

The substitution of the central potential distributions (13.11) into (a), equivalently relates the two carriers, electricity and current (b). The convective pair (13.15) accords to respective field relation (13.13a), and relative one (13.16) – to (13.14a). With respect to the field relations, each of these two sets lacks for a pair of relations. This fact is the consequence of the scalar form of the laws (13.12).

## 13.7 Field Transformations

The crosswise addition of the nominally similar fields, the *real* – from (13.13), and *equivalent* – from (13.14), gives the two *mutual inductions*:

$$B_+ = \varepsilon\mu(V - u) \times E, \qquad E_+ = -B \times (U - v) \tag{13.17}$$

These equations describe the same pair of interactions: electricity is the carrier and magnet the object in (a), and opposite – in (b). The convective terms arise at motion of dissimilar fields, and relative – of similar objects.

The case of a common detector moving at the speed $u = v$ – through the two EM fields, demands both equations (13.17). The addition of the inductions (13.17) to respective fields gives the *primary field transformations*:

$$B' = B + \varepsilon\mu V' \times E, \qquad E' = E + B \times U' \tag{13.18}$$

Here $V' = V{-}v$ and $U' = U{-}v$ are the speeds of the two fields, transformed in the relation to the common detector moving through them. These equations do not take into account the motion of the convective fields (13.13), nor the relative inductions in these fields. Added to the primary fields (13.18), these effects form the *secondary transformations*:

$$B'' = B' + \varepsilon\mu V_c' \times E_c, \qquad E'' = E' + B_c \times U_c' \tag{13.19}$$

Though the convective fields (13.13) are included into (13.17), their speeds are still unknown. Not only that this question has not been explicitly asked, but the resting convective fields were implicitly understood: $Vc = Uc = 0, or Vc' = Uc' = -v$. Thus defected transformations (13.19) were the starting bases for the *relativistic postulates*. The same theoretical gap had disabled in the former times the generalization of the incomplete kinetic law (13.4b). With gradual analysis of this problem, let us determine the wanted field speeds and formulate the law.

## 13.8 Kinetic Law

After elimination of the magnetic field from (13.9b) & (13.10b), with implicit supposition of its rest, there follows the conclusion of the torque on a moving dipole. On the other hand, if this field were moving together with its cause, the kinetic force acting on the object in common motion would be annulled. As the synthesis of the thesis and antithesis, we suppose the *transverse* field motion. With respect to this motion, the *radial* and *axial* force components are:

$$f_r = n[(U_c - v) \cdot V]r_0, \qquad f_a = -n[(U_c - v) \cdot r_0]V \tag{13.20}$$

Due to the transverse speed $U_c$, its scalar product with the axial speed $V$ annuls, and the radial force (a) turns into (13.4b). The annulment of the asymmetrical axial force (b), not noticed in the mentioned Trouton-Noble experiment – in the case of a moving dipole ($v = V$), demands the zero value of the expression in the brackets. This finally gives the wanted magnetic field speed (a):

$$U_c = Vcot\theta i_t, \qquad V_c = Ucot\theta i_t \tag{13.21}$$

The symmetric speed (b) of the convective electric field we introduce by analogy. $\theta$ is the angle between the given field and its speed. The circular field lines spread in the front and shrink behind a moving carrier, enabling thus its motion. The result (a) can be confirmed geometrically, as the transverse convective derivative at motion of a charged particle and its central potentials (13.11). These speeds are independent of the field objects, and so the two results (13.21) are general. (13.21a) applied to the poles independently moving in parallel turns the two *oblique* (13.20), into respective *orthogonal* components:

$$f_e = -nV^2cos\theta i_a, \qquad f_m = -nVvsin\theta i_t \tag{13.22}$$

Well, this is the central kinetic law in the form of Lorentz' force: $f = q(E + v \times B)$. At the two equal speeds ($v = V$), the vector sum accords to (13.4b).

The electric force (a) is odd around the transverse plane. In a line current, as the stream of plenty charges, the sum of such odd functions annuls. Therefore, it could not be noticed in technical practice. Respective axial electric field ($E = fe/q$) subtracted from the static central field, squeezes the result from this direction into ellipsoidal form. SRT ascribes this effect to the increased transverse fields, in ac-cord to respective artificial transformations. In fact, this theory was the direct consequence of the above theoretical misconception. Our resolution of the antinomy uproots the main reason for the primary introduction of SRT.

Though derived from the incomplete law (13.4a), the dynamic force (13.8) is resolved into the components, alike the kinetic force (13.22). Obviously, the two components (13.8), *inertial* and *centrifugal* ones, are in relation with *electric* and *magnetic* forces (13.22), respectively. Irrespective of their strict relation, these analogies say that the two longitudinal components – (13.8a) & (13.22a) – are of the *electro-dynamic*, and transverse ones – (13.8b) & (13.22b) – of *magneto-kinetic* natures. Their former names, (13.8) – as dynamic, and (13.22) – as kinetic, are thus surpassed. Though dynamic forces understand inertial and centrifugal components, the radiation fails at a centripetal charge acceleration. It is related with electric field, but the magnetic one keeps only the direction of motion, without any transfer of energy.

## 13.9 Differential Equations

On the bases of the space distributions of the central fields and potentials, in the functions of position and motion of their carriers, the following two differential sets are introduced in various ways. *Maxwell's equations* (a) relate the fields with carriers, and *gauge conditions* (b) – fields and potentials:

$$divE = \frac{Q}{\varepsilon}, \qquad E_s = -grad\Phi \tag{13.23}$$

$$curl B = \frac{\varepsilon\mu\partial E}{\partial t} + \mu J, \qquad B = curl A \tag{13.24}$$

$$curl E = \frac{\partial B}{\partial t}, \qquad E_d = -\frac{\partial A}{\partial t} \tag{13.25}$$

With respect to non-vortical static, and vortical dynamic fields, the same symbol ($E = E_s + E_d$) is used in (a). The *static equation* (13.23a) is some generalization of Gaussian theorem, as the generalized central static field (13.9a). The discrete charge is replaced by respective scalar field, $Q = \partial q / \partial v$, as the density of electricity representing the field sources. The elastic medium deformation, proportional to $\varepsilon$, partially compensates the initial charge, as the field source.

In the similar manner, the generalization of Ampere's theorem (13.9b), as the central distribution of the magnetic field, via the line integration along an infinite conductor, gives the *kinetic equation* (13.24a). The convectional current ($qV$) is thus substituted by respective density, $J = QV$, as the sum of the convective and conductive flows. This sum is supplemented by the field derivative, as a displacement current in dielectrics. Namely, electric field slightly displaces the bound electricity, positive in relation to negative polarities, disturbing thus their mutual position. This is well-known as the medium polarization, resulted from the process of dis-placing, as respective component of the electric current.

The *dynamic* equation (13.25a) is introduced by generalization of the integral law of induction, formulated on respective Faraday's empirical results. The three Maxwell's equations successively introduce the three EM fields: *static*, *kinetic* and *dynamic*. The second equation relies on the first, and the third – on the second. This explains their distinct forms, and asymmetry of the two electric and third magnetic fields. The forth Maxwell's equation, $div B = 0$, concerns free magnetic poles, as the supposed sources of respective field. As the result, it says that these poles do not exist, denying thus the initial supposition. In addition, this equation fails in respective pandanus between the gauge conditions, at right.

More directly than Maxwell's equations, gauge conditions express physical essences of EM forces: *static* – as elastic reaction of the compressible fluid, *kinetic* – as Bernoulli's transverse pressure loss, and *dynamic* one is nothing else than Newtonian force action law. In addition, their derivation is also simpler and more convincing. The two former conditions, static and kinetic, follow from the comparison of the central distributions. Radial derivative ($grad$) of the central potential (13.11a), gives respective electric field (13.9a). In the similar manner, the transverse derivative ($curl$) of the kinetic potential (13.11b) – uniformly directed, gives the magnetic field (13.9b). Curl applied to the *dynamic* condition, as the force action law, with substitution of the kinetic one, gives the dynamic equation (13.25a).

These two sets are in relation with the classical dilemma of the force action: by *successive transfer or directly at a distance*. In accord to the continuity principle, Maxwell advocated the former view. Though the dispute has not been reliably resolved, Maxwell's view has been accepted together with his equations, without any proof. Nobody has noticed that the speed of transfer was not included into these equations. The speed itself has never been determined, but the value c is understood. This view is transferred into quantum theory, where particles interact by continual barter of photons. However, this conception do not explain at all the interactions between particles, especially not their attraction.

According to this view, a particle would be an inexhaustible source emitting the fields and receiving such emissions from other particles. Unlikely, the other view concerns a particle only equipped by its fields stretched into surroundings, acting directly and instantly at each point. Really, EM forces are determined in practice from the field distributions, without elaborated theory of their propagation. In the final instance, the particles interact and manifest only by their fields. There is none a proof of existence of any hard particle body, distinct from its fields. A particle is the center, and may be something as a knot of its fields. This conclusion radically modifies the interpretation of the above differential sets.

The carriers at right sides of Maxwell's equations are nothing else than the differential features of the fields, indicated at left. Such mutual relation of the fields and potentials is more convincing. By the gauge conditions, the fields just describe the shapes of the potentials. In the tensor 4D calculus, the potentials are *vectors*, fields – *bi-vectors*, and carriers – *tri-vectors*. The numbers of components are also the same: 4-6-4. The components of 4D potential are associated to the four axes: static to *temporal*, and kinetic to *spatial* ones. EM fields, as the bi-vectors, belong to the six planes: electric – to *longitudinal* $(xt, yt, zt)$, and magnetic to *transverse* $(xy, yz, zx)$ ones, 'surnamed' in relation to the temporal axis.

## 13.10 EM Induction

By generalization of the two central fields (13.9), the two former Maxwell's equations are obtained, and by the field comparison – *rational* relation (13.13a). The dynamic equation (13.25a) and *empirical* relation (13.13b) rely on experience. The application of *div&curl* to the convective relations (13.13) gives two pairs of differential forms, wider than Maxwell's set. The comparison points to the space derivatives of field speeds, as the factors in the excessive terms. This fact demands the restriction of the relations to the homogeneous speeds, without rotation or deformation of the moving fields, behaving as the rigid structures stably oriented in space. Just such fields interact instantly at each point of the domain.

Apart from thus restricted form of the field motion, let us consider that of its direction. In accord to (13.17b), the relevant*mutual speed* and *induction* are perpendicular to the magnetic field. Their common surface also contains the *kinetic potential*, as the vortical field. The *transverse gradient* of the magnetic field, in the same surface, is perpendicular to the vector potential. Since the two mutually perpendicular vortical fields cannot exist on the same surface, the gradient is a *non-vortical* field. The motion in direction of the gradient causes the induction along vector-potential, and opposite. Of course, the *vortical* or *non-vortical* form of the induction accords to respective collinear field in each case.

Magnetic field does not move along the vector potential, or this motion does not cause any effect. Unlike the field, the object conductor can be moved or even rotated in any direction. In the relative EM induction, the two directions are relevant: the former along the field gradient, and latter along the vector-potential. Respective two types of induction are illustrated on Fig. 13.4.

Each of the two circuits consists of an instrument and its terminals, with an additional conductor sliding along the terminals. Instead of the primary conductor, the two secondary ones are moving, causing the relative inductions (13.14a). The transverse motion of the parallel conductor, along the field gradient, causes the vortical induction directed as the primary current or kinetic potential. The two opposite inductions, in the moving transverse conductor, play the role of the apparent non-vortical electric field symmetric to the primary current.

The former case may be generalized to the mutual motion of the primary and secondary conductors, thus changing the kinetic potential: $\partial A/\partial t = [(v - V) \cdot \nabla]A$. In accord to (13.25), this gives the real vortical induction. The latter case, however, does not obey the principle of relativity. The translation of the transverse conductor, together with its free electricity, represents the set of convectional longitudinal cur-rents, parallel to the primary one. In accord to (13.22b), this produces the transverse forces affecting the moving object electricity. The equivalent induced field does not exist out of the conductor itself. In addition, similar longitudinal motion of the current carrying conductor would not cause any effect.

In the Faraday's experiment (Fig. 13.5.) sliding contacts of the instrument terminals touch the center and rim of a conducting disc, rotating in the front of a cylindrical magnet, around the common axis. The thoroidal magnetic field (Fig. 13.2.), of the circular magnetization current on the magnet cover, is axially symmetric, without any gradient along the rotation. Free electricity inside the rotating disc forms the set of circular currents, attracted or repelled by the magnetization current. In the circuit and measuring instrument, this (in fact kinetic) effect gives an apparent impression of the radial non-vortical induction. This field, however, does not exist out of the rotating disc, and cannot be measured as such.

An additional Faraday's experiment is easily reducible to this one. With respect to the statistically neutral magnetic material, possible rotation of the magnet, together with the disc, does not change at all the magnetization current, nor the induction.

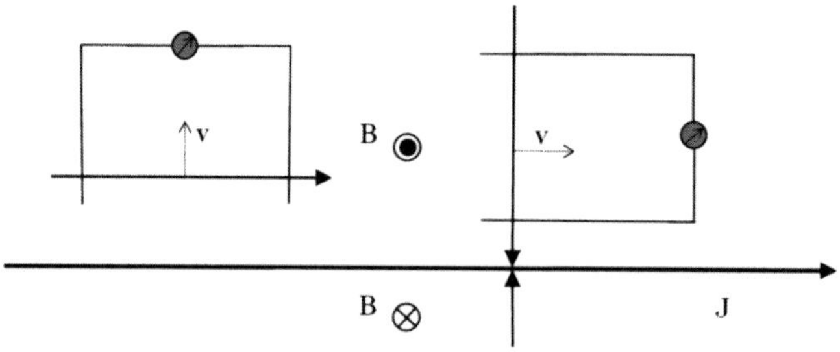


**Fig. 13.4** Vortical and 'non-vortical' inductions

This is the same as if the sliding contacts are reconnected directly to the magnet, as the sufficiently conducting material. Not only that this effect has been the enigma for EM theory up to these days, but this author himself was led astray, believing for many years in the real non-vortical induction. Being unable to ex-plain its essence, he is forced to the above reexamination.

## 13.11 EM Antinomies

Although Maxwell's equations, as the basic laws of EM theory, have been widely and successfully applied, their formalistic applications follow into a few antinomies, which solution demands the revision of some principal views. In this sense, the static equation (13.23a) applied to two typical surface charges – of a sphere (Fig. 13.1.) and of plane (Fig. 13.6.) – gives respective distributions in surrounding space, of the fields perpendicular to each of the two surfaces.

The central field apparently exists only out of the spherical volume, and symmetric field of a plane seems to be homogeneous. For the same surface charges, the former field is twice stronger than the latter, split into halves. The former field obeys inverse square function, and respective potential (13.11a) – inverse function. The latter potential is problematic. In accord to the homogeneous field, it would decrease by linear function, cut the abscissa at some distance, obtaining the negative values. The constant potential would mean the zero field. This antinomy demands some reexamination of the two field distributions.

If its radius grows into infinity, a finite sphere degenerates into infinite plane. There arises the question of the field redistribution during the limiting process, and of a physical reason for this event. The only possible compromise between the two extreme distributions is presented on Fig. 13.7. In accord to it, the central field is also equally divided into the two surface sides. Due to various field directions, the sum annuls in the center, and has some values in other internal points, maximal – close to the surface. Not only that the separate fields exist in the zero sum, but they continue to the opposite surfaces, break them and add to the local fields, with the sum twice stronger than the field of a charged plane.

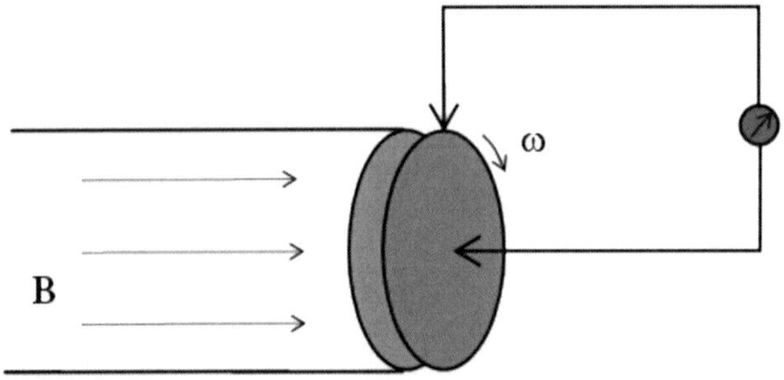


**Fig. 13.5** Faraday's rotational induction

The obtained relation between the two extreme field distributions results from the unity of the two surfaces. The locally observed flat surface is in fact the sphere of extremely long radius. Alike the equator on a globe, dividing the spherical surface into the two equal circles, with the poles as their opposite centers, the plane divides 3D space into the two hyper-spheres. This follows to the substitution of the flat – Euclidian, by the curved – Riemannian space, with circular dimensions. This view also explains the potential of the plane. In accord to the field on Fig. 13.7., the potential is only locally linear. At greater distances it decreases slowly, tending asymptotically to zero and approaching this value at infinity.

This explanation, however, calls in question the static equation (13.23a). Namely, the internal field lines (Fig. 13.7.) start from the charged surface, and terminate in the neutral space. Of course, this is the phenomenal vector sum, as the action of plenty separate fields. Keeping their individualities, these fields obey the static equation. This consideration in the whole points to undisturbed coexistence of more fields of the same nature at least, in the same locations. The numerous fields somehow pass by and cross each other, as if being displaced along the structural depts. In this sense, with the *three* spatial and *forth* temporal axes, a *structural dimension*, as the *fifth*, must be introduced. The following solution of an apparent antinomy of the kinetic Maxwell's equation, also speaks in favor of this idea.

The application of (13.24a) to a moving central field (Fig. 13.1.), via its convective derivative, $(V \cdot \nabla)E = -\partial E/\partial t$, finally gives the magnetic field (13.9b). However, the same procedure cannot be applied to a line current. If the vertical line on Fig. 13.6. were a line conductor, the horizontal arrows would represent electric field lines of the moving electricity forming the current. In accord to the above view,

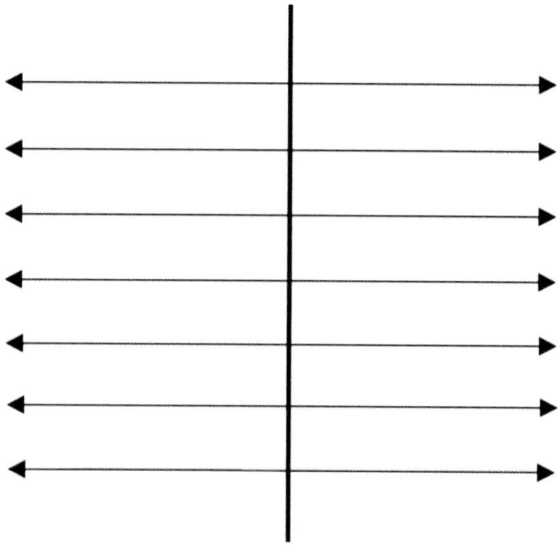

**Fig. 13.6** Euclidian field distribution

this field is separated from the opposite field of the resting polarity. There arises the additional problem, of the zero field gradient in direction of its motion. It can be surpassed only by separate treatment of elementary particles, their central fields and respective processes, with the final sum of the effects. This conclusion affirms the simultaneous coexistence of the multiple electric and magnetic fields in the same 3D location, being distributed along the structural dimension.

## 13.12 Structural Models

The motion of a charge, as the convectional current $qV$, is continued in space by displacement current $-\varepsilon(V \cdot \nabla)E$, in the surroundings. Axial gradient of the central field gives the *thoroidal vortex* of this current, Fig. 13.2. This vortex may be understood as the photon associated to the moving particle. Possible deceleration of the particle would release this photon fully or partially. Free from the carrying particle mass, it continues its own propagation at the speed $c$. In fact, it takes over the kinetic energy, as the energetic difference of the moving and resting particle. In accord to the function $r = r_0 g$, where $g$ is smaller than unit, the photon energy equals to that of the electric field between the two particle radii.

In analogy to the photon, the model of a charged particle may be also predicted. The quantity $c^2$ – in the static law (13.4a), against the product of the two speeds, in kinetic one (13.4b), points to the common speed of all the particles, along temporal axis. The same speed value points to similar structural model, as a *hyper-thoroidal*

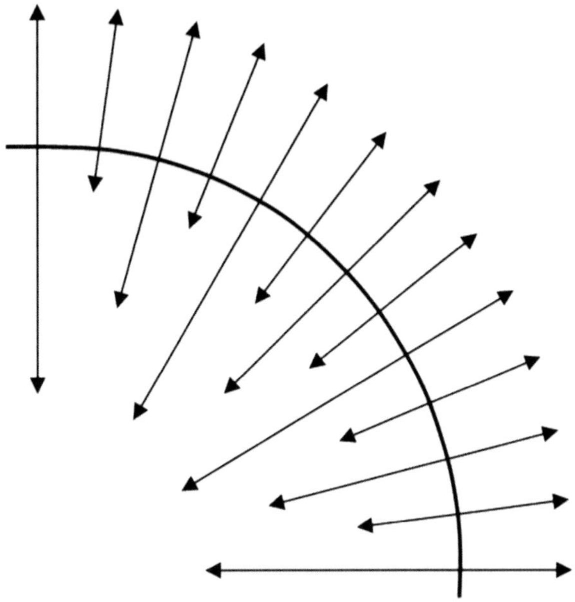

**Fig. 13.7** Riemannian field distribution

vortex of the quantum fluid flow, coaxial with temporal axis. Alike that of photon, this motion is the condition of the particle existence. The two references of propagation, of photons – through 3D space, and of particles – along the temporal axis, accord to the two structural levels along the fifth axis. They are substantial substrata of the two respective physical quantities: displacement current – of photons, and kinetic potential – of elementary charged particles.

Well, a hyper-thoroidal vortex, coaxial with t-axis, is the model of a charged particle. If Fig. 13.2. represents a positive, the opposite fluid flow forms the negative particle. These two models obey the known CPT symmetry (circulation-polarity-time). Each of the particles is symmetric in 3D space, but two opposite polarities are mutually anti-symmetric along the temporal axis. The alternation of one of the CPT parameters would alternate an additional. For instance, the reorientation of the temporal axis would alternate the polarities of all the particles. Irrespective of the theoretical basis of this quantum rule, it is fitted into our particle model. Moreover, this model offers the resolution of a sequence of scientific dilemmas. Let us announce at least some of these fundamental answers.

The boundary between the internal and external flows, as the cross-section of a circular line – on Fig. 13.2. – represents 3D particle surface. At a positive particle, the internal fluid flow adds to, and external one subtracts from the common speed. In accord to Bernoulli's law, these two results are manifest in 3D space (transverse to t-axis) by some disturbances of the pressure, as the static potential. This potential is negative inside, and positive out of the positive particle (Fig. 13.8.). This is the reason of the lesser radius and greater mass of a proton than the same electron measures, related by (13.7b). Unlike the current scientific views, proton would also be the elementary particle. The exceptional stabilities in separate states speak in favor of the elementary natures of the two basic charged particles.

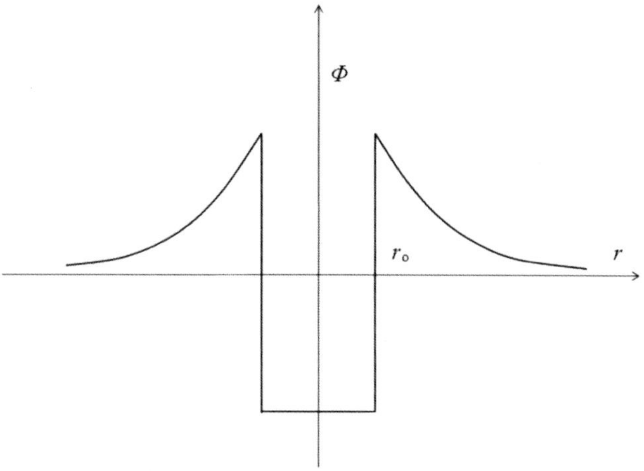


**Fig. 13.8** Elementary central potential

The common motion of all the particles along t-axis (at speed $c$) may be related with cosmic expansion. The cosmos itself would be reduced to a hipper-spherical wave, in accord to the circular form of its axes. If the temporal axis were a strait line, cosmos would expand into infinity. In order to avoid this undesirable process, the cosmology was looking for physical reasons for deceleration and possible redirection of the cosmic process. Without our idea of wave propagation, gravitational attraction between all the celestial bodies seemed to be such a cause. However, an estimation pointed to insufficiency of all the cosmic matter for such a reversal. To provide the needed gravitational attraction, some invisible *(dark) matter*, equivalent to the *massive quantum fluid*, has been introduced.

As the projection from the temporal into spatial domains, the speed of expansion depends on the t-axis form. In common with the wave flow, circular form of this axis would surpass the dead end of the former cosmology. The uniform motion along such an axis is projected to the cyclic cosmic pulsation between the two singularities in 4D space. In between, after *decelerated expansion*, there follows *accelerated contraction*. On the other hand, the parallel motion of all the matter, in accord to Bernoulli's law, points to gravitational attraction, as the consequence. The causal relation is thus reversed: instead of a determinative factor, gravitation may be a mere consequence of the cosmic flow. Apart from the opposite causality, the propagation need not be reversed for cosmic contraction.

This model itself is called in question by the recent remark of the increasing red shift of the light arrived from very distant super-nova stars. Instead of the expected deceleration, this fact is interpreted as the fantastic possibility of the *accelerated cosmic expansion*. In opposition to dark matter, its cause is ascribed to some *dark energy*, similar to the internal *pressure* of the compressible *quantum fluid*. Though the multiplication of indistinct concepts is none a scientific progress, this explanation has been widely accepted and crowned by Nobel prize. As if, the gravitational influence on the red shift is not taken into account. Its combination with collapsing process of mentioned stars may be the reason of the increasing red shift. This author himself fails in sufficient data for the final conclusion.

Some additional questions here arisen stay still open. Apart from determination of the temporal axis form, there are interrelations and final unity of various physical interactions, from the mechanical, via EM, up to the nuclear forces. Of course, these will be the themes of the future times and investigations.

## 13.13 Conclusion

With respect to very wide and long application of EM theory, all the facts being measurable had already been checked. Some crucial results, however, have so far not been convincingly explained. Apart from *reinterpretation* of these results, our text tries to *classify* conveniently all the quantities and their relations, with respective *arrangement* and *concise exposition* of the theory. Instead of the incomplete classical or *problematic* relativistic approaches, as the thesis and antithesis, this work may be understood as their *supplemented* synthesis.

The three classes of EM quantities have been considered: electric *charges* and *currents* – as the *carriers* of the static & dynamic *electric*, and kinetic *magnetic fields*, with the static – *scalar*, and kinetic *vector potentials*. In the reverse order, the three quantity classes are expressed by the vectors, bi-vectors and trivectors in 4D space. They are related in 3D space by a few types of equations. *Usual central laws* directly express EM forces, and *alternative ones* – respective energies. The quantities of the same ranks are related *algebraically*, and the order of *differential* equations equals to the difference of the related ranks. All the laws in common constitute a solid system of the complementary relations.

The general *kinetic central law* (13.22) is the palpable original contribution of this work. With explanation of a few problematic experiments, as Faraday's rotational induction and Trouton-Noble negative result, it removes respective theoretical dilemmas. The *secondary field transformations* (19) are also completed. The relativistic postulates and respective kinematical transformations, based on the incomplete field transformations, are finally disqualified. On the other hand, *Einstein's equation* and Lorentz' *mass function* are explained and affirmed. Instead of the *acceptance* or *rejection* of STR, its fundamental aspects are here separated from the mainstream speculations of this controversial theory.

EM processes are reduced to the mechanics of some *quantum fluid* in 4D space. The *compressibility*, *super-fluidity* and *inertia* of this fluid are the bases of the radial *static*, transverse *kinetic* and axial *dynamic* forces, respectively. All the forces depend on mutual *position* of interacting charges, the kinetic also depend on their *speeds*, and dynamic ones – on *acceleration* of one of them. The essences of *mass* and *electricity* are finally explained, with the main structural models, from a *photon* up to the full *cosmos*. The circular spatial axes are affirmed and explained by the wave cosmic process, which determines by itself *gravitational attraction*, with *unique* and *uniform lapse of time*, in accord to Galilean view.

## References

1. Mikovi, B.: ELECTRODYNAMICS, http://solair.eunet.rs/brami

# Chapter 14
# Advanced Adaptive Algorithms in 2D Finite Element Method of Higher Order of Accuracy

Pavel Karban, Ivo Doležel, František Mach, and Bohuš Ulrych

**Abstract.** Sophisticated methods of automatic adaptivity in finite element methods of higher order of accuracy are presented. The main attention is paid to hp-adaptivity techniques that exhibit the highest level of flexibility and exponential convergence of results. The technologies implemented in our own fully adaptive FEM codes Agros2D and Hermes2D based on a higher-order finite element method are illustrated by three typical examples.

## 14.1 Introduction

All advanced techniques for numerical solution of physical fields contain special algorithms for automatic adaptivity of discretization meshes (see, for example [1], [2], [3]. These algorithms are applied at the moment when the error of solution is higher than the acceptable tolerance. This error defined as the difference between the current numerical solution and exact solution is usually caused by locally rougher mesh, presence of one or more singular points, curvilinear boundaries or interfaces approximated by polygonal lines, etc. In all these cases, such errors must be identified in the course of computation and appropriate measures have to be taken for their fixing. This technology is also implemented in the codes Hermes2D [4] and Agros2D [5] that have been developed for a couple of years in our group. While Hermes is a library of numerical algorithms for monolithic and fully adaptive solution of systems of generally nonlinear and nonstationary partial differential equations (PDEs) based on the finite element method of higher order of accuracy, Agros is a powerful user's interface serving for preprocessing and postprocessing of the problems solved.

Pavel Karban · Ivo Doležel · František Mach · Bohuš Ulrych
University of West Bohemia, Faculty of Electrical Engineering,
Univerzitní 8, 306 14 Plzeň, Czech Republic
e-mail: {karban, idolezel, fmach, ulrych}@kte.zcu.cz

K. Kyamakya et al. (Eds.): Selected Topics in Nonlinear Dynamics, SCI 459, pp. 255–271.
DOI: 10.1007/ 978-3-642-34560-9_14 

## 14.2 Adaptivity Techniques in Agros and Hermes

The algorithms of automatic adaptivity implemented in Hermes and Agros are divided into the following principal groups:

- Refinement of elements in regions where the solution exhibits an unacceptable error. This way is called $h$-adaptivity while the original large finite element is split to several smaller elements, the degree of the polynomials replacing the real distribution of the investigated quantity in them remains the same. This is clear from Fig. 14.1, where both large element and smaller elements are described by polynomials of the same order.

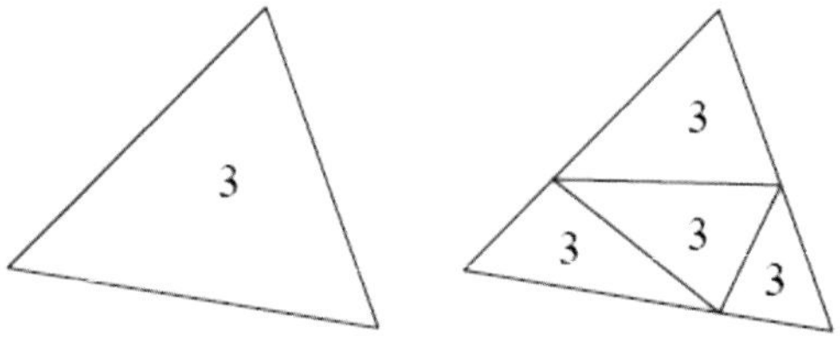


**Fig. 14.1** Typical application of $h$–adaptivity: polynomial orders in smaller elements remain unchanged

- Improvement of approximation of the investigated quantity. This way is called $p$-adaptivity the shapes of elements in the region do not change, but we increase the orders of the polynomial approximating the distribution of the investigated quantity. The situation is depicted in Fig. 14.2.

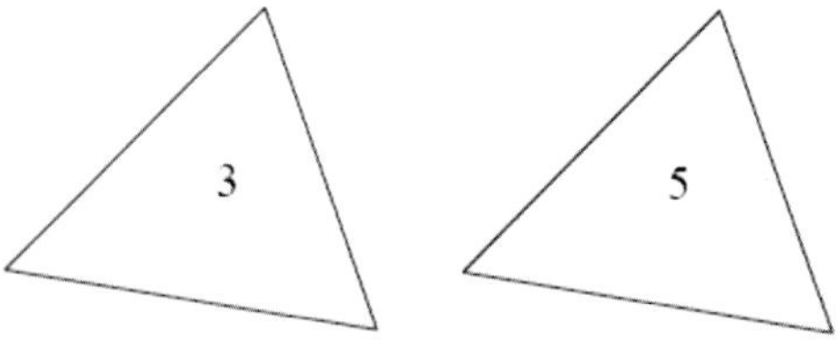


**Fig. 14.2** An application of $p$–adaptivity: polynomial order in the same element is increased

- The combination of both above ways is called $hp$-adaptivity, which belongs to the most flexible and powerful techniques characterized by an extremely fast (exponential) convergence of results. A typical possibility of its application is depicted in Fig. 14.3.
- Curvature of edges of selected elements adjacent to curvilinear interfaces and boundaries. As this technique is original and we do not know any commercial software (SW) that would use it (although, for example, ANSYS works with curvilinear elements that are, however, generated in a different manner), we will describe it in more detail. Agros2D discretizes 2D domains on the base of SW Triangle [6] that provides a high-quality triangular mesh. The corresponding input data for modeling curvilinear boundaries or interfaces in Triangle are given

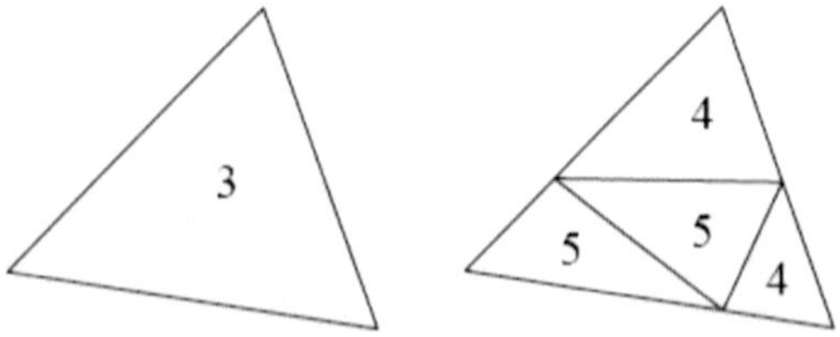


Fig. 14.3 Typical application of $hp$–adaptivity: the original element is divided into more smaller elements and polynomial order in them are increased

by a series of points lying on this line (together with the markers carrying information that these points belong to such a line) while the output is represented by a set of triangular elements (see Fig. 14.4). In the second step Agros2D repeats analyzing the curved lines and when any of the newly generated nodes approximating the curve (Fig. 14.4, right part) does not lie on it, it is automatically projected on the original arc. At the same time a special procedure determines the corresponding angles, such as angles $\alpha 1$ and $\alpha 2$ in the right part of Fig. 14.5.

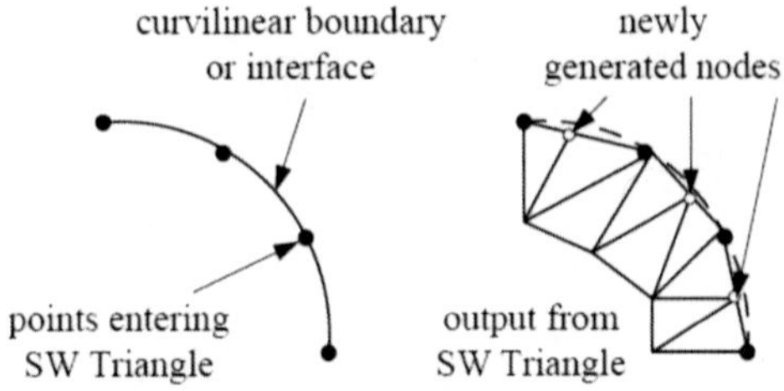


Fig. 14.4 Typical input data for SW Triangle (left) and output mesh (right)

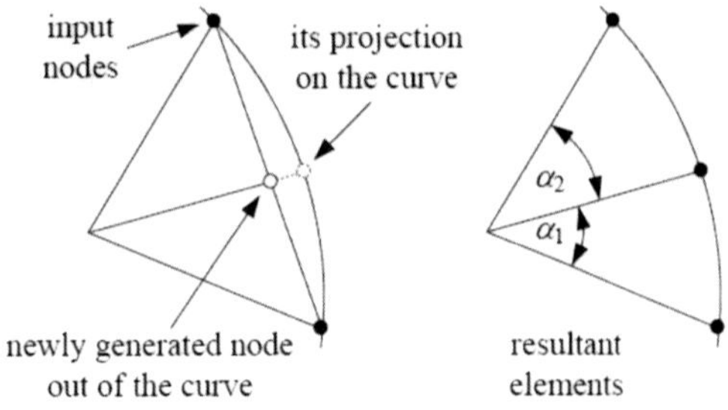


Fig. 14.5 Projection of a newly generated node on the curve (left) and resultant curvilinear elements formed in Agros2D (right)

In the course of numerical processing of the task the curvilinear elements are mapped back on normal triangles where all remaining operations (such as Gaussian numerical integration) are carried out and only in the final step postprocessing they are mapped again to the curvilinear elements.

The $h$-adaptivity is the simplest one and is implemented in practically all existing codes. The elements burdened by high errors are divided into several smaller elements, but there exist even other possibilities. The principal problems accompanying this type of adaptivity are the hanging nodes appearing on the interfaces between the refined elements and elements without refinements. What are the hanging nodes is clear from the right part of Fig. 14.6 (they do not represent vertexes of elements in not refined parts). These nodes must be handled with particular care, otherwise they may significantly contribute to the growth of the degrees of freedom of the problem solved [7].

The $p$-adaptivity is even simpler to implement, because the mesh remains unchanged. Only in the selected elements we enlarge the order of the corresponding approximating polynomials.

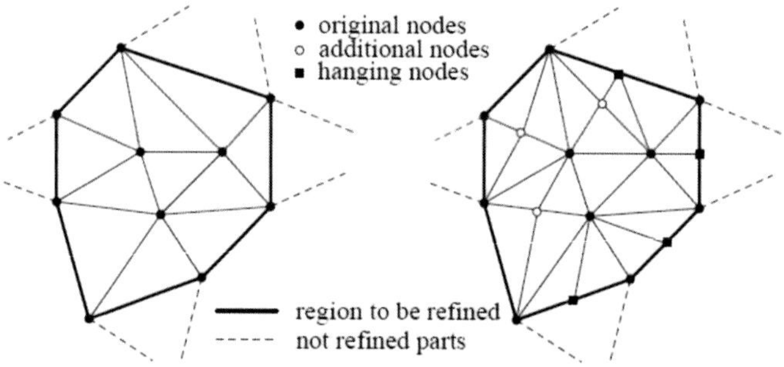


**Fig. 14.6** Refinement of the given area and generation of the hanging nodes

The $hp$-adaptivity represents the most complicated method and its implementation is highly nontrivial. On the other hand, it exhibits the exponential convergence of results and it was proven to be an extremely powerful tool just in the finite element methods of higher orders of accuracy. Although we can meet this method still very scarcely, its algorithms are already implemented in our SW Hermes2D and Agros2D.

The curved elements are applied in case of modeling curvilinear boundaries and interfaces.

In the present version, Agros2D also allows combining triangular and quadrilateral elements (these are desirable either in the regions with anisotropic materials or in regions where the field is expected sufficiently smooth).

## 14.3 Error of Solution

Algorithms of adaptivity start to be applied at the moment when some local error of solution is higher than the acceptable tolerance. Consider an equation

$$Lf = 0 \tag{14.1}$$

where $L$ denotes a differential operator and $f$ a function whose distribution over some domain $\Omega$ is to be found. If $f'$ is its approximation obtained by the numerical solution of (14.1), the absolute and percentage relative errors d and e are defined by the relations

$$\delta = f - f' \quad \eta = 100\left|\frac{\delta}{f}\right| \tag{14.2}$$

Even other quantities, however, may be used for the estimation of errors. These quantities are called norms (they are usually denoted by symbol $||e||$) and Hermes2D works with three of them:

- Basic "energetic" norm that is defined by the formula

$$||e|| = \left|\int_{\Omega} \delta(L\delta) d\Omega\right|^{\frac{1}{2}} \tag{14.3}$$

- $L^2$ norm defined by the relation

$$||e||_{L^2} = \left|\int_{\Omega} \delta^2 d\Omega\right|^{\frac{1}{2}} \tag{14.4}$$

- $H^1$ norm defined by the relation

$$||e||_{H^1} = \left|\int_{\Omega} (\delta^2 + grad\delta \cdot grad\delta) d\Omega\right|^{\frac{1}{2}} \tag{14.5}$$

Unfortunately, the exact solution $f$ is usually known only in very simple, analytically solvable cases. Moreover, there exists no general method that would provide a good estimation of the error for an arbitrary PDE (although for several classes of linear PDEs we can find it). Moreover, in the case of the $hp$ adaptivity the traditional error estimate (one number per element) is not enough; we must know its distribution over the whole element. In principle, it might be possible to obtain this information from estimates of local higher derivatives, but this approach is not very practical. That is why we work [8] with the reference solution $f_{ref}$ instead, that is obtained either by a refinement of the whole mesh ($h$-adaptivity), by enlargement of the polynomial degree ($p$-adaptivity) or by both above techniques ($hp$-adaptivity). In this manner we get the candidates for adaptivity even without knowledge of the exact solution $f$. The library Hermes2D works with very sophisticated and subtle tools based on the above considerations.

## 14.4 Illustrative Examples

In order to show the power of the indicated adaptive algorithms, we will illustrate their application by three different examples. In some of them we will compare the results with results obtained using other professional commercial codes.

### 14.4.1 Example I (hp Adaptivity)

The first example is inspired by the solution of the Schrödinger equation describing the interaction between two atoms. It can be found in a benchmark example collection [9] that serves for the comparison of capabilities of various existing codes. The equation in the form

$$\frac{1}{r}\frac{\partial}{\partial r}\left(r\frac{\partial u}{\partial r}\right)+\frac{u}{(r+\alpha)^4}=\frac{r-\alpha}{r(r+\alpha)^3}cos\left(\frac{1}{r+\alpha}\right),\quad r=\sqrt{x^2+y^2} \tag{14.6}$$

is solved on a unit square, whose low left corner is at the origin of the Cartesian coordinate system. Its particular analytical solution is

$$u=sin\left(\frac{1}{r+\alpha}\right) \tag{14.7}$$

and we shall assume the corresponding boundary conditions along the circumference of this unit square. The solution oscillates near the origin and the coefficient $\alpha$ is inversely proportional to the number of oscillations. For example, if $\alpha=1/10\pi$ we obtain ten oscillations and the resultant function $u$, evaluated exactly by SW Mathematica, is depicted in Fig. 14.7.

Figure 14.8 shows the meshes generated by code Agros2D after realizing the adaptive process. Its left part depicts a mesh obtained by using exclusively $h$-adaptive technique, while the right part contains a mesh obtained using fully

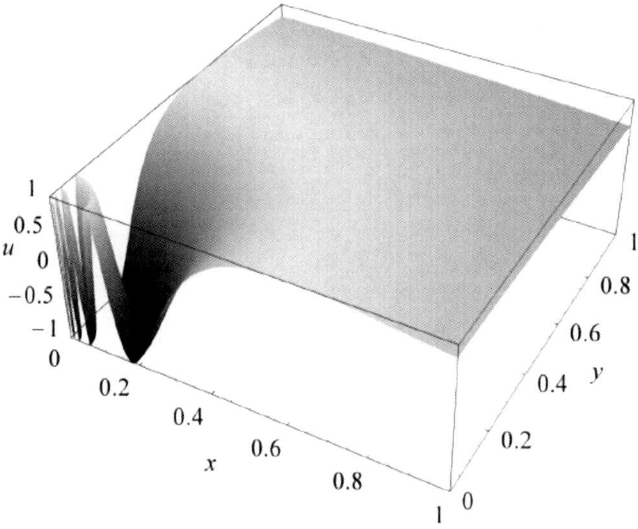


**Fig. 14.7** Distribution of function $u=sin(1/(r+a)$ for $a=1/10\pi$ (exact solution carried out using SW Mathematica)

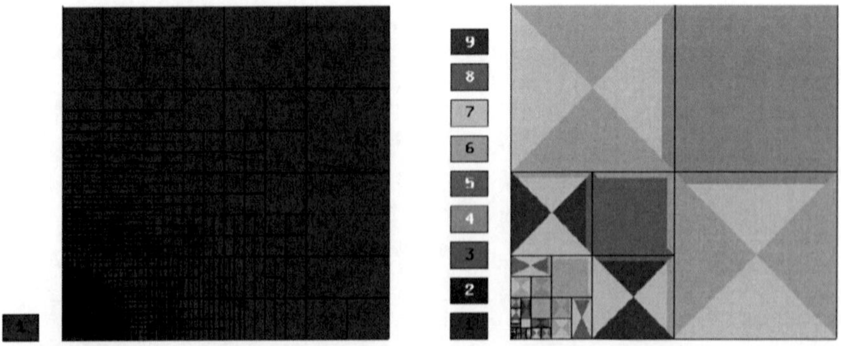


**Fig. 14.8** Discretization meshes for solving the problem: left—mesh obtained by using the h-adaptive technique (for $p = 1$), right—mesh obtained by using the hp-adaptive technique (the numbers in the rectangles show the degrees of the approximating polynomials

$hp$-adaptive technique. Both meshes satisfy the condition that the corresponding relative errors $\eta$ of solution do not exceed 0.01 %. It is obvious that the number of elements in the right-hand mesh (and, consequently, the number of the degrees of freedom (DOFs)) is much smaller, so that the $hp$-adaptivity is much more effective and the computation is characterized by a substantially shorter time and less intensive demands on the computer memory.

The numerical solution to (14.6) satisfying the condition $\eta \leq 0.01$ % obtained by Agros2D and Hermes2D is depicted in Fig. 14.9.

Figure 14.10 compares the adaptive techniques implemented in library Hermes2D with the techniques built in Comsol Multiphysics 3.5a. It is clear that Hermes2D is able to reach accuracy on the order of $10^{-2}$% with less than one tenth of

**Fig. 14.9** Numerical solution to (14.6) obtained by Agros2D and Hermes2D ($\eta \leq 0.01\%$)

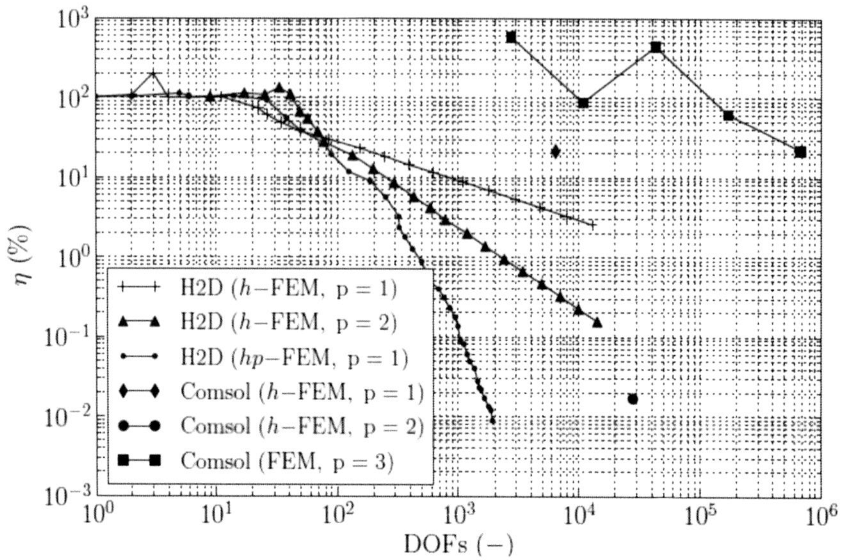


**Fig. 14.10** Comparison of adaptive techniques (relative error $\eta$ versus the number of DOFs; acronym H2D means Hermes2D

DOFs than the best version of Comsol ($h$-FEM, $p = 2$). The principal reason is that Comsol does not support the hanging nodes, which leads to a strong growth of the DOFs.

### 14.4.2 Example II (Curved Elements)

Application of the curvilinear elements will be illustrated on an example of induction heating of a cylindrical aluminum billet of diameter $d = 60mm$ rotating at a given angular frequency $n = 1500rpm$ in a stationary magnetic field generated by a system of permanent magnets, see Fig. 11. The axial length of the arrangement $l = 120mm$. The whole experimental stand is shown in Fig. 12.

The magnetic circuit is made of carbon steel CSN 12 040 [10], permanent magnets VMM10 (based on rare earths) exhibit the remanence $B_r = 1.41T$, and relative permeability in the second quadrant $\mu_r = 1.21$. All physical parameters of the billet and other materials are generally functions of the temperature [11]. The goal of the solution is to determine the time dependence of the average temperature of the billet.

The continuous mathematical model of the problem consists of two partial differential equations describing the distribution of magnetic and temperature fields. The first one, describing the magnetic field in terms of magnetic vector potential $A$, may be written in the form

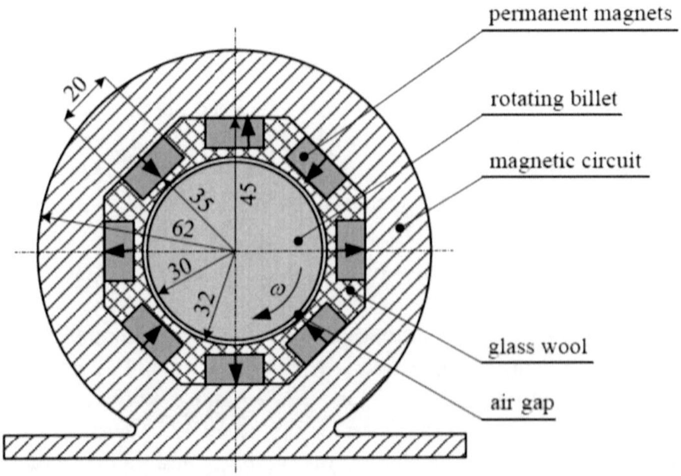


**Fig. 14.11** Cylindrical aluminum billet rotating in a system of permanent magnets (all dimensions given in mm)

**Fig. 14.12** View of the real experimental system

$$curl \left( \frac{1}{\mu} curl \mathbf{A} - \mathbf{H}_\mathrm{c} \right) - \gamma \mathbf{v} \times curl\ \mathbf{A} = \mathbf{0} \tag{14.8}$$

where $\gamma$ is the electrical conductivity, $\mathbf{v}$ is the local velocity of rotation, and $\mathbf{H}_\mathrm{c}$ stands for the coercive force of the permanent magnets. A sufficiently distant artificial boundary is characterized by the Dirichlet condition in the form $\mathbf{A} = \mathbf{0}$. The temperature field in the billet obeys the equation

$$div(\lambda gradT) = \rho c_p \left( \frac{\partial T}{\partial t} + \mathbf{v} \cdot gradT \right) - p_J \tag{14.9}$$

where $\lambda$ is the thermal conductivity, $\rho$ denotes the specific mass, and $c_p$ stands for the specific heat at a constant pressure. Finally, the symbol $p_J$ denotes the time average internal sources of heat (the specific Joule losses) determined from the formula $p_j = |\mathbf{J}_{eddy}^2/\gamma|$, where $\mathbf{J}_{eddy}$ is he density of currents induced in the billet. This is given by the relation $\mathbf{J}_{eddy} = \gamma \mathbf{v} \times curl\mathbf{A}$.

The task was solved numerically by the codes Agros2D and Hermes2D in the monolithic formulation (which means, that both magnetic and temperature fields were calculated simultaneously, so that the final numerical scheme was characterized by one stiffness matrix). We respected all possible nonlinearities, such as the magnetization characteristic of steel CSN 12 040 and temperature dependencies of all physical parameters of used materials. Figure 14.13 shows the discretization mesh (at the end of the process of adaptivity) used for computation of magnetic field in the system. The numbers in the rectangles denote the degrees of polynomials in particular elements. The billet is discretized using the mentioned curved elements (savings in DOFs are about 20% with respect to using normal triangular elements, compare with [11]). The regions in the neighborhood of the corners of the magnetic circuit representing the singular points are discretized by small triangles of low polynomial orders while places with expected smooth regions are covered by large triangles of high polynomial orders.

Figure 14.14 shows the distribution of magnetic field in the system for the nominal revolutions $n = 1500rpm$. Figure 14.15 shows the corresponding distribution of temperature after 120 s of heating. While the temperature of the billet reaches about $T \approx 150\,^\circ\mathrm{C}$ , due to the presence of good thermal insulation the permanent magnets remain practically cold (about 45 °C), so that there is no danger of deteriorating their physical properties because of overheating (maximum allowable temperature being about 80 °C).

The time evolution of the average temperature of the billet obtained by both modeling and measurement is depicted in Fig. 14.16. The figure contains two characteristics. The upper one corresponds to eight permanent magnet (see Fig. 14.12). The lower one was measured at the presence of only four magnets (instead of four remaining magnets the model contained four ferromagnetic poles of the same dimensions).

Finally, Fig. 14.17 shows the convergence curves when using triangular and curvilinear elements for different initial setting of approximating polynomials in particular cells of the mesh. Comparable results are obtained (after the process of adaptivity) for 1396 elements (some of them being curved) and 1824 purely triangular elements. The savings in DOFs in this case are about 240 %.

### *14.4.3 Example III (Curved Elements and Circular Points)*

Application of the curved elements will also be illustrated on another example that exhibits one more issue: a singular point. This means a point at which it is not

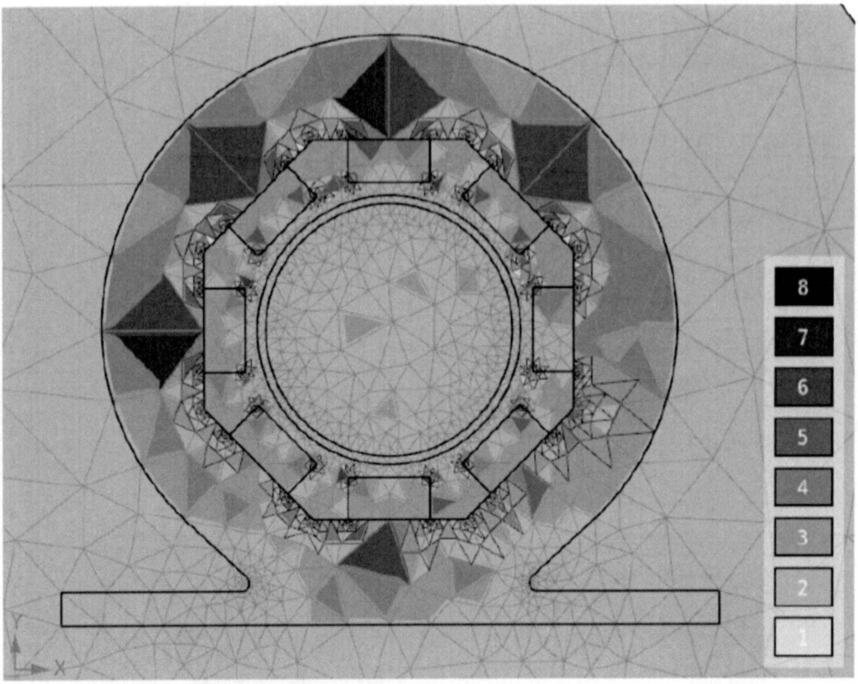


**Fig. 14.13** Discretization mesh: the cross section of the system is discretized using curved elements (light lines—before adaptivity, dark lines—after adaptivity, numbers in the right column show the orders of the corresponding elements

possible to define the normal and the gradient of the solution grows there over all limits.

Consider a spark gap in Fig. 14.18 consisting of two electrodes. The upper one (conical with a sharp point at its end) carries potential $\varphi = 1000V$, the lower one is spherical and grounded (i.e., $\varphi = 0$). The arrangement is considered axisymmetric and the task is to find the distribution of electric potential $\varphi$ in the system.

The viewpoint of modeling the problem is seemingly quite simple (electric field is described by the Laplace equation on a very simple domain). But this holds only to some extent. It turns out that common commercial codes are not able to provide correct values of potential near the peak of the cone and also its values near the sphere are somewhat distorted. The governing equation for the potential $\varphi$ in the air between the electrodes reads

$$\frac{1}{r}\frac{\partial}{\partial r}\left(r\frac{\partial \varphi}{\partial r}\right) + \frac{\partial^2 \varphi}{\partial z^2} = 0 \tag{14.10}$$

The boundary conditions are given by the prescribed potentials on the surfaces of the electrodes, and Neumann condition along a sufficiently distant artificial boundary.

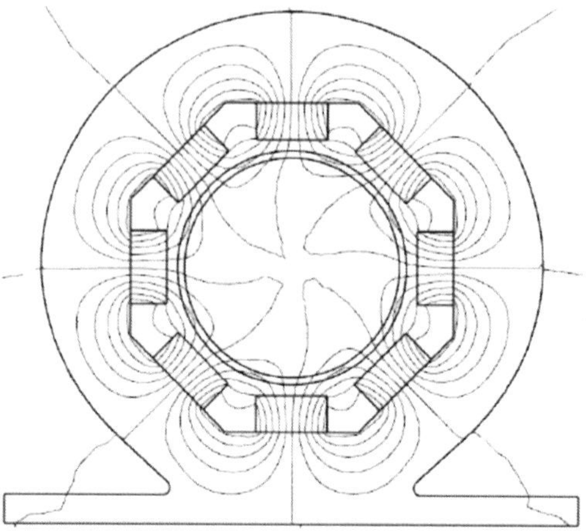

**Fig. 14.14** Distribution of magnetic field in the system for $n = 1500 rpm$

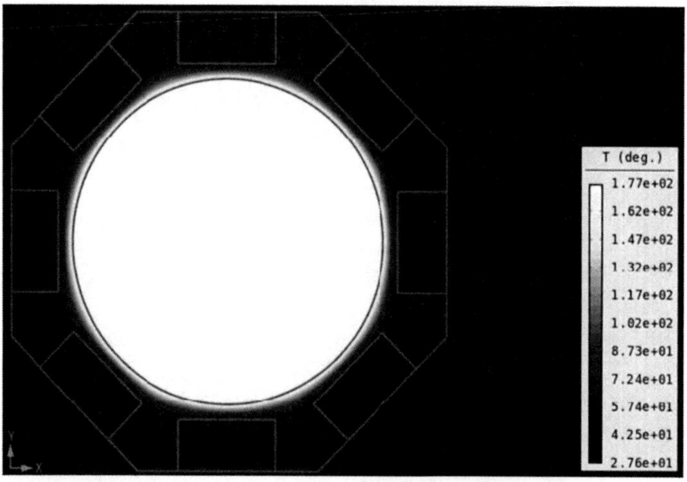


**Fig. 14.15** Distribution of the temperature in the area of the billet and permanent magnets after 120 s of heating

The problem was solved using several codes from which Comsol Multiphysics 3.5a and QuickField are commercial codes, while FEMM 4.2, Hermes2D and Agros2D are freely distributable applications.

Figure 14.19 shows some results obtained using Hermes2D and Agros2D. Its left part depicts the mesh (the light lines show the original mesh while the dark lines the final mesh after 25 steps of adaptivity and the numbers give the orders of the corresponding polynomials) and right part the distribution of potential $\varphi$ in the vicinity of both electrodes.

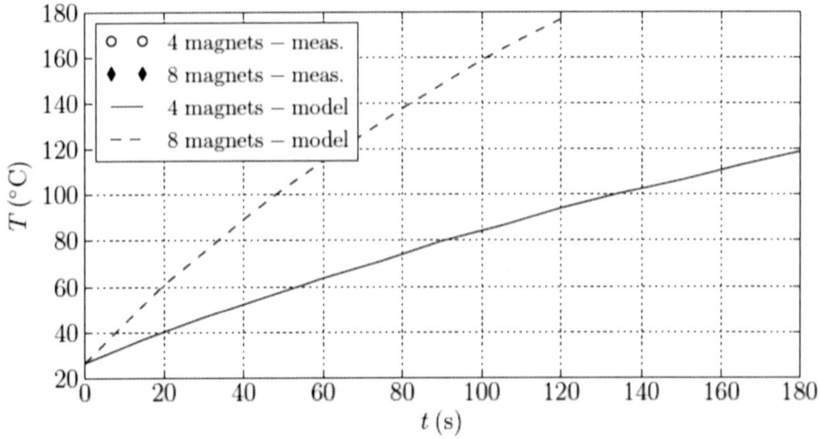


**Fig. 14.16** Time evolution of the average temperature of the billet for $n = 1500rpm$

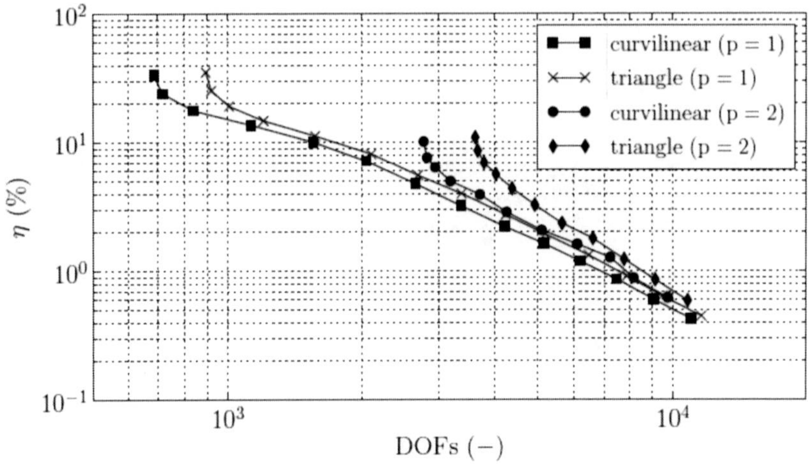


**Fig. 14.17** Convergence curves of the results: Agros2D, symbol $p$ denoting the initial degree of polynomials before starting of the adaptive process

It is clear that fine elements and hanging nodes were generated mostly in the region of the peak of the conical electrode. The spherical electrode is simulated by curvilinear elements, i.e., quite precisely (and the distribution of electric field in its vicinity, therefore, is modeled with a very high accuracy). Figure 14.20 depicts the dependence of the total electrostatic energy $W_e$ in the system calculated by different codes on the total number of the degrees of freedom (DOFs). The codes FEMM and QuickField only work with linear elements ($p = 1$, without adaptivity) and the

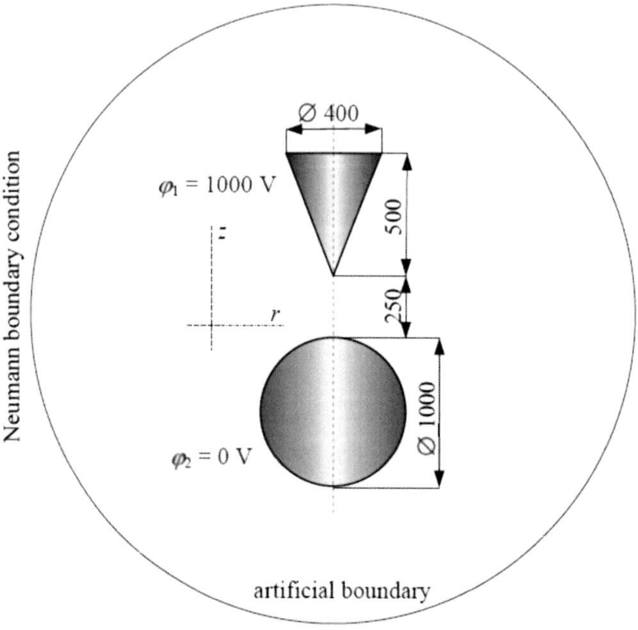


**Fig. 14.18** The investigated spark gap

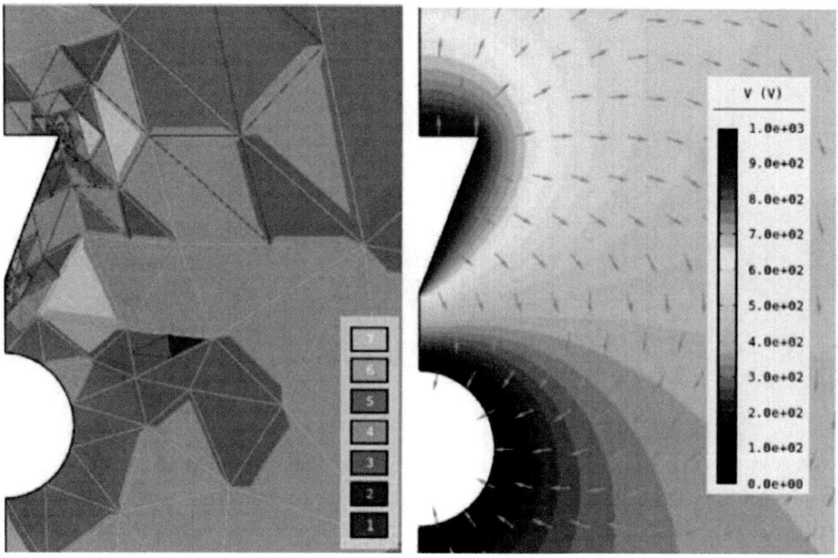


**Fig. 14.19** Left—original rougher mesh (white lines) and final mesh after adaptivity (dark lines), orders of elements in the column, right—distribution of potential and electric field near the electrodes (codes Agros2D and Hermes2D, $p = 1$, number of DOFs 1977, relative error of solution $\eta = 0.307\%$)

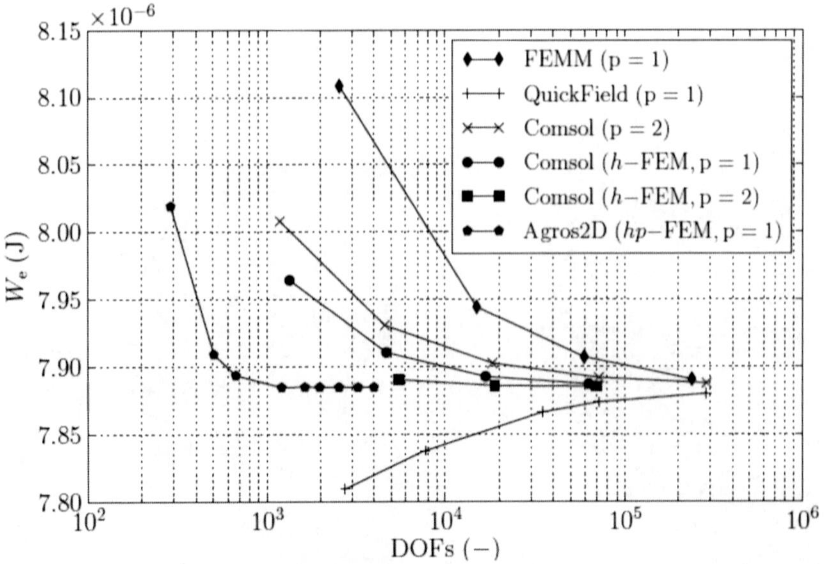


**Fig. 14.20** Dependence of total energy $W_e$ of electrostatic field in the system on the number of DOFs

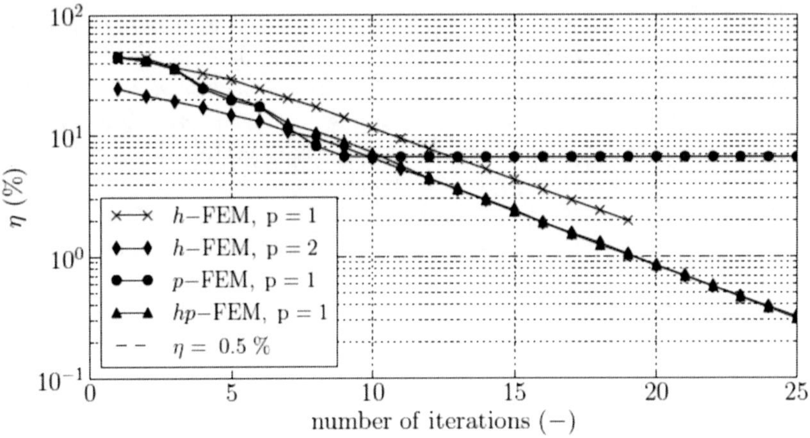


**Fig. 14.21** Comparison of various adaptive algorithms in Agros2D Fig.

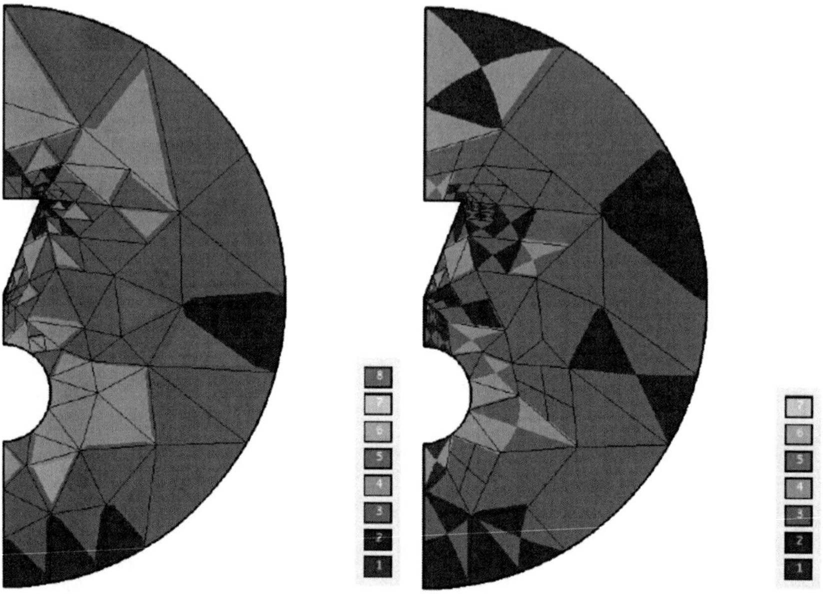

**Fig. 14.22** Two discretization meshes that provide results with relative error $\eta < 1\%$: purely triangular mesh - 1402 DOFs (left), combined mesh - 1186 DOFs (right)

results obviously converge very slowly. Faster is the convergence in Comsol Multiphysics, mainly with switched-on adaptivity. On the other hand, this code does not support the hanging nodes, so that much more elements are needed. Finally, Agros2D with hp-adaptivity starting from a rough mesh converges fast, with a substantially lower number of DOFs (their number is 1977 for relative error of solution $\eta = 0.307\%$). Thus, usage of Agros2D is much more effective.

Figure 14.21 shows analogous convergence curves obtained from Agros2D for several adaptive algorithms differing by the starting values of $p$. Again, the convergence of results based on the $hp$-adaptivity is the fastest. Finally, Fig. 14.22 shows a purely triangular mesh (left part) and mesh combining triangular and quadrilateral elements (right part). The second one provides the results of the same quality as the first one ($\eta < 1\%$), but the number of DOFs is lower by about 20%.

## 14.5 Conclusion

The methods of adaptivity described in the paper represent a powerful tool whose application leads to significant savings in DOFs in comparison with various available SW (by one order and more) even at a higher accuracy of the results.

Further improvement of these techniques can be achieved by using meshes that combine triangular, quadrilateral and curved elements, as quadrilateral elements seem to be effective mainly for modeling of domains with anisotropy. Nowadays, we are developing the corresponding algorithms and implementing them into the software Hermes and Agros.

**Acknowledgements.** This work was supported by the European Regional Development Fund and Ministry of Education, Youth and Sports of the Czech Republic (project No. CZ.1.05/2. 1.00/03.0094: Regional Innovation Centre for Electrical Engineering - RICE) and by the Grant project GACR P102/11/0498.

## References

1. Šolín, P., Dubcová, L., Kruis, J.: Adaptive hp-FEM with dynamical meshes for transient heat and moisture transfer problems. J. Comput. Appl. Math. 233(12), 3103–3112 (2010)
2. Dubcová, L., Šolín, P., Červený, J., Kůs, P.: Space and time adaptive two-mesh hp-FEM for transient microwave heating problems. Electromagnetics 30(1), 2340 (2010)
3. Castillo, L.E.G., Zubiaur, D.P., Demkowicz, L.F.: Fully automatic hp adaptivity for electromagnetics: Application to the analysis of H-plane and E-plane rectangular waveguide discontinuities. In: Proc. Microwave Symposium, Honolulu, HI, pp. 282–288 (June 2007)
4. Šolín, P., et al.: Code Hermes and manuals for its application, `http://hpfem.org/hermes/` (cited October 15, 2011)
5. Karban, P., et al.: Code Agros2D and manuals for its application, `http://agros2d.org` (cited October 15, 2011)
6. Shewchuk, J.R.: A two-dimensional quality mesh generator and delaunay triangulator Triangle and manual for its application, `http://www.cs.cmu.edu/quake/triangle.html` (cited October 15, 2011)
7. Šolín, P., Červený, J., Doležel, I.: Arbitrary-level hanging nodes and automatic adaptivity in the hp-FEM. Math. Comput. Simul. 77(1), 117–132 (2008)
8. Šolín, P., Segeth, K., Doležel, I.: Higher-Order Finite Element Methods. Chapman & Hall/CRC, Boca Raton, FL (2003)
9. Mitchell, F.M.: A Collection of 2D Elliptic Problems for Testing Adaptive Algorithms. NISTIR 7668 (2010)
10. Database of material parameters and their temperature dependencies, `http://www.jahm.com/` (cited October 15, 2011)
11. Karban, P., Mach, F., Doležel, I.: Induction heating of nonmagnetic cylindrical billets by rotation in magnetic field produced by static permanent magnets. Electrical Review 86(12), 53–56 (2010)

# Chapter 15
# SPICE Model for Fast Time Domain Simulation of Power Transformers, Exploiting the Ferromagnetic Hysteresis and Eddy-Currents

Lucian Mandache, Dumitru Topan, Mihai Iordache, and Ioana Gabriela Sirbu

**Abstract.** The paper proposes an effective time-domain modeling and simulation strategy of power transformers, using the SPICE circuit simulator. The nonlinear phenomena involved by the iron core are carefully considered, including the nonlinearity with saturation, static hysteresis and eddy currents. The principle of magnetic circuit modeling is based on analog lumped equivalent circuits and the SPICE implementation uses the principle of modularity. Such a module includes one transformer winding (either primary or secondary) and its core leg, implemented as a subcircuit with user-defined parameters. The method allows simulating normal operation modes, as well as critical transients and faulty conditions, the simulation result containing all electromagnetic quantities as time-domain functions. It is remarkable through its extremely short computation time and reasonable accuracy, it being conceived firstly as useful tool for design purposes, especially for repeated simulations required by optimization algorithms. A case study is presented to prove the feasibility, usefulness and accuracy of the proposed modeling and simulation approach, where the SPICE results are validated by experimental ones.

## 15.1 Introduction

The iron cores are basic components of transformers and inductors for wide application area, ranging from analog and digital microelectronics toward power converters and power systems. Because of the related phenomena involved by the iron core, as the nonlinearity with saturation, electromagnetic inertial behavior caused by the ferromagnetic hysteresis, anisotropy and induced eddy currents, a rigorous design of such electromagnetic devices is difficult. An optimal design of transformers should also consider the entire electromagnetic system, including the power supply and the

Lucian Mandache · Dumitru Topan · Ioana Gabriela Sirbu
University of Craiova, A.I. Cuza Str. 13, Craiova 200440, Romania

Mihai Iordache
University Politehnica of Bucharest, Spl. Independentei 313, Bucharest, 060042, Romania

K. Kyamakya et al. (Eds.): Selected Topics in Nonlinear Dynamics, SCI 459, pp. 273–286.
DOI: 10.1007/ 978-3-642-34560-9_15 

load, as well as all possible operation modes: normal operation mode (both steady-state and transients), malfunction modes or critical events.

This goal is only possible through CAD (Computer-Aided Design) tools, based on appropriate modeling and simulation approaches capable to provide the analysis of the entire equipment with adequate accuracy. Even though co-simulation solutions involving a FEM (Finite Element)-based simulator for the electromagnetic and thermal field analysis combined with a time-domain circuit simulator could offer the most accurate results [1,2], they are obviously the most costly regarding hardware and software requirements, computation effort and simulation time.

In such circumstances, in order to achieve reasonable design costs with reasonable accurate results, the concept of modeling the electromagnetic system (including the iron core) through equivalent diagrams with lumped circuits could be exploited successfully. Such an equivalent diagram is described by a mathematical model under the form of a nonlinear differential algebraic equation system (DAE). Although this concept has been promoted by many authors [4-10], it has not been sufficiently exploited yet. Depending on the analysis aim, the models based on analog lumped circuits have different degrees of accuracy. The simplest approaches do not consider the iron losses, but only the nonlinear behavior of the first-magnetization curve specified either by points or by analytic functions [10,11]. Other basic approaches consider the iron losses globally, with weak connection to the real physical phenomena [4,5]. Some authors developed more improved models starting from the specific physical phenomena which involve known core losses, as static hysteresis [6,8] or eddy currents [9].

We propose a more sophisticated but effective procedure for modeling and simulation of transformers supplied with harmonic or distorted voltages, the load being linear, nonlinear and/or time-dependent, taking into account, in the same time, the ferromagnetic nonlinearity with saturation, the static hysteresis and the eddy currents of the iron core. Since the corresponding mathematical model (DAE) is SPICE-compatible [3] the powerful software tool from SPICE family – ICAP4 from Intusoft co. [11] is exploited to perform the numerical analysis. Therefore, the entire electromagnetic system (including the chain power grid-transformer-load) is simulated in a unitary manner, and the time-consuming FEM-based analysis is avoided.

Although the general principle of this modeling and simulation procedure is known, our systematic study brings a significant improvement in terms of degree of generality and the ease of use. We conceived the modeling and simulation method to become a useful tool for researchers and designers, especially for repeated simulations required by optimization algorithms, where the shortening of the simulation time is paramount. Moreover, the chosen circuit simulator is remarkable through its accessibility among the specialists of academia and industry.

The modeling procedure considers the specific phenomena related to the wired ferromagnetic legs, explained by the Maxwell's theory. A concept of modularity has been exploited, in order to confer flexibility and ease of use. As absolute novelty, the main module is conceived as a subcircuit that combines: a ferromagnetic leg as magnetic field path, a winding spooled on it and the corresponding flux leakage path. Moreover, depending on the extern diagram, the winding of the model can play either the role of primary or secondary coil. Another module corresponds to transformer yokes or unwired legs, it being derived starting from the previously said model of the wired leg. It contains only a nonlinear, hysteretic and eddy current carrying iron core piece with flux leakage path. The air-gaps are modeled through linear resistances numerically equal to the corresponding magnetic reluctances.

The modeling procedure has a wide range of generality, being suitable to be extended to inductors, linear actuators and rotating motors with any normal operation mode (harmonic, distorted or switching-mode).

The section 15.2 describes briefly the modeling principles, while the SPICE implementation is shown in the section 15.3. An example of a single-phase transformer is given in the section 15.4, where the SPICE simulation results are judged in comparison with experimental results. Some other simulations were performed and the results are included here.

## 15.2 Modeling Principles

The simplest model of a ferromagnetic piece of cross section $S$ and length $l$ flowed by a magnetic flux assumed as uniform within the cross section (with the flux density $B$) consists in a voltage controlled nonlinear resistance flowed by a current numerically equal to the magnetic flux, the voltage across it being numerically equal to the magnetic force [9]. Its nonlinear characteristic $i(v)$ reproduces in scale the anhysteretic magnetization curve $B(H)$, commonly specified by the manufacturer of the ferromagnetic material; the scale factors depend on the cross-section and the length of the piece. Such a model does not consider the hysteresis effect, nor eddy currents.

The assumption related to the uniform distribution of the magnetic flux density within the cross section is in accordance to the real phenomenon for relatively small pieces of magnetic circuits. Moreover, it can be extended to power transformers, as proven by the experience [12]. If no uniform flux distribution is expected because of possible unusual shapes of core legs or yokes (e.g. as for water-cooled or forced air-cooled cores), in order to accomplish the assumption named before, the core leg model could be divided in smaller elements. We recommend deciding about splitting the core leg in elements with uniform flux density after a preliminary FEM-based analysis of the magnetic filed distribution, even for simple harmonic regime.

Although the initial magnetization characteristic is commonly specified by points as lookup table or graphic, for modeling and simulation purposes, to avoid convergence and stability problems during the numerical computation, an analytic representation is preferred, as continuous and derivable function. We tried successfully the Langevin approximation [13] that accomplishes the condition above and can be easily matched to the given lookup table. It is an implicit equation which provides the anhysteretic component of the magnetization:

$$M_{an}(H) = M_{sat}\left(coth\frac{H+\alpha M}{a} - \frac{a}{H+\alpha M}\right) \tag{15.1}$$

where the saturation magnetization $M_{sat}$ and the parameters $\alpha, a$can be established enough accurately starting from the characteristic specified in the datasheet. About the ferromagnetic hysteresis, the inertial curve of the irreversible component of the magnetization is exemplary described by the expression of the differential irreversible susceptibility $dM_{irr}/dH$ based on the theory of Jiles and Atherton [13,14]. In order to include the Jiles model in a time-domain circuit simulation program, the mathematical model must contain only derivatives with respect to time. Since $M_{irr}$ and $H$ are time-dependent, the derivative with respect to time of $M_{irr}$ includes the irreversible susceptibility developed by Jiles:

$$\frac{dM_{irr}}{dt} = \frac{M_{an} - M_{irr}}{ksgn\frac{dH}{dt} - \alpha(M_{an} - M_{irr})} \frac{dH}{dt} \tag{15.2}$$

where the constant $k$ is related to the width of the hysteresis cycle and the difference $M_{an} - M_{irr}$ gives the reversible component of the magnetization weighted with a subunit coefficient c correlated with the initial susceptibility [13]:

$$M_{rev} = c(M_{an} - M_{irr}) \tag{15.3}$$

The resulting flux density is therefore the effect of the reversible and irreversible magnetization:

$$M = M_{rev} + M_{irr}; \quad B = \mu_0(H + M) \tag{15.4}$$

Starting from the magnetic force as excitation quantity$u_m = Hl$, imposed by a given ampere-turns according to the Ampere's law, the corresponding computation chain of the hysteresis model joined to the Langevin representation of the initial magnetization curve is shown as block diagram in fig. 15.1.

Since the eddy-current losses can be easily assessed for common power transformers working at the grid frequency (starting from the specific power losses specified in the datasheet), the problem of eddy currents becomes more difficult in distorting operation modes, as under weak power quality conditions, at medium frequencies or in switching mode involved by power electronics applications. We developed

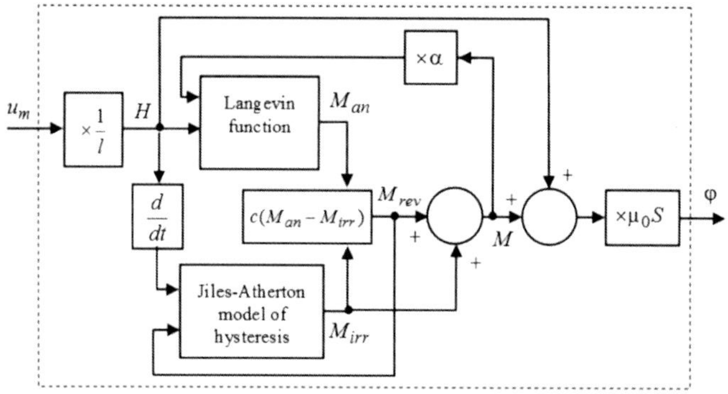


**Fig. 15.1** Computation chain of the hysteretic magnetization based on Jiles-Atherton model

previously an original model for eddy currents, in order to asses their instantaneous and mean power losses, as well as their parasitic magnetic field [9]. The principle is based on the computation of an equivalent eddy current whatever the shape or the iron core piece, and an equivalent resistance that dissipates the same power loss as in the real case. Massive ferromagnetic pieces, as well as silicon steel sheets have been treated. The calculus is based on the quantitative evaluation of the phenomena explained by the Maxwell's theory, according to the Faraday's law of induction (to compute the electromotive forces induced in the iron core by the time-variable magnetic field), the Ohm's law (to compute the induced eddy current), the Joule's law (to compute the power loss) and the Ampere's law (to compute the parasitic magnetic field). The equivalent eddy current in the time domain was obtained as:

$$i_{eddy}(t) = K\frac{p_0 \gamma l}{2\pi^2 f_0^2 B_0^2}\left(\frac{d\phi}{dt}\right) = K_{eddy}\frac{d\phi}{dt} \tag{15.5}$$

where $p_0$ is the specific power loss [W/kg] given in the datasheet for the reference frequency $f_0$ and the reference flux density $B_0$ in harmonic behavior. For silicon steel laminated sheets the reference conditions are usually 50 Hz and 1 T. K is a shape factor and $\gamma$is the mass density in kg/m3. If the eddy current mean power loss is computed in reference conditions, the equivalent eddy current resistance results:

$$R_{eddy} = \frac{1}{K^2}\frac{2\pi^2 f_0^2 B_0^2}{p_0 \gamma l S} \tag{15.6}$$

It dissipates accurately the actual power loss when it is flowed by the equivalent eddy current computed before, whatever the time-domain variation of the magnetic flux.

Therefore, the anhysteretic or hysteretic magnetization model shall be completed with additional elements in order to consider the eddy current, its power loss and its magnetic field, according to the computation chain shown in Fig. 15.2. The input quantity is the external ampere-turn, as primary source of the magnetic field, while the eddy current ampere-turn opposes to it.

A core piece carrying a coil is now considered, like a leg together with the primary winding of a transformer (Fig. 15.3a). All other windings are assumed in open-circuit and both the main flux $\phi$ and the leakage flux $\phi_l$ have been considered.

Obviously, they are fascicular magnetic fluxes. The Kirchhoff's voltage law for the primary circuit loop includes the components of the voltage across the winding of $N$ turns and resistance $R$, respectively the voltage drop across the resistance and the electromotive force induced by the total magnetic flux:

$$v = R_i + N\frac{d(\phi + \phi_l)}{dt} \tag{15.7}$$

The main magnetic flux flows through the ferromagnetic leg, while the leakage flux path can be assumed as a linear magnetic reluctance corresponding to the surrounding air. Therefore, an equivalent diagram of the structure of fig. 15.3a is conceived as in fig. 15.3b, where the model of the core piece includes both hysteresis and eddy-current effect. The magnetic reluctance corresponding to the leakage flux is noted $R_{ml}$.

The ampere-turn of the winding is modeled by the current controlled voltage source with the control resistance numerically equal to its number of turns. The second term of (15.7) is modeled as the current controlled voltage source, controlled with the time-derivative of the input current, numerically equal to the total magnetic flux $\varphi + \varphi_l$ . If the secondary winding is now considered, using similar notations as in (15.7) and fig. 15.3a, the term of the form $Nd(\varphi + \varphi_l)/dt$ represents the excitation of the secondary circuit, so that the corresponding expression of the voltage Kirchhoff's law is formally identical to (15.7), with the only change of the sign of the term $Ri$, as consequence of current sense inversion. Therefore, the model of fig. 15.3b remains valid for the assembly secondary winding - core leg. Since the mathematical model is a differential-algebraic equation system, the modeling procedure is compatible not only with SPICE, but also with other time-domain analysis methods based on state or semi-state equations [15,16].

## 15.3 SPICE Implementation

According to the principles exposed in the previous section, a complex model of a wired ferromagnetic piece has been developed and implemented as SPICE subcircuit (version ICAP4 from Intusoft). It is based on an equivalent lumped circuit diagram. It contains four terminals, two terminals of the electric circuit (winding terminals) and two terminals of the magnetic circuit. The same subcircuit can play either the role of primary or secondary.

Depending on the transformer construction, the model allows placing the primary and secondary on the same core leg or on different legs. The number of secondary windings is practically unlimited.

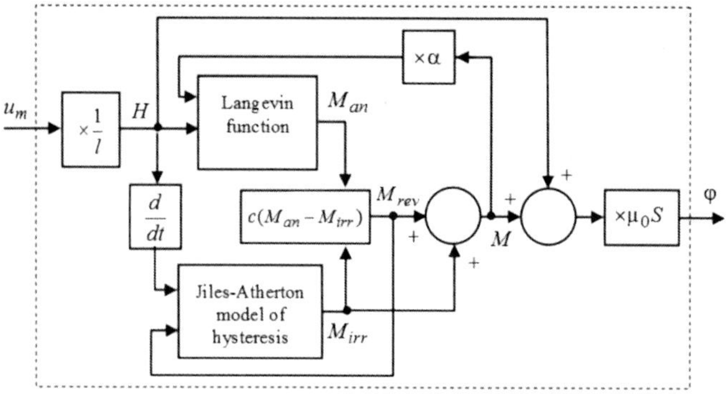


**Fig. 15.2** Computation chain of eddy currents and their effect in ferromagnetic core pieces

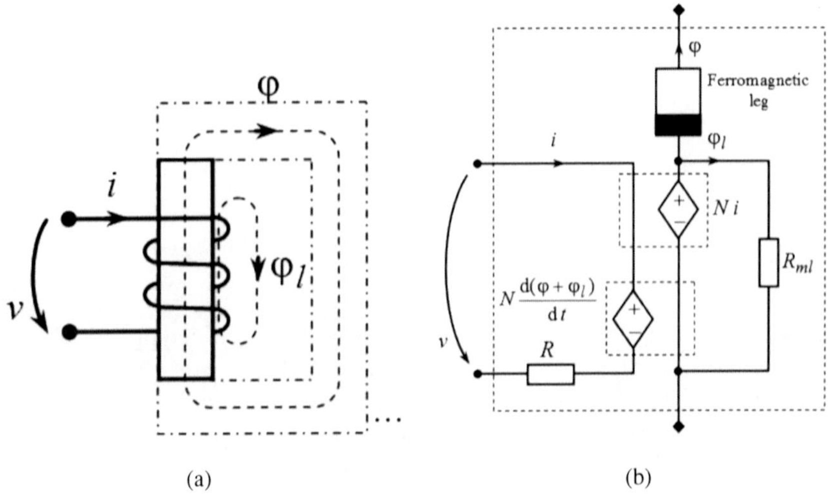


**Fig. 15.3** Model of a wired transformer leg

The subcircuit netlist (called LEDDY3) built according to the model above is detailed below. It implements the computation chains of Fig. 15.1 and Fig. 15.2 integrated in the model diagram of Fig. 15.3a:

```
*SYM=LEDDY
.SUBCKT LEDDY3  1B 2B 1M  2M {RCC=1.5, N=100, SFE=10e-4,
+LFE=0.1, BSAT=1.8,  GC=0.95, C=0.53, K=20, ALPHA=5e-5, A=35}
* Hysteresis model:
BB 1 2 I=3.1415*4e-7*{SFE}*(I(VH)+V(5))
GH 0 4 1 2 {1/LFE}
L2 4 9 1 IC=0
VH 9 0 0
BDMIRR 0 5 I=V(4)*(V(5)-V(6))/({C*K}*SGN(V(4))-
+{ALPHA}*(V(5)-V(6)))
BMREV 5 6 V={C}*(V(7)-V(6))
* Anhysteretic characteristic.
C1 6 0 1 IC=0
BAUX 8 0 V=(I(VH)+V(5)*{ALPHA})/{A}
R1 8 0 1e9
BMAN 7 0 V=ABS(V(8))>1e-3 ? {BSAT}/(3.1415*4e-7)*
+(1/TANH(V(8))-1/V(8))  : {BSAT}/(3.1415*4e-7)*V(8)/3
R2 7 0 1e9
* Eddy-current model:
VF 1A 1 0
FF 0 3A VF 1
LF 3A 0 1 IC=0
```

```
GE 0 4A 3A 0 {GC}
RE 4A 5A {2/3/GC}
VE 5A 0 0
HE 2 2M VE 1
* Model of eq. (7)
RC 1B 10 {RCC}
E1 10 3 3A 0 {N}
V1 3 2B
H1 1A 1M V1 {N}
.ENDS
```

The main subcircuit parameters are: the DC resistance of the winding, the winding number of turns, the cross section, the length and the shape factor of the ferromagnetic core leg, the saturation flux density, nearby other parameters related to the shape of the initial magnetization characteristic and static hysteresis, and the level of the equivalent eddy current, as described above. The quantities of interest, as the magnetic filed strength, magnetic flux density, equivalent eddy current, can be easily passed outside the subcircuit in order to be processed and judged according to the requirements of each particular case.

The subcircuit terminals 1B and 2B correspond to the electric circuit, while the terminals 1M and 2M correspond to the magnetic circuit. Four areas are distinguished depending on their roles, as it was emphasized by comments included in the subcircuit netlist. The subcircuit uses independent zero-voltage sources as ammeters (V1 for the winding current, VF for the magnetic flux, VH for the mag-netic field strength, VE for the equivalent eddy-current), and dependent voltage and current sources for modeling the equations (15.1)-(15.7) or to transfer some quantities from an area to another.

For legs or yokes without windings, a simplified version of this subcircuit has been created. Both subcircuits were included in the SPICE model library, in order to be called in future applications.

## 15.4 Example of Modeling and Simulation of a Single-Phase Power Transformer

To exemplify the usefulness of the developed model, we chose to simulate a single-phase voltage transformer in transient behavior using the SPICE implementation. The simulation results are compared with experimental results. To acquire with accuracy the experimental data, we used a power analyzer Chauvin Arnoux CA8352 with the sampling rate set at 19.2 kHz.

Summary description of the studied transformer designed and built for a particular application: rated voltage 220/120V, core type E+I (E20), silicon steel M4-0.27mm, main leg of 16mm2, primary winding of 420 turns.

We consider as relevant and expose here the case when the no-load transformer is connected to the grid at the moment of zero-crossing voltage, so that the peak value of the current achieves the highest value comparing to all other possible cases.

In Fig. 4 are shown the reference experimental results given directly by the power analyzer interface; both the voltage across the primary winding and the transient no-load current were acquired.

For the SPICE simulation, the model diagram of fig. 15.5 was used. It contains models like those described before for the wired central core leg (noted LEDDY3) and for the lateral legs (LEDDY0). The technological airgaps are represented as linear resistors (RGAP1 and RGAP2 on the figure, with the electric resistance numerically equal to the magnetic reluctance of the corresponding airgap) and the network parameters are modeled as RL impedance (LG and RG on the figure). The no-load operation mode is obtained with a high value of the load resistance (RLOAD). The diagram contains also testing points for currents and voltages. The test points for currents evaluate the currents through the additional zero-voltage independent voltage sources (V3 for the primary current, and V1 for the current numerically equal to the magnetic flux in the main leg). The voltage test points evaluate the voltages with respect to the ground of the points 6 (primary voltage) and 12 (this is numerically equal to the magnetic force across the main leg). The power source of 220V / 50Hz is modeled as an independent voltage source (VG on the figure) with no initial phase angle.

The transient analysis time covered ten complete cycles in order to obtain the damping of transient components. The maximum value of the integration time step was chosen $5\mu s$, corresponding to 40000 computation points.

The current given by the simulation is shown in Fig. 15.6, where the experimental curve was represented too for comparison. For more clarity, the first 4 cycles and the last cycle (the last cycle describes the steady-state operation mode) are shown separately at adequate representation scales. The simulation time was only 4.8 seconds on a PC of 2.2GHz/2GB RAM.

The accuracy of the simulation results, with acceptable deviation compared to actual (experimental) phenomena, proves the viability of the method.

Any electric or magnetic quantity, including instantaneous and mean power losses, can be also computed as simulation result and stored in the output file. This is extremely useful in design activities. As example, in Fig. 15.7 is shown a global image on the transient analysis described above, where the current (curve 2) is represented together with the voltage across the primary winding (curve 1) and the corresponding magnetization curve denoting a strong saturation (curve 3).

Contrarily, if the transformer is connected to the power grid at the peak value of the voltage (see curve 2 in Fig. 8), the transient behavior is extremely slight (see the current – curve 1), with no saturation (curve 3).

In Fig. 15.9 it is depicted the influence of the grid voltage on the primary no-load (magnetization) current, where the curve 1 corresponds to the rated value (slightly distorted), the curve 2 correspond to a 20% higher voltage (strongly distorted because of the iron core saturation level) and the curve 3 corresponds to a 20% lower voltage (quasi-sinusoidal). The RMS values of the currents are also given on the label attached to the figure. It must be emphasized that the simulation time for each studied case was below 10 seconds, on a common PC.

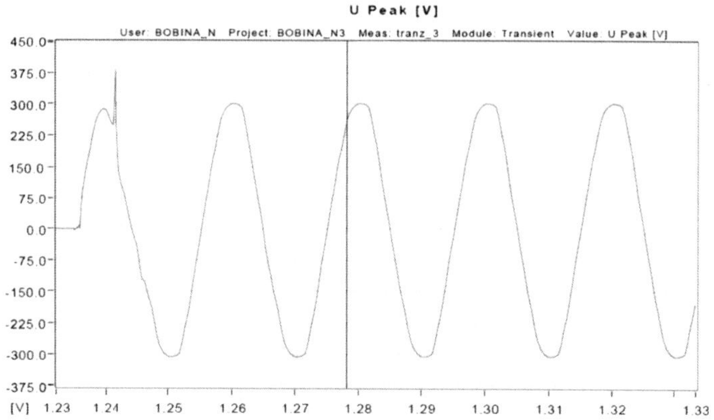


(a)

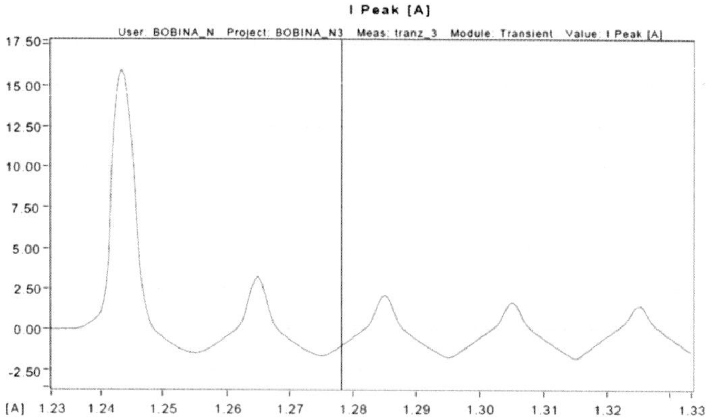


(b)

**Fig. 15.4** Rough experimental results acquired with a power analyzer: (a) – primary voltage; (b) – primary current in transient behavior

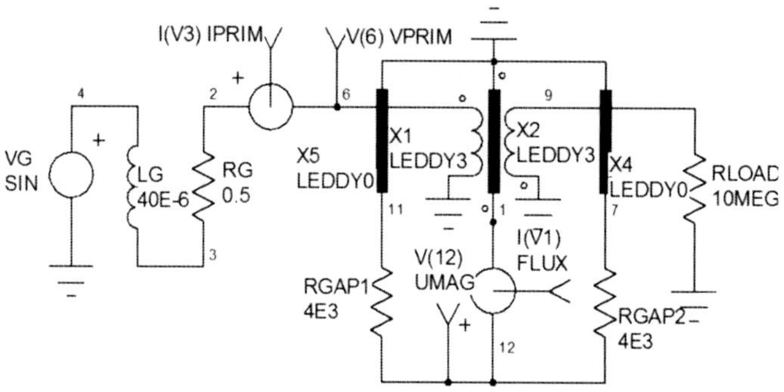


**Fig. 15.5** Example SPICE diagram

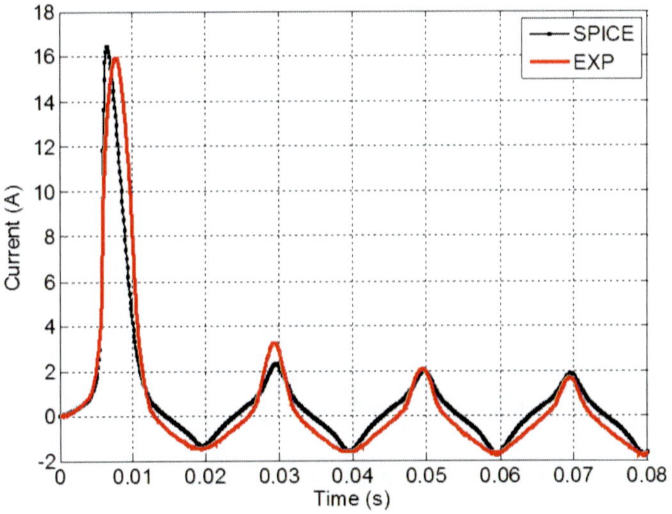


(a)

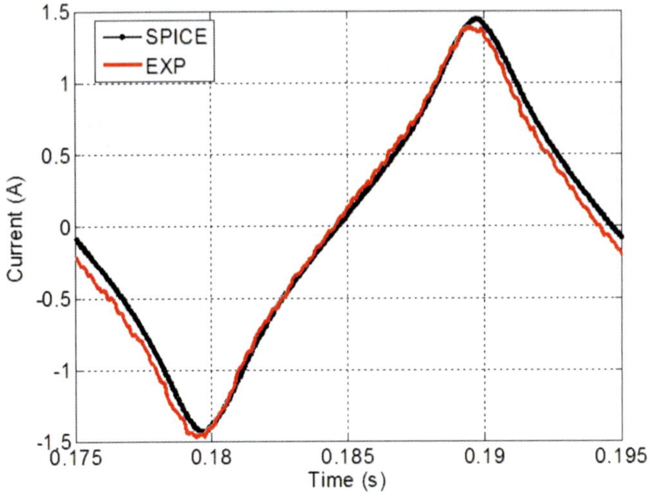


(b)

**Fig. 15.6** The primary no-load current as simulation/experimental result: (a) – the first four cycles after the connection to the power grid with detail on the peak value of the current; (b) – one cycle describing steady-state operation mode

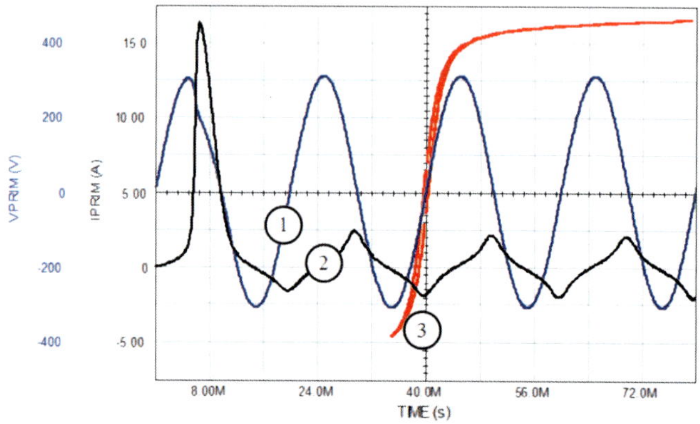


**Fig. 15.7** No-load transformer connected to the grid at the moment of zero-crossing voltage

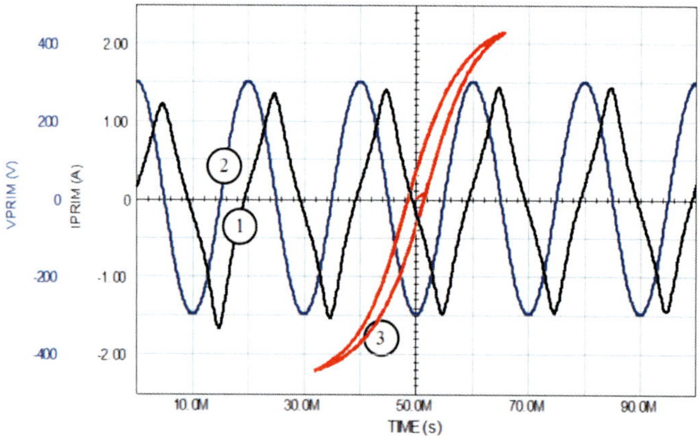


**Fig. 15.8** No-load transformer connected to the grid at the peak value of the voltage

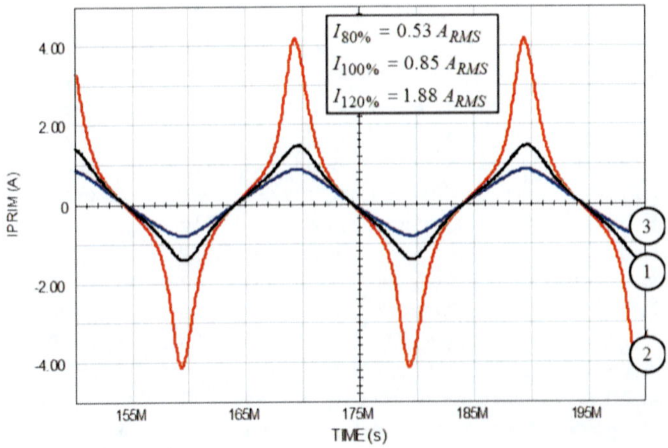


**Fig. 15.9** Influence of the grid voltage on the primary no-load current

## 15.5 Conclusion

The proposed modeling and simulation method is capable to offer complete information on conventional power transformers, as well as on those conceived for special applications, as in power electronics, due to the time-domain modeling. The complex nonlinear phenomena in ferromagnetic cores are also considered. The simplifying assumptions (see the uniformity of the flux density) are reasonable for the envisaged applications.

The flexibility and simplicity of implementation are obvious due to the modular structure of the model. The method is remarkable through the extremely short computation time, comparing to other modeling and simulation strategies, so that repeated simulations required by optimization algorithms can benefit by negligible costs. It is also robust and reliable of the point of view of the computation stability. Thus, the interaction of the studied transformer with the power grid and other neighboring devices can be easily considered to improve the accuracy of the simulation.

The method can be easily extended for other electromagnetic devices with soft iron core (e.g. filter inductors, electromagnetic actuators, excitation poles of DC or synchronous motors). To keep a reasonable accuracy, a preliminary FEM-based magnetic field analysis could be necessary before deciding about splitting the ferromagnetic core in elements with uniform flux density, as mentioned in the section 15.2.

**Acknowledgements.** This work was supported by The Romanian Ministry of Education, Research, Youth and Sport-UEFISCDI, project number 678/2009 PNII - IDEI code 539/2008.

## References

1. Costa, M.C., Nabeta, S.I., Cardoso, J.R.: Modified Nodal Analysis Applied to Electric Circuits Coupled with FEM in the Simulation of a Universal Motor. IEEE Transactions on Magnetics 36(4), 1431–1434 (2000)
2. Escarela-Perez, R., Melgoza, E., Alvarez-Ramirez, J.: Systematic Coupling of Multiple Magnetic Field Systems and Circuits Using Finite Element and Modified Nodal Analyses. IEEE Transactions on Magnetics 47(1), 207–213 (2011)
3. Mandache, L., Topan, D., Iordache, M., Sirbu, I.G.: SPICE model for effective and accurate time domain simulation of power transformers. In: 2011 Joint 3rd Int'l Workshop on Nonlinear Dynamics and Synchronization (INDS) & 16th Int'l Symposium on Theoretical Electrical Engineering (ISTET), Klagenfurt, July 25-27, pp. 1–6 (2011)
4. Chua, L.O., Stromsmoe, K.: Lumped-Circuit Models for Nonlinear Inductors Exhibiting Hysteresis Loops. IEEE Transactions on Circuit Theory CT-17(4), 564–574 (1970)
5. Chan, J.H., Vladimirescu, A., Gao, X.C., Liebmann, P., Valainis, J.: Nonlinear Transformer Model for Circuit Simulation. IEEE Transactions on Computer-Aided Design 10(4), 476–482 (1991)
6. Hsu, J.T., Ngo, K.D.T.: Subcircuit Modeling of Magnetic Cores with Hysteresis in PSpice. IEEE Transactions on Aerospace and Electronic Systems 38(4), 1425–1434 (2002)
7. Wilson, P.R., Ross, J.N., Brown, A.D.: Simulation of Magnetic Component Models in Electric Circuits Including Dynamic Thermal Effects. IEEE Transactions on Power Electronics 17(1), 55–65 (2002)
8. Thomas, D.W.P., Paul, J., Ozgonenel, O., Christopoulos, C.: Time-Domain Simulation of Non-linear Transformers Displaying Hysteresis. IEEE Transactions on Magnetics 42(7), 1820–1827 (2006)
9. Mandache, L., Topan, D.: Modeling and Time-Domain Simulation of Wired Ferromagnetic Cores for Distorted Regimes, Buletinul Institutului Politehnic din Iai, Univ. Tehnic Gh. Asachi; Tomul LIV (LVIII), fasc. 3 (2008), Electrotehnic, energetic, electronic, pp. 303–310; ISSN 1223-8139
10. Mandache, L., Topan, D., Iordache, M., Dumitriu, L., Sirbu, I.G.: On the time-domain analysis of analog circuits containing nonlinear inductors. In: 20th European Conference on Circuit Theory and Design, ECCTD 2011, August 29-31, pp. 98–101 (2011)
11. IsSpice4 Users Guide, Intusoft co., San Pedro, Ca. 90733-0710, USA (1995)
12. Pfutzner, H., Bengtsson, C., Booth, T., Loffler, F., Gramm, K.: Three dimensional flux distributions in transformer cores as a function of package design. IEEE Transactions on Magnetics 30(5), 2713–2727 (1994)
13. Jiles, D.C., Thoelke, J.B., Devine, M.K.: Numerical Determination of Hysteresis Parameters the Modeling of Magnetic Properties Using the Theory of Ferromagnetic Hysteresis. IEEE Transactions on Magnetics 28(1), 27–35 (1992)
14. Jiles, D.C., Atherton, D.L.: Ferromagnetic Hysteresis. IEEE Transactions on Magnetics MAG-19(5), 2183–2185 (1983)
15. Topan, D., Mandache, L., Suesse, R.: Advanced analysis of electric circuits. Wissenschaft Verlag Thringen (2011)
16. Iordache, M., Mandache, L., Perpelea, M.: Analyse numrique des circuits analogiques non linaires. Ed. Groupe Horizon, Marseille (2006)

# Chapter 16
# Hard-Coupled Modeling of Induction Shrink Fit of Gas-Turbine Active Wheel

Václav Kotlan, Pavel Karban, Bohuš Ulrych, Ivo Doležel, and Pavel Kůs

**Abstract.** Hard-coupled model of induction heating of a ferromagnetic disk is presented. The problem is described by three coupled partial differential equations (for the distribution of electromagnetic field, temperature field and field of thermoelastic displacements) whose coefficients are temperature-dependent functions. The system is solved numerically in the monolithic formulation by a fully adaptive finite element method of higher order of accuracy implemented into own codes Hermes and Agros. The methodology is illustrated by a typical example—heating of an active wheel of a high-speed gas turbine that is to be hot-pressed on a shaft with the aim of obtaining a shrink fit allowing transferring the given torque at the nominal revolutions. Evaluated and discussed are also the parameters of the heating process in transverse and longitudinal magnetic fields.

## 16.1 Introduction

Hot pressing belongs to widely spread industrial technologies used in a great number of industrial applications (for instance manufacturing of shrunk-on rings, crankshafts, tires of railway wheels, armature bandages in electrical machines, etc., see [1], [2]). The process of heating is mostly realized by gas or induction. Induction heating is characterized by an easy control of the intensity of heating and its local distribution, no chemical changes in the surface layers of the processed material, and no products of combustion.

Václav Kotlan · Pavel Karban · Bohuš Ulrych
University of West Bohemia, Faculty of Electrical Engineering,
Univerzitní 8, 306 14 Plzeň, Czech Republic
e-mail: {vkotlan,karban,ulrych}@kte.zcu.cz

Ivo Doležel · Pavel Kus
Academy of Sciences of the Czech Republic, Institute of Thermomechanics,
v.v.i, Dolejškova 5, 182 00 Praha 8, Czech Republic
e-mail: dolezel@it.cas.cz, pavel.kus@gmail.com

K. Kyamakya et al. (Eds.): Selected Topics in Nonlinear Dynamics, SCI 459, pp. 287–301.
DOI: 10.1007/ 978-3-642-34560-9_16 

In case of assembly of a shrink fit its external part (usually a chuck, a disk or a wheel) must be heated as long as the dilatation of its bore allows inserting the internal part (shaft, shank of a machine tool) into it. The system (particularly its heated part) is then cooled, which produces the shrink fit. Its purpose is usually to transfer a prescribed mechanical torque. The process is schematically depicted in Fig. 16.1.

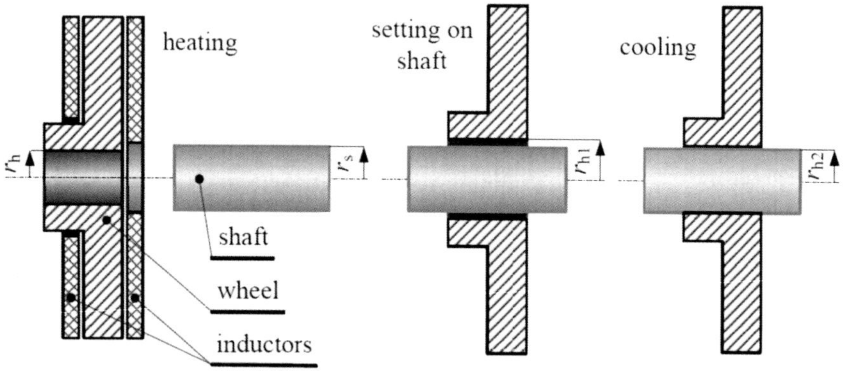


**Fig. 16.1** Schematic view of manufacturing a shrink fit

The radius of the internal hole of the wheel at cold is $r_h$, the radius of the shaft is $r_s$, and there holds $r_h < r_s$. After heating the wheel its hole enlarges its radius to $r_{h1} > r_s$. Then the wheel is put on the shaft and the whole system is cooled. After cooling the radii of hole and shaft are the same ($r_{h2}$) and there holds $r_h < r_{h2} < r_s$.

Manufacturing of shrink fits often requires a lot of energy. In order to minimize its consumption and also to anticipate the mechanical properties of the resultant fit, we should have detailed information about the complete process. And the most relevant tool is both the previous experience and computer modeling.

From the physical viewpoint, the process represents a triply coupled nonlinear and nonstationary problem characterized by a mutual interaction of three physical fields: magnetic field, temperature field and field of thermoelastic displacements. These fields (more or less influencing one another) are described by partial differential equations (PDEs) whose coefficients containing various material properties are temperature-dependent functions.

The above system was solved by several groups of authors working in the domain. They mostly used either a weakly coupled formulation (the three fields were solved independently of one another), see [3], [4], [5] or a mixed formulation, when magnetic field is solved independently, while the two remaining fields in the quasi-coupled or hard coupled formulation [6]. Usage of the weakly coupled formulation is acceptable just in the case of lower temperatures necessary for heating, as the material parameters in such a temperature range may be (with a small error) considered constant. The mixed formulation provides relatively good results even for higher temperature rises, but still there is a risk of unacceptable inaccuracies. And the temperature dependence of the relative permeability was considered only in a few cases (a serious reason is here the lack of experimental data).

The authors present the solution to the problem in the monolithic formulation that allows modeling of heating of the external part up to the Curie temperature or even higher. The algorithm is based on a fully adaptive finite element method of higher order of accuracy, which represents the basis of codes Hermes and Agros developed in our group. The algorithm is illustrated by an example of hot pressing of an active wheel on shaft of a high-speed gas turbine working at temperatures exceeding even 500 °C.

The aim of the paper is to propose a complete methodology of proposal of a shrink fit that satisfies all mechanical requirements and whose realization is (with respect to energy needed) as cheap as possible. More details are given in the next section, after explaining necessary particulars concerning the mechanical aspects of the problem.

## 16.2 Formulation of the Problem and Its Basic Analysis

In fact, the problem of pressing a wheel on a shaft based on the induction shrink fit is even more complex than it may seem from the previous section. For its proposal, we usually start from the following demands:

- The shrink fit has to transfer the required mechanical torque $M_m$.
- The maximum revolutions $n_{max}$ of the system wheel-shaft must be lower than the limit ("release") revolutions $n_{lim}$ (revolutions at which the centrifugal forces acting on the wheel produce such enlargement of its internal diameter that the fit becomes released).
- The pressure $p$ acting between both parts of the fit in any admissible operation regime must not exceed its maximum allowable value $p_{max}$.

Let us analyze the situation on a wheel of the simplest rectangular cross section as is depicted in Fig. 16.2. For the sake of simplicity we suppose that both the wheel and shaft are made of the same material (this case, however, is perhaps the most usual one in technical practice). The diameter of the bore in the wheel must always be smaller than the diameter of the shaft. Provided that the shaft is perfectly firm (which is an idealization of the real situation), the pressure p acting in the place of the fit is given by the formula

$$p = \frac{\Delta RE}{r_1} \frac{r_2^2 - r_1^2}{2r_2^2} \leq p_{max} \tag{16.1}$$

where E denotes the Young modulus of the material and $\Delta R$ denotes the interference (in Fig. 16.2 somewhat exaggerated for better visibility). Other dimensions are also obvious from Fig. 16.2. The pressure $p$ must satisfy the inequality

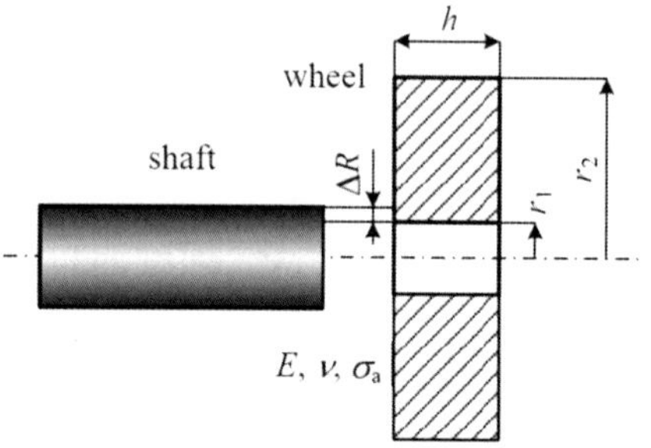


**Fig. 16.2** System shaft wheel before assembly

$$p\frac{2r_2^2}{r_2^2 - r_1^2} = \frac{\Delta RE}{r_1} \leq \frac{\sigma_a}{2} \Rightarrow \Delta R \leq \frac{\sigma_a r_1}{2E} \tag{16.2}$$

where $\sigma_a$ stands for the allowable stress of the material. The starting torque $M_{start}$ (that is always somewhat higher than the nominal torque $M_m$) is now given by the expression

$$M_{start} = 2\pi p r_1^2 h f_f \tag{16.3}$$

where $f_f$ stands for the coefficient of the static friction between the metals from which both wheel and shaft are produced. Finally, the limit revolutions $n_{lim}$ (in rev/min) follow from the formula

$$n_{lim} = \frac{60}{2\pi}\sqrt{\frac{4\Delta RE}{\rho r_1 r_2^2(3+v)}} \tag{16.4}$$

where $\rho$ denotes the specific mass of the wheel and $n$ is the Poisson number. Now we will explain the quantity $\Delta R$ called interference. As known, neither the radius of the shaft $r_s$ nor the radius of the bore in the wheel $r_h$ can be manufactured quite exactly. Each of these dimensions is characterized by a certain tolerance band, as is indicated in Fig. 16.3.

Suppose that the nominal radius of both the shaft and hole should be $r_1$. If we want to join both parts by a shrink fit, the real radius of the shaft must be little larger than $r_1$, while the real radius of the hole must be little smaller. Both mentioned radii lie between two bounds. The radius of the shaft $r_s \in< r_{s,min}, r_{s,max} >$ while the radius of the hole $r_h \in< r_{h,min}, r_{h,max} >$. Moreover, $r_{h,max} < r_1 < r_{s,min}$. Now it is clear that the real interference $\Delta R \in< \Delta R_{min}, R_{max} >$, where $\Delta R_{min} = r_{s,min} - r_{h,max}$ and $\Delta R_{max} = r_{s,max} - r_{h,min}$. And the conditions listed at the beginning of Section 16.2 must hold for both the limit values of the interference $\Delta R$, i.e., both for $\Delta R_{min}$ and $\Delta R_{max}$.

In the case of $\Delta R = \Delta R_{min}$ we must mainly check the starting torque $M_{start}$ and release revolutions $n_{lim}$, while there is practically no danger of exceeding the allowable pressure $p$. On the contrary, for $\Delta R = \Delta R_{max}$ the situation is opposite.

Consider an arrangement of an active wheel of a gas turbine and shaft depicted (in somewhat simplified manner) in Fig. 16.4. The disk is heated by two coils (indicated

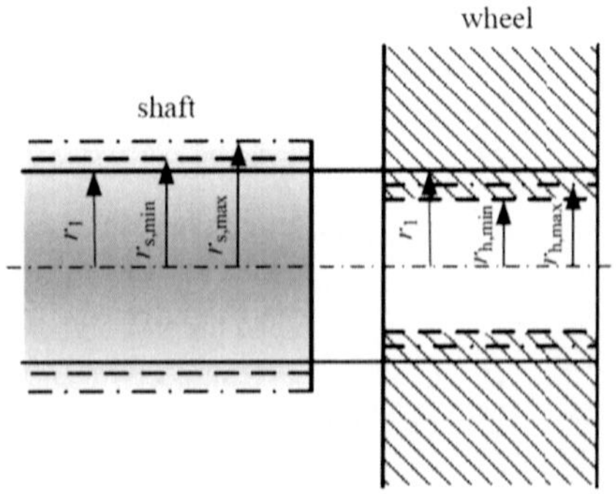


**Fig. 16.3** Tolerance bands for the shaft and wheel

in the right part of Fig. 16.4) in two variants, using transverse (up) and longitudinal (down) magnetic fields. The coils formed by a hollow massive copper conductor are adjusted to the surface of the heated part. In our case we will consider two conical plate-type coils, but even more sophisticated arrangements can be found in industrial practice.

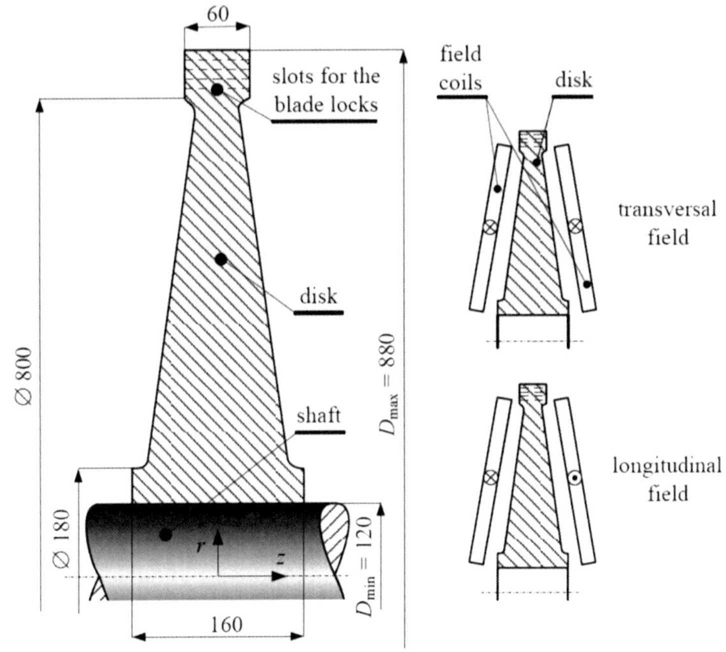


**Fig. 16.4** System disk–shaft (left, all dimensions in mm) and two possible arrangements of the field coils in the process of induction heating (right)

The aim of this paper is to carry out a complete study of preparation of the above shrink fit, consisting of the following steps:

- Finding the interference shaft-hole that allows transferring the prescribed torque within the range of the operation temperatures.
- Checking the nominal revolutions (with respect to the release revolutions) and operation pressure between the shaft and wheel.
- Estimation of the field currents in the inductors that produce a sufficient temperature rise of the wheel (that brings about the required dilatation of its hole) in an acceptable time period.
- Finding the principal characteristics describing the process of heating in the transverse and longitudinal magnetic fields,
- Evaluation of the results obtained and their discussion (with respect to the time of heating and efficiency).

## 16.3 Continuous Mathematical Model of the Process of Heating

Suppose that the material for the shaft and wheel is known (mainly its parameters $E$, $\nu$ and $\sigma_a$) and also the interference $\Delta R$. The process of induction heating of the wheel is then described by three partial differential equations (PDEs) describing the distributions of magnetic field, temperature field and field of thermoelastic displacements. These fields are mutually coupled through selected material parameters (that are generally functions of the temperature).

Electromagnetic field generated by the inductors in the wheel and its neighborhood is described by the well-known parabolic equation for the magnetic vector potential $\mathbf{A}$ in the form [7]

$$curl\left(\frac{1}{\mu}curl\mathbf{A}\right)+\gamma\frac{\partial\mathbf{A}}{\partial t}=\mathbf{J}_{ext} \tag{16.5}$$

where $\mu$ denotes the magnetic permeability, $\gamma$ is the electric conductivity and $\mathbf{J}_{ext}$ stands for the vector of the external harmonic current density in the inductors. Permeability $\mu$ is supposed to be a nonlinear function of not only flux density $B$, but also temperature $T$. But solution to equation (16.5) is practically unfeasible. The reason consists in the deep disproportion between the frequency $f$ (tens or hundreds Hz) of the field current and time of heating (tens of seconds or minutes). That is why the model was somewhat simplified using the assumption that the magnetic field is harmonic. In such a case it can be described by the Helmholtz equation for the phasor $\underline{\mathbf{A}}$ of the magnetic vector potential $\mathbf{A}$

$$curl\ culr\ \underline{\mathbf{A}}+j\omega\gamma\mu\underline{\mathbf{A}}=\mu\underline{\mathbf{J}}_{ext} \tag{16.6}$$

where $\omega$ is the angular frequency. But during the numerical processing of this equation, the magnetic permeability $\mu$ in any cell of the discretization mesh is assigned to the actual local value of magnetic flux density $B$. The conditions along the axis of the device and artificial boundary placed at a sufficient distance from the system are of the Dirichlet type $\underline{\mathbf{A}} = \underline{\mathbf{0}}$. The temperature field in the wheel is described by the heat transfer equation [8] in the form

$$div(\lambda\, grad\, T) = \rho c_p \frac{\partial T}{\partial t} - w \tag{16.7}$$

where $\lambda$ is the thermal conductivity, $\rho$ denotes the specific mass and $c_p$ stands for the specific heat at a constant pressure (all of these parameters are again temperature-dependent functions). Finally, symbol $w$ denotes the time average volumetric sources of heat that generally consist of the volumetric Joule losses $w_J$ due to eddy currents and magnetization losses $w_m$, so that

$$w = w_J + w_m \tag{16.8}$$

where

$$w_J = \frac{|\mathbf{J}_{eddy}|^2}{\gamma}, \quad \mathbf{J}_{eddy} = j\omega\gamma\underline{\mathbf{A}} \tag{16.9}$$

while losses $w_m$ (in the case that they are considered) are determined from the known measured loss dependence $w_m = w_m(|\mathbf{B}|)$ for the used material. The substantial advantage of the above approach consists in the fact that as the vector potential $\mathbf{A}$ in any element of this model is assumed harmonic, magnetic flux density $\mathbf{B}$ in every element is then also harmonic.

The boundary conditions for (16.7) should take into account convection and radiation, but their particular application depends on the case solved.

The solution of the thermoelastic problem was performed using the Lamé equation

$$(\varphi + \psi) grad\,(div\mathbf{u}) + \psi\Delta\mathbf{u} - (3\varphi + 2\psi)\alpha_T grad\, T + f = \mathbf{0} \tag{16.10}$$

where $\varphi$ and $\psi$ are coefficients associated with material parameters by the relations

$$\varphi = \frac{\nu E}{(1+\nu)(1-2\nu)}, \quad \psi = \frac{E}{2(1+\nu)} \tag{16.11}$$

Here, $\mathbf{u}$ represents the displacement vector, $\alpha_T$ denotes the coefficient of the linear thermal dilatability of material (which is, generally, a temperature-dependent function) and $f$ denotes the vector of the internal volumetric (electromagnetic, gravitational) forces. The application of the boundary conditions depends on the arrangement solved. Knowledge of the displacements $\mathbf{u}$ allows determining of the corresponding deformations, and mechanical strains and stresses of the thermoelastic origin in the wheel.

## 16.4 Numerical Solution

The numerical solution of the problem is realized by a fully adaptive higher-order finite element method [10], whose algorithms are implemented into codes Hermes

[11] and Agros [12]. Both codes have been developed in our group for a couple of years.

The codes written in C++ are intended for monolithic numerical solution of systems of generally nonlinear and nonstationary second-order partial differential equations whose principal purpose is hard-coupled modeling of complex physical problems. While Hermes is a library containing the most advanced procedures and algorithms for the numerical processing of the task solved, Agros represents a powerful preprocessor and postprocessor. Comprehensive information about them can be found on the corresponding www pages. Both codes are freely distributable under the GNU General Public License. The most important and in some cases quite unique features of the codes (that were used in the process of the numerical solution) follow:

- Solution of the system of PDEs is carried out monolithically, which means that the resultant numerical scheme is characterized by just one stiffness matrix. The PDEs are first rewritten into the weak forms whose numerical integration provides its coefficients. The integration is performed using the Gauss quadrature formulas.
- Fully automatic $hp$-adaptivity. In every iteration step the solution is compared with the reference solution (realized on an approximately twice finer mesh), and the distribution of error is then used for selection of candidates for adaptivity. Based on sophisticated and subtle algorithms the adaptivity is realized either by a subdivision of the candidate element or by its description by a polynomial of a higher order [13], [14].
- Each physical field can be solved on quite a different mesh that best corresponds to its particulars. This is of great importance, for instance, for respecting skin effect in the magnetic field, while the temperature field is usually smooth. Special powerful higher-order techniques of mapping are then used to avoid any numerical errors in the process of assembly of the stiffness matrix.
- In nonstationary processes every mesh can change in time, in accordance with the real evolution of the corresponding physical quantities.
- Easy treatment of the hanging nodes [15] appearing on the boundaries of subdomains whose elements have to be refined. Usually, the hanging nodes bring about a considerable increase of the number of the degrees of freedom (DOFs). The code contains higher-order algorithms for respecting these nodes without any need of an additional refinement of the external parts neighboring with the re-fined subdomain.
- Curved elements able to replace curvilinear parts of any boundary by a system of circular or elliptic arcs. These elements mostly allow reaching highly accurate results near the curvilinear boundaries with very low numbers of the DOFs.

## 16.5 Illustrative Example

A wheel is made of carbon steel (Czech make) CSN 12 040 and its dimensions are specified in Fig. 16.4). The most important physical parameters of the steel follow:

$E = 2.1 \times 10^{11} N/m^2, v = 0.3, \sigma_a = 2.75 \times 10^8 N/m2, \alpha_T = 1.25 \times 10^{-5}/K, \rho = 7650kg/m3, and f_f = 0.15$.

From (16.2) we immediately obtain the inequality $\Delta R \leq 39.3 \times 10^{-6}$ m. As the profile of the wheel (see Fig. 16.4)) is irregular, we have to determine the equivalent values of $r_2$ and $h(r_1 = 0.06m)$. From the viewpoint of the starting torque $M_{start}$, decisive is the length of the wheel bore, so that we accept $h = 0.16m$. Finally, the value $r_2$ is calculated from the condition that the axial cut through the wheel is rectangular and volume of the equivalent wheel is equal to the volume of the real wheel. In this way we obtain $r_2 = 0.3294m$. Furthermore, for the axisymmetric arrangement the vector of displacements **u** has components $(u_r, 0, u_z)$.

Using the above data we can determine the dependencies of mechanical values listed in section 16.2 on the interference $\Delta R$. Its value is determined as $\Delta R = min[u_r(r = r_1)]$ along the length of the hole, where $u_r$ denotes the radial component of the displacement vector **u**, see (16.10). Figure 16.5) shows the dependence $p = p(\Delta R)$ (eq. 16.1), Fig. 16.6 depicts the dependence $M_{start} = M_{start}(\Delta R)$ (eq. 16.3) and Fig. 16.7 contains the dependence $n_{lim} = n_{lim}(\Delta R)$ (eq. 16.4). The value of $\Delta R$ changes from 0 to $39.3 \times 10^{-6}m$, which is its maximum possible value with respect to the allowable stress.

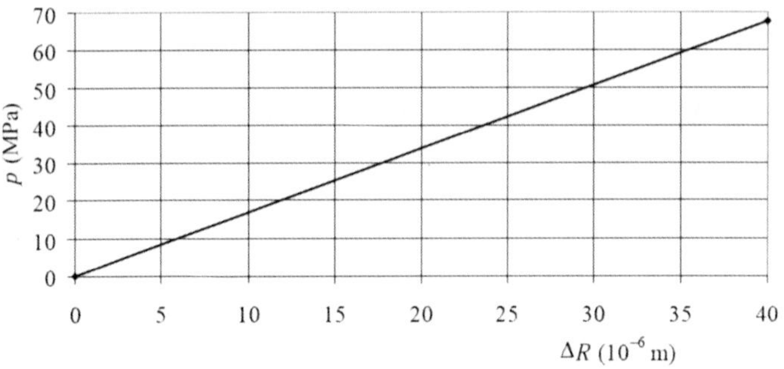


**Fig. 16.5** Dependence $p = p(\Delta R)$ of the investigated wheel

With respect that the nominal revolutions of such wheels do not exceed $n = 1500/min$, it seems to be reasonable to consider (with respect to the safety coefficient $k = 1.5$) $\Delta R \geq 10 \times 10^{-6}m$. Such values of $\Delta R$ also meet the demand laid on the transferred torque.

For the following electromagnetic, thermal and thermoelastic computations we will also need further important characteristics of the used carbon steel. For example, Fig. 16.8 shows its magnetization characteristic $B = B(H)$ and Figs. 16.9–16.11 depict

the temperature dependencies of its electric conductivity $\gamma$, thermal conductivity $\lambda$, and heat capacity $c_p$.

As told before, the magnetic permeability $\mu$ (more precisely its relative component $\mu_r$) is a function of magnetic flux density $B$ and temperature $T$. For every

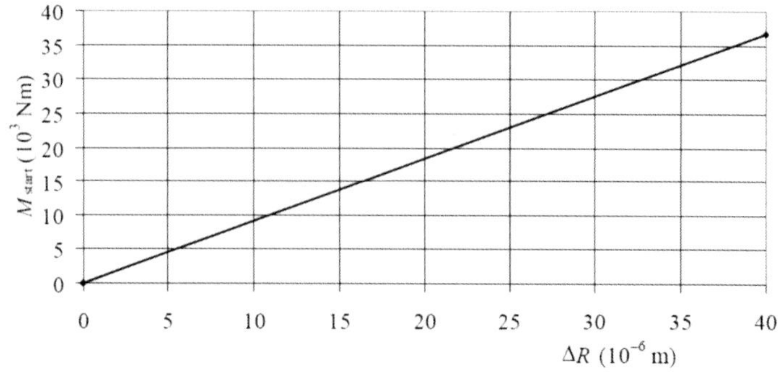


**Fig. 16.6** Dependence $M_{start} = M_{start}(\Delta R)$ of the investigated wheel

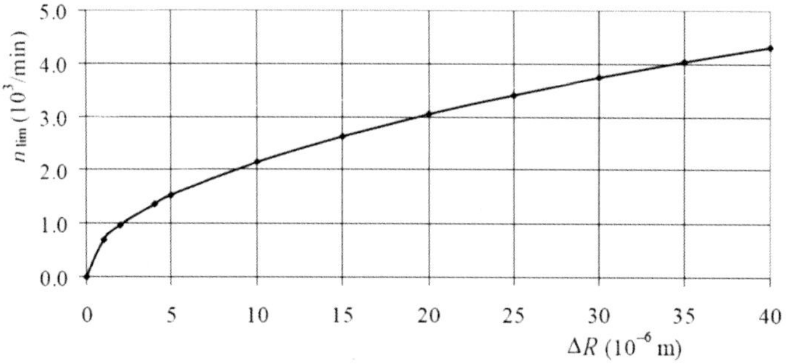


**Fig. 16.7** Dependence $n_{lim} = n_{lim}(\Delta R)$ of the investigated wheel

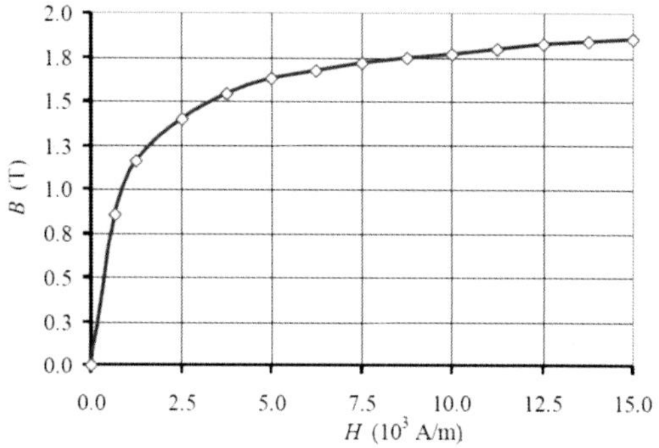


**Fig. 16.8** Magnetization curve of steel CSN 12 040

type of steel it must be found experimentally, which usually represents a serious difficulty. We introduce, therefore, an assumption that $\mu_r(B,T) = \mu_r(B,T_0)\phi(T)$, where $\mu_r(B,T_0)$ is the dependence of relative permeability $\mu_r$ on magnetic flux density $B$ at a given temperature $T_0$ (for example, $T_0 = 20\,^\circ$C) and function $\phi(T)$ is given by the relation

$$\begin{aligned} &for\ T_0 \le T \le T_C \quad \phi(T) = a - bT^2 \\ &for\ T_C \le T \qquad\quad \phi(T) = \frac{1}{\mu_r(B, T_0)} \end{aligned} \tag{16.12}$$

Here,

$$a = \frac{\mu_r(B,T_0)T_C^2 - T_0^2}{\mu_r(B,T_0)(T_C^2 - T_0^2)}, \quad b = \frac{\mu_r(B,T_0) - 1}{\mu_r(B,T_0)(T_C^2 - T_0^2)} \tag{16.13}$$

and TC is the Curie temperature (for the steel used its value is approximately 800 °C.

Both inductors are wound by a massive hollow copper conductor in an arrangement depicted in Fig. 16.12. Each of them has 15 turns and the conductors are intensively cooled by flowing water. The inductors are packed in glass wool (its physical parameters change with the temperature only very slightly, so that we consider $\gamma_{gw} = 0S/m, \lambda_{gw} = 0.04W/mK, \rho c_{p,gw} = 0.04824 \times 10^{-6} J/m^3K$) in order to reduce the thermal losses by convection and radiation from the heated wheel. This means that the wheel is thermally well insulated in the course of the whole process of heating. The temperature of the ambient air $T_0 = 30\,°C$ and average temperature of water flowing in the hollow conductors of the inductor $T_w = 50\,°C$.

For illustration, Fig. 16.13 shows the meshes (after adaptivity, frequency of the field current $f = 50Hz$) for computation of the temperature field (left) and thermoelastic displacements (right). They strongly differ from each other and also from the mesh (not depicted) used for the magnetic analysis. No radiation is assumed along the external boundary of the insulation system formed by glass wool.

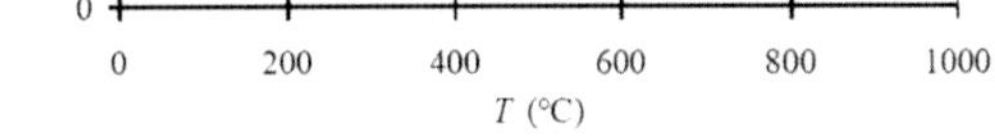


**Fig. 16.9** Dependence of electrical conductivity $\gamma$ on temperature $T$ for steel CSN 12

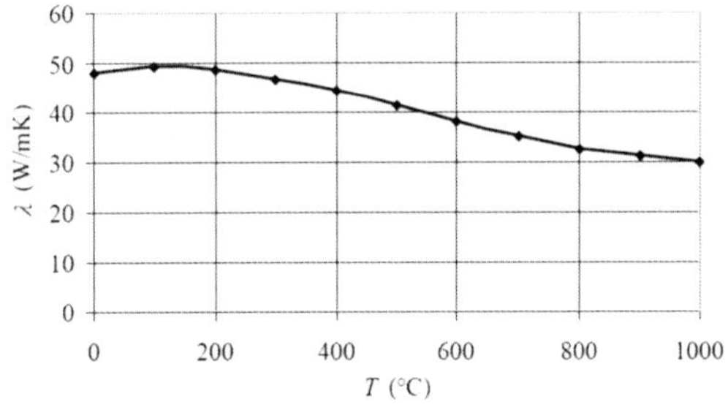


**Fig. 16.10** Dependence of thermal conductivity $\lambda$ on temperature $T$ for steel CSN 12

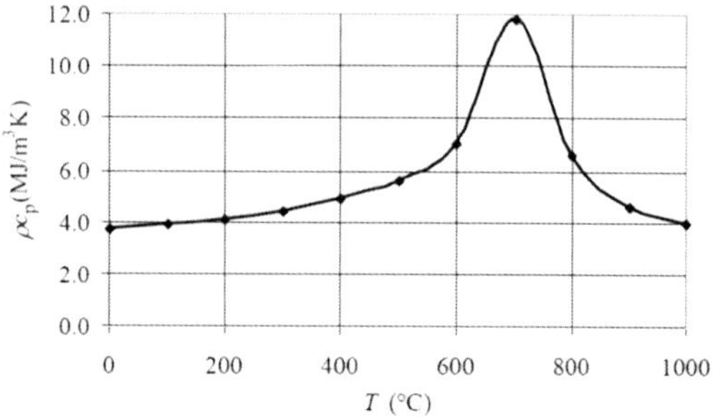


**Fig. 16.11** Dependence of heat capacity $\rho c_p$ on temperature $T$ for steel CSN 12 040

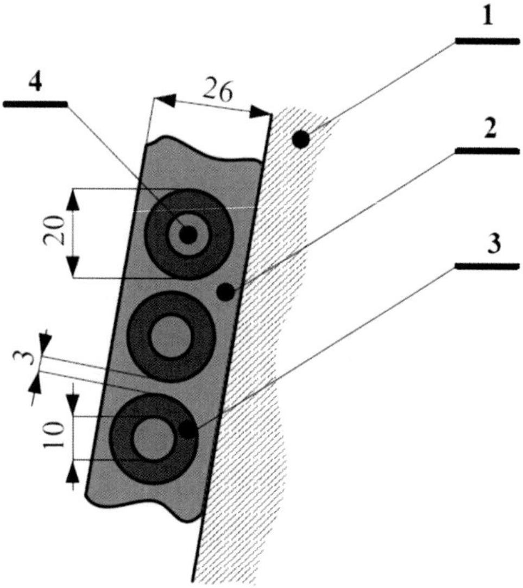


**Fig. 16.12** Arrangement of the inductor: **1**–heated wheel, **2**–thermal insulation (glass wool), **3**–hollow turns of the inductor, **4**–cooling water

The principal results representing the velocity of heating in transverse and longitudinal magnetic fields generated by field currents of the same value and analogous dependencies for the radial displacements are depicted in Figs. 16.14 and 16.15. It is clear, that from the viewpoint of the velocity of heating and total efficiency of the process the application of the transverse field is more favorable. From the graph we can choose an appropriate interference $\Delta R = u_r(r_{min})$ and using Figs. 16.5, 16.6 and 16.7 we can easily check the pressure $p$, starting torque $M_{start}$ and release revolutions $n_{lim}$.

Fig. 15 Resultant graphs for induction heating in longitudinal magnetic field: I. average temperature Tavrg in the wheel versus time, II. temperature of the inner surface T(r1) of the wheel versus time, III. radial dilatation ur(r1) of the inner surface of the wheel versus time for I =5 kA f =50 Hz

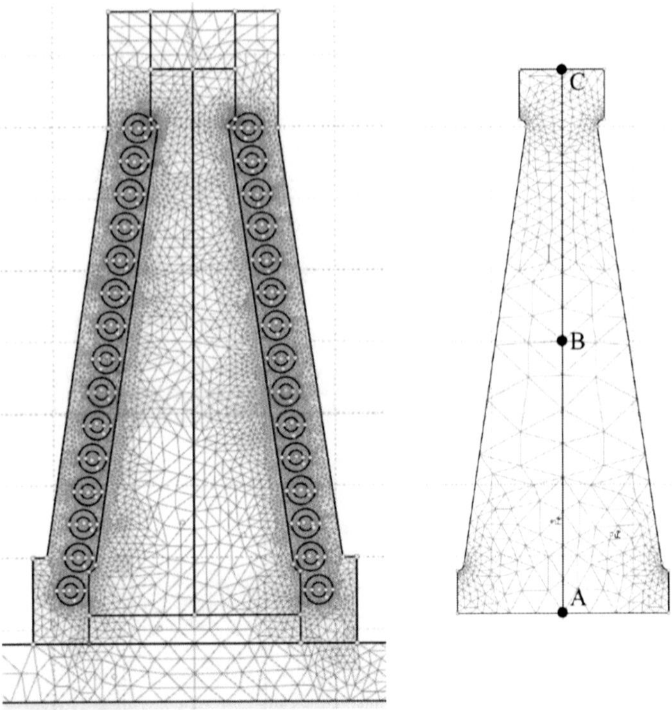


**Fig. 16.13** Discretization mesh (after adaptivity) used for computation of temperature field (left) and for computation of thermoelastic displacements (right) for frequency $f = 50Hz$

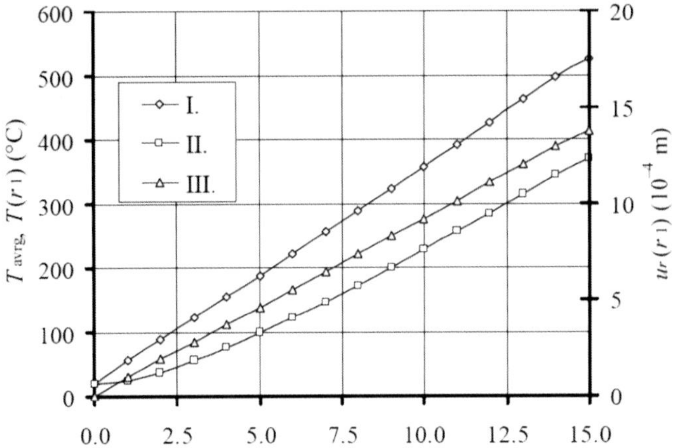


**Fig. 16.14** Resultant graphs for induction heating in transverse magnetic field: I. average temperature $T_{avrg}$ in the wheel versus time, II. temperature of the inner surface $T(r_1)$ of the wheel versus time, III. radial dilatation $u_r(r_1)$ of the inner surface of the wheel versus time for $I = 5kA$ $f = 50Hz$

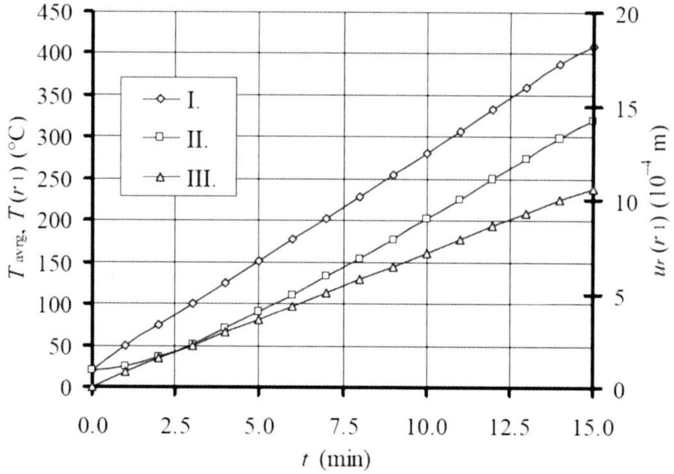


**Fig. 16.15** Resultant graphs for induction heating in longitudinal magnetic field: I. average temperature $T_{avrg}$ in the wheel versus time, II. temperature of the inner surface $T(r_1)$ of the wheel versus time, III. radial dilatation $u_r(r_1)$ of the inner surface of the wheel versus time for $I = 5kA$ $f = 50Hz$

## 16.6 Conclusion

For the investigated wheel, its heating in the transverse magnetic field exhibits higher velocity and better efficiency than in the longitudinal one. A substantial acceleration of the process is reached by using an appropriate thermal insulation along both sides of the wheel, that leads to a considerable reduction of thermal losses.

Next work in the field will be focused on further improvement of numerical algorithms and acceleration of the computations. After finding the temperature dependencies of the mechanical parameters ($E$, $\nu$ and $\alpha_T$) of the steel, they will also be included into the analysis.

**Acknowledgements.** This work was supported by the European Regional Development Fund and Ministry of Education, Youth and Sports of the Czech Republic (project No. CZ.1.05/2. 1.00/03.0094: Regional Innovation Center for Electrical Engineering - RICE), by the Grant projects GACR 102/09/1305 and GACR P102/11/0498, and by Research Plan MSM 6840770017. We wish also express our thanks to Prof. V. Laš from the Faculty of Applied Sciences of the University of West Bohemia in Pilsen for his valuable advices and commentaries.

## References

1. Cook, R.D.: Finite Element Modeling for Stress Analysis. Wiley, New York (1995)
2. Rudnev, V.I., Loveless, D., Cook, R., Black, M.: Handbook of Induction Heating. CRC Press, Boca Raton (2002)

3. Mun, C.S., Kwang, K.J., Kyo, J.H., Gyun, L.C.: Stress and thermal analysis coupled with field analysis of multi-layer buried magnet synchronous machine with a wide speed range. IEEE Trans. Magn. 41(5), 1632–1635 (2005)
4. Škopek, M., Ulrych, B., Doležel, I.: Optimized regime of induction heating of a disk before its pressing on shaft. IEEE Trans. Magn. 37(5), 3380–3383 (2001)
5. Kim, J.K., Kwak, S.Y., Cho, S.M., Jung, H.K., Chung, T.K., Jung, S.Y.: Optimization of multilayer buried magnet synchronous machine combined with stress and thermal analysis. IEEE Trans. Magn. 42(4), 1023–1026 (2006)
6. Pantelyat, M.: Coupled electromagnetic, thermal and elastic-plastic simulation of multi-impulse inductive heating. Int. J. Appl. Electromagn. Mech. 9(1), 11–24 (1998)
7. Kuczmann, M., Ivanyi, A.: The Finite Element Method in Magnetics. Akademiai Kiado, Budapest (2008)
8. Holman, J.P.: Heat Transfer. McGrawHill, New York (2002)
9. Boley, B., Wiener, J.: Theory of Thermal Stresses. Wiley, New York (1960)
10. Šolín, P., Segeth, K., Doležel, I.: Higher-Order Finite Element Methods. Chapman & Hall/CRC, Boca Raton, FL (2003)
11. Šolín, P., et al.: Code Hermes and manuals for its application, http://hpfem.org/hermes/ (cited October 15, 2011)
12. Karban, P., et al.: Code Agros2D and manuals for its application, http://agros2d.org (cited October 15, 2011)
13. Šolín, P., Dubcová, L., Kruis, J.: Adaptive hp-FEM with dynamical meshes for transient heat and moisture transfer problems. Appl. Math. 233(12), 3103–3112 (2010)
14. Dubcová, L., Šolín, P., Červený, J., Kůs, P.: Space and time adaptive two-mesh hp-FEM for transient microwave heating problems. Electromagnetics 30(1), 23–40 (2010)
15. Šolín, P., Červený, J., Doležel, I.: Arbitrary-level hanging nodes and automatic adaptivity in the hp-FEM. Math. Comput. Simul. 77(1), 117–132 (2008)

# Part IV
# Theory of Stability and Recent Trends

# Chapter 17
# Stability Analysis and Limit Cycles of High Order Sigma-Delta Modulators

Valeri Mladenov

**Abstract.** In this chapter we present an unified approach for study the stability and validation of potential limit cycles of one bit high order Sigma-Delta modulators. The approach is general because it uses the general form of a Sigma-Delta modulator. It is based on a parallel decomposition of the modulator and a direct nonlinear systems analysis. In this representation, the general N-th order modulator is transformed into a decomposition of low order, generally complex modulators, which interact only through the quantizer function. The developed conditions for stability and for validation of potential limit cycles are very easy for implementation and this procedure is very fast.

## 17.1 Introduction

Sigma-Delta modulation has become in recent years an increasingly popular choice for robust and inexpensive analog-to-digital and digital-to-analog conversion [1, 2]. Despite the widespread use of Sigma-Delta modulators theoretical understanding of Sigma-Delta concept is still very limited. This is a consequence of the fact that these systems are nonlinear, due to the presence of a discontinuous nonlinearity - the quantizer. Since the pioneering work of Gray and his co-workers beginning with [3], a number of researchers have contributed to the development of a theory of Sigma-Delta modulation based on the principles of nonlinear dynamics [4, 5, 6]. That work and references there, has succeeded in explaining many fundamentally nonlinear features of this system. The stability of high order interpolative Sigma-Delta ($\Sigma\Delta$) modulators based on nonlinear dynamics has been considered in a couple of papers [7, 8]. The authors present a technique, which in many cases simplifies the analysis. The technique involves a transformation of the state equations of a modulator into a form in which the individual state variables are essentially decoupled and interact

Valeri Mladenov
Dept of Theoretical Electrical Engineering, Technical University of Sofia
8, Kliment Ohridski St., Sofia, Bulgaria

K. Kyamakya et al. (Eds.): Selected Topics in Nonlinear Dynamics, SCI 459, pp. 305–327.
DOI: 10.1007/ 978-3-642-34560-9_17 

only within the quantizer function. In [9, 10, 11] and [12] a stability (in the sense of boundness of the states) analysis approach based on decomposition of the general N-th order modulator is presented. This decomposition is considered for all cases of poles of the transfer function of the modulator loop filter. Using this presentation the modulator could be considered as made up of N first order modulators, which interact only through the quantizer function. Based on this decomposition the stability conditions of high order modulators are extracted. They are determined by the stability conditions of each of the first order modulators but shifted with respect to the origin of the quantizer function. Limit cycles are well known phenomena that often appear in practical ΣΔ modulators. For data processing applications it is very important to predict and describe possible limit cycles. Main results concerning the limit cycles for low order Sigma-Delta modulators are presented in [6, 13, 14] and [15]. In [16, 17] authors use state space approach and present a mathematical framework for the description of limit cycles in 1-bit Sigma-Delta modulators for constant inputs. In [18] and [19] an approach for validation of potential limit cycles for high order modulators with constant input signals is presented. The approach is based on the same decomposition of the general N-th order modulator presented in [7, 9, 10, 11] and [12]. The conditions for the existence of limit cycles given in [18] and [19] are easily to be checked and they are basis of a searching procedure for possible limit cycles. In this contribution we do extend both techniques and present unified approach for study the stability and limit cycles of high order sigma-delta modulators. The study is organized as follows. In the next section we describe the parallel decomposition technique for different cases of poles of the loop filter transfer function. Then we present the stability analysis study for first and high order modulators together with an example. In Section 17.5 we present the limit cycle analysis and also give several examples to show the applicability of the presented techniques. The concluding remarks are given in the last section.

## 17.2 Parallel Decomposition of a Sigma Delta Modulator

The structure of a basic ΣΔ modulator is shown in Figure 17.1, and consists of a filter with transfer function $G(z)$ followed by a one-bit quantizer in a feedback loop. The system operates in discrete time.

The input to the loop is a discrete-time sequence $u(n) \in [-1,1]$, which is to appear in quantized form at the output. The discrete-time sequence $x(n)$ is the output

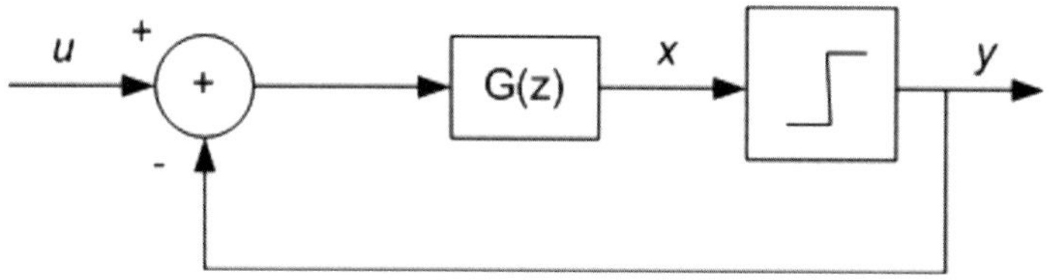


**Fig. 17.1** Basic structure of the sigma-delta modulator

of the filter and the input to the quantizer. Let us consider a N-th order modulator with a loop filter with a transfer function (TF) in the form

$$G(z) = \frac{a_1 z^{-1} + ... + a_N z^{-N}}{1 + d_1 z^{-1} + d_2 z^{-2} + d_N z^{-N}} \tag{17.1}$$

Suppose the transfer function has $N$ real distinct roots of the denominator. Then using partial fraction expansion we get

$$G(z) = \frac{a_1 z^{-1} + ... + a_N z^{-N}}{(1 - \lambda_1 z^{-1})...(1 - \lambda_N z^{-1})} = \frac{b_1 z^{-1}}{1 - \lambda_1 z^{-1}} + ... + \frac{b_N z^{-1}}{1 - \lambda_N z^{-1}} \tag{17.2}$$

where the coefficients $b_i, i = 1,2,...,N$ of the fractional components can be found easily using the well known formula $b_i = \left.\frac{(1-\lambda_i z^{-1})}{z^{-1}}\right|_{z=\lambda_i} G(z)$.

The corresponding block diagram of the modulator is given in Figure 17.2. Based on this presentation the state equations of the $\Sigma\Delta$ modulator are

$$\begin{aligned} x_k(n+1) =& \lambda_k x_k(n) + \left[u(n) - f\left(\sum_{i=1}^{N} b_i x_i(n)\right)\right] = \\ =& \lambda_k x_k(n) + \left[u(n) - f\left(b_k x_k(n) + \sum_{\substack{i=1 \\ i \neq k}}^{N} b_i x_i(n)\right)\right] \\ k =& 1,2,...,N \end{aligned} \tag{17.3}$$

where $\lambda_1, \lambda_2 \ldots, \lambda_N$ are poles (or modes) of the loop filter and the quantizer function $f$ is a sign function. Equation (17.3) also can be rewritten in the form

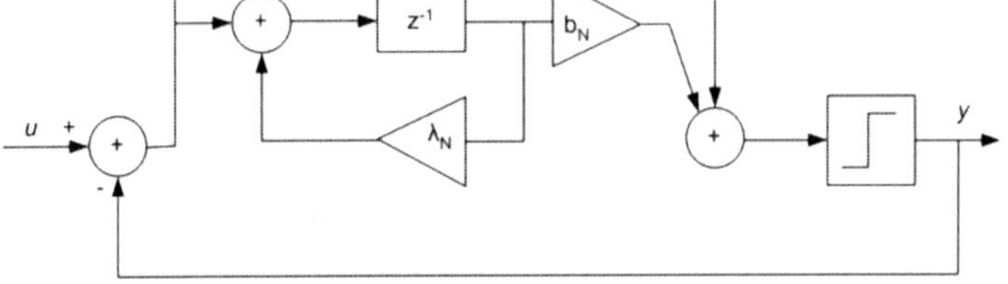


**Fig. 17.2** Block diagram of the modulator using parallel form of the loop filter

$$\begin{aligned} x_k(n+1) =& \lambda_k x_k(n) + \left[ u(n) - f\left( \sum_{i=1}^{N} b_i x_i(n) \right) \right] = \\ & \lambda_k x_k(n) + \left[ u(n) - f\left( \mathbf{b}^T \mathbf{x}(n) \right) \right] = \\ & \lambda_k x_k(n) + \left[ u(n) - y(n) \right] \quad k = 1,2,...,N \end{aligned} \tag{17.4}$$

where $\mathbf{b} = (b_1, b_2, ..., b_N)^T$ is the vector of fractional components coefficients and $\mathbf{x} = (x_1, x_2, ..., x_N)^T$ is the state vector. The above presentation indicates that high order modulators could be considered as built up of first order modulators, which interact only through the quantizer function. To simplify the notations, we will drop the indexes and will rewrite equation (17.3) in the following form

$$x_k(n+1) = \lambda x(n) + [u(n) - f(bx(n) + \alpha(n))] = \tag{17.5}$$

where

$$\alpha(n) = \sum_{\substack{i=1 \\ i \neq k}}^{N} b_i x_i(n) \tag{17.6}$$

and

$$y(n) = f\left( \sum_{i=1}^{N} b_i x_i(n) \right) = f\left( \mathbf{b}^T \mathbf{x}(n) \right) = \begin{cases} 1 & \mathbf{b}^T \mathbf{x}(n) \geq 0 \\ -1 & \mathbf{b}^T \mathbf{x}(n) < 0 \end{cases} \tag{17.7}$$

Equation (17.4) describes a first order shifted by $\alpha(n)$ modulator. A detailed analysis of the stability of these modulators will be presented in the next chapter.

In the general case the loop filter transfer function can have complex conjugated roots. Without loss of generality we will consider only one pair of complex conjugated roots. In this case (17.2) becomes

$$\begin{aligned} G(z) =& \frac{b_1 z^{-1}}{(1 - \lambda_1 z^{-1})} + ... + G_2(z) = \\ & \frac{b_1 z^{-1}}{(1 - \lambda_1 z^{-1})} + ... \frac{B_{N-1} z^{-1} + B_N z^{-2}}{1 - d_1 z^{-1} - d_2 z^{-2}} \end{aligned} \tag{17.8}$$

The denominator of the last part of (17.8) has a complex conjugated pair of roots. The main idea is to use a complex form of expansion of the last part of $G(z)$. Therefore (17.8) becomes

$$G(z) = \frac{b_1 z^{-1}}{(1 - \lambda_1 z^{-1})} + ... + \frac{b_{N-1} z^{-1}}{(1 - \lambda_{N-1} z^{-1})} + \frac{b_N z^{-1}}{1 - \lambda_N z^{-1}} \tag{17.9}$$

where

$$\begin{aligned} \lambda_{N-1} = \alpha + j\beta, \lambda_N = \alpha - j\beta \\ b_{N-1} = \delta - j\gamma, b_N = \delta + j\gamma \end{aligned} \tag{17.10}$$

i.e. $\lambda_{N-1}$, $\lambda_N$ and $b_{N-1}$, $b_N$ are complex conjugated numbers. Because of this we can use the same parallel presentation given in figure 2. However, the values of the last two blocks are complex. It should be stressed that the output signal of these two blocks is real. In order to make things more clear and without loss of generality we will consider only these blocks. They correspond to a second order $\Sigma\Delta$ modulator with complex conjugated poles of the loop filter transfer function $G(z)$. The block diagram of this modulator is given in figure 17.3.

Here both signals $x_1$ and $x_2$ are complex conjugated, namely

$$\begin{aligned} x_1(k+1) &= m(k+1) + jn(k+1) \\ x_2(k+1) &= m(k+1) - jn(k+1) \end{aligned} \tag{17.11}$$

Because of this the input of the quantizer is real i.e.

$$(\delta - j\gamma)x_1(k) + (\delta + j\gamma)x_2(k) = 2\delta m(k) + 2\gamma n(k) \tag{17.12}$$

As in the case of real poles, the modulator could be considered as two first order modulators interacting only through the quantizer function. The difference now is that the signals connected with both modulators are complex, but the input and output signals ($u$ and $y$) are the "true" signals of the modulator. This model will help us to make analysis simple. We will consider the state of the first order modulators as a point in a complex plane $(m,n)$. Depending on whether the input $2\delta m + 2\gamma n$ of the quantizer is positive or negative the state equation of the second order modulator could be described as follows:

$$\begin{aligned} x_1(k+1) &= (\alpha + j\beta)x_1(k) + [u(k) - 1], 2\delta m(k) + 2\gamma n(k) \geq 0 \\ x_2(k+1) &= (\alpha + j\beta)x_2(k) + [u(k) - 1], 2\delta m(k) + 2\gamma n(k) \geq 0 \end{aligned} \tag{17.13}$$

and

$$\begin{aligned} x_1(k+1) &= (\alpha + j\beta)x_1(k) + [u(k) + 1], 2\delta m(k) + 2\gamma n(k) < 0 \\ x_2(k+1) &= (\alpha + j\beta)x_2(k) + [u(k) + 1], 2\delta m(k) + 2\gamma n(k) < 0 \end{aligned} \tag{17.14}$$

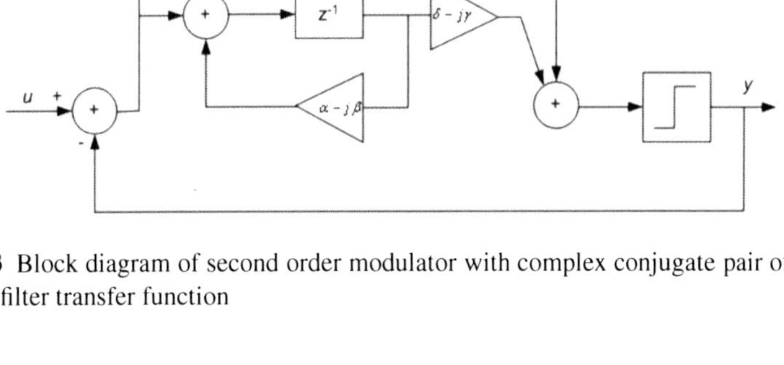


**Fig. 17.3** Block diagram of second order modulator with complex conjugate pair of roots of the loop filter transfer function

where $x_1$ and $x_2$ are given by (17.11). In fact $2\delta m + 2\gamma n$ is a line through the origin in the plane $(m,n)$ and depending on in what half the point $x_1$ is (because $x_1 = m + jn$), the description of the modulator is (17.13) or (17.14).

## 17.3 Stability of Shifted First Order Sigma-Delta Modulators

The shifted first order modulator is described by equation (17.5). Because of the ideal quantizer, the system can be viewed as two linear systems connected at point - $\alpha(n)/b$ and thus the equations describing the dynamics of the first order Sigma-Delta modulator from (17.5) are

$$\begin{aligned} x(n+1) &= \lambda x(n) + [u(n) - 1], x(n) \geq -\alpha(n)/b; b > 0 \\ x(n+1) &= \lambda x(n) + [u(n) + 1], x(n) < -\alpha(n)/b; b > 0 \end{aligned} \tag{17.15}$$

The fixed points of the system are $x' = \frac{u(n)-1}{1-\lambda}, x'' = \frac{u(n)+1}{1-\lambda}$.

In what follows we will consider the input signal $u(n)$ to be from the interval $u(n) \in [-\Delta u, \Delta u], \Delta u > 0$ and because of this the shift $\alpha(n)$ belongs to the interval $[-\Delta\alpha, \Delta\alpha], \Delta\alpha > 0$. The flow diagram of the system is given in Figure 17.4.

### *Stable Mode,* $\lambda < 1$

Depending on the parameters $b, \alpha(n)$ and input signal $u(n)$ the system can have two stable virtual fixed points (the case given in the figure) and a compact region exists between them (in fact this is an invariant set in state space, which has the property that all subsequent states lie in the original set for a certain class of input signals). For another set of parameters one of the virtual fixed points becomes a real fixed point. In each of the cases the system is stable but in the second one, there is

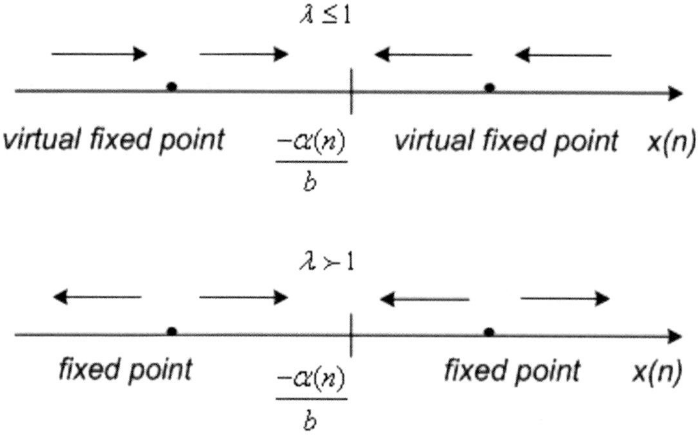


**Fig. 17.4** Flow diagrams of the first order system for the case of $\lambda \leq 1$ and $\lambda > 1$

no compact region. The system moves towards a single attractor at the stable fixed point. Anyway, if the initial condition is between the origin and the real fixed point of system (17.5) the state flow finishes at the equilibrium point (due to asymptotic movement to the single equilibrium point). It should be noted that this is not a desired Sigma-Delta modulator behavior. The Sigma-Delta modulator behavior appear when the first order system has two virtual fixed points and the state of (17.5) jumps between them. Thus the desired bitstream appear at the output of the quantizer.

## *Unstable Mode,* $\lambda > 1$

The stability in this case is connected with existence of a compact region between the unstable fixed points (not virtual). It is important to point out that $b > 0$. Otherwise the dynamic of the system is described by

$$\begin{aligned} x(n+1) &= \lambda x(n) + [u(n) - 1], x(n) < -\alpha(n)/b; b < 0 \\ x(n+1) &= \lambda x(n) + [u(n) + 1], x(n) \geq -\alpha(n)/b; b < 0 \end{aligned} \tag{17.16}$$

and it is easy to observe that the above system is always unstable, because at least one of the fixed points is virtual. Let's consider the map (17.5), given by (17.15) depicted in Figure 17.5. For a compact region (CR), to exists the fixed point should not be virtual i.e. $-\frac{\alpha(n)}{b} < \frac{u(n)-1}{(1-\lambda)}$ and $-\frac{\alpha(n)}{b} > \frac{u(n)+1}{(1-\lambda)}$

This should be true for the worst case i.e. $-\frac{\alpha(n)}{b} < \frac{\Delta u-1}{(1-\lambda)}$ and $-\frac{\alpha(n)}{b} > \frac{-\Delta u+1}{(1-\lambda)}$.
Taking into account that $(1-\lambda) < 0$ and $b > 0$ we get

$$\frac{b}{\lambda - 1}\Delta u - \frac{b}{(\lambda - 1)} < \alpha(n) < -\frac{b}{\lambda - 1}\Delta u + \frac{b}{(\lambda - 1)} \tag{17.17}$$

The second condition for the existence of a compact region is that it has to be included into the region between the fixed points i.e. the stable region. The maximum jump of the variable $x(n)$ from the Negative Half Line (NHL), with respect to $-\alpha(n)/b$, to the Positive Half Line (PHL), with respect to $-\alpha(n)/b$, is $[-\alpha(n)/b]\lambda + [u(n) + 1]$ and the maximum jump from PHL to NHL is $[-\alpha(n)/b]\lambda + [u(n) - 1]$. Hence in the worst case

$$-\frac{\alpha(n)}{b}\lambda + [\Delta u + 1] < \frac{\Delta u - 1}{(1 - \lambda)}, \quad -\frac{\alpha(n)}{b}\lambda + [-\Delta u - 1] > \frac{-\Delta u + 1}{(1 - \lambda)}$$

Solving the above inequalities with respect to $\alpha(n)$ we find that a compact region can only exist if $b > 0$ and

$$\begin{aligned} &b > 0 \\ &\frac{b}{\lambda - 1}\Delta u - \frac{b(2 - \lambda)}{\lambda(\lambda - 1)} < \alpha(n) < -\frac{b}{\lambda - 1}\Delta u + \frac{b(2 - \lambda)}{\lambda(\lambda - 1)} \end{aligned} \tag{17.18}$$

Because the above should be valid for all $y$ and for $y = 0$ as well then $(2-\lambda)/\lambda > 0$ or $\lambda < 2$, i.e. $1 < \lambda < 2$. Due to this $(2-\lambda)/\lambda < 1$ and hence if (17.18) is

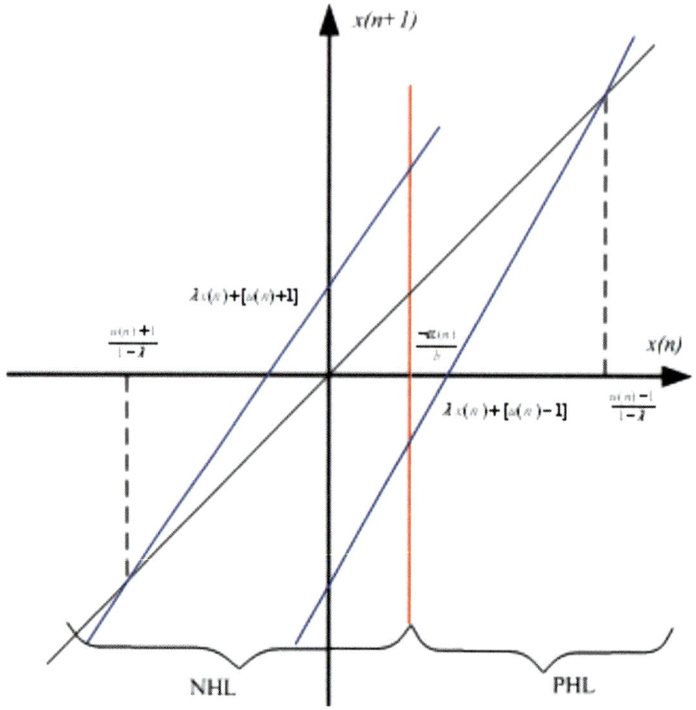


**Fig. 17.5** Map (17.5) given by (17.15) for the case of $\lambda > 1$

satisfied then (17.17) will be satisfied as well. Considering again these two conditions, the maximal shift of the input signal $\Delta u$, which ensures that the compact region is included into the region between the fixed points i.e. the stable region is given by

$$\Delta u < -\frac{\Delta\alpha(\lambda - 1)}{b} + \frac{2 - \lambda}{\lambda} \tag{17.19}$$

Note that condition (17.18) is a sufficient but not necessary condition. It has been derived for the worst case and if satisfied, the first order modulator is stable for the range of input signal given by (17.19). However, if (17.18) is not satisfied the modulator could be stable for certain input signal.

## 17.4 Stability of High Order Sigma-Delta Modulators

### *Stability of High Order Sigma-Delta Modulators with Real Poles*

Taking into account the parallel presentation given in Section 17.2, the stability of the high order Sigma-Delta modulator depends on the stability of each of the first order modulators. If all modes $\lambda_k$, are stable, i.e. $\lambda_k < 1$ then the corresponding high order Sigma-Delta modulator is stable in the sense of boundness of the states. If there exists even one unstable mode $\lambda_k$, i.e. $1 < \lambda_k < 2$, the stability conditions for shifted modulators given above should be applied. In this case the shift $\alpha_k(n)$ depends on the values of the other variables $x_i(n)$ i.e.

$$\lambda_k(n) = \sum_{\substack{i=1 \\ i \neq k}}^{N} b_i x_i(n) \tag{17.20}$$

From (17.18), we have

$$\begin{aligned} &\sum_{\substack{i=1 \\ i \neq k}}^{N} b_i x_i(n) < -\frac{b_k}{\lambda_k - 1}\Delta u + \frac{b_k(2-\lambda_k)}{\lambda_k(\lambda_k - 1)} \\ &\sum_{\substack{i=1 \\ i \neq k}}^{N} b_i x_i(n) > \frac{b_k}{\lambda_k - 1}\Delta u - \frac{b_k(2-\lambda_k)}{\lambda_k(\lambda_k - 1)}, \\ &k = 1,2,...,N \end{aligned} \tag{17.21}$$

The above should still be true when $x_k$ makes the maximal "jumps" into the PHL or into the NHL. Without loss of generality we will consider the first $p$ modes $\lambda_k$ of the high order Sigma-Delta modulator to correspond to $1 < \lambda_k < 2, k = 1,2,\ldots,p$ whereas the remaining $N-p$ modes correspond to $\lambda_k < 1, k = p+1,\ldots,N$. In this case only the first $p$ coefficients $b_k$ must be positive and the remaining $N-p$ coefficients could have any real value. The maximal "jumps" of the state variables corresponding to the first $p$ modes in the PHL and the NHL are $\frac{u(n)-1}{1-\lambda_k}$ and $\frac{u(n)+1}{1-\lambda_k}$, respectively (the fixed points of the system with respect to $x_k, k = 1,2,\ldots,p$). Similarly, the maximal "jumps" of the state variables corresponding to the last $N-p$ modes in the PHL and the NHL, are $\frac{u(n)+1}{1-\lambda_k}$ and $\frac{u(n)-1}{1-\lambda_k}$, respectively (the virtual or real fixed points of the system with respect to $x_k, k = p+1,\ldots,N$). Therefore from (17.21) for the worst case with respect to the input signal one can obtain

$$\begin{aligned} &\sum_{\substack{i=1 \\ i \neq k}}^{p} b_i \frac{-\Delta u - 1}{1-\lambda_i} + \sum_{i=p+1}^{N} |b_i| \frac{\Delta u + 1}{1-\lambda_i} < -\frac{b_k}{\lambda_k - 1}\Delta u + \frac{b_k(2-\lambda_k)}{\lambda_k(\lambda_k - 1)} \\ &\sum_{\substack{i=1 \\ i \neq k}}^{p} b_i \frac{\Delta u + 1}{1-\lambda_i} + \sum_{i=p+1}^{N} |b_i| \frac{-\Delta u - 1}{1-\lambda_i} > \frac{b_k}{\lambda_k - 1}\Delta u - \frac{b_k(2-\lambda_k)}{\lambda_k(\lambda_k - 1)} \\ &k = 1,2,...,p \end{aligned} \tag{17.22}$$

Note that we apply (17.22) only for the shifts connected to the first $p$ modulators. The other $N-p$ first order modulators are stable, because for their corresponding $\lambda_k, \lambda_k \leq 1, k = p+1,\ldots,N$. If there exists a region $[-\Delta u, \Delta u] \subseteq [-1,1]$, such that

$u \in [-\Delta u, \Delta u]$ and for this region conditions (17.22) are satisfied, then the Sigma-Delta modulator will be stable for all input signals from this region. Taking into account equation (17.22) we get

$$\left[\sum_{i=1}^{p}\frac{b_i}{\lambda_i-1}-\sum_{i=p+1}^{N}\frac{|b_i|}{\lambda_i-1}\right]\Delta u<\sum_{i=p+1}^{N}\frac{|b_i|}{\lambda_i-1}-\sum_{\substack{i=1\\i\neq k}}^{p}\frac{b_i}{\lambda_i-1}+\frac{b_k(2-\lambda_k)}{\lambda_k(\lambda_k)-1}$$

$$k=1,2,\dots,p \tag{17.23}$$

More detailed considerations of the above inequality shows that in order to ensure a consistent solution of (17.23) with respect to $\Delta u$

$$\sum_{\substack{i=1\\i\neq k}}^{p}\frac{b_i}{\lambda_i-1}-\sum_{i=p+1}^{N}\frac{|b_i|}{\lambda_i-1}-\frac{b_k(2-\lambda_k)}{\lambda_k(\lambda_k-1)}<0,\quad k=1,2,..,p \tag{17.24}$$

Hence the maximal shift of input signal $\Delta u$ ensuring the stability is given by

$$\Delta u<\frac{\sum_{i=p+1}^{N}\frac{|b_i|}{\lambda_i-1}-\sum_{\substack{i=1\\i\neq k}}^{p}\frac{b_i}{\lambda_i-1}+\frac{b_k(2-\lambda_k)}{\lambda_k(\lambda_k-1)}}{\sum_{i=1}^{p}\frac{b_i}{\lambda_i-1}-\sum_{i=p+1}^{N}\frac{|b_i|}{\lambda_i-1}},\quad k=1,2,\dots,p \tag{17.25}$$

Note that inequalities (17.25) should be valid simultaneously for each $k, k = 1,2,\dots,p$. Therefore, together with $b_k > 0, k = 1,2,\dots,p$, equation (17.24) gives the sufficient conditions for the stability of the Sigma-Delta modulator, namely

$$\frac{(2-\lambda_k)}{\lambda_k}\frac{b_k}{(\lambda_k-1)}>\sum_{\substack{i=1\\i\neq k}}^{p}\frac{b_i}{\lambda_i-1}-\sum_{i=p+1}^{N}\frac{|b_i|}{\lambda_i-1},\quad k=1,2,\dots,p \tag{17.26}$$

For the poles outside the unit circle, $k = 1,2,\dots,p$, we have that $(2-\lambda_k)/\lambda_k < 1$. This implies that the inequality, Eq. (17.26), can only hold for one value of $k$. Hence, Eq. (17.26) provides a sufficient condition for stability when $p = 1$ i.e. there is at most one unstable mode, and this sufficient condition cannot hold when there is more than one pole outside the unit circle. It is clear now that in the case of repeated poles $(\lambda_1,\dots,\lambda_m = \lambda)$ of the loop transfer function, the Sigma-Delta modulator is stable only when the corresponding modes are stable i.e. $\lambda \le 1$. Let us consider more precisely the case of identical poles. Without losing the generality we will consider that the pole $\lambda_1$ is repeated with order 2 i.e. $\lambda_1 = \lambda_2 = \lambda$. In this case (2) becomes

$$G(z)=\frac{b_1z^{-1}}{1-\lambda z^{-1}}+\frac{b_2z^{-2}}{(1-\lambda z^{-1})^2}+\dots+\frac{b_Nz^{-1}}{1-\lambda_Nz^{-1}} \tag{17.27}$$

And the state equations may be given as

$$\begin{aligned} x_1(n+1) &= \lambda x_1(n) + u(n) - sgn[b_1 x_1(n) + \sum_{i=2}^{N} b_i x_i(n)] \\ x_2(n+1) &= x_1(n) + \lambda x_2(n) \\ x_k(n+1) &= \lambda_k x_k(n) + u(n) - sgn[b_k x_k(n) + \sum_{\substack{i=1 \\ i \neq k}}^{N} b_i x_i(n)] \end{aligned} \tag{17.28}$$

$k = 3,..,N$

If $\lambda$ is an unstable mode, *i.e.* $1 < \lambda < 2$ then the corresponding first and second modulators should be stable in the sense of boundedness of the states. The first one can satisfy the conditions given by (17.18). The second one in fact is a linear system described by

$$x_2(n+1) = \lambda x_2(n) + x_1(n) \tag{17.29}$$

where the state variable $x_1$ could be considered as an input signal for this system. If $1 < \lambda < 2$ then all possible symbolic sequences represent admissible periodic orbits of $x_1$. Because of this, depending on the initial conditions a certain periodic orbit of $x_1$ could influence the instability in $x_2$.

## Stability of High Order Sigma-Delta Modulators with Complex Poles

In the particular case of two complex conjugated poles given in Figure 17.3, the dynamics of the Sigma-Delta Modulator is described by (17.13) or (17.14). The analysis of the behavior of both first order "complex" modulators is similar to the analysis of the first order "real" modulators, given in Section 17.3. Here we always should keep in mind that both modulators work cooperative, because their signals are conjugated. These modulators do not exist in the real Sigma-Delta modulator. They are introduced (like in the "real" case as well) to help us to carry out the analysis of the behavior of the whole system.

## Stable Mode, $|\lambda_1| = |\lambda_2| < 1$

In this case both modulators have two stable equilibrium points (in every half plane):

first modulator: $\frac{u-1}{1-\lambda_1}$ and $\frac{u+1}{1-\lambda_1}$ i.e. $\frac{(u-1)[(1-\alpha)+j\beta]}{(1-\alpha)^2+\beta^2}$ and $\frac{(u+1)[(1-\alpha)+j\beta]}{(1-\alpha)^2+\beta^2}$

second modulator: $\frac{u-1}{1-\lambda_2}$ and $\frac{u+1}{1-\lambda_2}$ i.e. $\frac{(u-1)[(1-\alpha)-j\beta]}{(1-\alpha)^2+\beta^2}$ and $\frac{(u+1)[(1-\alpha)-j\beta]}{(1-\alpha)^2+\beta^2}$

These fixed points could be virtual or real. Taking into account equations (17.12), (17.13) and (17.14), the fixed points of both modulators are "virtual" when $2\delta(1-\alpha)+2\gamma\beta>0$ and "non-virtual" when $2\delta(1-\alpha)+2\gamma\beta<0$. Both complex modulators are stable and the second order modulator is stable as well. As was mentioned in section 17.3, the Sigma-Delta modulator behavior appears when the first order system has two virtual fixed points and the states of (17.13), (17.14) jump between them. Thus the desired bitstream appears at the output of the quantizer. According to [12], in the general case, when the last two first order modulators are "complex", i.e. correspond to a stable complex conjugated pair of roots; condition (17.26) has the form

$$\frac{(2-\lambda_1)}{\lambda_1}\frac{b_1}{(\lambda_1-1)} > -\sum_{i=2}^{N-2}\frac{|b_i|}{\lambda_i-1}+\frac{2|\delta(1-\alpha)+\gamma\beta|}{(1-\alpha)^2+\beta^2} \tag{17.30}$$

and the maximal range of input signal $\Delta u$ ensuring the stability is expressed by

$$\Delta u < \frac{\sum_{i=2}^{N-2}\frac{|b_i|}{\lambda_i-1}+\frac{2|\delta(1-\alpha)+\gamma\beta|}{(1-\alpha)^2+\beta^2}+\frac{b_1(2-\lambda_1)}{\lambda_1(\lambda_1-1)}}{\frac{b_1}{\lambda_1-1}-\sum_{i=2}^{N-2}\frac{|b_i|}{\lambda_i-1}+\frac{2|\delta(1-\alpha)+\gamma\beta|}{(1-\alpha)^2+\beta^2}} \tag{17.31}$$

## Unstable Mode, $|\lambda_1| = |\lambda_2| > 1$

In this case both modulators have two unstable fixed points (in every half plane). Depending on parameters, these points could be "non-virtual" or "virtual". In the case of virtual fixed points, both "complex" modulators are unstable and the whole system is unstable. In the case of real fixed points, the possibility for Sigma-Delta modulator behavior is connected with the existence of a compact region in the complex plane.

To summarize the results on stability of high order Sigma-Delta modulators from this section, we have the following: 1. Any Sigma-Delta modulator comprised entirely of parallel sections with poles inside the unit circle is inherently stable. 2. Any Sigma-Delta modulator with only real poles is guaranteed to be stable if (17.26) holds and (17.25) provides the maximum input for stability. Equation (17.26) also implies that the sufficient conditions for stability are violated if at least 2 real poles are outside the unit circle. 3. Any Sigma-Delta modulator comprised entirely of parallel sections with poles inside the unit circle and one complex conjugate pair inside the unit circle is inherently stable. 4. Any Sigma-Delta modulator comprised entirely of parallel sections with some real poles outside the unit circle and one complex conjugate pair inside the unit circle is guaranteed to be stable if (17.30) holds, and (17.31) provides the maximum input for stability. Equation (17.30) also implies that the sufficient conditions for stability are violated if at least 2 real poles are outside the unit circle. It should be emphasized, that present theoretical study includes only the cases considered above; real poles not equal to 1, or complex poles inside the unit circle. To demonstrate the applicability of the presented conditions we consider a $\Sigma\Delta$ modulator with the following loop filter transfer function

$$G(z) = \frac{b_1 z^{-1}}{1-\lambda_1 z^{-1}} + \frac{2r\cos\theta z^{-1} - r^2 z^{-2}}{1-2r\cos\theta z^{-1} + r^2 z^{-2}}, \quad \lambda_1 > 1$$

In this case $\lambda_2 = \alpha + j\beta, \quad \lambda_3 = \alpha - j\beta; \quad b_2 = \delta - j\gamma, \quad b_3 = \delta + j\gamma$ where $\alpha = r\cos\theta, \quad \beta = r\sin\theta, \quad \delta = r\cos\theta; \quad \gamma = r\frac{\cos 2\theta}{2\sin\theta}$, Then the stability condition becomes

$$\frac{(2-\lambda_1)}{\lambda_1}\frac{b_1}{(\lambda_1 - 1)} > \frac{2r\cos\theta - r^2}{1-2r\cos\theta + r^2} \tag{17.32}$$

Let's consider two different modulators with the following set of parameters: $r = 0.9, \theta = 15°, \lambda_1 = 1.05, b_1 = 0.5$ and $r = 0.9, \theta = 15°, \lambda_1 = 1.05, b_1 = 1$. One can simulate numerically the behavior of both modulators and could observe that the first modulator is unstable, whereas the second one is stable for a certain range of input signal (given by (17.31)) because stability condition (17.32) is satisfied.

## 17.5 Analysis of Limit Cycles in High Order Sigma-Delta Modulators

In what follows, the following case will be considered:

1. The input signal $u = u(n)$ is constant from interval [-1, 1]

$$u = const., u \in [-1, 1] \tag{17.33}$$

2. The poles of the loop filter $\lambda_1, \lambda_2, \ldots, \lambda_n$ are in the unit circle

$$(\forall_{k=1}^{N} : |\lambda_k| < 1) \tag{17.34}$$

One of the important observations in [4] is that the case of pairs of complex conjugated poles can be considered on similar way with the help of presentation given in Figure 17.2. However, in this case the corresponding signals and coefficients are complex conjugated. As it has been stressed the contribution of the state variables corresponding to every pair of complex conjugated poles, to the input of the quantizer is real. In the next investigations we are going to skip also the case of real repeated roots. This assumption is practical, because it is very difficult to have this case due to unavoidable noise in every sigma-delta modulator realization. Without loss of generality we will consider the case with real distinct poles. Thus the discrete time sequence for state variables $x_1, x_2, \ldots, x_N$ is given by:

$$
\begin{aligned}
x_k(1) =&\lambda_k x_k(0) + [u(0) - y(0)],\\
x_k(2) =&\lambda_k x_k(1) + [u(1) - y(1)] = \lambda_k^2 x_k(0) +\\
&+ [u(0) - y(0)]\lambda_k^1 + [u(1) - y(1)]\\
\ldots\ldots&\\
x_k(n) =&\lambda_k x_k(n-1) + [u(n-1) - y(n-1)] =\\
&\lambda_k^n x_k(0) + [u(0) - y(0)]\lambda_k^{n-1} + [u(1) - y(1)]\lambda_k^{n-2} + \ldots\\
&[u(n-2) - y(n-2)]\lambda_k^1 + [u(n-1) - y(n-1)] =\\
&\lambda_k^n x_k(0) + \sum_{i=0}^{n-1} [u(i) - y(i)]\lambda_k^{n-i-1}\\
k = 1,2,&\ldots,N
\end{aligned}
\tag{17.35}
$$

The limit cycles correspond to periodic solutions in time domain. The periodic solutions can be observed at the output of the modulator as repetitive sequences of 1's and -1's. Let's consider a periodic sequence $y(0), y(1), \ldots, y(M-1)$ with length $M$ at the output of the modulator. In this case $y(M) = y(0), y(M+1) = y(1), \ldots, y(2M-1) = y(M-1)$, etc. Every periodic output sequence corresponds to a periodic sequence in the states i.e. every state variable $x_k$ is periodic. This can be observed easily if we write the state variable $x_k$ after $L$ periods.

$$
x_k(L.M) = \lambda_k^{L.M} x_k(0) + \sum_{i=0}^{LM-1} [u(i) - y(i)]\lambda_k^{LM-i-1} \qquad k = 1,2,\ldots,N
\tag{17.36}
$$

Taking into account that every $[u(i) - y(i)]$ is the same after each $M$ samples, (17.36) can be rewritten as

$$
\begin{aligned}
x_k(L.M) =&\lambda_k^{L.M} x_k(0) + \sum_{i=0}^{LM-1} [u(i) - y(i)]\lambda_k^{LM-i-1}\\
&\lambda_k^{L.M} x_k(0) + \sum_{p=0}^{L-1} \lambda_k^{p.M} \left( \sum_{i=0}^{M-1} [u(i) - y(i)]\lambda_k^{M-i-1} \right) =\\
&\lambda_k^{L.M} x_k(0) + \frac{1 - \lambda_k^{LM}}{1 - \lambda_k^{M}} \left( \sum_{i=0}^{M-1} [u(i) - y(i)]\lambda_k^{M-i-1} \right)\\
k = 1,2,&\ldots,N
\end{aligned}
\tag{17.37}
$$

The above is correct, because $\sum_{p=0}^{L-1} \lambda_k^{p.M}$ is a partial sum of the first $L$ terms of a geometric series with value $\frac{1-\lambda_k^{LM}}{1-\lambda_k^{M}}$. If $|\lambda_k| < 1$, for every $L$ that is large enough (after enough time) $x_k(L.M) = \frac{1}{1-\lambda_k^{M}} \left( \sum_{i=0}^{M-1} [u(i) - y(i)]\lambda_k^{M-i-1} \right)$, i.e. $x_k(L.M)$ does not depend on $L$. This means repetition of the value of state $x_k$ after every

$M$ instances, i.e. the states are periodic. If $|\lambda_k| > 1$, from (17.37) follows that the boundness of the states is ensured if

$$x_k(0) = \frac{1}{1-\lambda_k^M}\left(\sum_{i=0}^{M-1}[u(i)-y(i)]\lambda_k^{M-i-1}\right) \tag{17.38}$$

and thus $x_k(L.M) = x_k(0)$. This means that the initial condition with respect to $x_k$, should be taken in accordance with (17.38), in order to ensure stability of the solution. This fits with the results in concerning the stability of high order modulators when $\lambda_k > 1$. If $\lambda_k = 1$, $x_k(L.M) = x_k(0)$ for every $L$ and every $x_k(0)$, because at the periodic orbit $\sum_{i=0}^{M-1}[u-y(i)] = 0$ for constant input signal $u$. This actually means that periodicity with respect to $x_k$ is ensured. In the case of complex pair of poles the results are similar, but the initial conditions connected with the complex conjugated pair of poles are also complex conjugated. We should stress again that the contribution of the state variables corresponding to these poles, to the input of the quantizer is real. The obtained results have been derived without matching the time sequence of the states $x_k(0), x_k(1), \ldots, x_k(M-1), k = 1,2,\ldots N$ with the time sequence of the output signal $y(0), y(1), \ldots, y(M-1)$ in the framework of one period. In fact to have a valid output sequence $y(0), y(1), \ldots, y(M-1)$ condition (17.7) should be satisfied. Thus,

$$\begin{aligned}
&\left(\sum_{k=1}^{N} b_k x_k(n)\right) = (\mathbf{b}^T\mathbf{x}(n)) \geq 0, \quad if \quad y(n) = 1\\
&\left(\sum_{k=1}^{N} b_k x_k(n)\right) = (\mathbf{b}^T\mathbf{x}(n)) < 0, \quad if \quad y(n) = -1\\
&n = 1,2,\ldots,M
\end{aligned} \tag{17.39}$$

or

$$\begin{aligned}
&\sum_{k=1}^{N} b_k\lambda_k^n x_k(0) \geq -\sum_{k=1}^{N} b_k\left(\sum_{i=0}^{n-1}[u(i)-y(i)]\lambda_k^{n-i-1}\right), \quad if \quad y(n) = 1\\
&\sum_{k=1}^{N} b_k\lambda_k^n x_k(0) < -\sum_{k=1}^{N} b_k\left(\sum_{i=0}^{n-1}[u(i)-y(i)]\lambda_k^{n-i-1}\right), \quad if \quad y(n) = -1\\
&n = 1,2,\ldots,M
\end{aligned} \tag{17.40}$$

Hence

$$\begin{aligned}
&\sum_{k=1}^{N} b_k\lambda_k^n x_k(0) \geq -\sum_{i=1}^{n-1}\left([u(i)-y(i)]\sum_{k=1}^{N} b_k\lambda_k^{n-i-1}\right), \quad if \quad y(n) = 1\\
&\sum_{k=1}^{N} b_k\lambda_k^n x_k(0) < -\sum_{i=1}^{n-1}\left([u(i)-y(i)]\sum_{k=1}^{N} b_k\lambda_k^{n-i-1}\right), \quad if \quad y(n) = -1\\
&n = 1,2,\ldots,M
\end{aligned} \tag{17.41}$$

In the case of a complex pair of poles $\lambda_i, \lambda_{i+1}$ the result has the same form. It should be noted that the left and right parts of inequalities (17.41) are real. The strategy for searching the limit cycles that correspond to a given output sequence of 1's and -1's with arbitrary length $M$ is based on finding the appropriate initial conditions $x_k(0), k = 1,2,\ldots,N$ with respect to state variables that ensure periodicity after the first period. Afterward the validity of the corresponding output sequences has to be checked. To simplify conditions (17.41) for validation of a given limit cycle connected with the corresponding vector of initial conditions $x_k(0), k = 1,2,\ldots,N$ obtained by (17.38), we are going, substitute (17.38) in all conditions (17.41). Thus taking into account (17.7) we get

$$\begin{aligned}
&\sum_{k=1}^{N} b_k\lambda_k^n \frac{1}{1-\lambda_k^M}\left(\sum_{i=0}^{M-1}[u-y(i)]\lambda_k^{M-i-1}\right) \geq -\sum_{i=0}^{n-1}\left([u-y(i)]\sum_{k=1}^{N} b_k\lambda_k^{n-i-1}\right),\\
&if \quad y(n)=1\\
&\sum_{k=1}^{N} b_k\lambda_k^n \frac{1}{1-\lambda_k^M}\left(\sum_{i=0}^{M-1}[u-y(i)]\lambda_k^{M-i-1}\right) < -\sum_{i=0}^{n-1}\left([u-y(i)]\sum_{k=1}^{N} b_k\lambda_k^{n-i-1}\right),\\
&if \quad y(n)=-1\\
&n=1,2,\ldots,M
\end{aligned} \tag{17.42}$$

Conditions (17.42) can be combined in one, multiplying both sides by $y(n)$ that is either 1 or -1.

$$\begin{aligned}
&y(n).\sum_{k=1}^{N} b_k\lambda_k^n \frac{1}{1-\lambda_k^M}\left(\sum_{i=0}^{M-1}[u-y(i)]\lambda_k^{M-i-1}\right) \geq\\
&-y(n).\sum_{i=0}^{n-1}\left([u-y(i)]\sum_{k=1}^{N} b_k\lambda_k^{n-i-1}\right)\\
&n=1,2,\ldots,M
\end{aligned} \tag{17.43}$$

The above inequalities (17.43) can be developed further as follows

$$\begin{aligned}
&y(n).\sum_{k=1}^{N} b_k\lambda_k^n \frac{1}{1-\lambda_k^M}\left(\sum_{i=0}^{M-1}[u-y(i)]\lambda_k^{M-i-1}\right) \geq\\
&-y(n).\sum_{k=1}^{N} b_k\sum_{i=0}^{n-1}[u-y(i)]\lambda_k^{n-i-1}\\
&n=1,2,\ldots,M
\end{aligned} \tag{17.44}$$

or

$$y(n).\sum_{k=1}^{N} b_k\lambda_k^n \frac{u}{1-\lambda_k^M}(1+\lambda_k+\lambda_k^2+...+\lambda_k^{M-1})-$$
$$y(n).\sum_{k=1}^{N} b_k\lambda_k^n \frac{u}{1-\lambda_k^M}(y(0)\lambda_k^{M-1}+y(1)\lambda_k^{M-2}+...+y(M-2)\lambda_k+y(M-1)) \geq$$
$$-y(n)\sum_{k=1}^{N} b_k u.(1+\lambda_k+\lambda_k^2+...+\lambda_k^{n-1})+$$
$$+y(n).\sum_{k=1}^{N} b_k(y(0)\lambda_k^{n-1}+y(1)\lambda_k^{n-2}+...+y(n-2)\lambda_k+y(n-1))$$
$$n=1,2,...,M \tag{17.45}$$

Taking into account that the value of the sum $(1+\lambda_k+\lambda_k^2+...+\lambda_k^{M-1})$ is $\frac{1-\lambda_k^M}{1-\lambda_k}$ and the value of the sum $(1+\lambda_k+\lambda_k^2+...+\lambda_k^{n-1})$ is $\frac{1-\lambda_k^n}{1-\lambda_k}$, inequalities (17.45) become

$$y(n).\sum_{k=1}^{N} b_k\lambda_k^n \frac{u}{1-\lambda_k}+y(n)\sum_{k=1}^{N} b_k u\frac{1-\lambda_k^n}{1-\lambda_k} \geq$$
$$y(n).\sum_{k=1}^{N} b_k\lambda_k^n \frac{u}{1-\lambda_k^M}(y(0)\lambda_k^{M-1}+y(1)\lambda_k^{M-2}+...+y(M-2)\lambda_k+y(M-1))+$$
$$+y(n).\sum_{k=1}^{N} b_k(y(0)\lambda_k^{n-1}+y(1)\lambda_k^{n-2}+...+y(n-2)\lambda_k+y(n-1))+$$
$$n=1,2,...,M \tag{17.46}$$

Thus we get

$$y(n).u.\sum_{k=1}^{N} b_k\frac{1}{1-\lambda_k} \geq$$
$$y(n).\sum_{k=1}^{N} \frac{b_k}{1-\lambda_k^M}(y(0)\lambda_k^{n+M-1}+y(1)\lambda_k^{n+M-2}+...+y(M-2)\lambda_k^{n+1}+y(M-1)\lambda_k^n)+$$
$$+y(n).\sum_{k=1}^{N} b_k(y(0)\lambda_k^{n-1}+y(1)\lambda_k^{n-2}+...+y(n-2)\lambda_k+y(n-1))+$$
$$n=1,2,...,M \tag{17.47}$$

Further investigations on (17.47) leads to

$$
\begin{aligned}
&y(n).u.\sum_{k=1}^{N} b_k \frac{1}{1-\lambda_k} \geq \\
&y(n).\sum_{k=1}^{N} b_k \Big[\Big(\frac{\lambda_k^{n+M-1}}{1-\lambda_k^M} + \lambda_k^{n-1}\Big)y(0) + \Big(\frac{\lambda_k^{n+M-2}}{1-\lambda_k^M} + \lambda_k^{n-2}\Big)y(1) + \\
&\dots + \Big(\frac{\lambda_k^{M}}{1-\lambda_k^M} + 1\Big)y(n-1) + \frac{\lambda_k^{M-1}}{1-\lambda_k^M}y(n) + \dots + \frac{\lambda_k^{n+1}}{1-\lambda_k^M}y(M-2) + \frac{\lambda_k^{n}}{1-\lambda_k^M}y(M-1)\Big] \\
&n = 1,2,\dots,M
\end{aligned}
$$

and hence

$$
\begin{aligned}
&y(n).u.\sum_{k=1}^{N} b_k \frac{1}{1-\lambda_k} \geq \\
&y(n).\sum_{k=1}^{N} b_k \Big[\Big(\frac{\lambda_k^{n-1}}{1-\lambda_k^M}\Big)y(0) + \Big(\frac{\lambda_k^{n-2}}{1-\lambda_k^M}\Big)y(1) + \\
&\dots + \Big(\frac{1}{1-\lambda_k^M}\Big)y(n-1) + \frac{\lambda_k^{M-1}}{1-\lambda_k^M}y(n) + \\
&\dots + \frac{\lambda_k^{n+1}}{1-\lambda_k^M}y(M-2) + \frac{\lambda_k^{n}}{1-\lambda_k^M}y(M-1)\Big] \\
&n = 1,2,\dots,M
\end{aligned}
$$

Therefore, with respect to the output bitstream sequence $y(0),y(1),\dots,y(M-1)$ the inequalities that have to be satisfied are $M$ linear inequalities in form

$$
\begin{aligned}
&y(n).u.\sum_{k=1}^{N} b_k \frac{1}{1-\lambda_k} \geq \left(y(n).\sum_{k=1}^{N} \frac{b_k\lambda_k^{n-1}}{1-\lambda_k^M}\right) y(0) + \\
&+\left(y(n).\sum_{k=1}^{N} \frac{b_k\lambda_k^{n-2}}{1-\lambda_k^M}\right) y(1) + \dots + \left(y(n).\sum_{k=1}^{N} \frac{b_k}{1-\lambda_k^M}\right) y(n-1) + \\
&+\left(y(n).\sum_{k=1}^{N} \frac{b_k\lambda_k^{M-1}}{1-\lambda_k^M}\right) y(n) + \dots + \left(y(n).\sum_{k=1}^{N} \frac{b_k\lambda_k^{n+1}}{1-\lambda_k^M}\right) y(M-2) + \\
&+\left(y(n).\sum_{k=1}^{N} \frac{b_k\lambda_k^{n}}{1-\lambda_k^M}\right) y(M-1) \\
&n = 1,2,\dots,M
\end{aligned}
\tag{17.48}
$$

Inequalities (17.48) have a geometrical interpretation. In the $M$ dimensional space of the output sequences $y(0),y(1),\dots,y(M-1)$ with length $M$, every output bitstream of 1's and -1's is a vertex of the $M$ dimensional hypercube in this space. Such a vertex represents a possible limit cycle if it is on the corresponding side of all $M$ hyperplanes, given by (17.48) that are equivalent to this vertex.

In extended form the inequalities (17.48) could be rewritten as follows. For $n = 1$

$$y(1).u.\sum_{k=1}^{N} b_k \frac{1}{1-\lambda_k} \geq \left(y(1).\sum_{k=1}^{N} \frac{b_k}{1-\lambda_k^M}\right) y(0) + \left(y(1).\sum_{k=1}^{N} \frac{b_k \lambda_k^{M-1}}{1-\lambda_k^M}\right) y(1) +$$
$$\ldots + \left(y(1).\sum_{k=1}^{N} \frac{b_k \lambda_k^2}{1-\lambda_k^M}\right) y(M-2) + \left(y(1).\sum_{k=1}^{N} \frac{b_k \lambda_k}{1-\lambda_k^M}\right) y(M-1)$$

For $n = 2$

$$y(1).u.\sum_{k=1}^{N} b_k \frac{1}{1-\lambda_k} \geq \left(y(2).\sum_{k=1}^{N} \frac{b_k \lambda_k}{1-\lambda_k^M}\right) y(0) + \left(y(2).\sum_{k=1}^{N} \frac{b_k}{1-\lambda_k^M}\right) y(1) +$$
$$+ \left(y(2).\sum_{k=1}^{N} \frac{b_k \lambda_k^{M-1}}{1-\lambda_k^M}\right) y(2) + \ldots +$$
$$+ \left(y(2).\sum_{k=1}^{N} \frac{b_k \lambda_k^3}{1-\lambda_k^M}\right) y(M-2) + \left(y(2).\sum_{k=1}^{N} \frac{b_k \lambda_k^2}{1-\lambda_k^M}\right) y(M-1)$$
$$n = 1,2,\ldots,M$$

For $n = M, y(M) = y(0)$, because the limit cycle is with length $M$

$$y(0).u.\sum_{k=1}^{N} b_k \frac{1}{1-\lambda_k} \geq \left(y(0).\sum_{k=1}^{N} \frac{b_k \lambda_k^{M-1}}{1-\lambda_k^M}\right) y(0) +$$
$$+ \left(y(0).\sum_{k=1}^{N} \frac{b_k \lambda_k^{M-2}}{1-\lambda_k^M}\right) y(1) + \ldots + \left(y(0).\sum_{k=1}^{N} \frac{b_k}{1-\lambda_k^M}\right) y(M-1)$$

Taking into account that at the limit cycle $y(0) = y(M), y(1) = y(M+1), \ldots, y(M-1) = y(2M-1), y(M) = y(2M) = y(0)$ or $y(p) = y(p+M)$ for $p = 1,2,\ldots,M-1$, we can rewrite conditions (17.48) in more general form:

$$y(n).u.\sum_{k=1}^{N} \frac{b_k}{1-\lambda_k} \geq \left(y(n).\sum_{k=1}^{N} \frac{b_k \lambda_k^{M-1}}{1-\lambda_k^M}\right) y(n) +$$
$$+ \left(y(n).\sum_{k=1}^{N} \frac{b_k \lambda_k^{M-2}}{1-\lambda_k^M}\right) y(n+1) + \left(y(n).\sum_{k=1}^{N} \frac{b_k \lambda_k^{M-3}}{1-\lambda_k^M}\right) y(n+2) + \ldots +$$
$$\left(y(n).\sum_{k=1}^{N} \frac{b_k \lambda_k}{1-\lambda_k^M}\right) y(n+M-2) + \left(y(n).\sum_{k=1}^{N} \frac{b_k}{1-\lambda_k^M}\right) y(n+M-1)$$
$$n = 1,2,\ldots,M \tag{17.49}$$

This result simplify conditions (17.41) for validation of a given limit cycle connected with an output bitstream $y(0), y(1), \ldots, y(M-1)$ with length $L$, because directly incorporates the values of the bitsteram sequence, the constant input signal $u$ and the parameters of parallel presentation of the loop filer of the sigma-delta modulator considered. It should be stressed that the coefficients $\sum_{k=1}^{N}\frac{b_k}{1-\lambda_k}$, $\sum_{k=1}^{N}\frac{b_k\lambda_k^{M-1}}{1-\lambda_k^M}$, $\sum_{k=1}^{N}\frac{b_k\lambda_k^{M-2}}{1-\lambda_k^M}, \ldots, \sum_{k=1}^{N}\frac{b_k}{1-\lambda_k^M}$ are common for all inequalities and thus the conditions (17.49) could be checked very easy.

For better understanding the validation formulas (17.49) we are going to present a particular case for verification of limit cycles with length $M = 4$ for a sigma-delta modulator with a third order loop filter $N = 3$. In this case formulas (17.49) become

$$y(1).u.\sum_{k=1}^{3}\frac{b_k}{1-\lambda_k} \geq \left(y(1).\sum_{k=1}^{3}\frac{b_k\lambda_k^3}{1-\lambda_k^4}\right)y(1) + \left(y(1).\sum_{k=1}^{3}\frac{b_k\lambda_k^2}{1-\lambda_k^4}\right)y(2)+$$
$$+\left(y(1).\sum_{k=1}^{3}\frac{b_k\lambda_k}{1-\lambda_k^4}\right)y(3) + \left(y(1).\sum_{k=1}^{3}\frac{b_k}{1-\lambda_k^4}\right)y(0)$$
$$y(2).u.\sum_{k=1}^{3}\frac{b_k}{1-\lambda_k} \geq \left(y(2).\sum_{k=1}^{3}\frac{b_k\lambda_k^3}{1-\lambda_k^4}\right)y(2) + \left(y(1).\sum_{k=1}^{3}\frac{b_k\lambda_k^2}{1-\lambda_k^4}\right)y(3)+$$
$$+\left(y(2).\sum_{k=1}^{3}\frac{b_k\lambda_k}{1-\lambda_k^4}\right)y(0) + \left(y(1).\sum_{k=1}^{3}\frac{b_k}{1-\lambda_k^4}\right)y(1)$$
$$y(3).u.\sum_{k=1}^{3}\frac{b_k}{1-\lambda_k} \geq \left(y(3).\sum_{k=1}^{3}\frac{b_k\lambda_k^3}{1-\lambda_k^4}\right)y(3) + \left(y(3).\sum_{k=1}^{3}\frac{b_k\lambda_k^2}{1-\lambda_k^4}\right)y(0)+$$
$$+\left(y(3).\sum_{k=1}^{3}\frac{b_k\lambda_k}{1-\lambda_k^4}\right)y(1) + \left(y(3).\sum_{k=1}^{3}\frac{b_k}{1-\lambda_k^4}\right)y(2)$$
$$y(0).u.\sum_{k=1}^{3}\frac{b_k}{1-\lambda_k} \geq \left(y(0).\sum_{k=1}^{3}\frac{b_k\lambda_k^3}{1-\lambda_k^4}\right)y(0) + \left(y(0).\sum_{k=1}^{3}\frac{b_k\lambda_k^2}{1-\lambda_k^4}\right)y(1)+$$
$$+\left(y(0).\sum_{k=1}^{3}\frac{b_k\lambda_k}{1-\lambda_k^4}\right)y(2) + \left(y(0).\sum_{k=1}^{3}\frac{b_k}{1-\lambda_k^4}\right)y(3)$$

Based on considerations here, given periodic output sequence of 1's and -1's with arbitrary length $M$, corresponds to a limit cycle if the inequalities (17.49) are satisfied. The application of the approach considered consists of checking inequalities (17.49) for every possible output sequence of 1's and -1's with length $M$. The number of these sequences is $2^M$. Developed conditions (17.49) are $M$ inequalities for every output sequence. Because the coefficients $\sum_{k=1}^{N}\frac{b_k}{1-\lambda_k}, \sum_{k=1}^{N}\frac{b_k\lambda_k^{M-1}}{1-\lambda_k^M}, \sum_{k=1}^{N}\frac{b_k\lambda_k^{M-2}}{1-\lambda_k^M}, \ldots, \sum_{k=1}^{N}\frac{b_k}{1-\lambda_k^M}$ are common for all inequalities, conditions (17.49) are checked very fast and easy. This result accelerates the validation check for the limit cycles in the general case considered in this section. To demonstrate applicability of the new conditions (17.49), we consider a second order sigma-delta modulator with the following loop filter transfer function [4]

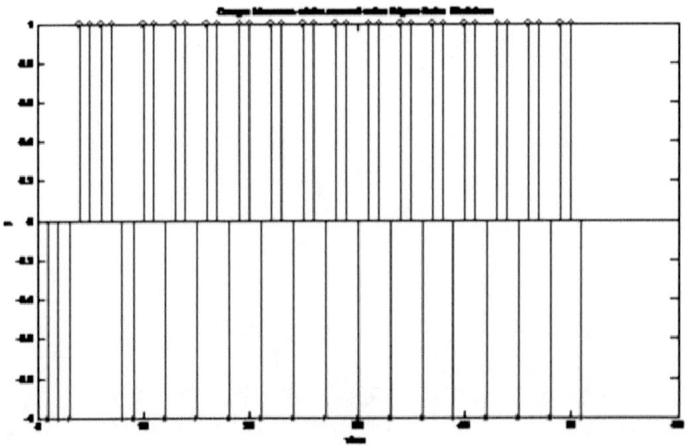

**Fig. 17.6** Output bitstream of the second order Sigma-Delta modulator with loop filter transfer function given by (17.50)

$$G(z) = \frac{2r\cos\theta z^{-1} - r^2 z^{-2}}{1 - 2r\cos\theta z^{-1} + r^2 z^{-2}} \qquad (17.50)$$

In this case $\lambda_1 = \alpha + j\beta$, $\lambda_2 = \alpha - j\beta$, $b_1 = \delta - j\gamma$, $b_2 = \delta + j\gamma$, where $\alpha = r.\cos\theta$, $\beta = r.\sin\theta$, $\delta = r.\cos\theta$, $\gamma = r.(\cos 2\theta)/(2\sin\theta)$. When $r = 0.9$ and $\theta = 30°$, $\lambda_1 = 0.7794 + j0.4500$, $\lambda_2 = 0.7794 - j0.4500$, $b_1 = 0.7794 - j0.4500$, $b_2 = 0.7794 + j0.4500$ and the initial conditions that lead to a periodic output sequence $y(0) = 1$, $y(1) = 1$, $y(2) = -1$ without transient are $x_1(0) = 0.779 - j0.123$, $x_2(0) = 0.779 + j0.123$ [18], [19]. When the constant input is $u = 0.4$, the periodic output sequence with length $M = 3$ is $y(0) = 1, y(1) = 1, y(2) = -1$, and can be detected in Figure 17.6.

In this case the inequalities (17.49) have the form

$$y(1).u.\sum_{k=1}^{2}\frac{b_k}{1-\lambda_k} \geq \left(y(1).\sum_{k=1}^{2}\frac{b_k\lambda_k^2}{1-\lambda_k^3}\right)y(1) + \left(y(1).\sum_{k=1}^{2}\frac{b_k\lambda_k}{1-\lambda_k^3}\right)y(2) + \left(y(1).\sum_{k=1}^{2}\frac{b_k}{1-\lambda_k^3}\right)y(0)$$

$$y(2).u.\sum_{k=1}^{2}\frac{b_k}{1-\lambda_k} \geq \left(y(2).\sum_{k=1}^{2}\frac{b_k\lambda_k^2}{1-\lambda_k^3}\right)y(2) + \left(y(2).\sum_{k=1}^{2}\frac{b_k\lambda_k}{1-\lambda_k^3}\right)y(0) + \left(y(2).\sum_{k=1}^{2}\frac{b_k}{1-\lambda_k^3}\right)y(1)$$

$$y(3).u.\sum_{k=1}^{2}\frac{b_k}{1-\lambda_k} \geq \left(y(3).\sum_{k=1}^{2}\frac{b_k\lambda_k^2}{1-\lambda_k^3}\right)y(0) + \left(y(3).\sum_{k=1}^{2}\frac{b_k\lambda_k}{1-\lambda_k^3}\right)y(1) + \left(y(3).\sum_{k=1}^{2}\frac{b_k}{1-\lambda_k^3}\right)y(2)$$

where $y(3) = y(0)$.

The coefficients $\sum_{k=1}^{2}\frac{b_k}{1-\lambda_k} = 2.9816$, $\sum_{k=1}^{2}\frac{B_k\lambda_k^2}{1-\lambda_k^3} = 0.4775$, $\sum_{k=1}^{2}\frac{b_k\lambda_2}{1-\lambda_k^3} = 1.0578$, $\sum_{k=1}^{2}\frac{b_k}{1-\lambda_k^3} = 1.4463$ are common for the above 3 inequalities that are satisfied for the output bitstream sequence $y(0) = 1, y(1) = 1, y(2) = -1$.

$$y(1).u.\sum_{k=1}^{2}\frac{b_k}{1-\lambda_k} - \left[\left(y(1).\sum_{k=1}^{2}\frac{b_k\lambda_k^2}{1-\lambda_k^3}\right)y(1) + \left(y(1).\sum_{k=1}^{2}\frac{b_k\lambda_k}{1-\lambda_k^3}\right)y(2) + \left(y(1).\sum_{k=1}^{2}\frac{b_k}{1-\lambda_k^3}\right)y(0)\right]$$
$$= 0.3267 \geq 0$$

$$y(2).u.\sum_{k=1}^{2}\frac{b_k}{1-\lambda_k} - \left[\left(y(2).\sum_{k=1}^{2}\frac{b_k\lambda_k^2}{1-\lambda_k^3}\right)y(2) + \left(y(2).\sum_{k=1}^{2}\frac{b_k\lambda_k}{1-\lambda_k^3}\right)y(0) + \left(y(2).\sum_{k=1}^{2}\frac{b_k}{1-\lambda_k^3}\right)y(1)\right]$$
$$= 0.8340 \geq 0$$

$$y(3).u.\sum_{k=1}^{2}\frac{b_k}{1-\lambda_k} - \left[\left(y(3).\sum_{k=1}^{2}\frac{b_k\lambda_k^2}{1-\lambda_k^3}\right)y(0) + \left(y(3).\sum_{k=1}^{2}\frac{b_k\lambda_k}{1-\lambda_k^3}\right)y(1) + \left(y(3).\sum_{k=1}^{2}\frac{b_k}{1-\lambda_k^3}\right)y(2)\right]$$
$$= 1.1037 \geq 0$$

## 17.6 Conclusions

In this chapter we present an unified approach for study the stability and validation of potential limit cycles of one bit high order Sigma-Delta modulators. The approach is general because it uses the general form of a Sigma-Delta modulator. It is based on a parallel decomposition of the modulator and a direct nonlinear systems analysis. In this representation, the general $N-th$ order modulator is transformed into a decomposition of low order, generally complex modulators, which interact only through the quantizer function. The developed conditions for stability and for validation of potential limit cycles are very easy for implementation and this procedure is very fast. The reported results can be elaborated further for some particular cases, and investigating the possibilities to skip the check of some output bitstream sequences and thus to accelerate extra the limit cycle validation procedure. Furthermore it is an open problem how to use the developed conditions for Sigma-Delta modulators design, i.e. to design a Sigma-Delta modulator working on a desired limit cycle.

**Acknowledgements.** This work is supported by N.W.O. visitor travel grant Nr. 040.11.312 for 2012.

## References

1. Candy, J.C., Temes, G.C.: Oversampling Delta-Sigma Data Converters. IEEE Press, New York (1992)
2. Norsworthy, S.R., Schreier, R., Temes, G.C.: Delta-Sigma Data Converters. IEEE Press, New York (1997)
3. Gray, R.M.: Oversampled sigma-delta modulation. IEEE Trans. Commun. 35, 481–489 (1987)
4. Feely, O., Chua, L.O.: The effect of integrator leak in Sigma-delta modulation. IEEE Trans. Circuits and Systems 38, 1293–1305 (1991)
5. Feely, O., Chua, L.O.: Nonlinear dynamics of a class of analog-to-digital converters. Int. J. of Bifurcation and Chaos 2, 325–340 (1992)
6. Feely, O.: A tutorial introduction to non-linear dynamics and chaos and their application to sigma-delta modulators. Int. J. Circuit Theory and Applications 25, 347–367 (1997)
7. Steiner, P., Yang, W.: A framework for analysis of high-order sigma-delta modulators. IEEE Transactions on Circuits and Systems II: CAS II 44, 1–10 (1997)

8. Steiner, P., Yang, W.: Stability of high order sigma-delta modulators. In: International Symposium on Circuits and Systems, ISCAS 1996, vol. 3, pp. 52–55 (1996)
9. Mladenov, V., Hegt, H., Roermund, A.V.: Stability analysis of high order sigma-delta modulators. In: Proc. of the 15th European Conference on Circuit Theory and Design ECCTD 2001, pp. I-313 – I-316. Helsinki University of Technology, Finland (2001)
10. Mladenov, V., Hegt, H., van Roermund, A.: On the Stability of High Order Sigma-Delta Modulators. In: Proceedings of the 8th IEEE International Conference on Electronics, Circuits and Systems, ICECS 2001, Malta, pp. 1383–1386 (2001)
11. Mladenov, V., Hegt, H., van Roermund, A.: On the Stability Analysis of High Order Sigma-Delta Modulators. An International Journal on Analog Integrated Circuits and Signal Processing 36(1-2), 47–55 (2003)
12. Mladenov, V., Hegt, H., van Roermund, A.: On the Stability Analysis of Sigma-Delta Modulators. In: Proceedings of the 16th European Conference on Circuit Theory and Design, ECCTD 2003, Cracow, Poland, September 1-4, pp. I-97– I-100 (2003)
13. Friedman, V.: The structure of the limit cycles in sigma delta modulation. IEEE Transactions on Communications 36, 972–979 (1988)
14. Mann, S., Taylor, D.: Limit cycle behavior in the double-loop bandpass sigma delta a/d converter. IEEE Transactions on Circuits and Systems-II 46, 1086–1089 (1999)
15. Hyun, D., Fischer, G.: Limit cycles and pattern noise in single-stage single-bit delta-sigma modulators. IEEE Trans. Circuits and Systems I 49, 646–656 (2002)
16. Reefman, D., Reiss, J., Janssen, E., Sandler, M.: Description of limit cycles in sigma delta modulators. IEEE Trans. on CAS 52(6), 1211–1223 (2005)
17. Reiss, J.D., Sandler, M.B.: Detection and removal of limit cycles in sigma delta modulators. IEEE Transactions on Circuits and Systems I: Regular Papers 55(10), 3119–3130 (2008)
18. Mladenov, V.: A Method for Searching the Limit Cycles of High Order Sigma-Delta Modulators. In: Proceedings of the 19th European Conference on Circuit Theory and Design, ECCTD 2009, Antalya, Turkey, August 23-27, pp. 543–546 (2009)
19. Mladenov, V.: A Method for Validation the Limit Cycles of High Order Sigma-Delta Modulators. In: Proceedings of the 3rd International Workshop on Nonlinear Dynamics and Synchjronization, INDS 2011, Klagenfurt, Austria, July 25-27, pp. 234–238 (2011)

# Chapter 18
# Stability Analysis of Vector Equalization Based on Recurrent Neural Networks

Mohamad Mostafa, Werner G. Teich, and Jürgen Lindner

**Abstract.** Since the pioneer work of Hopfield on the computational capabilities of recurrent neural networks (RNNs), they have been applied to solve classification and optimization problems in many scientific disciplines. This can be done, either by using conventional training algorithms like back propagation through time, or by investigating the Lyapunov stability of these RNNs and comparing the corresponding Lyapunov function with the cost function of the optimization problem to be solved. The later method is especially interesting in the field of engineering because no training phase is needed, which is always associated with computational effort and time. In this chapter we focus on an application of RNNs in communications engineering, namely the vector equalization. The importance of this procedure arises from the fact that there is no need for training. The parameters of the RNN to act as vector equalizer can be obtained by investigating the stability properties of these networks and by choosing a suitable activation function, which will be the core of this work.

**Keywords:** Recurrent neural networks, stability analysis, vector equalization.

## 18.1 Organization of the Chapter

In Section 2 we introduce the vector-valued transmission model and present the problem of vector equalization. In Section 3 we discuss the recurrent neural networks (RNNs) with the corresponding state-space equations and revisit the Lyapunov theory on stability. The stability analysis of RNNs with time-invariant activation functions is considered in Section 4. Section 5 is dedicated to analyze the optimum activation function for the vector equalizer based on RNNs. In Section 6 we present the stability analysis of the RNN for time-variant activation functions in detail. The comparison between local and global stable vector equalizer

Mohamad Mostafa · Werner G. Teich · Jürgen Lindner
Institute of Communications Engineering, Albert-Einstein-Allee 43, 89081 Ulm, Germany
e-mail: mohamad.mostafa@uni-ulm.de

K. Kyamakya et al. (Eds.): Selected Topics in Nonlinear Dynamics, SCI 459, pp. 329–349.
DOI: 10.1007/ 978-3-642-34560-9_18 

based on RNNs is discussed in Section 7. We finish the chapter with a conclusion in Section 8.

The following notation is needed throughout the chapter. Vectors are underlined once, matrices twice. $(\cdot)^T$, $(\cdot)^H$ and $|\cdot|$ denote the transpose, conjugate transpose and the absolute value of a matrix or a vector. A matrix $\underline{\underline{B}} \geq 0$ ($\underline{\underline{B}} > 0$) means it is positive semidefinite (positive definite).

We restrict ourself to real-valued RNNs without hidden neurons. We will point out, where a result is also valid for complex-valued RNNs. Extension to complex-valued RNNs is in progress. Parts of this chapter have been published in ( [17], [18], [19]).

## 18.2 Vector-Valued Transmission Model

The block vector-valued transmission model for linear modulation schemes without channel coding is shown in Fig. 18.1 and is described as follows [13], [14]:

$$\underline{\tilde{x}} = \underline{\underline{r}} \cdot \underline{x} + \underline{n_e} \tag{18.1}$$

- $\underline{x}$ is the transmit vector of size $(n \times 1)$
- $\underline{\tilde{x}}$ is the receive vector of size $(n \times 1)$
- $\underline{\check{x}}$ is the soft-valued decided vector of size $(n \times 1)$ at the output of the vector equalizer, cf. Fig. 18.1
- $\underline{\hat{x}}$ is the decided vector of size $(n \times 1)$, cf. Fig. 18.1
- $\underline{n_e}$ is the colored noise vector of size $(n \times 1)$ with correlation matrix

$$\underline{\underline{\Phi}} = \frac{N_0}{2} \cdot \underline{\underline{r}} \tag{18.2}$$

$N_0$ is the single-sided noise power spectral density
- $\underline{\underline{r}}$ is the discrete-time channel matrix of size $(n \times n)$. It is a symmetric[1] $\underline{\underline{r}} = \underline{\underline{r}}^T$ and positive semidefinite matrix $\underline{\underline{r}} \geq 0$ because it is a correlation matrix. This property arises from the use of the channel matched filter at the receiver. $\underline{\underline{r}}$ depends on the transmission scheme (basic wave forms) and the channel. For coherent transmission (the case we are considering) a perfect knowledge of the channel impulse response and thus the discrete-time channel matrix $\underline{\underline{r}}$ at the receiver side is required
- the channel matrix $\underline{\underline{r}}$ contains all physical properties of the transmission model.
- $\forall i \in \{1,2,\cdots,n\}$, $\forall j \in \{1,2,\cdots,M\}$, $M = 2^s$, $s \in \mathbb{N}/\{0\}$ :
  $x_i \in \mathbb{A}_x = \{a_1, a_2, \cdots, a_M\}$, $a_j \in \mathbb{R}$. In this case there are $M^n$ possible transmit vectors[2].

The vector-valued transmission model depicted in Fig. 18.1 is a general model and fits to different transmission schemes like OFDM (orthogonal frequency division

[1] For complex-valued case it is a hermitian matrix $\underline{\underline{r}} = \underline{\underline{r}}^H$.

[2] For complex-valued case $a_j \in \mathbb{C}$.

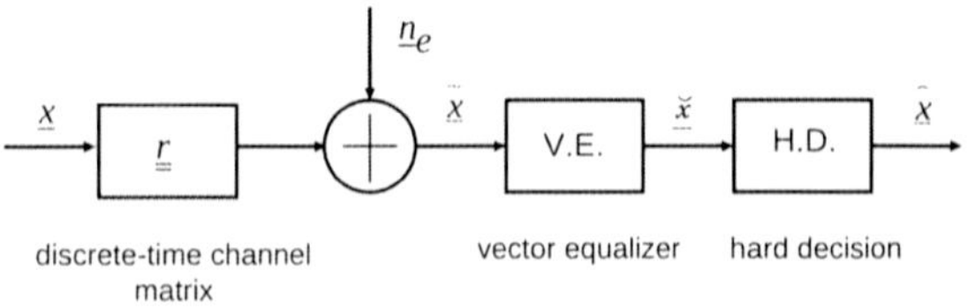


**Fig. 18.1** Block vector-valued transmission model for linear modulation schemes without channel coding

multiplexing), CDMA (code division multiple access), MC-CDMA (multi carrier CDMA), MIMO (multiple input multiple output), ... etc [13]. All Information about the physical background of the model in Fig. 18.1 is contained in $\underline{\underline{r}}$. More details about the connection between this model and the physical model of a transmission scheme can be found in [13]. The vector equalizer in Fig. 18.1 acts as a classifier.

The maximum-likelihood vector equalizer represents the optimum vector equalizer[3] and its functionality can be described as follows: For each received vector $\underline{\tilde{x}}$ it calculates the distance between the received vector $\underline{\tilde{x}}$ and all possible transmit vectors $\underline{\xi}$ (finite set with cardinality $M^n$) and then decides in favor of the minimum distance as follows[4] [16]:

$$\begin{aligned}\Gamma(\underline{\xi}) &= -2\cdot\underline{\tilde{x}}^T\cdot\underline{\xi}+\underline{\tilde{\xi}}^T\cdot\underline{\underline{r}}\cdot\underline{\tilde{\xi}}\\ \hat{x} &= \arg\min_{\underline{\xi}}\left\{\Gamma(\underline{\xi})\right\}\end{aligned} \tag{18.3}$$

In the distance calculation the correlation of the noise Eq. (18.2) has to be taken into account. This leads to the Mahalanobis metric [6] as the proper distance measure.

If the channel matrix $\underline{\underline{r}}$ is fully occupied, the complexity of the optimum equalizer increases in general exponentially with the length of the vectors in Eq. (18.1) and is too complex to be implemented for a realistic $n$ [26]. Therefore research efforts have been concentrated on the development of suboptimum vector equalizers, which have lower complexity and near-optimum performance.

One suboptimum vector equalizer with remarkable performance and moderate complexity is the RNN. For a properly defined RNN this can be explained by the equivalence between the Lyapunov function (in the case of local stability) and the cost function of the optimum vector equalizer Eq. (18.3) [13]. In this case there is no need for a learning phase, in contrast to the usual strategy when using artificial neural networks (ANN) for optimization problems. This represents a very attractive feature.

Having the channel matrix $\underline{\underline{r}}$ the weight coefficients of the RNN can be set up directly. This connection will be explained in the following sections.

---

[3] Optimum in the sense that it leads to the minimum number of erroneous decisions.

[4] In complex-valued case it is $\Gamma(\underline{\xi}) = -2\cdot\Re\{\underline{\tilde{x}}^H\cdot\underline{\xi}\}+\underline{\tilde{\xi}}^H\cdot\underline{\underline{r}}\cdot\underline{\tilde{\xi}}$.

## 18.3 Recurrent Neural Networks

In this section we introduce the structure and the dynamical behavior of both discrete-time (serial and parallel update) and continuous-time RNNs. We revisit also the stability theorem based on Lyapunov functions.

### *18.3.1 Discrete-Time RNNs*

Figure 18.2 shows a discrete-time RNN with parallel update. $\underline{v}$ is the output, $\underline{u}$ the inner state, $\underline{e}$ the external input, $\varphi(\cdot)$ the activation function, $w_{ij}$ the weight coefficient from the output of the $j^{th}$ neuron to the input of the $i^{th}$ neuron, $w_{i0}$ the weight coefficient of the $i^{th}$ external input, $n$ the number of neurons in the network. $\underline{v},\underline{u},\underline{e} \in \mathbb{R}^n, \underline{\underline{w}}, \underline{\underline{w}}_0 \in \mathbb{R}^{n\times n}, \underline{\underline{w}}_0 = \text{diag}\{w_{10}, w_{20}, \cdots, w_{n0}\}$.

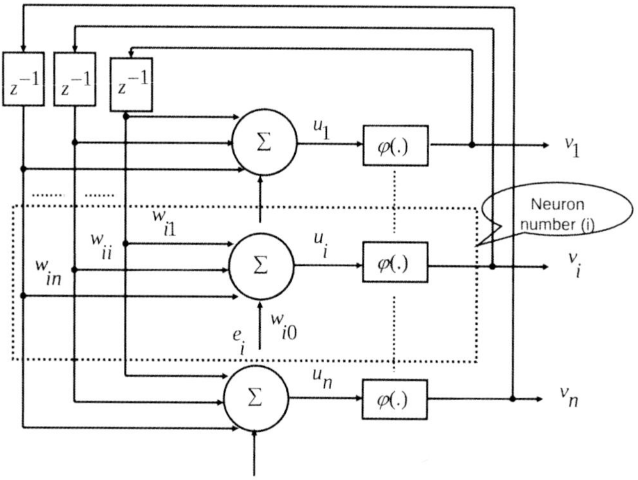


**Fig. 18.2** Discrete-time recurrent neural network, $\underline{v}$ is the output, $\underline{u}$ the inner state, $\underline{e}$ the external input, $\varphi(\cdot)$ the activation function, $w_{ij}$ the weight coefficient from the output of the $j^{th}$ neuron to the input of the $i^{th}$ neuron, $w_{i0}$ the weight coefficient of the $i^{th}$ external input, $n$ the number of neurons in the network. $\underline{v},\underline{u},\underline{e} \in \mathbb{R}^n, \underline{\underline{w}}, \underline{\underline{w}}_0 \in \mathbb{R}^{n\times n}, \underline{\underline{w}}_0 = \text{diag}\{w_{10}, w_{20}, \cdots, w_{n0}\}$.

In the parallel update case all neurons are updated in one step $(k)$. The dynamical behavior in this case is described as follows:

$$\begin{aligned} \underline{u}(k+1) &= \underline{\underline{w}} \cdot \underline{v}(k) + \underline{\underline{w}}_0 \cdot \underline{e} \\ \underline{v}(k) &= \varphi[\underline{u}(k)] \end{aligned} \tag{18.4}$$

In the serial update case one neuron is updated every step $(\rho)$. The dynamical behavior of the discrete-time RNN with serial update, assuming the $j^{th}$ neuron is being updated, is described as follows:

$$u_j(\rho+1) = \sum_{i=1}^{n} w_{ji} \cdot v_i(\rho) + w_{j0} \cdot e_j$$
$$v_j(\rho) = \varphi[u_j(\rho)],\ j \in \{1,2,\cdots,n\} \tag{18.5}$$

Depending on Eq. (18.5) we define the output vector before and after updating the $j^{th}$ neuron i.e. the transition from the discrete-time index $(\rho)$ to $(\rho+1)$ as follows:

$$\underline{v}(\rho) = [v_1(\rho),\cdots,v_{j-1}(\rho),v_j(\rho),v_{j+1}(\rho),\cdots,v_n(\rho)]^T$$
$$\underline{v}(\rho+1) = [v_1(\rho),\cdots,v_{j-1}(\rho),v_j(\rho+1),v_{j+1}(\rho),\cdots,v_n(\rho)]^T \tag{18.6}$$

We notice that $\forall i \neq j : v_i(\rho+1) = v_i(\rho)$. This definition will be useful during the stability proof.

*Remark 18.1.* In this case $n$ steps are required to update all neurons.

*Remark 18.2.* The order of the update in this case may influence the performance of the RNN.

### 18.3.2 Continuous-Time RNNs

The dynamical behavior of the continuous-time RNN, cf. Fig. 18.3, is described as follows:

$$\underline{\underline{\tau}} \cdot \frac{\mathrm{d}\underline{u}(t)}{\mathrm{d}t} = -\underline{u}(t) + \underline{\underline{w}} \cdot \underline{v}(t) + \underline{\underline{w}}_0 \cdot \underline{e}$$
$$\underline{v}(t) = \varphi[\underline{u}(t)]$$
$$\underline{\underline{\tau}} = \mathrm{diag}\{\tau_1,\tau_2,\cdots,\tau_n\} \tag{18.7}$$

$\tau_i = R_i \cdot C_i$ is the time constant of the $i^{th}$ neuron in the continuous-time model.

*Remark 18.3.* All variables in Eq. (18.7) are like those in Eq. (18.4).

*Remark 18.4.* The activation function in Eq. (18.4,18.5,18.7) is applied element-wise.

### 18.3.3 Stability Analysis Based on Lyapunov Functions

**Definition 18.1.** The equilibrium states $v_{eq}$ of a continuous-time dynamical system $\frac{\mathrm{d}v(t)}{\mathrm{d}t} = g[v(t)]$ fulfill $g[v_{eq}] = 0$.

**Definition 18.2.** The fixed points $v_f$ of a discrete-time dynamical system $v(k+1) = g[v(k)]$ fulfill $g[v_f] = v_f$.

*Remark 18.5.* If the continuous map $\lambda(\underline{u}) : \underline{u} \to \underline{\underline{w}} \cdot \varphi[\underline{u}] + \underline{\underline{w}}_0 \cdot \underline{e}$ is bounded then the RNNs have at least one fixed point i.e. the activation function $\varphi$ must be bounded [15].

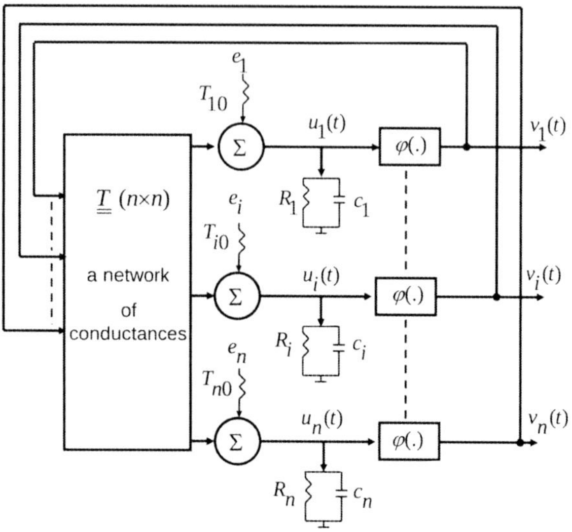


**Fig. 18.3** Continuous-time recurrent neural network, $\underline{v}$ is the output, $\underline{u}$ the inner state, $\underline{e}$ the external input, $\varphi(\cdot)$ the activation function, $w_{ij} = \frac{T_{ij}}{T_i}$ the weight coefficient from the output of the $j^{th}$ neuron to the input of the $i^{th}$ neuron, $w_{i0} = \frac{T_{i0}}{T_i}$ the weight coefficient of the $i^{th}$ external input, $n$ the number of neurons in the network. $\underline{v}, \underline{u}, \underline{e} \in \mathbb{R}^n, \underline{\underline{w}}, \underline{\underline{w}}_0 \in \mathbb{R}^{n\times n}, \underline{\underline{w}}_0 = \text{diag}\{w_{10}, w_{20}, \cdots, w_{n0}\}$.

The equilibrium state of a continuous-time dynamical system (defined as $\underline{v}_{eq}$ in the following) is asymptotically stable if in a small neighborhood of $\underline{v}_{eq}$ (the domain of attraction $\mathfrak{B}$) there exists a positive definite function $E(\underline{v})$. The derivative of $E(\underline{v})$ with respect to the time along the dynamics is negative definite in that region [5]. A scalar function $E(\underline{v})$ that satisfies these requirements is called a strict Lyapunov function for the equilibrium state $\underline{v}_{eq}$. In other words $E(\underline{v})$ has to fulfill:

1. $E(\underline{v})$ has continuous partial derivatives with respect to the elements of the state vector $\underline{v}$
2. $E(\underline{v}_{eq}) = 0$
3. $E(\underline{v}) > 0$ if $\underline{v} \in \mathfrak{B}/\{\underline{v}_{eq}\}$
4. $\frac{\mathrm{d}E(\underline{v})}{\mathrm{d}t} < 0$ if $\underline{v} \in \mathfrak{B}/\{\underline{v}_{eq}\}$
5. $\frac{\mathrm{d}E(\underline{v})}{\mathrm{d}t} = 0$ if $\underline{v} = \underline{v}_{eq}$.

If the domain of attraction $\mathfrak{B}$ is unlimited (unbounded), then the system possesses only one equilibrium and is globally stable. For local stability results we focus on the asymptotic stability of the equilibrium $\underline{v}_{eq} = \underline{0}$, for other equilibrium points we just have to shift the equation of motion of the dynamical system and the Lyapunov function to that equilibrium point. We adopt this procedure for global stability results. For discrete-time dynamical systems we replace the differentiation $\frac{\mathrm{d}E(\underline{v})}{\mathrm{d}t}$ with $E[\underline{v}(k+1)] - E[\underline{v}(k)]$.

**Definition 18.3.** The discrete-time RNN reaches a limit cycle of length $T_c \in \mathbb{N}$ means:

$$E[\underline{v}(k)] = E[\underline{v}(k+T_c)]$$

## 18.4 Stability Analysis of RNNs with Time-Invariant Activation Functions

In this section we introduce some known results related to the stability of the RNNs with time-invariant activation functions. To do so, we begin with the following definition:

**Definition 18.4.** A set of real-valued functions $f(x)$ satisfying the following conditions is said to be class $F^{(1)}$

- $f(x)$ is continuously differentiable with respect to $x$
- $f(0) = 0$
- $f(x)$ is a bounded function i.e. there exists some $h$ such that $|f(x)| \leq h < \infty$
- $f(x)$ is a monotonically increasing function i.e. $\frac{\mathrm{d}f(x)}{\mathrm{d}x} > 0$

From the above listed conditions we conclude that:

- there exists a positive real-valued $l_1$ such that $\forall x, y \in \mathbb{R}, x \neq y : l_1 \geq \frac{f(x)-f(y)}{x-y}$ i.e. $\frac{\mathrm{d}f(x)}{\mathrm{d}x}$ is over bounded
- $f(x)$ is invertible

The inverted function $f^{-1}(x)$ has the following properties [23]:

- $f^{-1}(x)$ is continuously differentiable with respect to $x$
- $f^{-1}(0) = 0$
- $f^{-1}(x)$ is a monotonically increasing function i.e. $\frac{\mathrm{d}f^{-1}(x)}{\mathrm{d}x} > 0$
- There exists a positive real-valued $l_2$ such that $\forall x, y \in \mathbb{R}, x \neq y : l_2 \leq \frac{f^{-1}(x)-f^{-1}(y)}{x-y}$ and $l_2 = \frac{1}{l_1}$ i.e. $\frac{\mathrm{d}f^{-1}(x)}{\mathrm{d}x}$ is under bounded.

**Theorem 18.1.** *The discrete-time RNN with parallel update, cf. Fig. 18.2, Eq. (18.4), is locally asymptotically stable with maximum length of cycles equals to two if:*

- *the activation function* $\varphi(x) \in F^{(1)}$
- *there exists a diagonal positive definite matrix* $\underline{\underline{D}} = diag\{d_1, d_2, \cdots, d_n\} > 0$ *such that* $\underline{\underline{D}} \cdot \underline{\underline{w}} = \{\underline{\underline{D}} \cdot \underline{\underline{w}}\}^T$.

*Proof.* The Lyapunov function in this case is given by:

$$E[\underline{v}(k)] = -\underline{v}^T(k+1) \cdot \underline{\underline{D}} \cdot \underline{\underline{w}} \cdot \underline{v}(k) - [\underline{v}^T(k+1) + \underline{v}^T(k)] \cdot \underline{\underline{D}} \cdot \underline{\underline{w}}_0 \cdot \underline{e} + \sum_{l=1}^{n} d_l \cdot \left\{ \int_0^{v_l(k+1)} \varphi^{-1}(\zeta) d\zeta + \int_0^{v_l(k)} \varphi^{-1}(\zeta) d\zeta \right\} \tag{18.8}$$

This function is monotonically decreasing along the discrete-time dynamics of the network i.e. $E[\underline{v}(k+1)] - E[\underline{v}(k)] \leq 0$. The proof can be found in [3].

**Theorem 18.2.** *The discrete-time RNN with serial update, cf. Eq. (18.5) and Eq. (18.6), is locally asymptotically stable with maximum length of cycles equals to one if:*

- *the activation function* $\varphi(x) \in F^{(1)}$
- *there exists a diagonal positive definite matrix* $\underline{\underline{D}} = diag\{d_1, d_2, \cdots, d_n\} > 0$ *such that* $\underline{\underline{D}} \cdot \underline{\underline{w}} = \{\underline{\underline{D}} \cdot \underline{\underline{w}}\}^T$
- *the diagonal elements of the weight matrix are non-negative*

*Proof.* The Lyapunov function in this case is given by:

$$E[\underline{v}(\rho)] = -\frac{1}{2}\underline{v}^T(\rho) \cdot \underline{\underline{D}} \cdot \underline{\underline{w}} \cdot \underline{v}(\rho) - \underline{v}^T(\rho) \cdot \underline{\underline{D}} \cdot \underline{\underline{w}}_0 \cdot \underline{e} + \sum_{l=1}^{n} d_l \cdot \int_0^{v_l(\rho)} \varphi^{-1}(\zeta) d\zeta \tag{18.9}$$

This function is monotonically decreasing along the discrete-time dynamics of the network i.e. $E[\underline{v}(\rho+1)] - E[\underline{v}(\rho)] \leq 0$. The proof can be found in [3].

**Theorem 18.3.** *The continuous-time RNN, cf. Fig. 18.3 and Eq. (18.7), is locally asymptotically stable if:*

- *the activation function* $\varphi(x) \in F^{(1)}$
- *there exists a diagonal positive definite matrix* $\underline{\underline{D}} = diag\{d_1, d_2, \cdots, d_n\} > 0$ *such that* $\underline{\underline{D}} \cdot \underline{\underline{w}} = \{\underline{\underline{D}} \cdot \underline{\underline{w}}\}^T$

*Proof.* The Lyapunov function in this case is given by:

$$E[\underline{v}(t)] = -\frac{1}{2}\underline{v}^T(t) \cdot \underline{\underline{D}} \cdot \underline{\underline{w}} \cdot \underline{v}(t) - \underline{v}^T(t) \cdot \underline{\underline{D}} \cdot \underline{\underline{w}}_0 \cdot \underline{e} + \sum_{l=1}^{n} d_l \cdot \int_0^{v_l(t)} \varphi^{-1}(\zeta) d\zeta \tag{18.10}$$

This function is monotonically decreasing along the continuous-time dynamics of the network. The proof can be found in[5] [12].

**Theorem 18.4.** *The fixed points of the discrete-time RNN with serial and parallel updating are the same.*

*Proof.* This theorem has been proven for $\varphi(x) = \text{sign}(x)$ in [1]. A generalization to any $\varphi \in F^{(1)}$ is similar and is omitted here for lack of space.

*Remark 18.6.* In the original proof of theorems (18.1-18.3) it is assumed that $\underline{\underline{D}} = \underline{\underline{I}}$ and $\underline{\underline{w}} = \underline{\underline{w}}^T$. However, the generalization to $\underline{\underline{D}} \neq \underline{\underline{I}}$ is only a minor modification of the proof and we will see later that this assumption is very useful.

*Remark 18.7.* If $\varphi(\cdot) = \tanh(\cdot)$ then theorem (18.3) coincides with the work of Hopfield presented in [9].

[5] The original proof is for complex-valued RNNs

*Remark 18.8.* If $\varphi(x) = \tanh(\beta \cdot x)$ and $\beta \to \infty$ then theorem (18.2) coincides with the work of Hopfield presented in [8].

## 18.5 Analyzing The Optimum Activation Function

In this section we introduce the optimum activation function for the vector equalizer based on RNN and we prove that the optimum activation function belongs to $F^{(1)}$. At the end of this chapter we derive the connection between the Lyapunov function of the RNNs Eq. (18.8-18.10) and the maximum likelihood function Eq. (18.3).

### *18.5.1 The Optimum Activation Function*

In[6] [1], [21] the optimum estimate $\check{x}$ of a single-valued symbol $x \in \mathbb{A}_x = \{a_1, a_2, \cdots, a_M\}$, $a_i \in \mathbb{R}$, which is disturbed by real-valued additive white Gaussian noise has been obtained using the mean squared error $J = E\{|x - \check{x}|^2 | \tilde{x}\}$. The minimum of $J$ can only be reached if $\check{x}$ is a continuous value (soft estimate), i.e. $\check{x}$ is not restricted to the discrete-valued symbol alphabet $\mathbb{A}_x$ see Fig. 18.4.

This problem can be treated as a problem of parameter estimation [4], [21] and in the case of a real-valued symbol alphabet $\mathbb{A}_x$ leads to the optimum estimate:

$$\check{x} = \theta_{opt}(\tilde{x}) = \frac{\sum_{i=1}^{M} a_i \exp\{-\frac{\beta}{2} a_i^2\} \exp\{\beta a_i \tilde{x}\}}{\sum_{i=1}^{M} \exp\{-\frac{\beta}{2} a_i^2\} \exp\{\beta a_i \tilde{x}\}} \tag{18.11}$$

where $\sigma^2 = \frac{1}{\beta}$ is the power of the real-valued additive white Gaussian noise.

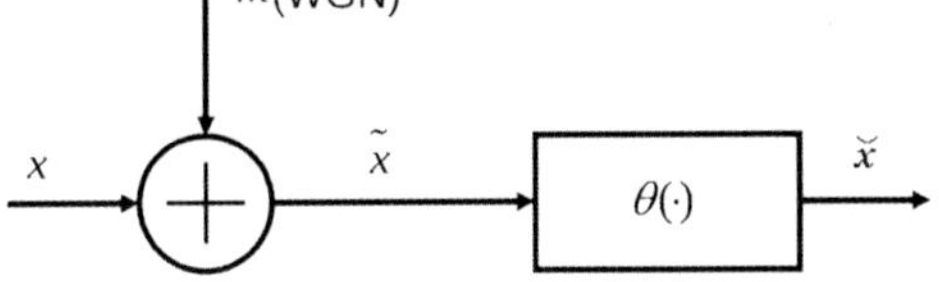


**Fig. 18.4** Parameter estimation problem

This function $\theta_{opt}(\cdot)$ has been widely used as activation function in case of using RNNs as multiuser detector or vector equalizer [2], [11], [20], [21], [24], [25]. We will see later that $\beta$ is nothing else than the slope of the function $\theta_{opt}(\cdot)$. There are many methods to define $\beta$ when using the function $\theta_{opt}(\cdot)$ as activation function for the vector equalizer based on RNNs. One method is to assume it to stay constant during the evolution process. Another one is to assume it to be time-variant during the iteration process. The later approach leads to a

[6] The estimation has been done for complex-valued symbols and noise.

time-varying activation function. Therefore we need to analyze the stability under this time-varying condition. One but not the only way of this time-variation is to update $\sigma^2$ (the power of the additive white Gaussian noise plus the power of the interference) after each iteration. Another possibility is to let the slope $\beta$ to be increased during the iteration process. We restrict ourself to symbol alphabets $\mathbb{A}_x = \{a_{-\frac{M}{2}}, a_{-\frac{M}{2}+1}, \cdots, a_{-1}, a_1, \cdots, a_{\frac{M}{2}-1}, a_{\frac{M}{2}}\}$, which fulfill:

- Symbol alphabets of length $M$: $M = 2^s$, $s \in \mathbb{N}/\{0\}$
- Real-valued symbol alphabets: $\forall i \in \{-\frac{M}{2}, -\frac{M}{2}+1, \cdots, -1, 1, \cdots, \frac{M}{2}-1, \frac{M}{2}\} : a_i \in \mathbb{R}$
- Symmetric symbol alphabets: $\forall i \in \{1, 2, \cdots, \frac{M}{2}\} : a_i = -a_{-i}$
- Equal distanced symbol alphabets: $\forall i \in \{-\frac{M}{2}, -\frac{M}{2}+1, \cdots, -1, 1, \cdots, \frac{M}{2}-1\}$ : $a_{i+1} = a_i + d_h$, $d_h \in \mathbb{R}^+$.

Using these conditions we can rewrite Eq. (18.11) as follows:

$$\theta_{opt}(\tilde{x}) = \frac{\sum_{i=1}^{\frac{M}{2}} a_i \exp[-\frac{\beta}{2} a_i^2] \sinh(\beta a_i \tilde{x})}{\sum_{i=1}^{\frac{M}{2}} \exp[-\frac{\beta}{2} a_i^2] \cosh(\beta a_i \tilde{x})} \tag{18.12}$$

*Remark 18.9.* If $a_1 = 1$ and $\forall i \in \left\{-\frac{M}{2}, -\frac{M}{2}+1, \cdots, \frac{M}{2}-1\right\} : \alpha_i = \frac{a_{i+1}+a_i}{2}$ it can be shown that:

$$\theta_{opt}(\tilde{x}) \approx \sum_{i=-\frac{M}{2}}^{\frac{M}{2}-1} \tanh[\beta \cdot (\tilde{x} - \alpha_i)]$$

The larger is the slope $\beta$, the better is the approximation.

*Remark 18.10.* For binary phase shift keying (BPSK) $\mathbb{A}_x = \{-1, +1\}$:

$$\theta_{opt}(\tilde{x}) = \tanh(\beta \cdot \tilde{x})$$

## *18.5.2 Properties of the Optimum Activation Function*

It is easy to prove that the optimum activation function $\theta_{opt}(\cdot)$ in Eq. (18.11,18.12) fulfills the following conditions:

- its domain is $\mathbb{R}$ and it is continuously differentiable with respect to its argument
- $\theta_{opt}(0) = 0$
- it is a bounded function $\theta_{opt}(\cdot) \in [\min\{\mathbb{A}_x\}, \max\{\mathbb{A}_x\}]$
- it is a monotonically increasing function $\frac{\mathrm{d}\theta_{opt}(x)}{\mathrm{d}x} > 0$
- $\max\left\{\frac{\theta_{opt}(x)-\theta_{opt}(y)}{x-y}\right\} = \max\left\{\frac{\mathrm{d}\theta_{opt}(x)}{\mathrm{d}x}\right\} \approx \beta$. The larger is $\beta$, the better is the approximation

- the optimum activation function is invertible
- $\theta_{opt}^{-1}(\cdot)$ is continuously differentiable with respect to its argument
- $\theta_{opt}^{-1}(0) = 0$
- it is a monotonically increasing function $\frac{\mathrm{d}\theta_{opt}^{-1}(x)}{\mathrm{d}x} > 0$
- $\min\left\{\frac{\theta_{opt}^{-1}(x)-\theta_{opt}^{-1}(y)}{x-y}\right\} = \min\left\{\frac{\mathrm{d}\theta_{opt}^{-1}(x)}{\mathrm{d}x}\right\} \approx \frac{1}{\beta}$. The larger is $\beta$, the better is the approximation
- $\beta >> 1 \Rightarrow \int_0^x \theta_{opt}^{-1}(\zeta)d\zeta$ is a strictly convex function [7]
- $\beta << 1 \Rightarrow \int_0^x \theta_{opt}^{-1}(\zeta)d\zeta$ is a strongly convex function [7]

We conclude that $\theta_{opt}(\cdot) \in F^{(1)}$ and the theorems (18.1-18.4) are valid if $\varphi(\cdot) = \theta_{opt}(\cdot)$.

*Example 18.1.* Figures 18.5 and 18.6 show those properties using a 4ASK symbol alphabet $\mathbb{A}_x = \{-3,-1,1,3\}$.

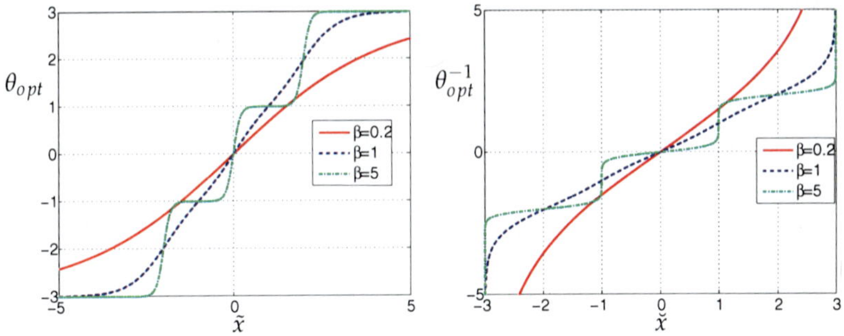


**Fig. 18.5** The activation function and the inverted activation function

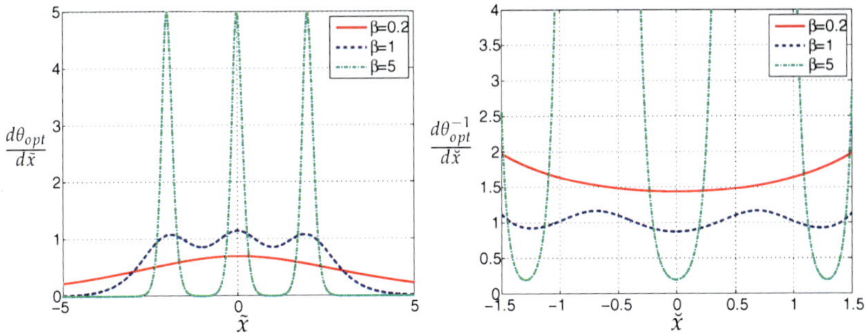


**Fig. 18.6** Derivative of the activation function and the inverted activation function

Interpreting the integration as calculation of surface, it is easy from Fig. 18.5 to see that:

$$\begin{aligned} \beta_1 > \beta_2 &\Rightarrow \int_0^x \theta^{-1}_{opt,\beta_1}(\zeta)d\zeta \leq \int_0^x \theta^{-1}_{opt,\beta_2}(\zeta)d\zeta \\ &\Rightarrow \int_0^x \theta^{-1}_{opt,\beta_2}(\zeta)d\zeta - \int_0^x \theta^{-1}_{opt,\beta_1}(\zeta)d\zeta \geq 0 \\ &\Rightarrow \frac{\partial[\int_0^x \theta^{-1}_{opt,\beta}(\zeta)d\zeta]}{\partial\beta} \leq 0 \end{aligned} \tag{18.13}$$

*Remark 18.11.* For large slopes $\beta$ no difference can be seen in the activation function. For $\beta \rightarrow \infty$,$\theta_{opt}(\cdot)$ tends to a staircase function.

### 18.5.3 Lyapunov Function vs. Maximum Likelihood Function

The basic idea behind the ability of using RNNs as vector equalizer without training is the correspondence between the Lyapunov function and the Mahalanobis metric. Let us compare Eq. (18.3) and Eq. (18.9):

$$\Gamma(\underline{\xi}) = \underline{\tilde{\xi}}^T \cdot \underline{\underline{r}} \cdot \underline{\tilde{\xi}} - 2 \cdot \underline{\tilde{\xi}}^T \cdot \underline{x}$$

$$E[\underline{v}(\rho)] = -\frac{1}{2}\underline{v}^T(\rho) \cdot \underline{\underline{D}} \cdot \underline{\underline{w}} \cdot \underline{v}(\rho) - \underline{v}^T(\rho) \cdot \underline{\underline{D}} \cdot \underline{\underline{w}}_0 \cdot \underline{e} + \sum_{l=1}^{N} d_l \cdot \int_0^{v_l(\rho)} \varphi^{-1}(\zeta)d\zeta$$

We are looking for the discrete-valued vector $\underline{\xi}$ which minimizes $\Gamma(\underline{\xi})$. The RNN is minimizing the Lyapunov function (energy function) $E[\underline{v}(\rho)]$. If we compare $\Gamma(\underline{\xi})$ with $E[\underline{v}(\rho)]$ term by term assuming $\underline{\underline{r}}_d = \text{diag}\{r_{11}, r_{22}, \cdots, r_{nn}\}$ we conclude:

$$\begin{aligned} \underline{\underline{D}} = \underline{\underline{r}}_d,\ \underline{\underline{w}}_0 = \underline{\underline{r}}_d^{-1} &\Rightarrow \underline{\underline{D}} \cdot \underline{\underline{w}}_0 = \underline{\underline{I}} \\ \underline{e} = \underline{\tilde{x}},\ \underline{\underline{w}} &= \underline{\underline{r}}_d^{-1} \cdot \left\{\underline{\underline{r}}_d - \underline{\underline{r}}\right\} \\ \varphi(\cdot) = \theta_{opt}(\cdot) \end{aligned} \tag{18.14}$$

After reaching a fixed point:

$$\underline{\xi} = \text{DECI}\{\underline{v}\}$$

The vector equalizer based on RNNs is a suboptimum equalizer because:

- the third term in the Lyapunov function does not have any correspondence in the Mahalanobis metric

- to fulfill the third stability condition of the discrete-time RNN with serial update ( cf. theorem 18.2) the weight matrix $\underline{\underline{w}}$ is forced to have zero diagonal elements. Therefore the first two terms in both Lyapunov function and maximum likelihood metric are not exactly coinciding, except for BPSK
- we are looking for the global minima of the Mahalanobis metric but the RNN finds in general only the local minima.

## 18.6 Stability Analysis of RNNs with Time-Variant Activation Functions

Previous work showed that vector equalizer based on RNNs set up with the rules in Eq. (18.14), where the slope of the activation function $\beta$ is constant during the iteration process result in a good performance for channels with small to moderate interference. To be able to avoid local minima, a linear increasing slope has been suggested and better results could be achieved. On the other hand the stability proof in theorems (18.1-18.3) is not valid any more. Therefore we provide in this section conditions on the time-variant activation function, for which the stability of the RNNs can be proven.

### 18.6.1 Discrete-Time RNNs with Parallel Update

**Theorem 18.5.** *The discrete-time RNN with parallel update, cf. Fig. 18.2 and Eq. (18.4), is locally asymptotically stable with maximum length of cycles equals to two if:*

- *the slope of the activation function $\beta$ is non-decreasing during the iteration process*
- *the activation function $\varphi(\cdot) = \theta_{opt}(\cdot) \in F^{(1)}$*
- *there exists a diagonal positive definite matrix $\underline{\underline{D}} = diag\{d_1, d_2, \cdots, d_n\} > 0$ such that $\underline{\underline{D}} \cdot \underline{\underline{w}} = \{\underline{\underline{D}} \cdot \underline{\underline{w}}\}^T$.*

*Proof.* The Lyapunov function in this case is given by:

$$E[\underline{v}(k)] = -\underline{v}^T(k+1) \cdot \underline{\underline{D}} \cdot \underline{\underline{w}} \cdot \underline{v}(k) - [\underline{v}^T(k+1) + \underline{v}^T(k)] \cdot \underline{\underline{D}} \cdot \underline{\underline{w}}_0 \cdot \underline{e} + \sum_{l=1}^{n} d_l \cdot \left\{ \int_0^{v_l(k+1)} \varphi^{-1}_{\beta(k+1)}(\zeta)d\zeta + \int_0^{v_l(k)} \varphi^{-1}_{\beta(k)}(\zeta)d\zeta \right\} \tag{18.15}$$

We proof in the following that $E[\underline{v}(k)]$ is monotonically decreasing with time. For this reason we define:

$$\Delta E = E[\underline{v}(k)] - E[\underline{v}(k-1)]$$

Using the conditions above and the mean value theorem [22] we find:

$$\Delta E = -\left\{\underline{v}^T(k+1) - \underline{v}^T(k-1)\right\} \cdot \underline{\underline{D}} \cdot \underline{\underline{w}} \cdot \underline{v}(k) - \left\{\underline{v}^T(k+1) - \underline{v}^T(k-1)\right\} \cdot \underline{\underline{D}} \cdot \underline{\underline{w}}_0 \cdot \underline{e}$$
$$+ \sum_{l=1}^{n} d_l \cdot \left\{ \int_0^{v_l(k+1)} \varphi^{-1}_{\beta(k+1)}(\zeta) d\zeta - \int_0^{v_l(k-1)} \varphi^{-1}_{\beta(k-1)}(\zeta) d\zeta \right\}$$
$$\Delta E = -\left\{\underline{v}^T(k+1) - \underline{v}^T(k-1)\right\} \cdot \underline{\underline{D}} \cdot \underline{u}(k+1) + \sum_{l=1}^{n} d_l \cdot \left\{ \int_{v_l(k-1)}^{v_l(k+1)} \varphi^{-1}_{\beta(k+1)}(\zeta) d\zeta \right\}$$
$$- \sum_{l=1}^{n} d_l \cdot \left\{ \underbrace{\int_0^{v_l(k-1)} \varphi^{-1}_{\beta(k-1)}(\zeta) d\zeta - \int_0^{v_l(k-1)} \varphi^{-1}_{\beta(k+1)}(\zeta) d\zeta}_{I_l} \right\}$$

Because $\beta(k)$ is non-decreasing then according to
Eq. (18.13): $\forall l \in \{1,2,\cdots,n\} : I_l \geq 0 \Rightarrow \sum_{l=1}^{n} d_l \cdot I_l \geq 0$

$$\Delta E = -\left\{\underline{v}^T(k+1) - \underline{v}^T(k-1)\right\} \cdot \underline{\underline{D}} \cdot \{\underline{u}(k+1) - \underline{u}_0\} - \sum_{l=1}^{n} d_l \cdot I_l$$
$$\underline{u}_0 \in ]\min\{\underline{u}_1, \underline{u}(k+1)\}, \max\{\underline{u}_1, \underline{u}(k+1)\}[$$
$$\underline{u}_1 = \varphi^{-1}_{\beta(k+1)}[\underline{v}(k-1)]$$

Because of the special properties of the activation function Fig. 18.5 $\Rightarrow \Delta E \leq 0$ and the equality holds in two cases:

- limit cycle of length two $\underline{v}(k-1) = \underline{v}(k+1) \neq \underline{v}(k)$
- fixed point $\underline{v}(k-1) = \underline{v}(k) = \underline{v}(k+1)$

for:

- $\beta(k)$ reaches a constant value
- $\beta(k)$ reaches a large value such that increasing it makes no difference on the activation function $\varphi_\beta(\cdot)$

□

*Remark 18.12.* We can imagine a way to avoid the limit cycle of length two by calculating the next value of the slope $\beta$ dependent on its previous values and previous outputs of the RNN.

## 18.6.2 Discrete-Time RNN with Serial Update

**Theorem 18.6.** *The discrete-time RNN with serial update, cf. Eq. (18.5) and Eq. (18.6), is locally asymptotically stable with maximum length of cycles equals to one if:*

- *the slope of the activation function $\beta$ is non-decreasing during the iteration process*
- *the activation function $\varphi(\cdot) = \theta_{opt}(\cdot) \in F^{(1)}$*

- *there exists a diagonal positive definite matrix* $\underline{\underline{D}} = diag\{d_1, d_2, \cdots, d_n\} > 0$ *such that* $\underline{\underline{D}} \cdot \underline{\underline{w}} = \{\underline{\underline{D}} \cdot \underline{\underline{w}}\}^T$
- *the diagonal elements of the weight matrix are non-negative.*

*Proof.* The Lyapunov function in this case is given by:

$$E[\underline{v}(\rho)] = -\frac{1}{2}\underline{v}^T(\rho) \cdot \underline{\underline{D}} \cdot \underline{\underline{w}} \cdot \underline{v}(\rho) - \underline{v}^T(\rho) \cdot \underline{\underline{D}} \cdot \underline{\underline{w}}_0 \cdot \underline{e} + \sum_{l=1}^{n} d_l \cdot \int_0^{v_l(\rho)} \varphi^{-1}_{\beta(\rho)}(\zeta) d\zeta \tag{18.16}$$

We proof in the following that $E[\underline{v}(\rho)]$ is monotonically decreasing with time assuming the $j^{th}$ neuron, $j \in \{1, 2, \cdots, n\}$, has been updated i.e. $v_l(\rho+1) = v_l(\rho)$ if $l \neq j$. For this reason we define:

$$\Delta E = E[\underline{v}(\rho+1)] - E[\underline{v}(\rho)]$$

Using the conditions listed above and the mean value theorem [22] we find:

$$\begin{aligned}\Delta E = &- d_j \cdot \{v_j(\rho+1) - v_j(\rho)\} \cdot u_j(\rho+1) - \frac{1}{2} \cdot d_j \cdot w_{jj} \cdot \{v_j(\rho+1) - v_j(\rho)\}^2 \\ &+ \sum_{l=1}^{n} d_l \cdot \left\{ \int_0^{v_l(\rho+1)} \varphi^{-1}_{\beta(\rho+1)}(\zeta) d\zeta - \int_0^{v_l(\rho)} \varphi^{-1}_{\beta(\rho)}(\zeta) d\zeta \right\}\end{aligned}$$

$$\begin{aligned}\Delta E = &- d_j \cdot \{v_j(\rho+1) - v_j(\rho)\} \cdot \{u_j(\rho+1) - u_{j,0}\} - \frac{1}{2} \cdot d_j \cdot w_{jj} \cdot \{v_j(\rho+1) - v_j(\rho)\}^2 \\ &- \sum_{l=1}^{n} d_l \cdot \left\{ \underbrace{\int_0^{v_l(\rho)} \varphi^{-1}_{\beta(\rho)}(\zeta) d\zeta - \int_0^{v_l(\rho)} \varphi^{-1}_{\beta(\rho+1)}(\zeta) d\zeta}_{I_l} \right\}\end{aligned}$$

Because $\beta(\rho)$ is non-decreasing then according to Eq. (18.13): $\forall l \in \{1, 2, 1 \cdots, n\}$ : $I_l \geq 0 \Rightarrow \sum_{l=1}^{n} d_l \cdot I_l \geq 0$

$$\begin{aligned}\Delta E = &- d_j \cdot \{v_j(\rho+1) - v_j(\rho)\} \cdot \{u_j(\rho+1) - u_{j,0}\} \\ &- \frac{1}{2} \cdot d_j \cdot w_{jj} \cdot \{v_j(\rho+1) - v_j(\rho)\}^2 - \sum_{l=1}^{n} d_l \cdot I_l \\ u_{j,0} \in &\,]\min\{u_{j,1}, u_j(\rho+1)\}, \max\{u_{j,1}, u_j(\rho+1)\}[ \\ u_{j,1} = &\ \varphi^{-1}_{\beta(\rho+1)}[v_j(\rho)]\end{aligned}$$

Because of the special properties of the activation function Fig. 18.5 $\Rightarrow \Delta E \leq 0$ and the equality holds if $v_j(\rho) = v_j(\rho+1)$ and $\beta_{(\rho)} = \beta_{(\rho+1)}$ or $\beta(\rho)$ is so large such that increasing it makes no difference for the activation function. A fixed point will be reached if $\forall l \in \{1, 2, \cdots, n\} : \Delta E = 0$. □

### 18.6.3 Continuous-Time RNN

**Theorem 18.7.** *The continuous-time RNN, cf. Fig. 18.3 and Eq. (18.7), is locally asymptotically stable if:*

- *the slope of the activation function $\beta$ is non-decreasing during the iteration process*
- *the activation function $\varphi(\cdot) = \theta_{opt}(\cdot) \in F^{(1)}$*
- *there exists a diagonal positive definite matrix $\underline{\underline{D}} = diag\{d_1, d_2, \cdots, d_n\} > 0$ such that $\underline{\underline{D}} \cdot \underline{\underline{w}} = \{\underline{\underline{D}} \cdot \underline{\underline{w}}\}^T$.*

*Proof.* The Lyapunov function in this case is given by:

$$E[\underline{v}(t)] = -\frac{1}{2}\underline{v}^T(t) \cdot \underline{\underline{D}} \cdot \underline{\underline{w}} \cdot \underline{v}(t) - \underline{v}^T(t) \cdot \underline{\underline{D}} \cdot \underline{\underline{w}}_0 \cdot \underline{e} + \sum_{l=1}^{n} d_l \cdot \int_0^{v_l(t)} \varphi_{\beta(t)}^{-1}(\zeta) d\zeta \tag{18.17}$$

We proof in the following that $\frac{dE[\underline{v}(t)]}{dt}$ is monotonically decreasing with time.

$$\frac{\mathrm{d}E[\underline{v}(t)]}{\mathrm{d}t} = -\frac{\mathrm{d}\underline{v}^T(t)}{\mathrm{d}t} \cdot \underline{\underline{D}} \cdot \underline{\underline{\tau}} \cdot \frac{\mathrm{d}\underline{u}(t)}{\mathrm{d}t} + \sum_{l=1}^{n} d_l \cdot \underbrace{\frac{\partial}{\partial \beta}\left\{\int_0^{v_l} \varphi_\beta^{-1}(\zeta) d\zeta\right\}}_{\leq 0 \text{ according to Eq. (18.13)}} \cdot \frac{\mathrm{d}\beta(t)}{\mathrm{d}t}$$

$$\Rightarrow \frac{dE[\underline{v}(t)]}{dt} \leq 0$$

An equilibrium point is reached if $\frac{\mathrm{d}v(t)}{\mathrm{d}t} = 0$ and $\beta(t)$ reaches a constant or a large value. □

*Remark 18.13.* In the proof of the theorems (18.5-18.7) it has been assumed that all neurons have the same slope $\beta$. This can easily be generalized to the case, where different neurons have different slopes. Important is that these slopes are non-decreasing with respect to time.

## 18.7 Global vs. Local Stability for Vector Equalizer Based on RNN

A RNN can be globally or locally stable depending on the fulfilled inherent stability conditions. When applying the RNN to solve practical optimization problems, the RNN is usually designed to have an unique equilibrium and to be globally asymptotically stable to avoid spurious responses or the problem of local minima [10]. In this section we show that for a vector equalizer based on RNN, the global stability conditions can be achieved easily. However, this leads to a long latency time, i.e. a longer processing time is required. On the other hand, local stability conditions offer more free parameters to optimize its performance. We show also, in which situations both stability cases are especially interesting.

### 18.7.1 Discrete-Time RNN with Parallel Update

**Theorem 18.8.** *The discrete-time RNN with parallel update, cf. Fig. 18.2, Eq. (18.4), is globally asymptotically stable if:*

- *the activation function* $\varphi(\cdot) = \theta_{opt}(\cdot) \in F^{(1)}$
- *there exists a diagonal positive definite matrix* $\underline{\underline{P}}$ *such that:*

$$\underline{\underline{L}}^{-1} \cdot \underline{\underline{P}} \cdot \underline{\underline{L}}^{-1} - |\underline{\underline{w}}|^T \cdot |\underline{\underline{P}}| \cdot |\underline{\underline{w}}| > 0 \tag{18.18}$$

*where* $\underline{\underline{L}} = diag\{\beta_1, \beta_2, \cdots, \beta_n\}$.

*Proof.* The corresponding Lyapunov function (shifted to the global fixed point) is given by:

$$E[\underline{Z}(k)] = \underline{Z}^T(k) \cdot \underline{\underline{P}} \cdot \underline{Z}(k) \tag{18.19}$$

$\underline{Z}(k)$ is the difference between the output of the discrete-time RNN at discrete time $k$ and its global equilibrium. The proof can be found in[7] [27].

### 18.7.2 Continuous-Time RNN

**Theorem 18.9.** *The continuous-time RNN, cf. Fig. 18.3, Eq. (18.7), is globally stable if:*

- *the activation function* $\varphi(\cdot) = \theta_{opt}(\cdot) \in F^{(1)}$
- *there exists a diagonal positive definite matrix* $\underline{\underline{P}}$ *such that:*

$$\underline{\underline{P}} \cdot \underline{\underline{\tau}}^{-1} \cdot \left\{ \underline{\underline{L}}^{-1} - |\underline{\underline{w}}| \right\} > 0 \tag{18.20}$$

*where* $\underline{\underline{L}} = diag\{\beta_1, \beta_2, \cdots, \beta_n\}$.

*Proof.* The corresponding Lyapunov function (shifted to the global fixed point) is given by:

$$E[\underline{Z}(t)] = \underline{Z}^T(t) \cdot \underline{\underline{L}} \cdot \underline{\underline{P}} \cdot \underline{Z}(t) \tag{18.21}$$

$\underline{Z}(t)$ is the difference between the output of the continuous-time RNN at time instant $t$ and its global equilibrium. The proof can be found in[8] [27].

### 18.7.3 Discussion

As mentioned before, the RNNs with a unique equilibrium and globally asymptotical stability are preferred in general, when applying RNNs to solve optimization problems.

In the following we show, what that means, when applying the RNNs as vector equalizer. At the end of this section we will be able to have a better understanding of

[7] The original proof is valid for complex-valued RNN also.

[8] The original proof is valid for complex-valued RNN also.

a hybrid scheme of global and local stability, which has already been applied for the RNN vector equalizer. Let us begin with the globally stable, continuous-time RNN Eq. (18.7). In this case Eq. (18.20) must be fulfilled to guarantee that the RNN is globally stable.

A direct solution to fulfill Eq. (18.20) is to assume:
$\forall i \in \{1,2,\cdots,n\}: \; p_i = p > 0, \; \tau_i = \tau > 0 \Rightarrow$

$$\underline{\underline{P}} \cdot \underline{\underline{\tau}}^{-1} \cdot \left\{ \underline{\underline{L}}^{-1} - |\underline{\underline{w}}| \right\} > 0 \Leftrightarrow \frac{p}{\tau} \cdot \left\{ \underline{\underline{L}}^{-1} - |\underline{\underline{w}}| \right\} > 0$$
$$\Leftrightarrow \underline{\underline{L}}^{-1} - |\underline{\underline{w}}| > 0$$

$\underline{\underline{L}}^{-1} - |\underline{\underline{w}}|$ is a square matrix with positive diagonal elements and negative off-diagonal elements. To force this matrix to be positive definite, we must increase its diagonal elements $L_i^{-1} = \beta_i^{-1}$ i.e. decreasing $\beta_i$. This leads to activation functions $\varphi_i(\cdot)$ with very small slopes, where a huge evolution time (multiple of $\tau$) is required to reach the global minimum. In other words, global stability requires activation functions with small slopes but it does not demand explicitly any conditions on the weight matrix itself.

Similar results can be obtained when analyzing the globally stable, discrete-time RNN Eq. (18.4). In this case Eq. (18.18) must be fulfilled to guarantee that the RNN is global stable.

A direct solution to fulfill Eq. (18.18) is to assume:
$\forall i \in \{1,2,\cdots,n\}: \; p_i = p > 0$

$$\underline{\underline{L}}^{-1} \cdot \underline{\underline{P}} \cdot \underline{\underline{L}}^{-1} - |\underline{\underline{w}}|^T \cdot \underline{\underline{P}} \cdot |\underline{\underline{w}}| > 0 \Leftrightarrow$$
$$p \cdot \left\{ \underline{\underline{L}}^{-2} - |\underline{\underline{w}}|^T \cdot |\underline{\underline{w}}| \right\} > 0 \Leftrightarrow$$
$$\underline{\underline{L}}^{-2} - |\underline{\underline{w}}|^T \cdot |\underline{\underline{w}}| > 0$$

As in the previous case, one direct solution to fulfill the condition above is to decrease $\beta_i$.

The correspondence between the cost function of the optimum vector equalization and the Lyapunov function of the RNN (in the case of local stability) depends on the slope $\beta_i$. The smaller the slope, the smaller is the similarity. This means, even if the RNN is globally stable, the Lyapunov function does not match to the Mahalanobis distance any more. Unlike the global stability conditions, the local stability does not demand any conditions on the slope of the activation function $\beta_i$ but it demands a symmetric weight matrix (for $\underline{\underline{D}} = \underline{\underline{I}}$). In the context of vector equalization the global stability becomes interesting, if the symmetry property of the weight matrix is lost (because of electronic elements non accuracy for continuous-time RNN as example).

A combination of global and local stability of the RNN vector equalizer (at least for discrete-time RNN) can be used. In this case the iteration process begins with a small slope $\beta$ (good assumption for global stability). The value of the slope increases after each iteration (transition from global to local stability)[9] [1].

## 18.8 Conclusion

In this chapter we reviewed the problem of vector equalization and how to apply a RNN to solve it as suboptimum scheme without training phase. We analyzed the properties of the optimum activation function for the vector equalizer based on RNNs and we generalized the known results of the stability of RNNs to a class of time-varying activation functions. These functions emerge from the problem of parameter estimation and have been used intensively in conjunction with the application of RNNs as multiuser detector or vector equalizer in wireless communications.

In this paper we analyzed also the global and local stability of the discrete-time and continuous-time RNN in the context of a vector equalizer. We studied the impact of the stability conditions on the equalization process. We showed that the global stability is interesting if the weight matrix is not symmetric. This might be the case because of electronic elements non-accuracy in case of continuous-time RNN as example. On the other hand global stability in general leads to long latency time and degraded performance at the receiver. This means a longer time is required to equalize each received vector because of the small slope of the activation function.

Local stability demands for symmetric weight matrices. The optimum slope and the evolution time have to be found by simulations.

We showed that a vector equalizer based on RNNs with increasing slope of the activation function during the iteration process can be interpreted as a scheme with a transition from global stability to local stability. On the other hand local stability still holds with increasing slope.

**Acknowledgements.** Financial support by the Deutsche Forschungsgemeinschaft (DFG project Li 65912-1) is gratefully acknowledged.

## References

1. Engelhart, A.: Vector detection techniques with moderate complexity. Dissertation, university of Ulm, institute of information technology, VDI Verlag GmbH, Düsseldorf (2003)
2. Engelhart, A., Teich, W.G., Lindner, J., Jeney, G., Imre, S., Pap, L.: A Survey of multiuser/multisubchannel detection schemes based on recurrent neural networks. In: Wireless Communications and Mobile Computing, special issue on Advances in 3G Wireless Networks, vol. 2(3), pp. 269–284. John Wiley & Sons, Ltd. (2002)
3. Fogelman-Soulié, F., Mejia, C., Goles, E., Martinez, S.: Energy functions in neural networks with continuous local functions. Complex Systems 3, 269–293 (1989)

[9] The stability with increased slope has been proven in previous sections.

4. Frey, T., Reinhardt, M.: Signal estimation for interference cancellation and decision feedback equalization. In: Proceeding IEEE Vehicular Technology Conference, VTC 1997, pp. 155–159 (1997)
5. Haykin, S.: Neural Networks: A Comprehensive Foundation, pp. 545–548. Macmillan College Publishing Company, Inc., USA (1994)
6. Hill, T., Lewicki, P.: Statistics: Methods and applications: A comprehensive reference for science, industry and data mining. StatSoft, Inc., USA (2006)
7. Hiriart-Urruty, J.-B., Lemaréchal, C.: Fundamentals of convex analysis, pp. 110–117. Springer, USA (2001)
8. Hopfield, J.J.: Neural networks and physical systems With emergent collective computational abilities. Proceeding of Natural Academic Science 79, 2554–2558 (1982)
9. Hopfield, J.J.: Neurons with graded response have collective computational properties like those of two-state neurons. Proceeding of Natural Academic Science 81, 3088–3092 (1984)
10. Hu, S., Wang, J.: Global stability of a class of discrete-time recurrent neural networks. IEEE Transactions on Circuits and Systems I: Fundamental Theory and Applications 49(8), 1104–1117 (2002)
11. Kechriotis, G.I., Manolakos, E.S.: Hopfield neural networks implementation of the optimal CDMA multiuser detector. IEEE Transactions on Neural Networks 7(1), 131–141 (1996)
12. Kuroe, Y., Hashimoto, N., Mori, T.: On energy function for complex-valued neural networks and its applications. In:Proceeding of the 9th International Conference on Neural Information Processing, ICONIP 2010, vol. 3, pp. 1079–1083 (2002)
13. Lindner, J.: MC-CDMA in the context of general multiuser/multisubchannel transmission methods. European Transactions on Telecommunications 10(4), 351–367 (1999)
14. Lindner, J.: Informationsübertragung: Grundlagen der Kommunikationstechnik. ch.8. Springer, Heidelberg (2005)
15. Luenberger, D.G.: Optimization by vector space method. John Wiley, NY (1969)
16. Lupas, R., Verdu, S.: Near-far resistance of multiuser detectors in asynchronous channels. IEEE Transactions on Communications 38(4), 496–508 (1990)
17. Mostafa, M., Teich, W.G., Lindner, J.: A modified recurrent neural network as vector detector. In: Proceeding of Asia-Pacific Conference on Circuits and Systems, APCCAS 2010, pp. 620–623 (2010)
18. Mostafa, M., Teich, W.G., Lindner, J.: Stability analysis of recurrent neural networks with time-varying activation functions. In: Proceeding of the International Workshop on Nonlinear Dynamical Systems, INDS 2011, pp. 239–244 (2011)
19. Mostafa, M., Teich, W.G., Lindner, J.: Global vs. local stability for recurrent neural networks as vector equalizer. In: International Conference on Signal Processing and Communication Systems ICSPCS (2011)
20. Sgraja, C., Engelhart, A., Teich, W.G., Lindner, J.: Combined equalization and decoding for general BFDM packet transmission schemes. In: Proceeding of the 1st international OFDM-Workshop, pp. 1–6 (1999)
21. Sgraja, C., Engelhart, A., Teich, W.G., Lindner, J.: Equalization with recurrent neural networks for complex-valued modulation schemes. In: Proceeding of the 3rd Workshop Kommunikationstechnik, pp. 7–12 (1999)
22. Sahoo, P.K., Riedel, T.: Mean value theorems and functional equations. ch.4. World Scientific, USA (1998)

23. Spivak, M.: Calculus on Mainfolds, pp. 35–37. Addison-Wesley (1965)
24. Teich, W.G., Seidl, M.: Code division multiple access communications: Multiuser detection based on a recurrent neural network structure. In: Proceeding of the International Symposium on Spread Spectrum Techniques and Applications, ISSSTA 1996, vol. 3, pp. 979–984 (1996)
25. Teich, W.G., Engelhart, A., Schlecker, W., Gessler, R., Pfleiderer, H.-J.: Towards an efficient hardware implementation of recurrent neural network based multiuser detection. In: 6th International Symposium on Spread Spectrum Techniques and Applications ISSSTA 2000, pp. 662–665 (2000)
26. Verdu, S.: Computational complexity of optimum multiuser detection. Algorithmica, 303–312 (1989)
27. Yoshida, M., Mori, T.: Complex-valued neural networks: Utilizing high-dimensional parameterss, pp. 104–114. IGI Global (2009)

# Chapter 19
# Stability of Linear Circuits with Interval Data: A Case Study

Zygmunt A. Garczarczyk

**Abstract.** In this chapter we study the problem of checking stability of a linear lumped electric circuits with interval data that model uncertainties of their element parameters (passive element values R, L, C and controlled source coefficients k). For such circuits the problem is concerned with examination of the eigenvalues of interval matrix. Presented approach is based on checking stability of symmetric interval matrix associated with the state matrix and is based on some interval analysis results. The method is not complex and in some cases we can determine circuit stability. We illustrate the applicability of studied approach by means of two numerical examples.

## 19.1 Introduction

Much of modern system theory adresses problems involving uncertainty. The mathematical model of a system might have various physical parameters (as, for example coefficients of friction, spring constants, capacitances, inductances, etc.) whose values are specified only within given intervals. The effects of system parameter uncertainties on system performance are important aspect of system design. A fundamental problem in system theory is concerned with stability of a given linear system. Stability analysis of systems with parametric uncertainties has attracted much attention in recent years. Motivated by the celebrated Kharitonov's Theorem and Edge Theorem, a number of papers have concentrated on robust stability of polynomial and matrix families with more complex uncertainty structures [2]-[4], [7], [8], [10], [12], [16]-[18].

The objective of this chapter is to present the results of our studies based on interval analysis techniques on checking stability of a linear lumped electric circuits with

Zygmunt A. Garczarczyk
Faculty of Electrical Engineering, Silesian Technical University
ul. Akademicka 10, 44-100 Gliwice, Poland
e-mail: Zygmunt.Garczarczyk@polsl.pl

K. Kyamakya et al. (Eds.): Selected Topics in Nonlinear Dynamics, SCI 459, pp. 351–358.
DOI: 10.1007/ 978-3-642-34560-9_19 

interval data that model uncertainties of their element parameters (passive element values R, L, C and controlled source coefficients k).

## 19.2 Problem Statement

A most applicable characterization of engineering systems is through their state equations. For linear time-invariant circuits the characterization is commonly expressed by normal form of the first order differential state equations which can be written in terms of the vector p of uncertain parameters. We consider

$$\frac{dx(t)}{dt} = A(p)x(t) + B(p)u(t) \tag{19.1}$$

and call $A(p)$ and $B(p)$ an uncertain or interval matrices. Their entries are not known exactly and are represented by some intervals. The following example will demonstrate this fact.

EXAMPLE 1. Consider the circuit of Fig. 19.1.

A state-space equation of the circuit is given by

$$\frac{d}{dt}\begin{bmatrix} u_1 \\ u_2 \end{bmatrix} = \begin{bmatrix} -\frac{2}{RC} & -\frac{k+1}{RC} \\ -\frac{1}{RC} & -\frac{1}{RC} \end{bmatrix} \begin{bmatrix} u_1 \\ u_2 \end{bmatrix} + \begin{bmatrix} \frac{2}{RC} & 0 \\ \frac{1}{RC} & -\frac{1}{C} \end{bmatrix} \begin{bmatrix} e \\ j \end{bmatrix}$$

Assume that the capacitors are exactly given as $C = 1$, and resistances of resistors and gain of the voltage - controlled voltage source are done as interval numbers, viz. $R = [0.935, 1.01]$ and $k = [0.7, 0.4]$. They form the uncertain vector $p = (R, k)$. With the numerical values substituted, the matrices on the right-hand side become

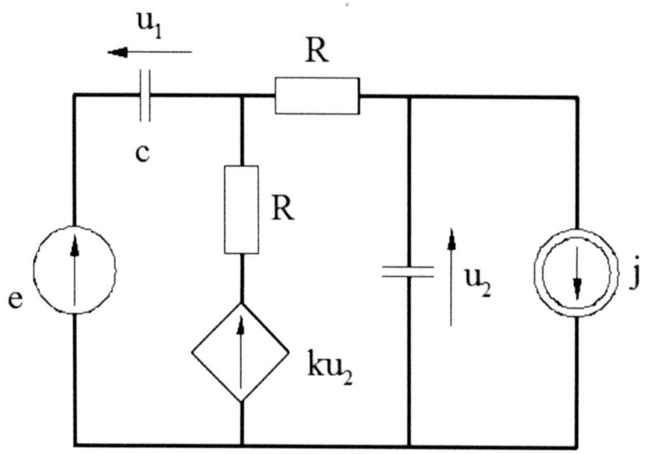


**Fig. 19.1** Linear time-invariant circuit for Example 1

$$A(p) = \begin{bmatrix} [-2.14,-1.96] & [-1.86,-1.67] \\ [-1.07,-0.98] & [-1.07,-0.98] \end{bmatrix}$$

$$B(p) = \begin{bmatrix} [1.96,-2.14] & 0 \\ [0,98,1.07] & -1 \end{bmatrix}$$

It's seen that in stability analysis of circuits, the stability of an interval (uncertain) matrices, equivalent to a set of matrices mathematically, is of concern.

It is well known that linear time-invariant continuous-time system is stable if and only if all the eigenvalues of the matrix $A(p)$ have negative real parts. $A(p)$ is an interval matrix i.e. is a real matrix in which all elements are known only within certain closed intervals. In precise terms, an $n \times n$ interval matrix $A^I = A(p)$ is a set of real matrices defined by

$$A^I = \{A = [A_{ij}] : b_{ij} \leq a_{ij} \leq c_{ij}, i,j = 1,2,..,n\} \tag{19.2}$$

An interval matrix $A^I$ is said to be stable if $\forall A \in A^I$ , all the eigenvalues of A are in the open left-half of the complex plane, i.e., $Re\{\lambda_i(A)\}$, where $\lambda(A)$ denotes an eigenvalues of $A$. The problem of stability of such matrices is much more complicated than the analogous one for systems described by their characteristic polynomial, as suggested by the fact that Kharitonov's Theorem for interval polynomials does not hold for interval matrices.

## 19.3 Stability Of Interval Matrices

First we introduce some notations. For a square real matrix $A = |a_{ij}|$ we denote the transpose by $A^T$, its absolute value as the matrix $|A| = (|a_{ij}|)$ and the spectral radius by $\rho(A)$, $\rho(A) = max|\lambda_i(A)|, i = l, ..., n$. Matrix is called symmetric if $A = A^T$. Symmetric matrices are known to have all eigenvalues real. W shall denote by $\sigma(A)$ the spectrum of $A$ i.e. the set of all $\lambda(A)$. Matrix inequalities, as $A < B$ or $A > B$, are to be understood componentwise. Let $A_c$ and $\Delta$ be real $n \times n$ matrices, $\Delta \geq 0$. The set of matrices

$$A^I = [A_c - \Delta, A_c + \Delta] = \{A : A_c - \Delta \leq A \leq A_c + \Delta\} \tag{19.3}$$

is called an interval matrix. Elements of center matrix $A_c$ and radius matrix $\Delta$ are defined following

$$[a_{cij}] = \frac{1}{2}(b_{ij} + c_{ij}) \tag{19.4}$$

$$[\Delta_{ij}] = \frac{1}{2}(c_{ij} - b_{ij}) \tag{19.5}$$

$A^I$ is said to be symmetric if both $A_c$ and $\Delta$ are symmetric. With each interval matrix $A^I = [A_c - \Delta, A_c + \Delta$ we shall associate the symmetric interval matrix

$$A^I_s = [A'_c - \Delta', A'_c + \Delta'] \tag{19.6}$$

where

$$A_c' = \frac{1}{2}(A_c + A_c^T) \tag{19.7}$$

$$\Delta' = \frac{1}{2}(\Delta + \Delta^T) \tag{19.8}$$

Obviously, if $A \in A^I$, then $\frac{1}{2}(A_c + A_c^T) \in A_s^I$ , and $A^I$ is symmetric if and only if $A^I = A_s^I$.

Stability properties of associated symmetric interval matrix are closely related to stability of any interval matrix $A^I$. In his paper, Rohn [15] exploiting Bendixon's theorem pointed out that if $A_s^I$ is stable, then $A^I$ is also stable. Unfortunately, the converse implication is generally not valid. For stability verification it's also useful result showing that, if for symmetric interval matrix $A_s^I$ center matrix $A_c'$ is stable and

$$\rho\left(|A_c'^{-1}|\Delta'\right) < I \tag{19.9}$$

holds, then $A_s^I$ is stable. Because of roundoff errors in computations of the inverse matrix, for practical purposes condition (19.9) may be replaced by following one

$$\rho\left(|I - QA_c'^{-1}| + |Q|\Delta'\right) < I \tag{19.10}$$

where $I$ is the unit matrix and $Q \approx A_c'^{-1}$.

The problem of finding the convex hull of the eigenvalues of a symmetric interval matrices has been recently studied by Hertz [10]. The maximal eigenvalue of eigenvalues of these interval matrices coincides with the maximal eigenvalue of a set containing $2^{n-1}$ symmetric vertex matrices. It's seen the complexity of evaluation the maximal eigenvalues grows up immediately with a dimension of state equations.

Basing on the above results simpler procedure for testing the stability or instability can be suggested:

STEP 1. Evaluate center matrix $A_c$ and radius matrix $\Delta$ with element defined by (19.4), (19.5). If $A_c$ and $\Delta$ are symmetric go STEP 3.
STEP 2. Using (19.7), (19.8) calculate symmetric matrices $A_c'$ and $\Delta$.
STEP 3. Test stability of the center matrix $A_c'$. If $A_c'$ is stable go to STEP 4 else go to STEP 6.
STEP 4. If condition (19.9) (respectively (19.10)) is fulfilled go to STEP 5 else go to STEP 6.
STEP 5. Stop. The state-space equation (19.1) is stable.
STEP 6. Stop. No decision concerning the stability or the instability of (19.1) can be made.

If the procedure terminates in STEP 6 no decision concerning the instability of the set of linear circuits with interval parameters can be made. This follows from the fact the stability problem of symmetric interval matrices is imbedded in the broader stability problem of any interval matrices. Therefore the instability of matrix (19.6) does not necessarily entail instability of matrix (19.3). In such case another stability condition must be tried (cf. for example [3], [12], [16], [18]).

## 19.4 Computational Aspects

Problem of testing robust stability of linear circuits is formulated, from the mathematical viewpoint, as the matrix eigenvalue problem. There are many numerical methods for asymmetric and symmetric eigenvalue problem, like QR method, Jacobi method, power method, Lanczos method etc. [9], [14]. Some of these techniques are appropriate when only a few eigenvalues are desired. Although computing the matrix eigenvalues may not be simply it's relatively easy to estimate some of them. In the foregoing procedure in STEP 3 and STEP 4 it's necessary only to calculate the spectral radius of given matrix to decide about its stability. This task can be done with use of the power method.

Given a matrix A, the power method is defined by the iterations

$$x_k = Ax_{k-1}, k = 1, 2, ... \tag{19.11}$$

where $x_0$ is the starting guess. The iterations converge to an eigenvector corresponding to the $\Lambda(A)$ with largest magnitude i.e. $\rho(A)$. The spectral radius estimate is calculated as the Rayleigh quotient

$$\rho(A) = \lim_{k\to\infty} \frac{x_k^T A x_k}{x_k^T x_k} \tag{19.12}$$

Furthermore notice that condition (19.9) (resp.(19.10)) is formulated for nonnegative or positive matrix $C = |A_C^{'-1}|\Delta' \geq 0$.

For estimation $\rho(C)$ we can exploit the Perron-Froebenius Theorem [8]. For example, if $C > 0$ and if $x$ is Perron vector of $C$ then

$$\rho(C) = \sum_{i=1}^{n}\sum_{j=1}^{n} c_{ij} x_j \tag{19.13}$$

Recall that $x_1 + x_2 + ... + x_n = 1$ by definition. According to the Brauer's theorem [17],[18] we can simply estimate the bounds for spectral radius

$$\begin{aligned} s + m(h-1) \leq \rho(C) &\leq S - m\left(1 - \frac{1}{g}\right) \\ g &= \frac{S - 2m + \sqrt{S^2 - 4m(S-s)}}{2(2(s-m))} \\ h &= \frac{-s + 2m + \sqrt{S^2 - 4m(S-s)}}{2m} \end{aligned} \tag{19.14}$$

where

$$m = \min_{i,j} c_{ij}$$

$$s = \min_i S_i, \quad S = \max_i S_i, \quad S_i = \sum_{j=1}^{n} c_{ij}$$

## 19.5 Numerical Experiments

To illustrate properties of studied approach we consider two numerical examples.

EXAMPLE 2. For the introductory circuit of Fig. 19.1 we shall check stability using the foregoing procedure. Matrices of centers and radiuses of the interval entries of matrix $A^I = A(p)$ are following

$$A_c = \begin{bmatrix} -2.05 & -1.765 \\ -1.025 & -1.025 \end{bmatrix}, \Delta = \begin{bmatrix} 0.09 & 0.097 \\ 0.045 & 0.045 \end{bmatrix}$$

Associated symmetric center matrix $A_c'$ and radius matrix $\Delta'$ are

$$A_c' = \begin{bmatrix} -2.05 & -1.395 \\ -1.395 & -1.025 \end{bmatrix}, \Delta' = \begin{bmatrix} 0.09 & 0.071 \\ 0.071 & 0.045 \end{bmatrix}$$

Spectrum of $A_c'$ is equal $\rho(A_c') = (-3.023, -0.051)$ thus $A_c'$ is stable. Because $\rho\left(|A_c'^{-1}|\Delta'\right) = 0.486 < 1$ circuit of Fig.1 is stable.

EXAMPLE 3. Consider stability of the circuit of Fig. 19.2. The state-space equations are done as

$$\frac{d}{dt}\begin{bmatrix} i \\ u_1 \\ u_2 \end{bmatrix} = \begin{bmatrix} 0 & -\frac{1}{L} & 0 \\ -\frac{1-k}{C_1} & -\frac{1}{RC_1} & \frac{1}{RC_1} \\ \frac{k}{C_2} & \frac{1}{RC_2} & -\frac{1}{RC_2} \end{bmatrix} \begin{bmatrix} i \\ u_1 \\ u_2 \end{bmatrix}$$

For numerical data $R = [2, 2.5], L = 1, C1 = [0.5, 1], C2 = [1, 2], k = [0.2, 0.4]$ matrix $A(p) = A^I$ is following

$$\begin{bmatrix} 0 & -1 & 0 \\ [0.6, 1.6] & [-1, 0.4] & [0.4, 1] \\ [0.1, 0.4] & [0.2, 0.5] & [-0.5, -0.2] \end{bmatrix}$$

Center matrix $A_c$ and radius matrix $\Delta$ are

$$A_c = \begin{bmatrix} 0 & -1 & 0 \\ 1.1 & -0.7 & 0.7 \\ 0.25 & 0.35 & 0.35 \end{bmatrix}$$

$$\Delta = \begin{bmatrix} 0 & 0 & 0 \\ 0.5 & 0.3 & 0.3 \\ 0.15 & 0.15 & 0.15 \end{bmatrix}$$

For associated symmetric interval matrix $A_s^I$ we obtain

$$A_c' = \begin{bmatrix} 0 & 0.05 & 0.125 \\ 0.05 & -0.7 & 0.525 \\ 0.125 & 0.525 & -0.35 \end{bmatrix}$$

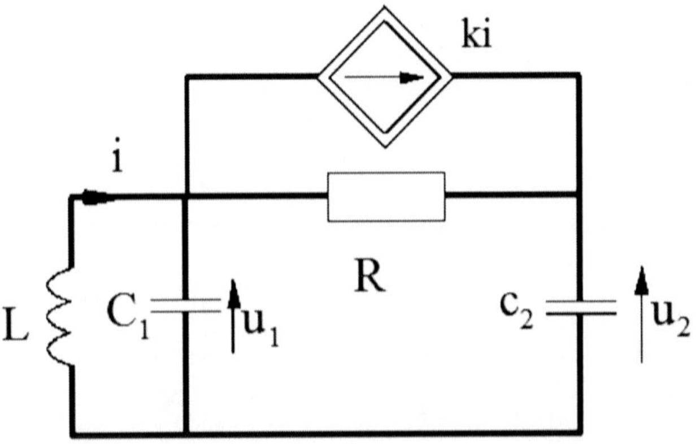


**Fig. 19.2** Linear time-invariant circuit for Example 3

$$\Delta' = \begin{bmatrix} 0 & 0.25 & 0.075 \\ 0.25 & 0.3 & 0.225 \\ 0.075 & 0.225 & 0.15 \end{bmatrix}$$

$A'_c$ is stable because spectrum is equal

$$\sigma(A'_c) = (-2.022, -0.252, -0.057)$$

Spectral radius $\rho\left(|A_c'^{-1}|\Delta'\right) = 0.847 < 1$ and circuit of Fig. 19.2 is stable. For another set of data $(R = [0.4, 0.5], L = 1, C_1 = 1, C_2 = [1,2], k = [0.2, 0.4])$ no conclusion concerning the stability or the instability of circuit was made.

## 19.6 Final Remarks

Checking stability of linear time-invariant systems with uncertain parameters is a key problem of system theory. It has received considerable attention in recent years especially after the seminal work of Kharitonov. In this chapter, we have presented a simple procedure for testing robust stability of linear circuits. It's seen from numerical experiments that procedure works effectively for some interval state matrix. Future studies should be focused on improving and tailoring foregoing procedure to the underlying robust stability structure.

## References

1. Alefeld, G., Herzberger, J.: Introduction to interval computations. Academic Press, New York (1983)
2. Argoun, M.B.: On sufficient conditions for the stability of interval matrices. Int. J. Control 44, 1245–1250 (1986)

3. Barmish, B.R., Kang, H.I.: A survey of extreme point results for robustness of control systems. Automatica 29, 13–35 (1993)
4. Barmish, B.R.: New tools for robustness of linear systems. Macmillan Publ. Comp., New York (1994)
5. Brauer, A.: The theorems of Ledermann and Ostrowski on positive matrices. Duke. Math. J. 24, 265–274 (1957)
6. Carotenuto, L., et al.: Computational methods to analyze the stability of interval matrices. IEE Proc. Control Theory Appl. 151, 669–674 (2004)
7. Chen, J.: Sufficient conditions on stability of interval matrices: connections and new results. IEEE Trans. Autom. Control 37, 541–544 (1992)
8. Franze, G., et al.: New inclusion criterion for the stability of interval matrices. IEE. Proc. Control Theory Appl. 153, 478–482 (2006)
9. Golub, G.H., Van Loan, C.F.: Matrix computations, 3rd edn. The Johns Hopkins University Press, Baltimore (1996)
10. Hertz, D.: The extreme eigenvalues and stability of real symmetric interval matrices. IEEE Trans. Autom. Control 37, 532–535 (1992)
11. Horn, R.A., Johnson, C.A.: Matrix analysis. Cambridge University Press, Cambridge (1985)
12. Kogan, J.: Robust stability and convexity. Springer, London (1995)
13. Minc, H.: Nonnegative matrices. Wiley, New York (1988)
14. Press, W.H., et al.: Numerical recipes - the art. of scientific computing, 3rd edn. Cambridge University Press, Cambridge (2007)
15. Rohn, J.: Positive definiteness and stability of interval matrices. SIAM J. Matrix Anal. Appl. 15, 175–184 (1994)
16. Yedavalli, R.K.: Stability of interval matrices: another sufficient condition. Int. J. Control 43, 767–772 (1986)
17. Xu, S.J., et al.: Robustness analysis of interval matrices based on Kharitonov's theorem. IEEE Trans. Autom. Control 43, 273–278 (1998)
18. Zeheb, E.: Conditions for Hurwitz stability of interval matrices. In: Proc. ISCAS 1994, pp. 125–127 (1994)

# Part V

# Further Application Area- Optimization, Data Mining, Pattern Recognition and Image Processing

# Chapter 20
# Data Reconciliation and Bias Estimation in On-Line Optimization

Moufid Mansour

**Abstract.** The reliability of measured data, which can be subject to both systematic and random errors, is of great importance for the monitoring and evaluation of process performance and the determination of control action. This Chapter presents and assesses bias estimation (as a type of systematic error) technique and data reconciliation methods for the detection, estimation and elimination of biases and random errors respectively. It is shown how these methods can be successfully employed within an on-line Integrated System Optimisation and Parameter Estimation (ISOPE) scheme for the determination of the process optimum, despite the existence of model-reality differences and measurement errors. The performance of the resulting scheme is demonstrated by application to a two tank CSTR system.

## 20.1 Introduction

In recent times, established techniques of Integrated System Optimization and Parameter Estimation (ISOPE) have been seen to be successfully applied in the on-line process optimization situation when model-reality differences exist (see for example Ellis et al., 1988 and Roberts and Williams, 1981). The ISOPE approach includes process measurements as part of the procedure and has been seen to perform well, and obtain the real process optimum, when employed with faithful measurements. However, in many practical situations, errors in process measurements may exist, and it is this setting of the ISOPE algorithm that is being considered. Measurements from process plants are seldom error free, as they are prone to contain both random and systematic errors. Systematic errors are caused by non-random events such as process leaks, biases in instrument measurements, malfunction of instruments and inadequate accounting of departures from steady state conditions. Random errors arise from chance occurrences and are generally normally distributed. In this

Moufid Mansour
Faculty of Electronics and Computer Science, University of Sciences and Technology BP. 32, El-Allia, Algiers, Algeria
e-mail: `M.Mansour@city.ac.uk`

K. Kyamakya et al. (Eds.): Selected Topics in Nonlinear Dynamics, SCI 459, pp. 361–372.
DOI: 10.1007/ 978-3-642-34560-9_20 

Chapter, we aim at investigating a technique for Bias Estimation (BE) to tackle biases as a type of systematic error and methods of Data Reconciliation (DR), to deal with random errors, and incorporate them within an ISOPE scheme.

## 20.2 Data Reconciliation

Data reconciliation (also called validation), allows state estimation and measurement correction problems to be addressed in a global way. The aim here is to obtain error-free values of available measurements, and also yield consistent and complete estimates of all the process state variables as well as unmeasured process parameters. Data reconciliation is based on measurement redundancy (Arora et al., 2002). By definition, a redundant measurement is a measurement which the value can be calculated based on other measurements. Data reconciliation uses measurement redundancies with the fact that at least some information about the process is known (a priory) and exploits it together with the relationships that exist between the measurements to correct them. The yielded measurements are accurate and more reliable. As a result, the reconciled values are of a lower variance compared to original raw measurements.

The main drawback in the data reconciliation problem is the incompleteness of the measurement sets. Usually, process variables cannot all be obtainable or available for measurement. This is due to different reasons; among them we mention cost considerations and technical unfeasibility. Therefore, we sometimes proceed to estimating the value of some unmeasured variables by way of mass, energy and component balances (Mah et al., 1976).

The general assumptions lying behind steady-state data reconciliation methods and their software implementations are given as follows (Kim et al., 1997):

1. The process is stationary: The system is at steady-state.
2. The measurement error is Gaussian with zero mean value, and known variances. This signifies that the measurements are not affected by any gross error.

Therefore, the data reconciliation problem can be stated as follows (Bagajewicz, 2003): Given a set of measurement values of a subset state variables, it is desired to obtain the best estimators of these measured state variables and as many unmeasured variables as possible.

### 20.2.1 *Types of Errors*

There are two types of experimental errors: systematic errors and random errors Systematic errors affect the accuracy of the measurement. In fact, in the absence of other types of errors, repeated measurements on the same variable yield the same result that differs from the true or accepted value by the same amount. Common sources of systematic errors are faulty calibration of measuring instruments, biases in instrument measurements, inadequate accounting of departures from steady-state operations and/or inaccurate process models. On the other hand, random errors are errors that affect the precision of the measurement. In the absence of other types of errors, repeated measurements on the same variable yield a result that fluctuates around the true or accepted value. Random errors come from the randomness of the measurements, such a process noise and they are normally distributed.

### 20.2.2 Brief History

Data reconciliation has been around for several years now. It has been used as a mean for obtaining accurate and reliable data in process plants. The earliest work reported in the literature is probably that of Kuehn and Davison (1961), where the authors presented a formulation of the data reconciliation problem and a method based on Lagrange Multipliers in order to solve the steady-state data reconciliation problem. Gelb (1974) used Kalman Filtering successfully to recursively smooth measurement data and estimate parameters In dynamic cases. However, both steady-state and dynamic studies were developed for linear systems only. Therefore, modifications were made in order to handle a more general form: nonlinear systems. Knepper and Gorman (1980) proposed a method in which successive linearization of the system's nonlinear equations was used. It was based on the application of the analytical solution for the linearly constrained data reconciliation problem. In a comparison study, Jang et al. (1986) concluded that nonlinear programming gives better results in terms of response to changes in parameters and robustness in the presence of modelling errors and strong nonlinearities. In the same context, Liebman and Edgar (1988) demonstrated that nonlinear programming yield improved reconciliation estimates compared to successive linearization. In 1983, Crowe et al. (1983) proposed a method based on matrix projection to reconcile process flows. They used a Chi-square test based on the inverse of the reduced Hessian. In 1987, Narasimhan and Mah (1987) introduced their well known Generalise Likelihood Ratio (GLR) method for gross error detection. This method based on the likelihood ratio statistical test was capable of detecting, identifying, estimating and eliminating a wide variety of gross errors. In the same paper, they proposed a strategy for identifying multiple gross errors namely the Serial Compensation strategy. Based on the Chi-square test, a linear combination technique that identifies equivalent gross errors was derived by Rollins et al. (1996). A review of important results for gross error detection prior to 1996 is available in Crowe (1996), for steady-state systems and Albuquerque and Kramer (1995) for dynamic systems.

Based on a bivariate distribution function constructed using the maximum likelihood principle, Tjoa and Biegler (1991) presented a method for combined data reconciliation and gross error detection applied to steady-state processes. Because the assumption that all data reconciliation algorithms are based on, is that the measurements are taken at steady-state, and as processes are generally never in a steady-state, means not only random errors but also process variations are averaged with good measurements. This issue was addressed in many publications (Narasimhan, 1984, Holly et al., 1989, and Abu-el-zeet et al., 2000). In the year 2000, Narasimhan and Jordache (2000) published a book, which provides a systematic and comprehensive treatment of data reconciliation and gross error detection techniques.

The literature is rich with excellent review papers: Mah (1982); Tamhane and Mah (1985); Mah (1990); Madron (1992); and Crowe (1996).

### *20.2.3 The Benefits of Data Reconciliation*

The benefits gained from data reconciliation in the chemical and process industry are numerous. We mention (Arora et al., 2002):

- Improvement of measurement layout.
- Fewer routine analyses.
- Reduced frequency of sensor calibration (only faulty sensors need to be calibrated).
- Removal of systematic measurement errors.
- Systematic improvement of process data.
- A clear picture of plant operating condition.
- Reduced measurement noise for key variables.

Moreover, monitoring through data reconciliation leads to early detection of sensor deviation and equipment performance degradation, actual plant balances for accounting and performance follow-up, safe operation closer to the process limits and improved quality and performance at the process level.

### *20.2.4 Recent Developments and Software Packages*

People both from academia and industry, are being attracted to the area of data reconciliation. Hundreds of articles have been published, several books have been written and a couple of industrial software packages exist at the present moment (Bagajewicz, 2003).

Recent developments in the field aim at combining online data acquisition with data reconciliation, where reconciled data are displayed in control rooms in parallel with raw measurements. Departure between reconciled and measured data can trigger alarms and analysis of time variation of those corrections can draw attention to drifting sensors that need recalibration (Arora et al., 2002).

Amongst the software packages developed to date, we note: PRECISE, from Ok-Solutions, VALI, which is a data reconciliation and data validation software available from BELSIM s.a. We also mention RAGE, which is a software package for data reconciliation and gross error detection developed by the Chemical Engineering Department (IIT Madras).

In the next subsection we present the mathematical structure of the steady-state data reconciliation problem and the different means used to solve it. After that, bias estimation technique is given to demonstrate one type of systematic errors. A comprehensive case study that illustrates the use of the above tools is provided in

Section 5. The case study demonstrates the treatment of random errors as well as biases using the appropriate techniques on a two CSTR system.

### 20.2.5 Formulation of the Data Reconciliation Problem

Data Reconciliation (DR) is a necessary operation for obtaining accurate and consistent data in process plants by forcing them to obey natural laws such as material and energy balances so that, ultimately, the material and, if considered, the energy balances are satisfied exactly (Abu-el-Zeet et al., 2000). Generally speaking, the data reconciliation problem can be formulated as a constrained optimization problem. That is, as a least squares estimation problem if the measurements contain random errors only. This will be the case here as any prior biases have already been removed. Let $\varepsilon$ be a vector of random measurement errors:

$$\varepsilon = y_m - y_{true} \tag{20.1}$$

where $y_m$ is the vector of measured process variables, and $y_{true}$ denotes the vector of true values of measured variables. If these errors are considered to be normally distributed with zero mean, and a covariance matrix, $V$ , the data reconciliation problem can be defined as a least squares estimation problem:

$$\begin{aligned} &Minimise : F(y_m, y_{true}) \\ F(y_m, y_{true}) &= \frac{1}{2}(y_m, y_{true})^T V^{-1} (y_m, y_{true}) \\ &subject\ to : h(y_{true}) = 0 \end{aligned} \tag{20.2}$$

where $h$ is a set of algebraic equality constraint equations, and $V$ is the variance-covariance matrix, where each element $V_{ii}$ is $\sigma_i^2 (i = 1, m)$, and is assumed to be the same for all data sets.

The above problem can be solved using several approaches.

#### 20.2.5.1 Nonlinear Programming (NLP)

The problem of equation (20.2) can for instance be solved by any nonlinear programming technique. Often Sequential Quadratic Programming (SQP) is used as it requires the fewest function evaluations. In this case, upper and lower bounds on the measured variables are added, so problem (20.2) can be more generalized. These upper and lower bounds can be considered as an extra inequality constraint.

#### 20.2.5.2 Quadratic Programming

In the special case where the equality constraint equations are linear, or can be linearized if they are almost linear, problem (20.2) can be reduced to an unconstrained Quadratic Programming problem (QP) that can be solved analytically. We write then,

$$h(y_{true}) = Ay_{true} = 0 \tag{20.3}$$

where A is the Jacobian of the constraint equations and the solution is obtained by the use of Lagrange multipliers and is given by (Abu-el-zeet, 2000):

$$y_{true} = y_m - VA^T(AVA^T)^{-11}\delta \tag{20.4}$$

where $\delta$ is the residual of the unsatisfied balances and is given by:

$$\delta = A\varepsilon = Ay_m \tag{20.5}$$

#### 20.2.5.3 Successive Linearization

A shortcoming of the linear solution is that the solution does not necessarily satisfy the non-linear constraints. In successive linearization, the linear problem is iterated until an optimal point is obtained satisfying the non-linear constraints. As in the linear solution method, the advantage of successive linearization is its relative simplicity and fast calculation.

## 20.3 Bias Estimation

When both random and biases are present on process measurements, bias estimation techniques are firstly applied to the measurements to eliminate, or reduce, the non-random errors (Narasimhan and Jordache, 2000).

In the special case where the locations of the biased variables are known a priori, bias can be estimated as a parameter (McBrayer and Edgar, 1995). This methodology is appropriate here and the procedure is to solve the following non-linear programming (NLP) problem:

$$\begin{aligned} &Min : J(\bar{y},\hat{b}) \\ &subject\ to : \\ &f(\bar{y}) = 0 \end{aligned} \tag{20.6}$$

$$\begin{aligned} &\bar{y}_{l,i} \le \bar{y}_i \le \bar{y}_{u,i}\ \forall i, \\ &\hat{b}_{l,i} \le \hat{b}_i \le \hat{b}_{u,i}\ \forall i, \end{aligned} \tag{20.7}$$

Where:

$$J(\bar{y},\hat{b}) = \left(\frac{\bar{y}_1 - \left(y_{m1} - \hat{b}_1\right)}{\sigma_1}\right)^2 + \left(\frac{\bar{y}_2 - \left(y_{m2} - \hat{b}_2\right)}{\sigma_2}\right)^2 + \left(\frac{\bar{y}_3 - \left(y_{m3} - \hat{b}_3\right)}{\sigma_3}\right)^2 \tag{20.8}$$

where $y_{mi}$ is the $i^{th}$ measured variable, $\bar{y}_i$ is the $i^{th}$ estimate, $\sigma_i$ is the measurement noise standard deviation of the $i^{th}$ measured variable and $\hat{b}_i$ is the estimate of bias on the $i^{th}$ measured variable. It should be noted that the bias, $\hat{b}_i$ , is also included in the inequality constraints. This allows for physical limits on the range of admissible biases.

## 20.4 ISOPE and the Inclusion of Data Reconciliation and Bias Estimation

The ISOPE algorithm is a model-based system optimization technique that was developed to overcome model-reality differences and generate the system optimum (Roberts, 1979). The basic form of ISOPE is discussed here in connection with BE and DR but extended versions can readily be employed if the situation demands (Ellis et al., 1988).

As process measurements are being used within the algorithm, there are, inevitably, difficulties in that measurements are likely to be contaminated by various types of errors. By applying BE and DR techniques to the measurements, it would be hoped that the performance of the ISOPE algorithm could be improved.

The ISOPE algorithm addresses the general non-linear programming problem, where * refers to the real process (Mansour and Ellis, 2003):

$$\underset{c}{Min} Q(c, y_*) \tag{20.9}$$

subject to:

$$y_* = F_*(c) \tag{20.10}$$

$$g(y_*) \leq 0 \tag{20.11}$$

$$c_{min} \leq c \leq c_{max} \tag{20.12}$$

where, $c$ and $y_*$ are the controls and outputs, respectively, of the process. The general form of the ISOPE algorithm, with error free output measurements can be seen in Ellis et al., (1988) and Roberts, (1979).

With the inclusion of Bias Estimation and DR, the ISOPE algorithm takes the following form:

1. Apply the current control, $c_k$, to the real process and, after an appropriate settling time, obtain steady-state measurements, $y_{*k}$. Where, $k$ is the iteration and $y_{*k}$ is the error burdened output measurement.
2. Apply BE and DR techniques as required to the measured outputs, $y_{*k}$, to yield the error free outputs, $y_{*k}$.
3. The process model is given by:

$$y = F(c, \alpha) \tag{20.13}$$

where, $\alpha$ are free model parameters. Assuming that measurements of all out-puts are available, (20.10) and (20.13) can be used in a simple estimation procedure to determine the model parameters, $\alpha$:

$$y = F(c,\alpha) = F_*(c) \tag{20.14}$$

This estimation procedure also has the benefit of satisfying one of the necessary system optimality conditions (Ellis et al., 1988).

4. Solve the modified model-based optimization problem given by:

$$\begin{aligned} &\underset{c}{Min} Q(c,\alpha) - \lambda^T c \\ &y = F(c,\alpha) \\ &g(c,\alpha) \leq 0 \end{aligned} \tag{20.15}$$

Where,

$$\lambda = \left[\frac{\partial^T F}{\partial c} - \frac{dF_*}{dc}\right]\left[\frac{\partial^T F}{\partial \alpha}\right]\left[\frac{\partial Q}{\partial \alpha}\right] \tag{20.16}$$

$\lambda$ is termed a modifier and arises from the necessary optimality conditions, of the system optimization problem (Ellis et al., 1988, Roberts and Williams, 1981, Rob-erts, 1979).

5. In order to control convergence of the algorithm, the new control $\hat{c}_k$ , obtained from the model-based problem of (20.15), is not directly applied to the system. Instead, the following under-relaxation scheme is used to provide updated controls, $\hat{c}_{k+1}$ , for the process :

$$c_{k+1} = c_k + K(\hat{c}_k - c_k) \tag{20.17}$$

where $K$ is a relaxation gain matrix, and governs the actual changes made to the real process inputs from one iteration to another. Its purpose is to ensure that excessive alterations are not made.

The above steps are repeated until satisfactory convergence is obtained. Convergence occurs when no further improvement is observed. In other words, when the new control is no longer a better candidate than the previous one.

## 20.5 Application to a Continuous Stirred Tank Reactor System

The ISOPE algorithm, using the Bias Estimation (BE) and Data Reconciliation (DR) schemes, is now applied, under simulation, to a Continuous Stirred Tank Reactors (CSTR) system (Garcia and Morari, 1981) which has two tanks connected in cascade (Figure 20.1).

An exothermic autocatalytic reaction takes place in the reactors with interaction taking place in the units in both directions due to a recycle of 50% of the product stream into the first reactor. Regulatory controllers are used to control the temperature in both reactors.

The reaction is:

$$A + B \underset{k-}{\overset{k+}{\rightleftharpoons}} 2B \tag{20.18}$$

The system has four outputs which are the concentrations of the two components $A$ and $B$ in the two tanks, i.e.: $y = (C_{a1}, C_{b1}, C_{a2}, C_{b2}$ . In our example, the concentrations of species $B$ in both tanks $C_{b1}$ and $C_{b2}$ are to be monitored for steady-state identification. Temperatures in the two tanks, $T_1$ and $T_2$ are the set-points.

The simulations were started from the same initial operating point given by $T_1 = 307K$ and $T_2 = 302K$ yielding the following steady-state output values of the concentration of product $B$ in the two tanks 1 and 2, $C_{b1}(0) = 0.0516[kmol/m^3]$ , and $C_{b1}(0) = 0.0586[kmol/m^3]$ .

Measurement noise was simulated as normally distributed with zero mean. The value of the variance-covariance matrix was chosen to be:

$$V = \begin{bmatrix} \sigma_1^2 & 0 \\ 0 & \sigma_2^2 \end{bmatrix} \tag{20.19}$$

where $\sigma_1$ is the standard deviation for the variable $C_{b1}$ and was chosen to be 5% of the nominal value, $\sigma_2$ is the standard deviation for $C_{b2}$ and was of a value of 5%. These values were chosen as they represent typical values in many realistic situations.

The aim here is to maximize the concentration of component $B$ in tank 2, giving the objective function, (20.9), as:

$$Q(c, y_*) = -C_{b2} \tag{20.20}$$

The controls, at the real process optimum, obtained directly from the real process equations, are: $T_1 = 312K$ and $T_2 = 310.2K$ With corresponding output concentration measurements $C_{b1}(0) = 0.0644[kmol/m^3]$ , and $C_{b1}(0) = 0.0725[kmol/m^3]$ The model adopted is of the approximate linear form [1]:

$$V = \begin{bmatrix} C_{b1} \\ C_{b2} \end{bmatrix} = \begin{bmatrix} a_{11} & a_{12} \\ a_{21} & a_{22} \end{bmatrix} \begin{bmatrix} T_1 \\ T_2 \end{bmatrix} + \begin{bmatrix} \alpha_1 \\ \alpha_2 \end{bmatrix} \tag{20.21}$$

where $\alpha_1$ and $\alpha_2$ are the free model parameters to be estimated and $a_{ij} i, j = 1,2$ are also model parameters which are assigned the values of the process derivatives. The real process derivatives, $df/dc$ , may be estimated by such techniques as applying control perturbations or, as is the case here, by the use of Broydon's method (Fletcher, 1980). The estimation of the process derivatives may be made by such techniques for the calculation of the modifier, $\lambda$, in (20.16). The procedure is implemented, using a MATLAB/SIMULINK software plat-form, following the steps described in Section 20.4. Initially, to illustrate the performance of the ISOPE algorithm, without any BE or DR being implemented, both measurements were subject to 5% additive noise. Figures 20.2(a) shows the measurements, while Figure 20.2(b) shows the controls not converging to the correct real optimum. This is due to flawed data measurements producing erroneous model parameters. The sample extreme case of gross error, represented here is in the form of measurement biases. With noise present on the measurements as well, it is observed with BE and DR applied, as described in Sections 20.2 and 20.3, to the real process measurements. Figure

20.3(a) shows, due to the introduction of BE and DR, error free measurements being obtained. Having error free measurements available, the ISOPE algorithm is now able to function correctly, as can be seen in Figure 20.3(b), where the controls converge the correct real process optimum. Figures 20.3(a) and 20.3(b) demonstrate the effectiveness of the BE and DR procedures to enable the ISOPE algorithm to perform successfully. Less extreme situations, such as when only noise is present on the measurements, have also been seen to be handled satisfactorily by the BE and DR procedures [10].

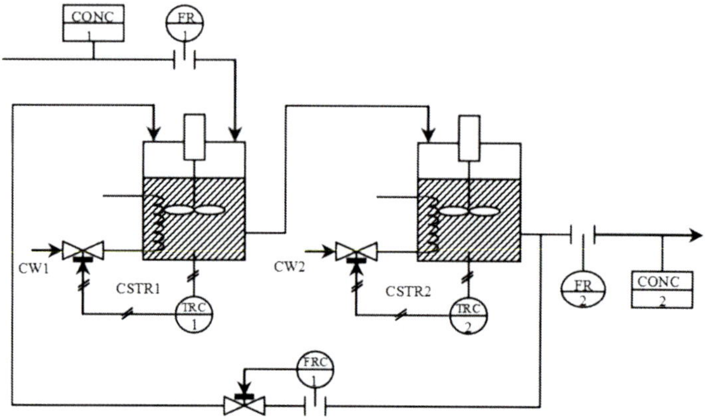


**Fig. 20.1** Continuous Stirred Tank Reactor System

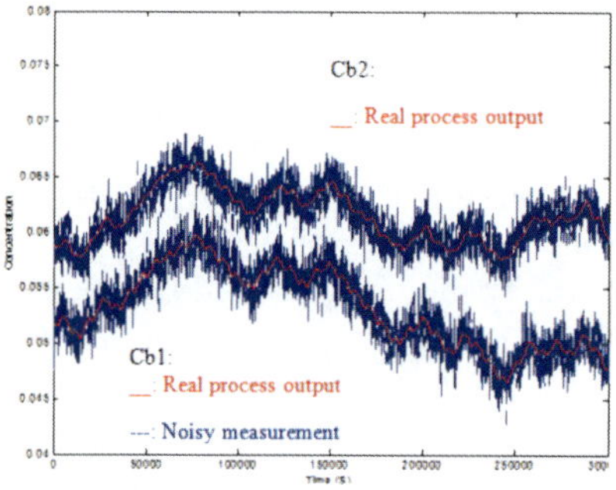


(a) ISOPE Process Outputs and Measurements

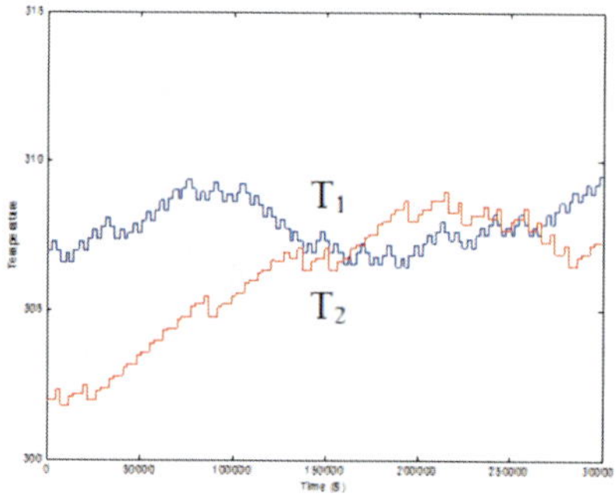


(b) ISOPE Control Trajectories

**Fig. 20.2** Noise applied without BE and DR

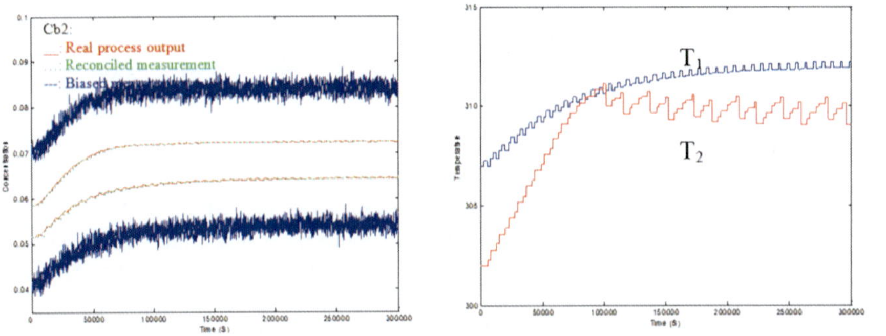


(a) ISOPE Process Outputs and Measurements with Implementation of BE and DR

(b) ISOPE Control Trajectories with Implementation of BE and DR

**Fig. 20.3** Multiple Biases and Noise Present

## 20.6 Conclusion

In the on-line process optimization problem, when measurements are subject to gross errors and/or noise, it has been seen how techniques of Bias Estimation (BE) and Data Reconciliation (DR) can be employed in conjunction with an algorithm for Integrated System Optimisation and Parameter Estimation (ISOPE) to enable the real process optimum to be found. This is in the situation when there also exists model-reality differences.

These techniques have been demonstrated for the on-line optimization of a two tank CSTR system. Despite the presence of multiple biases and noise on the process measurements, together with an unfaithful model, the real process optimum was seen to be achieved.

Thus, these techniques of BE and DR are therefore eminently suitable for the often encountered on-line optimization situation, when measurements are subject to errors, yet still enable the process optimum to be achieved.

## References

1. Abu-el-Zeet, Z.H.: Optimisation Techniques for Advanced Process Supervision and Control. PhD Thesis, City University, London, EC1V 0HB, U.K (2000)
2. Abu-el-Zeet, Z.H., Becerra, V.M., Roberts, P.D.: Data Reconciliation and Steady-State Detection Applied to a Chemical Process. In: UKACC International Conference (Control 2000), September 4-7, University of Cambridge, Cambridge (2000)
3. Albuquerque, J.S., Biegler, L.T.: Data Reconciliation and Gross-Error Detection for Dynamic Systems. AIChE Journal 42(10) (1996)
4. Arora, N., Biegler, L.T., Heyen, G.: Optimization Formulations for Data Reconciliation. Computer Aided Process Engineering (2002)
5. Bagajewicz, M.J.: Data Reconciliation and Instrumentation Upgrade. Overview and Challenges. In: FOCAPO (Foundations of Computer Aided Process Operations), Coral Springs, FL, USA (2003)

6. Crowe, C.M.: Data reconciliation-progress and challenges. Journal of Process Control 6(2/3), 89–98 (1996)
7. Ellis, J.E., Kambhampati, C., Sheng, G., Roberts, P.D.: Approaches to the Optimizing Control Problem. Int. J. of Systems Science 19, 1969–1985 (1988)
8. Fletcher, R.: Practical Methods of Optimization. Wiley Interscience (1980)
9. Garcia, C.E., Morari, M.: Optimal Operation of Integrated processing Systems. AICHE Journal 27, 960–968 (1981)
10. Gelb, A.: Applied Optimal Estimation. MIT Press, Cambridge (1974)
11. Jang, S.S., Joseph, B., Mukai, H.: Comparison of two approaches to on-line parameter and state estimation of nonlinear systems. Ind. Engng. Chem. Process. Des. Dev. 25, 809–814 (1986)
12. Kim, I.W., Kang, M.S., Park, S., Edgar, T.F.: Robust Data Reconciliation and Gross Error Detection: The Modified MIMT Using NLP. Computers and Chemical Engineering 21(7), 775–782 (1997)
13. Liebman, M.J., Edgar, T.F.: Data reconciliation for nonlinear processes, Paper Presented at the AIChE Annual Meeting, Washington, DC (1988)
14. Madron, F.: Process plant performance: Measurement and data processing for optimization and retrofits. Ellis Horwood, Chichester (1992)
15. Mah, R.S.H.: Design and Analysis of Process Performance Monitoring Systems. In: Edgar, T.F., Seborg, D.E. (eds.) Chemical Process Control II, Engineering Foundation, New York, p. 525 (1982); Proceedings of the Engineering Foundation Conference, Sea Island, Georgia, January 18-23 (1981)
16. Mah, R.S.H.: Chemical Process Structures and Information Flows. ch. 8,9. Butterworths, Stoneham (1990)
17. Mah, R.S.H., Stanley, G.M., Downing, D.M.: Reconciliation and rectification of process flow and inventory data. Ind. Engng Chem. Process. Des. Dev. 15, 175–183 (1976)
18. Mah, R.S.H., Tamhane, A.C.: Detection of Gross Errors in Process Data. AIChE Journal 28, 828–830 (1982)
19. McBrayer, K.F., Edgar, T.F.: Bias Detection and Estimation in Dynamic Data Reconciliation. Journal of Process Control 5, 285–289 (1995)
20. Mansour, M., Ellis, J.E.: Comparison of Methods for Estimating Real Process Derivatives in On-line Optimization. Applied Mathematical Modelling 27, 275–291 (2003)
21. Mansour, M., Ellis, J.E.: Methodology of On-line Optimisation applied to a Chemical Reactor. Applied Mathematical Modelling 32, 170–184 (2008)
22. Narasimhan, S., Mah, R.S.H.: Generalized Likelihood Ratio Method for Gross Error Identification. AIChE Journal 33(9), 1514–1521 (1987)
23. Narasimhan, S., Jordache, C.: Data Reconciliation and Gross Error Detection: An Intelligent Use of Process Data. Gulf Publishing Company (2000)
24. Roberts, P.D.: An Algorithm for Steady-State System Optimisation and Parameter Estimation. Int. J. of Systems Science 10, 719–734 (1979)
25. Roberts, P.D., Williams, T.W.C.: On an Algorithm for Combined System Optimisation and Parameter Estimation. Automatica 17, 199–209 (1981)
26. Rollings, D.K., Cheng, Y., Devanathan, S.: Intelligent Selection of Hypothesis tests to Enhance Gross Error Identification. Comp. and Chem. Eng. 20(5), 517–530 (1996)
27. Tamhane, A.C., Mah, R.S.H.: Data Reconciliation and Gross Error Detection in Chemical Process Networks. Technometrics 27(4), 409–422 (1985)
28. Tjoa, I.B., Biegler, L.T.: Simultaneous Strategies for Data Reconciliation and Gross Error Detection of Nonlinear Systems. Computers and Chemical Engineering 15(10), 679–690 (1991)

# Chapter 21
# Image Edge Detection and Orientation Selection with Coupled Nonlinear Excitable Elements

Atsushi Nomura, Yoshiki Mizukami, Koichi Okada, and Makoto Ichikawa

**Abstract.** This chapter presents an image-processing algorithm for edge detection and orientation selection with discretely coupled nonlinear elements. The algorithm utilizes the nonlinear characteristic of the FitzHugh-Nagumo model and arranges the elements on an image grid system. The model is described with a pair of ordinary differential equations with activator and inhibitor variables, and exhibits mono-stable excitability. It was previously found that a grid system consisting of mono-stable nonlinear elements self-organizes pulses at crossing points between an initial activator distribution and a threshold level. In particular, the imposition of strong inhibitory coupling on the grid system causes stationary pulses at the crossing points. The algorithm presented here focuses on the phenomenon in which the grid system self-organizes stationary pulses at the crossing points. In addition, the algorithm introduces anisotropic coupling strength into the grid system; the coupling strength is decided according to the difference between the gradient direction of the inhibitor distribution and the specific orientation. An experimental section demonstrates the results of edge detection and orientation selection for artificial and real images.

## 21.1 Introduction

Nonlinear elements are common in nature and exhibit interesting temporal behavior such as nonlinear excitation and oscillation [1]. Very classical researches done

Atsushi Nomura
Faculty of Education, Yamaguchi University, Japan
e-mail: anomura@yamaguchi-u.ac.jp

Yoshiki Mizukami
Graduate School of Science and Technology, Yamaguchi University, Japan

Koichi Okada
Center for the Promotion of Higher Education, Yamaguchi University, Japan

Makoto Ichikawa
Faculty of Letters, Chiba University, Japan

K. Kyamakya et al. (Eds.): Selected Topics in Nonlinear Dynamics, SCI 459, pp. 373–392.
DOI: 10.1007/ 978-3-642-34560-9_21 

in physiology showed that a nerve axon responds to an external stimulus and its state traces a nonlinear process of excitation and inhibition [2]. A chemical reaction system can also show nonlinear excitation or oscillation processes [3].

A system consisting of coupled nonlinear elements exhibits further interesting phenomena, such as self-organized pattern-formation processes. A reaction-diffusion system refers to a system of diffusely or continuously coupled nonlinear elements [1]. Some reaction-diffusion systems self-organize spiral waves and target waves traveling in space while others self-organize stationary periodic waves from initial uniform distribution.

Several researchers have found that a reaction-diffusion system has functions of information processing in addition to the function of information transmission. In a chemical reaction-diffusion system, an external stimulus initiates a target wave. The wave travels in space and reaches another point in space. If we artificially stimulate a point in the space of the reaction-diffusion system, and if we observe another point, we can find the traveling wave passing the observation point after a finite duration of time. This means that an information transmission system with traveling waves is established between the two points [4]. Another chemical reaction-diffusion system self-organizes waves in an area or along edges of an image-brightness distribution projected onto the system [5]. Thus, we can understand that a reaction-diffusion system also has the image-processing functions of edge detection and segmentation.

Image-processing functions can also be provided with a grid system of discretely coupled nonlinear excitable or oscillatory elements. A typical nonlinear element has activator and inhibitor variables; the activator variable excites the state of the element and the inhibitor variable inhibits the element from exciting. A grid system named LEGION (Locally Excitatory Globally Inhibitory coupled Oscillatory Network) performs image segmentation [6, 7]. We previously found that a grid system of nonlinear excitable elements with strong inhibitory coupling could also perform edge detection and segmentation [8–10].

This chapter presents an algorithm for edge detection and orientation selection with a two-dimensional grid system of discretely coupled nonlinear excitable elements. According to our previous findings [8–10], the algorithm utilizes a grid system consisting of nonlinear excitable elements. When a particular element has isotropic, strong inhibitory coupling to its neighboring elements, the system performs edge detection. When a particular element has anisotropic inhibitory coupling to its neighboring elements, the system performs orientation selection. A reaction-diffusion model with anisotropic inhibitory coupling originally demonstrated oriented periodic pattern formation in a biological system [11]. The algorithm presented here is obtained by replacing the anisotropic diffusion of the reaction-diffusion model with anisotropic discrete coupling. An experimental section shows that the algorithm indeed works for edge detection and orientation selection for artificial binary images and real gray-level images. In addition, the section demonstrates quantitative performance of the algorithm for noisy artificial binary images, in comparison with two representative existing algorithms proposed by Marr and Hildreth [12] and by Canny [13].

## 21.2 Background

### *21.2.1 Coupled Nonlinear Elements*

Biological systems perform information transmission with pulses stably propagating in space [1]. For example, a nerve axon has a mechanism of transmitting information from one point to another point with pulses. Hodgkin and Huxley [2] examined the mechanism with physiological experiments, and presented a mathematical model describing the response of the axon to an external stimulus. In addition, they derived a partial differential equation describing pulse transmission along the axon.

FitzHugh [14] derived a simplified model so as to qualitatively simulate behavior of the Hodgkin-Huxley model. On the one hand, the Hodgkin-Huxley model has four variables and is highly complex; on the other, the FitzHugh model has only two variables of activator and inhibitor. Thus, he simplified the Hodgkin-Huxley model by retaining its qualitative characteristics of excitation and inhibition. Nagumo et al. [15] almost simultaneously presented a similar simplified model, and implemented a circuit system simulating pulse transmission along a nerve axon. Nowadays, the FitzHugh-Nagumo model refers to that derived by FitzHugh [14] and also by Nagumo et al. [15].

A system of coupled nonlinear elements such as those modeled by the FitzHugh-Nagumo model and the Hodgkin-Huxley model causes stable pulses and their robust propagation. It is difficult to explain such a robust pulse transmission phenomenon with linear elements. There are many types of nonlinearity depending on biological systems and their functions. Among them, the FitzHugh-Nagumo model in particular has been studied with theoretical and numerical analyses.

A system of diffusely coupled nonlinear elements becomes a reaction-diffusion system, which also brings a wide variety of pattern-formation phenomena such as spiral and target waves propagating in two- or three-dimensional space. The Belousov-Zhabotinsky reaction system self-organizes spiral and target waves [3], and a biological amoeba system also self-organizes such waves for transmitting information about environmental changes [16]. The reaction-diffusion mechanism of diffusely coupled nonlinear elements causes these pattern-formation phenomena.

Some of reaction-diffusion systems with strong inhibitory diffusion generate localized and stationary periodic patterns, which simulate morphogenesis in hydras and patterns of markings on fish skin. Turing was the first to propose a model of reaction-diffusion and its condition for explaining morphogenesis [17]; later, Gierer and Meinhardt proposed more realistic models [18]. Besides these early theoretical results, laboratory experiments have shown that several real chemical and biological systems self-organize such localized and stationary periodic patterns [19,20].

Besides the chemical and biological systems stated above, there exist artificial systems such as circuit systems exhibiting nonlinear excitation and oscillation. Chua and Matsumoto proposed a circuit system exhibiting nonlinearity [21]; the system is now called the Chua circuit. Later, the Chua circuit was combined with the cellular neural network approach; Chua also proposed the cellular neural network approach as a framework implementing discretely coupled elements on a circuit system [22]. Thus, we can now implement reaction-diffusion systems including the FitzHugh-Nagumo type not only with numerical computation, but also with circuit systems by combining the Chua circuit with the cellular neural network approach. Several Chua circuit systems have successfully demonstrated traveling and stationary waves.

### *21.2.2 Edge Detection*

An edge refers to a position having rapid brightness change on image space. Marr and Hildreth proposed an edge-detection algorithm that located zero-crossing points in second derivatives of image brightness distributions [12]. Their algorithm is composed of a Gaussian filter for noise reduction and a Laplacian filter for the second derivatives. The brightness distribution of a strongly blurred image intersects with a weakly blurred one at its inflection point. Thus, the algorithm has another version composed of two Gaussian filters with different spatial spreads, instead of the Laplacian-of-the-Gaussian filter.

The application of a Gaussian filter to image brightness distribution results in a blurred image, which is also the case with a diffusion equation. The solution of the diffusion equation becomes a convolution of the Gaussian function and its initial condition [23]. If the diffusion equation has the initial condition of image brightness distribution, its solution becomes the same as the output of the Gaussian filter applied to the image brightness distribution. Thus, we can utilize the diffusion equation instead of the Gaussian filter in image processing.

By accepting the application of diffusion equations to image processing, several researchers have come to propose applying anisotropic diffusion equations to edge detection [24]. Since an isotropic diffusion equation blurs image brightness distribution almost everywhere, it tends to remove meaningful image structures, such as corner points required for later processes of image understanding and recognition. By modulating its diffusion coefficient adaptively for original image brightness distribution, the researchers have tried to solve the difficulty of preserving meaningful image structures.

The cellular neural network approach [22] was combined with edge detection algorithms utilizing diffusion equations, and therefore brought further practical applications in circuit systems. Isotropic and anisotropic diffusion equations were reformulated with templates of the cellular neural network approach [25].

Part of the research on image processing was originally motivated by biological vision. For example, the Mach bands effect, which is a kind of edge enhancement phenomenon in human and biological vision, has attracted much attention from practitioners of image processing as well as vision researchers. Barlow et al. presented a mathematical model for explaining the phenomenon, by modeling interactions among discretely coupled visual receptor units in the lateral eye of Limulus [26]. The important point in the model is the nonlinear and inhibitory interactions. In addition, it is known that a biological visual system has orientation selectivity. Several researchers have presented models for explaining the selectivity; one of them explains that some asymmetry causes orientation selectivity [27].

Motivated by previous research work on biological vision and nonlinear phenomena, several researchers have proposed image processing algorithms such as segmentation and edge detection. Wang and Terman [6] focused on the biological nonlinear response described with the FitzHugh-Nagumo model. They arranged the FitzHugh-Nagumo nonlinear oscillator elements on an image grid system, and coupled them with local excitation and global inhibition; the LEGION network refers to a grid system. We ourselves presented a grid system of discretely coupled FitzHugh-Nagumo nonlinear excitable elements, and reported that the grid system is applicable to edge detection and segmentation [8–10]. We imposed strong inhibitory coupling on the grid system; this is a key point for self-organizing stable pulses indicating edges.

This chapter presents an edge detection and orientation selection algorithm by employing our previous grid system consisting of the FitzHugh-Nagumo elements. The algorithm for edge detection is the same as that of our previous algorithm designed for binary image edge detection [8–10]. In addition, by introducing anisotropic inhibitory coupling into the grid system, we propose to perform orientation selection. The anisotropic coupling is motivated by a reaction-diffusion model simulating the oriented periodic patterns of markings on fish skin [11].

## 21.3 FitzHugh-Nagumo Elements on a Grid System

### 21.3.1 FitzHugh-Nagumo Element

The FitzHugh-Nagumo nonlinear element qualitatively simulates the temporal response of a nerve axon to a stimulus [14, 15]. A mathematical model explaining the response of the element is described with a pair of time-evolving ordinary differential equations with an activator variable $u(t)$ and an inhibitor variable $v(t)$, as follows:

$$\frac{\mathrm{d}u}{\mathrm{d}t} = \frac{1}{\epsilon}\{u(u-a)(1-u)-v\}, \tag{21.1}$$

$$\frac{\mathrm{d}v}{\mathrm{d}t} = u - bv, \tag{21.2}$$

in which $\mathrm{d}/\mathrm{d}t$ is a temporal derivative; $\epsilon$ is a positive small constant ($0 < \epsilon \ll 1$); $a$ is a constant and $b$ is a positive constant ($0 < b$). Equation (21.1) has cubic nonlinearity. The initial conditions for $u(t)$ and $v(t)$ are as follows:

$$u(0) = u_0, \quad v(0) = v_0. \tag{21.3}$$

Phase plane analysis (see eg Appendix 1 in [1]) visualizes the solution trajectory of the FitzHugh-Nagumo model described with Eqs. (21.1) and (21.2), as shown in Fig. 21.1. The phase plane is partitioned into four domains, depending on the positive or negative sign of $\mathrm{d}u/\mathrm{d}t$ and $\mathrm{d}v/\mathrm{d}t$. Let us consider an initial condition $(u,v) = (u_0,v_0)$ denoted by $\mathrm{P}_0$. Since the point $\mathrm{P}_0$ is located in the domain of

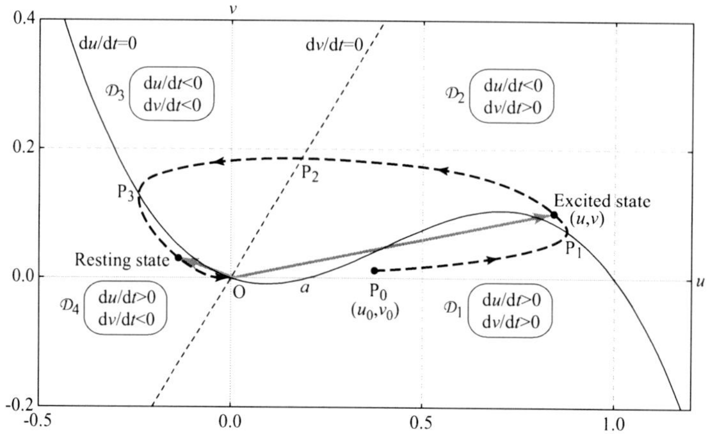


**Fig. 21.1** Phase plane for the FitzHugh-Nagumo model of Eqs. (21.1) and (21.2). Null-clines denoted by $\mathrm{d}u/\mathrm{d}t = 0$ and $\mathrm{d}v/\mathrm{d}t = 0$ partition the plane into four domains according to the combination of positive and negative signs of $\mathrm{d}u/\mathrm{d}t$ and $\mathrm{d}v/\mathrm{d}t$; $\mathcal{D}_1 = \{(u,v)|\mathrm{d}u/\mathrm{d}t > 0, \mathrm{d}v/\mathrm{d}t > 0\}$, $\mathcal{D}_2 = \{(u,v)|\mathrm{d}u/\mathrm{d}t < 0, \mathrm{d}v/\mathrm{d}t > 0\}$, $\mathcal{D}_3 = \{(u,v)|\mathrm{d}u/\mathrm{d}t < 0, \mathrm{d}v/\mathrm{d}t < 0\}$ and $\mathcal{D}_4 = \{(u,v)|\mathrm{d}u/\mathrm{d}t > 0, \mathrm{d}v/\mathrm{d}t < 0\}$. For example, an initial solution $(u,v) = (u_0, v_0)$ denoted by $\mathrm{P}_0$ traces the trajectory $\mathrm{P}_0 \rightarrow \mathrm{P}_1 \rightarrow \mathrm{P}_2 \rightarrow \mathrm{P}_3 \rightarrow \mathrm{O}$ as time proceeds. An excited state refers to areas having a large value $u$ around $\mathrm{P}_1$, and a resting state refers to areas having a small value $u$ around $\mathrm{P}_3$ and O.

$\mathrm{d}u/\mathrm{d}t > 0$ and $\mathrm{d}v/\mathrm{d}t > 0$, the solution $(u,v)$ traces the trajectory denoted by $\mathrm{P}_0 \rightarrow \mathrm{P}_1$ and enters an excited state having a large value of $u$. After the solution enters the area having $\mathrm{d}u/\mathrm{d}t < 0$ and $\mathrm{d}v/\mathrm{d}t > 0$, the variable $u(t)$ begins to decrease as time proceeds. Thus, the solution traces the trajectory along $\mathrm{P}_1 \rightarrow \mathrm{P}_2 \rightarrow \mathrm{P}_3$, and returns to the resting state having a small value of $u$. At the final stage, the solution converges to the stable steady state located at the origin O via the trajectory along $\mathrm{P}_3 \rightarrow \mathrm{O}$, as suggested by the linear stability analysis. Thus, the origin is the only globally stable steady state, in the case of Fig. 21.1. Let us consider a situation in which the model is in the stable steady state O in Fig. 21.1. If the model receives an external stimulus beyond the threshold level $a$, it again enters the excited state and traces the similar trajectory $\mathrm{P}_0 \rightarrow \mathrm{P}_1 \rightarrow \mathrm{P}_2 \rightarrow \mathrm{P}_3 \rightarrow \mathrm{O}$. Thus, this model describes what is called an excitable element.

When the solution $(u,v)$ traces the trajectory along $\mathrm{P}_0 \rightarrow \mathrm{P}_1$, the solution $u$ increases with acceleration. This is because Eq. (21.1) has the cubic nonlinearity with the large velocity $1/\varepsilon$. In contrast to this, after the element enters an excited state, the variable $v$ also becomes large, resulting in $\mathrm{d}u/\mathrm{d}t < 0$. A large value of the variable $v$ inhibits the variable $u$ from increasing. Thus, the variable $u$ is called an activator and the variable $v$ is called an inhibitor.

The behavior of the FitzHugh-Nagumo model depends on the settings of the parameters $a$ and $b$. If the model has one or two stable steady states, it becomes a mono-stable or bi-stable excitable system, respectively. In the bi-stable system an

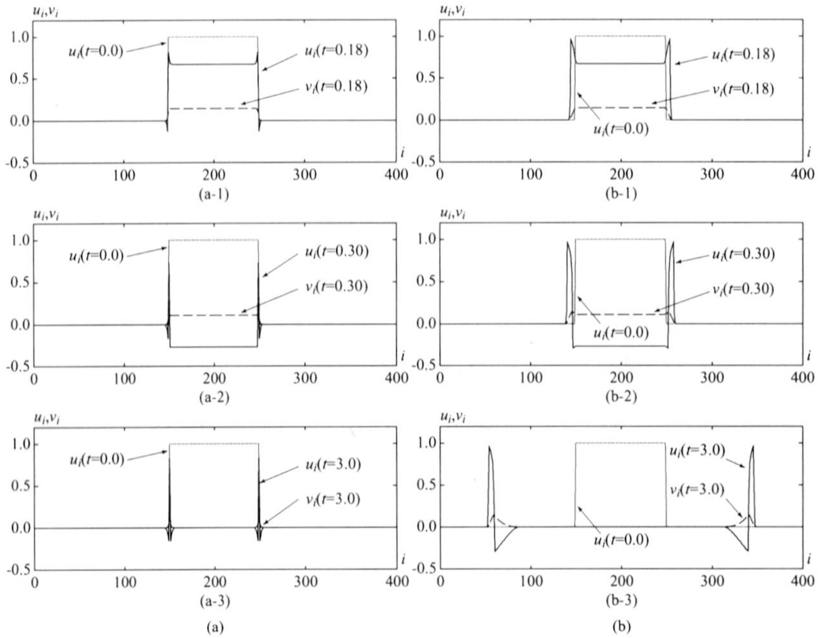


**Fig. 21.2** One-dimensional results for coupled elements described with Eqs. (21.4) and (21.5). The parameter settings were as follows: (a) $C_u = 2.0$ and $C_v = 6.0$ and (b) $C_u = 6.0$ and $C_v = 2.0$; other parameter settings were fixed at $a = 0.05, b = 1.0, \varepsilon = 1.0 \times 10^{-3}, \delta t = 1.0 \times 10^{-4}$ and $L = 400$; $\delta t$ is the finite difference for numerical computation of Eqs. (21.4) and (21.5). The horizontal axes denote the index number $i$ and the vertical axes do values of $u_i$ and $v_i$. Solid gray lines indicate the initial condition for $u_i$; the initial condition $v_i$ was zero for all of the elements. Figures (a-1) and (b-1) show the results obtained at $t = 0.18$, Figs. (a-2) and (b-2) show those at $t = 0.30$, and Figs. (a-3) and (b-3) show those at $t = 3.0$.

external stimulus triggers the switch of the two stable steady states. If the model has no stable steady state, it becomes an oscillatory system, in which the state of the system autonomously oscillates as time proceeds. A model having two stable steady states works for a bi-stable nonlinear excitable element and that exhibiting the oscillatory behavior works for a nonlinear oscillatory element. On the one hand, Wang and Terman utilized the FitzHugh-Nagumo type nonlinear oscillatory elements for image segmentation [6, 7]; on the other hand, we utilize nonlinear mono-stable excitable elements in this chapter.

### 21.3.2 Coupled Elements

Kurata et al. found that a pair of FitzHugh-Nagumo excitable elements coupled with strong inhibition self-organizes in a localized pattern [10]. Its one-dimensionally extended version is as follows:

$$\frac{du_i}{dt} = \frac{1}{\epsilon}\{u_i(u_i - a)(1 - u_i) - v_i\} + C_u\{(u_{i+1} - u_i) - (u_i - u_{i-1})\} \quad (21.4)$$

$$\frac{dv_i}{dt} = u_i - bv_i + C_v\{(v_{i+1} - v_i) - (v_i - v_{i-1})\}, \quad (21.5)$$

in which the index number $i = 1, \cdots, L$ identifies each of the elements, $(u_i, v_i)$ is a solution of the $i$-th element, and $C_u$ and $C_v$ denote the coupling strength among neighboring elements. In Eqs. (21.4) and (21.5), replacements of the discrete coupling terms with diffusion coupling terms bring a one-dimensional FitzHugh-Nagumo type reaction-diffusion system. Note that Eq. (21.4) employs inhibitory coupling; however, the original FitzHugh-Nagumo type reaction-diffusion system does not employ inhibitory diffusion coupling [14, 15].

Figure 21.2 shows results of numerical computation of Eqs. (21.4) and (21.5). On the one hand, Fig. 21.2(a) shows that the two pulses are self-organized and remain at the edges of the initial one-dimensional distribution; on the other hand, Fig. 21.2(b) shows that the two pulses self-organized at the edges and traveled in a one-dimensional domain. The ratio between two values of the coupling strength $C_u$ and $C_v$ caused the difference between the two results of Figs. 21.2(a) and 21.2(b), that is, $C_v/C_u = 3$ in Fig. 21.2(a) and $C_v/C_u = 1/3$ in Fig. 21.2(b). Thus, we can understand that the pulses remain at fixed points due to the strong inhibitory coupling.

## 21.4 Algorithm

We present an algorithm for edge detection and orientation selection with the FitzHugh-Nagumo type nonlinear mono-stable excitable elements. The algorithm utilizes a grid system consisting of mono-stable excitable elements; it imposes strong inhibitory couplings among neighboring elements. We substituted an image brightness distribution to the initial conditions of the activator variables of the elements, and initiated computation of ordinary differential equations governing states of the elements. After a certain duration of time, the grid system self-organized a pulse pattern as excited pulses remaining at the edges, as shown in Section 21.3.2. Thus, edge detection could be implemented by finding excited activators in excited states. In addition, we introduced anisotropy into the inhibitory coupling for strengthening the propagation of the inhibiting effect at a specific orientation. The strong inhibitory coupling allows only elements located along the edge lines perpendicular to the orientation.

In Section 21.4.1 we present the algorithm for edge detection. In Section 21.4.2 we extend the algorithm so as to deal with both edge detection and orientation selection by introducing anisotropic inhibitory coupling.

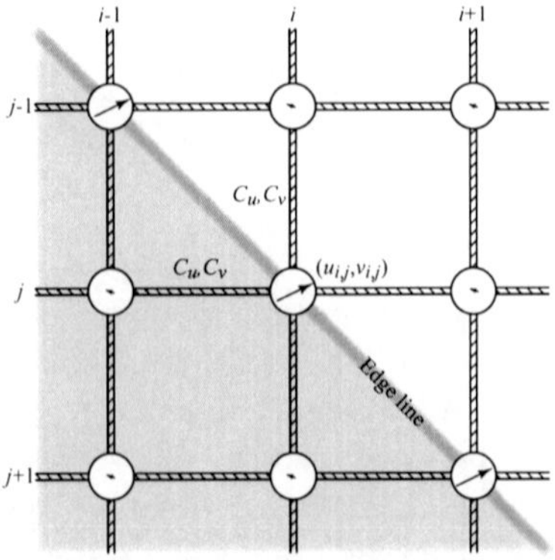


**Fig. 21.3** Two-dimensional grid system consisting of the FitzHugh-Nagumo nonlinear elements of Eqs. (21.6) and (21.7). The coupling strength is $C_u$ for the activator $u_{i,j}$ and $C_v$ for the inhibitor $v_{i,j}$, in which $(i,j)$ denotes a grid point on the grid system $\mathcal{L}_i \times \mathcal{L}_j$. See Eq. (21.9) for the boundary conditions of the grid system.

### *21.4.1 Edge Detection Algorithm with a Two-Dimensional Grid System*

Figure 21.3 shows a grid system utilized in the algorithm. Let $\mathcal{L}_i \times \mathcal{L}_j$ be a two-dimensional grid system; the number of grid points is $L_i \times L_j = |\mathcal{L}_i \times \mathcal{L}_j|$. At each grid point $(i,j) \in \mathcal{L}_i \times \mathcal{L}_j$, we arrange the FitzHugh-Nagumo element governed by

$$\begin{aligned}\frac{\mathrm{d}u_{i,j}}{\mathrm{d}t} &= \frac{1}{\epsilon}\left\{u_{i,j}(u_{i,j}-a)(1-u_{i,j}) - v_{i,j}\right\} \\ &\quad + C_u\left\{(u_{i+1,j}-u_{i,j}) - (u_{i,j}-u_{i-1,j})\right\} \\ &\quad + C_u\left\{(u_{i,j+1}-u_{i,j}) - (u_{i,j}-u_{i,j-1})\right\}, \end{aligned} \qquad (21.6)$$

$$\begin{aligned}\frac{\mathrm{d}v_{i,j}}{\mathrm{d}t} &= u_{i,j} - bv_{i,j} \\ &\quad + C_v\left\{(v_{i+1,j}-v_{i,j}) - (v_{i,j}-v_{i-1,j})\right\} \\ &\quad + C_v\left\{(v_{i,j+1}-v_{i,j}) - (v_{i,j}-v_{i,j-1})\right\}. \end{aligned} \qquad (21.7)$$

The second and third terms of the right-hand sides in Eqs. (21.6) and (21.7) couple four neighboring elements located at $(i+1,j), (i-1,j), (i,j+1)$ and $(i,j-1)$ with the element located at $(i,j)$.

The pixel value of an input image $I$ at the coordinate $(i,j)$, $I_{i,j}, (i,j) \in \mathcal{L}_i \times \mathcal{L}_j$, is fed to the excitation element as the initial value $u_{i,j}(0)$ of the grid system. Thus, the initial conditions for $u_{i,j}$ and $v_{i,j}$ are as follows:

$$u_{i,j}(0) = I_{i,j}, \quad v_{i,j}(0) = 0. \qquad (21.8)$$

This is because the FitzHugh-Nagumo elements coupled with strong inhibition self-organize pulses at the edge positions in the initial condition of $u_i$, as shown in the one-dimensional case of Fig. 21.2. The boundary conditions for $u_{i,j}$ are as follows:

$$u_{0,j} = u_{1,j},\quad u_{L_i,j} = u_{L_i+1,j},\quad u_{i,0} = u_{i,1},\quad u_{i,L_j} = u_{i,L_j+1}. \tag{21.9}$$

The boundary conditions for $v_{i,j}$ are the same as those of Eq. (21.9).

With the initial conditions of Eq. (21.8) and the boundary conditions of Eq. (21.9), we carried out numerical computation of the coupled FitzHugh-Nagumo excitable elements. After a sufficient duration of time, the grid system self-organized pulses on the edge lines (see Fig. 21.3). Thus, by finding excited activators in excited states, we could detect edges. The edge map $\mathcal{M}(t)$ at time $t$ is as follows:

$$\mathcal{M}(t) = \{(i,j)|u_{i,j}(t) > 1/2\}, \tag{21.10}$$

in which we judge the state of an element with the threshold level $1/2$; if $u_{i,j} > 1/2$, we estimate the element to be in an excited state and the position $(i,j)$ to be an edge.

## 21.4.2 Algorithm for Edge Detection and Orientation Selection

As described in Section 21.3.2 and Section 21.4.1, the grid system self-organizes a pulse pattern at edges of an initial activator distribution. Strong inhibitory coupling is the necessary condition for organizing the pulses, and weak inhibitory coupling does not allow such pulses to remain at edge positions. The adoption of anisotropic inhibitory coupling to the grid system is expected to realize orientation selection of edge lines. Only elements having strong inhibitory coupling to neighboring elements can maintain the excitation state. The strength of the inhibitory coupling is given based on the directional difference between the gradient of the inhibitor variables and the specific orientation to be selected.

Let us introduce anisotropic inhibitory coupling strength $A(\theta_{i,j})$ into inhibitor distribution on the grid system, as follows:

$$\begin{aligned}\frac{dv_{i,j}}{dt} &= u_{i,j} - bv_{i,j}\\ &+C_v\left\{A(\theta_{i+1/2,j})\cdot(v_{i+1,j}-v_{i,j}) - A(\theta_{i-1/2,j})\cdot(v_{i,j}-v_{i-1,j})\right\}\\ &+C_v\left\{A(\theta_{i,j+1/2})\cdot(v_{i,j+1}-v_{i,j}) - A(\theta_{i,j-1/2})\cdot(v_{i,j}-v_{i,j-1})\right\},\end{aligned} \tag{21.11}$$

in which the interpolations $A(\theta_{i+1/2,j})$ and $A(\theta_{i,j+1/2})$ are

$$A(\theta_{i+1/2,j}) = \frac{A(\theta_{i+1,j}) + A(\theta_{i,j})}{2},\quad A(\theta_{i,j+1/2}) = \frac{A(\theta_{i,j+1}) + A(\theta_{i,j})}{2}. \tag{21.12}$$

The direction $\theta_{i,j}$ of $A(\theta_{i,j})$ denotes the gradient direction of the inhibitor distribution $v_{i,j}$, as defined by

$$\theta_{i,j} = \tan^{-1}\left(\frac{v_{i,j+1}(t) - v_{i,j-1}(t)}{v_{i+1,j}(t) - v_{i-1,j}(t)}\right). \tag{21.13}$$

The anisotropic inhibitory coupling strength $A(\theta_{i,j})$ is defined by

$$A(\theta_{i,j}) = 1/\sqrt{1 - \delta_v \cos 2(\theta_{i,j} - \phi)}, \tag{21.14}$$

in which $\delta_v$ denotes the strength of the inhibitory coupling and takes the range of $0 \leq \delta_v < 1$; if $\delta_v = 0$, $A(\theta_{i,j})$ becomes 1.0 and thus Eq. (21.11) returns to the isotropic coupling of Eq. (21.7). In Eq. (21.14), $A(\theta_{i,j})$ is designed to strengthen the inhibitory coupling around the area having a similar direction between the gradient direction $\theta_{i,j}$ and the specific orientation $\phi$. Thus, the orientation $\phi \pm \pi/2$ indicates the edge orientation to be selected.

Shoji et al. [11] presented a reaction-diffusion model for explaining oriented periodic patterns of markings on fish skin. The earlier work done by Kondo and Asai showed that the Turing type periodic pattern appears on fish skin [20]; a reaction-diffusion model self-organizes the periodic pattern under strong inhibitory diffusion. Shoji et al. proposed to introduce anisotropy into the diffusion coefficient to explain the oriented periodic patterns. The formulation of Eqs. (21.11) and (21.14) originated from the reaction-diffusion model presented by Shoji et al. [11], and was derived from the replacement of anisotropic diffusion with anisotropic discrete coupling.

**Algorithm 1.** for edge detection and orientation selection.

**for all** $(i,j) \in \mathcal{L}_i \times \mathcal{L}_j$ **do**
  $u_{i,j} \leftarrow I_{i,j}$, $v_{i,j} \leftarrow 0$ ▷ Set the initial conditions with Eq. (21.8).
**end for**
$t \leftarrow 0$
**while** $t < L_t$ **do**
  **for all** $(i,j) \in \mathcal{L}_i \times \mathcal{L}_j$ **do**
    Update $u_{i,j}$ with Eq. (21.6). ▷ With the boundary conditions of Eq. (21.9).
    Compute $\theta_{i,j}$ with Eq. (21.13).
    Compute $A(\theta_{i,j})$ with Eq. (21.14).
    Update $v_{i,j}$ with Eq. (21.11).
  **end for**
  $t \leftarrow t + \delta t$
**end while**
$\mathcal{M} \leftarrow \emptyset$
**for all** $(i,j) \in \mathcal{L}_i \times \mathcal{L}_j$ **do**
  **if** $u_{i,j} > 1/2$ **then**
    $\mathcal{M} \leftarrow \mathcal{M} \cup \{(i,j)\}$ ▷ Update an edge or orientation-selected map with Eq. (21.10).
  **end if**
**end for**

Algorithm 1 describes the algorithm of edge detection and orientation selection for an image-brightness distribution $I_{i,j}$, $(i,j) \in \mathcal{L}_i \times \mathcal{L}_j$. When $\delta_v = 0$, the algorithm provides an edge map; when $\delta_v > 0$, the algorithm provides an orientation-selected map for specific orientation $\phi$. For computing ordinary differential equations, we utilized a finite difference $\delta t$ for temporal discretization. The algorithm computes Eqs. (21.6) and (21.11) iteratively until $t = L_t$.

## 21.5 Experimental Results and Discussion

### *21.5.1 Examples of Edge Detection and Orientation Selection*

This section presents experimental results of edge detection and orientation selection for artificial and real images. Algorithm 1 performed edge detection with the parameter of $\delta_v = 0.0$ and orientation selection with $\phi = 0.0, \pi/4, \pi/2, 3\pi/4$ and $\delta_v = 0.95$. The FitzHugh-Nagumo nonlinear element has the parameter $a$ in Eq. (21.1); the parameter works as a threshold level for an initial condition of $u = u_0$. The algorithm first divides the initial distribution into two levels higher or lower than the threshold level $a$. Then, it performs edge detection or orientation selection for areas segmented by $a$. Thus, we first confirmed that Algorithm 1 works for the artificial binary image having brightness values $I_{i,j} \in \{0,1\}$, and then tried to apply the algorithm to a real image.

Figure 21.4 shows an artificial binary image and the results of edge detection and orientation selection. The algorithm with $\delta_v = 0.0$ provided a good result of edge detection; it successfully detected edges around corner points and in tiny areas; in particular, it fully detected edges as seen in the word "University" and the abbreviation "PDE". In the results of orientation selection shown in Figs. 21.4(c)∼21.4(f), the algorithm almost successfully provided maps of vertical and horizontal orientation selection. In particular, it detected vertical lines with $\phi = 0.0$ and horizontal lines with $\phi = \pi/2$ along the edges of cells and small rectangles. On the orientation $\pi/4$ and $3\pi/4$, although the algorithm provided almost well detected maps, it partly generated noisy results that can be seen as dotted lines. For example, see the edge lines along the triangle in Figs. 21.4(e) and 21.4(f).

Figure 21.5 shows a real image and its results of edge detection and orientation selection. Although the algorithm with $\delta_v = 0.0$ detected edges along the contours of high-contrast objects such as window frames and tables, it failed to detect the edges of the gray horizontal bar clearly seen against the white wall. This is because the algorithm detected edges for areas segmented with the threshold level $a$. (Recall that an element initiated with $u_0 > a$ can enter an excited state; that with $u_0 < a$ cannot enter the excited state, but directly returns to a resting state, as described in Section 21.3.1.) Figures 21.5(g) and 21.5(h) show two situations in one-dimensional distributions of image brightness values $I_{i,j}$ and obtained solutions $u_{i,j}$ as well as the threshold level $a = 0.1$. For example in Fig. 21.5(g), while the algorithm successfully detected edges around $j = 190$ and $j = 227$, it failed to detect those around $j = 45$ and $j = 56$, at which there was the horizontal bar on the wall. On the one hand,

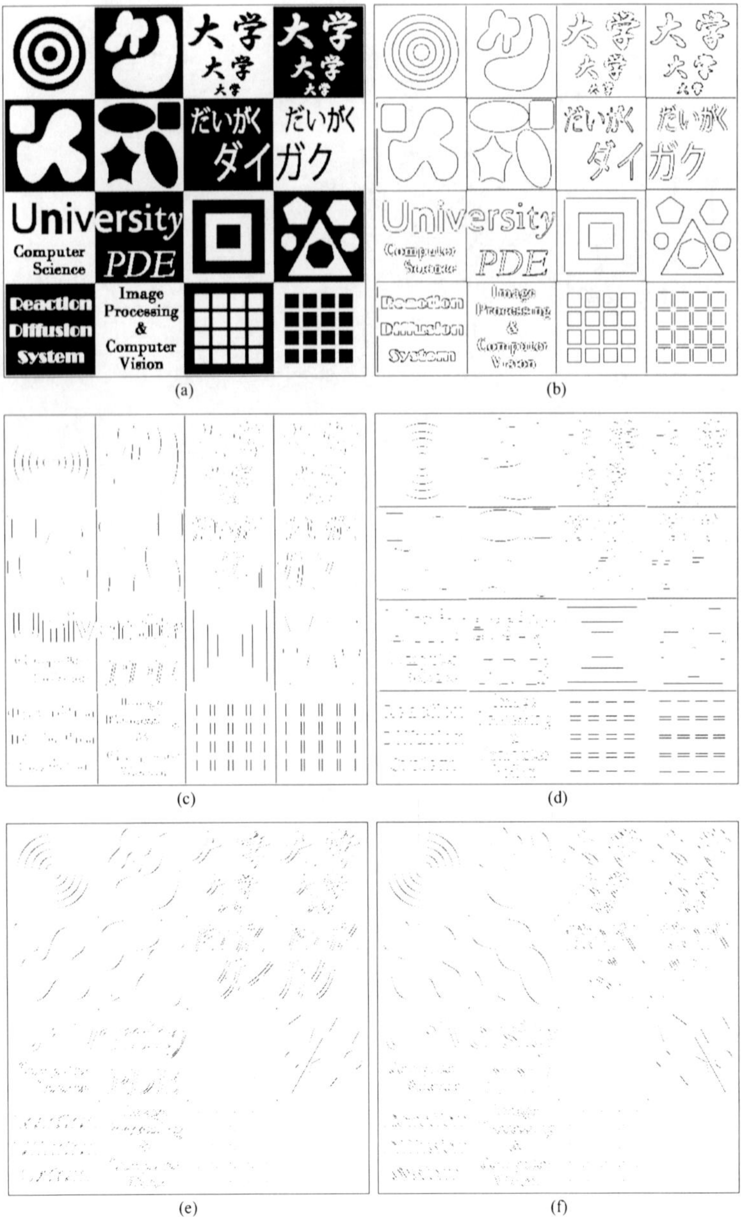


**Fig. 21.4** Results of edge detection and orientation selection for an artificial binary image. Figure (a) shows an artificial binary image with the size of $L_i \times L_j = 500 \times 500$ pixels. Algorithm 1 provided an edge map (b) for the image (a); its parameter settings on coupling were $C_u = 2.0, C_v = 6.0$ and $\delta_v = 0.0$. Algorithm 1 provided orientation-selected maps with (c) $\phi = 0.0$, (d) $\phi = \pi/2$, (e) $\phi = \pi/4$ and (f) $\phi = 3\pi/4$; its parameter settings on coupling were $C_u = C_v = 2.0$ and $\delta_v = 0.95$. Other parameter settings were fixed across (b)$\sim$(f) as $\delta t = 1.0 \times 10^{-4}, a = 0.10, b = 1.0, \varepsilon = 1.0 \times 10^{-3}$ and $L_t = 5.0$.

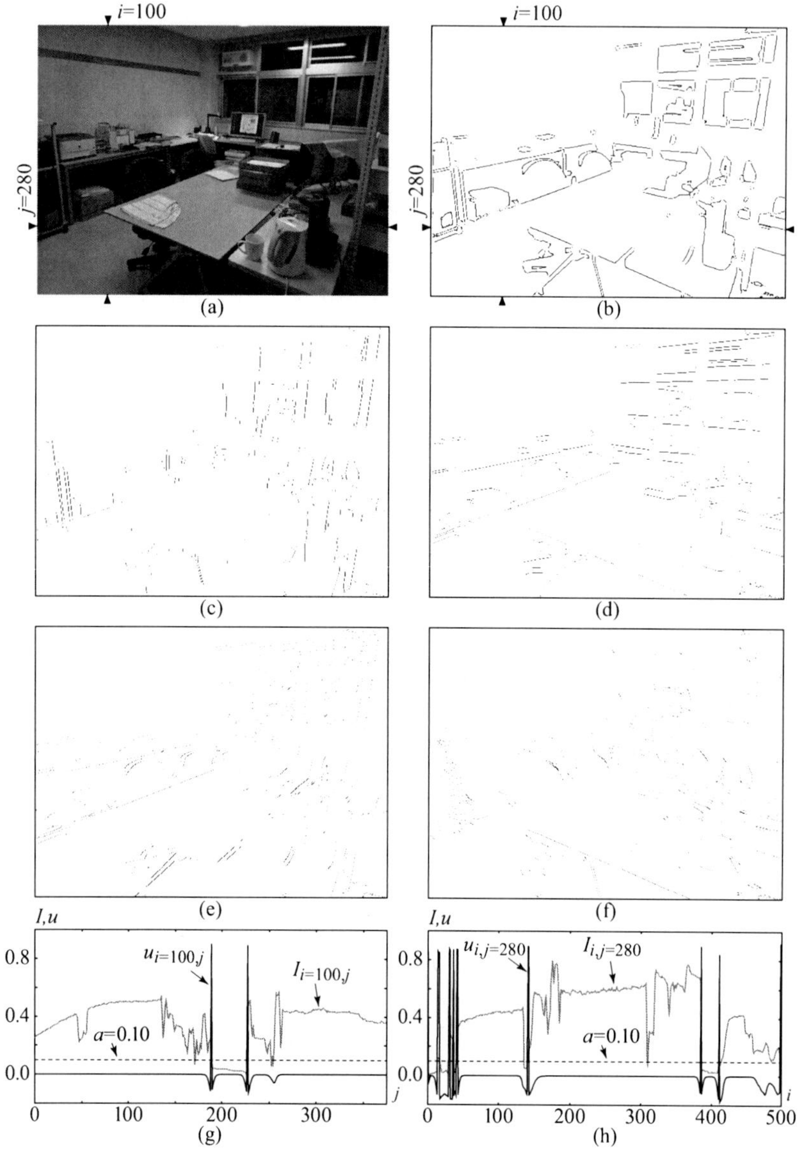


**Fig. 21.5** Results of edge detection and orientation selection for a real image. Figure (a) shows the real image with the size of $L_i \times L_j = 500 \times 375$ pixels and 256 brightness levels (eight bits quantization). Algorithm 1 with the parameter settings of $C_u = 4, C_v = 12$ and $\delta_v = 0.0$ provided an edge map (b) for the image (a). Algorithm 1 with the parameter settings of $C_u = C_v = 2.0$ and $\delta_v = 0.95$ provided orientation-selected maps with (c) $\phi = 0.0$, (d) $\phi = \pi/2$, (e) $\phi = \pi/4$ and (f) $\phi = 3\pi/4$. Other parameter settings were the same as those in Fig. 21.4. Figure (g) shows one-dimensional distributions of image-brightness values $I_{i=100,j}$ and obtained solutions $u_{i=100,j}$; Fig. (h) shows those of $I_{i,j=280}$ and $u_{i,j=280}$. Two-dimensional distributions of $I_{i,j}$ and $u_{i,j}$ are shown in Figs. (a) and (b). The parameter $a$ was chosen as $a = 0.1$ in the edge detection (b).

the image brightness distribution does not cross the threshold level $a$ at $j = 45 \sim 56$; on the other hand, it crosses the level at $j = 190$ and $j = 227$. Figure 21.5(h) also shows a similar situation around $i = 308 \sim 320$. An edge of a table exists around $i = 308 \sim 320$; in spite of that, the algorithm did not detect the edge, as also shown in Fig. 21.5(b). Thus, except for the low-contrast edges not crossing the threshold level, we confirmed that the algorithm almost correctly detected edges also for the real image. When turning our attention to results of vertical and horizontal orientation selection, we could find orientation selected lines. However, for example, along the edges of the window frames in Fig. 21.5(c), the algorithm provided results that can be seen as broken and dotted lines, as also confirmed in Fig. 21.4.

Let us consider the reason why the algorithm provided orientation selection results that can be seen as dotted lines and broken ones in both Figs. 21.4 and 21.5. For example, when focusing on the result along the edge lines of the window frames in Fig. 21.5(d), we can observe the edge lines not lying on an exact horizontal orientation, but lying on orientation slightly slanted from the horizontal one. A series of short horizontal lines can approximate a straight line slightly slanted from the horizontal orientation. Thus, we understand that the broken lines obtained for slightly slanted lines are reasonable as an approximation in the orientation selection. However, when turning our attention to highly slanted edge lines of the tables in Fig. 21.5(d), we can find dotted-line like results as a horizontal orientation; that is, the algorithm detected the small dots for the highly slanted edge lines. The results of the small dots are considered as over-detection and are unacceptable as approximation of the highly slanted edge lines. We believe that this would be avoidable with optimal parameter settings and with a coupling scheme that takes into account not only four neighboring adjacent elements, but also those located diagonally on the grid, or more global ones. Ways of estimating the optimal parameter settings and coupling with more global elements are future research topics.

### 21.5.2 *Quantitative Performance Evaluation on Edge Detection*

This section presents quantitative performance evaluation of Algorithm 1 consisting of coupled FitzHugh-Nagumo (FHN) elements, in comparison with two representative algorithms proposed by Marr and Hildreth [12] and Canny [13]. The algorithm proposed by Marr and Hildreth utilizes the difference-of-two-Gaussians filter, and is called the DoG algorithm. An output of a Gaussian filter is generally equivalent to a solution of a diffusion equation [23]. Thus, Eqs. (21.6) and (21.7) without the reaction terms of $[u(u-a)(1-u)-v]/\varepsilon$ and $u - bv$ approximate two Gaussian filters. We implemented the DoG algorithm by solving two diffusion equations having different diffusion coefficients, on which we imposed a weak excitatory coupling $C_u = 4.0$ and a relatively strong inhibitory one $C_v = 26.1$, as suggested by Marr and Hildreth. We estimated the duration of time in computing the diffusion equations so as to achieve the best performance for each image (in most cases, we estimated it at $L_t = 0.6$). For the other algorithm proposed by Canny [13], we utilized a computer program code provided on a website by Heath et al. [28, 29].

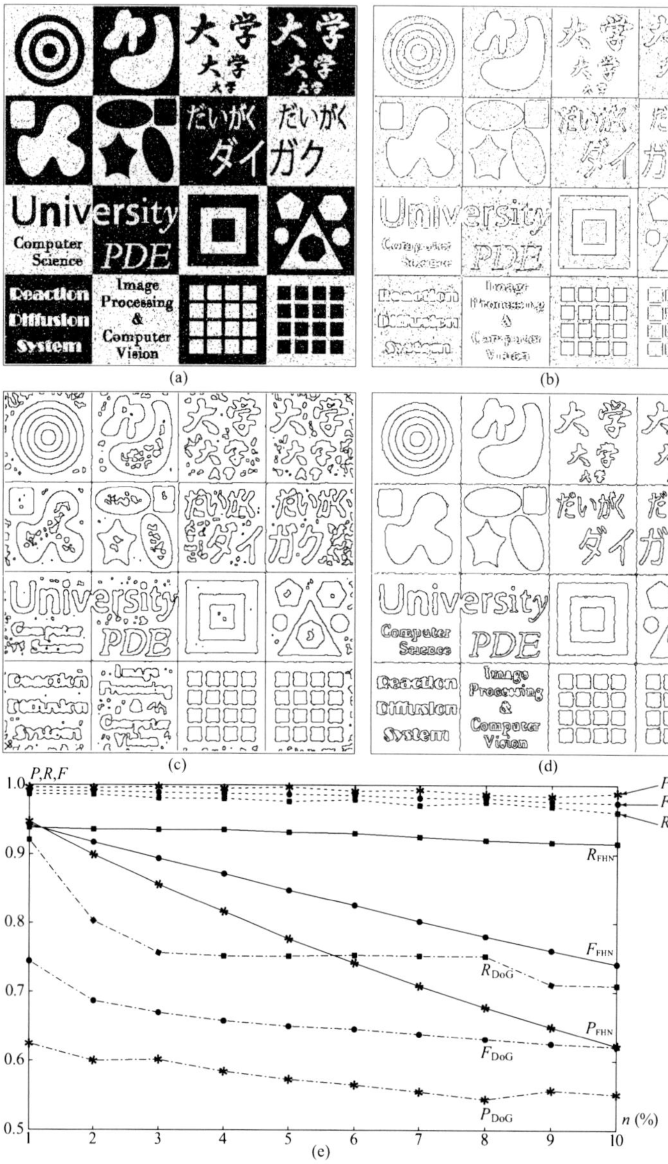


**Fig. 21.6** Edge detection results for noisy binary images. Figure (a) shows a noisy image generated by adding $n = 5$ % dot noise to the binary image of Fig. 21.4(a). Algorithm 1 consisting of coupled FitzHugh-Nagumo (FHN) elements provided an edge map (b) with the same parameter settings as those of Fig. 21.4(b); the DoG algorithm proposed by Marr and Hildreth with the difference-of-two-Gaussians filter [12] provided the edge map (c); the Canny algorithm [13, 28] provided the edge map (d). Figure (e) shows the dependence on the noise ratio $n$ (%) of the three algorithms; $P_{FHN}$, $R_{FHN}$ and $F_{FHN}$ show the dependence of Algorithm 1; $P_{DoG}$, $R_{DoG}$ and $F_{DoG}$ show the dependence of the DoG algorithm; and $P_{Canny}$, $R_{Canny}$ and $F_{Canny}$ show the dependence of the Canny algorithm. See Eq. (21.15) for the measures $P, R$ and $F$.

The Canny algorithm has three parameters: $\sigma$ and two threshold levels; we chose their parameter settings so as to achieve the best performance among the combinations of $\sigma = 0.2, 0.4, \cdots, 2.0$, the lower threshold level $= 0.1, 0.2, \cdots, 0.8$ and the higher threshold level $= 0.2, 0.3, \cdots, 0.9$ for each image.

We applied the three algorithms: Algorithm 1, the DoG algorithm and the Canny algorithm to noisy artificial binary images, and evaluated their performance quantitatively with three measures. The noisy images were generated by adding binary noise to the artificial binary image shown in Fig. 21.4(a); the noise ratio $n$ (%) refers to the ratio of the number of binary noise pixels to that of all the pixels $L_i \times L_j$. Figure 21.6(a) shows one of the noisy images ($n = 5$ %), and Figs. 21.6(b)~21.6(d) show edge maps provided by the three algorithms for the noisy image of Fig. 21.6(a). The following F-measure computed from precision and recall measures $P$ and $R$ [30] evaluates the performance of the algorithms for each noisy artificial image:

$$P = \frac{|\mathcal{M}_t \cap \mathcal{M}_o|}{|\mathcal{M}_o|}, \quad R = \frac{|\mathcal{M}_t \cap \mathcal{M}_o|}{|\mathcal{M}_t|}, \quad F = \frac{PR}{\alpha P + (1-\alpha)R}, \tag{21.15}$$

in which $\mathcal{M}_t$ is the true edge map and $\mathcal{M}_o$ is the obtained one. The parameter $\alpha$ defines the balance between the two measures $P$ and $R$; we fixed $\alpha$ at $\alpha = 0.5$, which means equal balance. Larger values of $P$, $R$ and $F$ indicate better performance. Figure 21.6(e) shows the dependence of the three algorithms on the noise ratio $n$.

Let us discuss the results of Fig. 21.6. In the result of Algorithm 1, the measure $P_{\text{FHN}}$ decreases almost linearly as the noise ratio $n$ increases, while the measure $R_{\text{FHN}}$ is almost constant ($R_{\text{FHN}} \simeq 0.95$), and the measure $F_{\text{FHN}}$ also decreases linearly. Since the recall measure $R$ evaluates the ratio of the number of detected true edges $|\mathcal{M}_t \cap \mathcal{M}_o|$ to that of the true edges $|\mathcal{M}_t|$, the constant large value of $R_{\text{FHN}}$ implies that the algorithm always successfully detected most of the true edges. In contrast to this, since the precision measure $P$ evaluates the ratio of the number of detected true edges to that of the detected edges $|\mathcal{M}_o|$, the linearly decreasing trend implies that the algorithm detected false edges and the number of false edges increased as the noise ratio $n$ increased. From these considerations, the decreasing trends of $P_{\text{FHN}}$ and $F_{\text{FHN}}$ suggest that the algorithm falsely detected noise pixels as edges, and thus clearly showed the noise vulnerability of Algorithm 1. This is also recognizable with the obtained edge map shown in Fig. 21.6(b). However, as also recognized in Fig. 21.6(b) simultaneously, the algorithm successfully removed noise in areas having a higher brightness level. This may provide a clue for improving the noise-robustness of the algorithm. The DoG algorithm provided the worst edge detection results among the three algorithms. The Canny algorithm achieved the best performance among them, and its performance is almost constant or only slightly decreases as the noise ratio $n$ increases. Thus, the Canny algorithm is very robust to noise, when its parameter settings are adaptively chosen.

For confirming the convergence of Algorithm 1 and the DoG algorithm, we evaluated the edge detection processes of the algorithms applied to the image of Fig. 21.6(a). Figure 21.7 shows temporal changes of the three measures. For

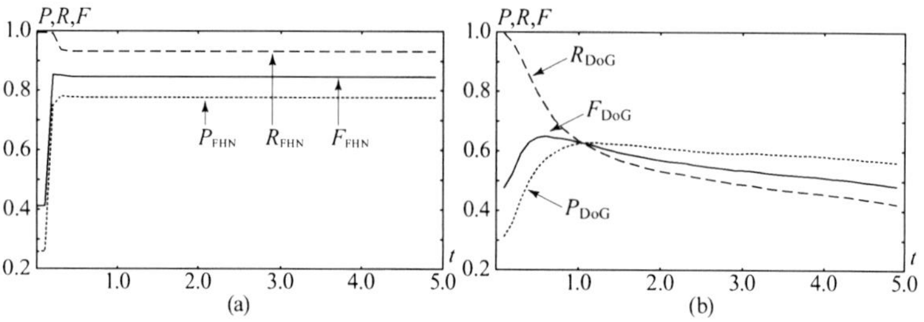


**Fig. 21.7** Temporal changes of the three measures $P, R$ and $F$ for edge detection processes of Algorithm 1 consisting of coupled FitzHugh-Nagumo (FHN) elements and the DoG algorithm proposed by Marr and Hildreth with the difference-of-two-Gaussians filter [12]. Figure (a) shows the temporal changes of $P_{\mathrm{FHN}}, R_{\mathrm{FHN}}$ and $F_{\mathrm{FHN}}$ for the edge detection process of Algorithm 1 applied to the noisy image ($n = 5$ %) of Fig. 21.6(a). Figure (b) shows the temporal changes of $P_{\mathrm{DoG}}, R_{\mathrm{DoG}}$ and $F_{\mathrm{DoG}}$ for the edge-detection process of the DoG algorithm applied to the same image. See Figs. 21.6(b) and 21.6(c) for their obtained edge maps.

Algorithm 1, the temporal changes rapidly converged to almost constant values before $t = 1.0$, as shown in Fig. 21.7(a). For the DoG algorithm, the three measures dynamically changed in the range of $t < 1.0$, and decreased almost monotonically in the range of $t > 1.5$, as shown in Fig. 21.7(b); the algorithm achieved the best performance at $t = 0.6$. Thus, how to estimate the optimal duration of time or the stopping time for the best performance is a problem yet to be solved; this is the stopping time evaluation problem [31] or the termination problem [7]. We believe that the FitzHugh-Nagumo type nonlinearity of Eqs. (21.1) and (21.2) works to avoid the termination problem in Algorithm 1.

## 21.6 Conclusion

This chapter presented an algorithm for edge detection and orientation selection with a grid system consisting of coupled nonlinear excitable elements. Several biological phenomena caused by strong inhibitory coupling or long-range inhibition [17, 18, 20, 26] motivated us to utilize the FitzHugh-Nagumo model [14, 15] as a nonlinear excitable element and to couple elements with strong inhibition. The two conditions of strong inhibitory coupling and strong nonlinearity induced stationary pulses remaining at positions having rapid changes across a threshold level in their initial activator distribution [10]. Thus, our edge detection algorithm utilized the phenomenon in which the grid system self-organizes a pulse pattern at edges in an initial condition. In addition, by introducing anisotropic inhibitory coupling into the grid system, we developed an edge detection algorithm so as to perform orientation selection as well. The idea of introducing the anisotropy came from the model that explained oriented periodic patterns of markings on fish skin [11].

We experimentally applied the algorithm of edge detection and orientation selection to artificial binary and real images. The algorithm provided almost correct results of edge detection, except that it did not detect edges with low brightness changes in the real gray level image. We also performed orientation selection with four different orientation settings. Although the algorithm provided fairly reasonable results for orientation selection, it also provided dotted-line or broken-line edges. We finally compared the quantitative performance among our algorithm and two representative previous algorithms proposed by other researchers [12, 13] for artificially generated noisy binary images. Although the algorithm presented here did not achieve the best performance among the three algorithms, it performed better than the previous algorithm proposed by Marr and Hildreth [12].

There exist several future research topics for the present algorithm. The algorithm couples an element with its horizontally and vertically nearest four neighboring adjacent elements on the grid system. For more reliable and accurate orientation selection, we will study additional diagonal or more global coupling schemes. We are expecting that this will bring more precise orientation selection than the four orientation selections in the present algorithm. As also suggested by the results of performance comparison, the algorithm has noise vulnerability; in order to solve the vulnerability, we are now reconsidering the initial conditions provided for the FitzHugh-Nagumo elements. In addition to these algorithmic improvements, from an engineering point of view, it would be interesting to apply the algorithm to a cellular neural network approach for hardware implementation, developing various visual functions, and building a biologically inspired visual system. From a scientific point of view, we hope that the resulting artificial visual system provides some hints for understanding the human visual system and visual functions therein.

**Acknowledgements.** The present work was supported in part by a Grant-in-Aid for Scientific Research (C) (No. 23500278) from the Japan Society for the Promotion of Science.

## References

1. Murray, J.D.: Mathematical Biology. Springer, Berlin (1989)
2. Hodgkin, A.L., Huxley, A.F.: A quantitative description of membrane current and its application to conduction and excitation in nerve. J. Physiol. 117, 500–544 (1952)
3. Zaikin, A.N., Zhabotinsky, A.M.: Concentration wave propagation in two-dimensional liquid-phase self-oscillating system. Nature 225, 535–537 (1970)
4. Busse, H., Hess, B.: Information transmission in a diffusion-coupled oscillatory chemical system. Nature 244, 203–205 (1973)
5. Kuhnert, L., Agladze, K.I.: Krinsky VI Image processing using light-sensitive chemical waves. Nature 337, 244–247 (1989)
6. Wang, D.L., Terman, D.: Locally excitatory globally inhibitory oscillatory networks. IEEE Trans. Neural Netw. 6, 283–286 (1995)
7. Chen, K., Wang, D.L.: A dynamically coupled neural oscillator network for image segmentation. Neural Netw. 15, 423–439 (2002)
8. Nomura, A., Ichikawa, M., Miike, H., Ebihara, M., Mahara, H., Sakurai, T.: Realizing visual functions with the reaction-diffusion mechanism. J. Phys. Soc. Jpn. 72, 2385–2395 (2003)
9. Ebihara, M., Mahara, H., Sakurai, T., Nomura, A., Osa, A., Miike, H.: Segmentation and edge detection of noisy image and low contrast image based on a reaction-diffusion model. J. IIEEJ 32, 378–385 (2003)

10. Kurata, N., Kitahata, H., Mahara, H., Nomura, A., Miike, H., Sakurai, T.: Stationary pattern formation in a discrete excitable system with strong inhibitory coupling. Phys. Rev. E 79, 056203 (2009)
11. Shoji, H., Iwasa, Y., Mochizuki, A., Kondo, S.: Directionality of stripes formed by anisotropic reaction-diffusion models. J. Theor. Biol. 214, 549–561 (2002)
12. Marr, D., Hildreth, E.: Theory of edge detection. Proc. Roy. Soc. Lond. Ser. B, Biol. Sci. 207, 187–217 (1980)
13. Canny, J.: A computational approach to edge detection. IEEE Trans. Patt. Anal. Mach. Intell. 8, 679–698 (1986)
14. FitzHugh, R.: Impulses and physiological states in theoretical models of nerve membrane. Biophys. J. 1, 445–466 (1961)
15. Nagumo, J., Arimoto, S., Yoshizawa, S.: An active pulse transmission line simulating nerve axon. Proc. IRE 50, 2061–2070 (1962)
16. Tyson, J.J., Alexander, K.A., Manoranjan, V.S., Murray, J.D.: Spiral waves of cyclic AMP in a model of slime mold aggregation. Phys. D 34, 193–207 (1989)
17. Turing, A.M.: The chemical basis of morphogenesis. Phil. Trans. Roy. Soc. Lond. Ser. B, Biol. Sci. 237, 37–72 (1952)
18. Gierer, A., Meinhardt, H.: A theory of biological pattern formation. Kybernetik 12, 30–39 (1972)
19. Castets, V., Dulos, E., Boissonade, J., De Kepper, P.: Experimental evidence of a sustained standing Turing-type nonequilibrium chemical pattern. Phys. Rev. Lett. 64, 2953–2956 (1990)
20. Kondo, S., Asai, R.: A reaction-diffusion wave on the skin of the marine angelfish Pomacanthus. Nature 376, 765–768 (1995)
21. Matsumoto, T.: A chaotic attractor from Chua's circuit. IEEE Trans. Circ. Syst. 31, 1055–1058 (1984)
22. Chua, L.O., Hasler, M., Moschytz, G.S., Neirynck, J.: Autonomous cellular neural networks: a unified paradigm for pattern formation and active wave propagation. IEEE Trans. Circ. 42, 559–577 (1995)
23. Koenderink, J.J.: The structure of images. Biol. Cybern. 50, 363–370 (1984)
24. Perona, P., Malik, J.: Scale-space and edge detection using anisotropic diffusion. IEEE Trans. Patt. Anal. Mach. Intell. 12, 629–639 (1990)
25. Rekeczky, C.: CNN architectures for constrained diffusion based locally adaptive image processing. Int. J. Circ. Theor. Appl. 30, 313–348 (2002)
26. Barlow, R.B.: Jr, Quarles DA Jr Mach bands in the lateral eye of *Limulus*. J. Gen. Physiol. 65, 709–730 (1975)
27. Ferster, D., Koch, C.: Neuronal connections underlying orientation selectivity in cat visual cortex. Trends in Neurosci. 10, 487–492 (1987)
28. Heath, M., Sarkar, S., Sanocki, T., Bowyer, K.: `http://marathon.csee.usf.edu/edge/edge_detection.html`
29. Heath, M.D., Sarkar, S., Sanocki, T., Bowyer, K.W.: A robust visual method for assessing the relative performance of edge-detection algorithms. IEEE Trans. Patt. Anal. Mach. Intell. 19, 1338–1359 (1997)
30. Martin, D.R., Fowlkes, C.C., Malik, J.: Learning to detect natural image boundaries using local brightness, color, and texture cues. IEEE Trans. Patt. Anal. Mach. Intell. 26, 530–549 (2004)
31. Mrázek, P., Navara, M.: Selection of optimal stopping time for nonlinear diffusion filtering. Int. J. Comp. Vis. 52, 189–203 (2003)

# Chapter 22
# Consecutive Repeating State Cycles Determine Periodic Points in a Turing Machine

Michael Stephen Fiske

**Abstract.** The Turing machine is studied with new methods motivated by the notion of recurrence in classical dynamical systems theory. The state cycle of a Turing machine is introduced. It is proven that each consecutive repeating state cycle in a Turing machine determines a unique periodic configuration (point) and vice versa. This characterization is a *periodic point* theorem for Turing machines. A Turing machine is defined to be periodic if it has at least one periodic configuration or it only has halting configurations. Using the notion of a prime directed edge and a mathematical operation called edge pattern substitution, a search procedure finds consecutive repeating state cycles. If the Turing machine is periodic, then this procedure eventually finds each periodic point or this procedure determines that the machine has only halting configurations. New mathematical techniques are demonstrated such as edge pattern substitution and prime directed edge sequences that could be quite useful in the further study of the aperiodic Turing machines. The aperiodicity appears to play an integral role in the undecidability of the Halting problem.

## 22.1 Introduction

New results are achieved here by analyzing the Turing machine from a dynamical systems point of view. Classical dynamical systems theory has been successful by first understanding the periodic behavior and then studying more general recurrent behavior (as in [3], [4], [7], and [11] ). This paper follows the classical approach, by finding a notion of recurrence that characterizes periodic configurations of a Turing machine.

In the next section, the Turing machine is briefly reviewed and then some definitions are presented for studying its recurrent behavior. The Turing machine is represented as $(Q, A, \eta)$ where $Q$ is a finite set of states, $A$ is a finite alphabet and

Michael Stephen Fiske
Aemea Institute, San Francisco, CA, 94129
e-mail: mf@aemea.org

K. Kyamakya et al. (Eds.): Selected Topics in Nonlinear Dynamics, SCI 459, pp. 393–417.
DOI: 10.1007/ 978-3-642-34560-9_22 

the program $\eta$ is a function $\eta : Q \times A \to Q \times A \times \{-1,+1\} \cup \{\mathcal{H}\} \times A \times \{0\}$. A Turing machine configuration $(q,k,T)$ is a triplet, where $q$ is the current state of the machine, the function $T : Z \to A$ represents the tape, where $Z$ is the integers and $k$ is an integer that is the tape head location. A machine starts program execution at configuration $(q,k,T)$ where $T(k) = \alpha$ and $\eta$ determines the next configuration according to three cases:

1. $(r,k-1,S)$ if $\eta(q,\alpha) = (r,\beta,-1)$
2. $(r,k+1,S)$ if $\eta(q,\alpha) = (r,\beta,+1)$
3. $(\mathcal{H},k,S)$ if $\eta(q,\alpha) = (\mathcal{H},\beta,0)$

such that for all three cases the new tape $S(j) = T(j)$ whenever $j \neq k$ and $S(k) = \beta$.

Case 1 means that the machine moves to state $r$, replaces alphabet symbol $\alpha$ with symbol $\beta$ at tape square $k$ and then moves the tape head left $-1$ to tape square $k-1$. Case 2 means the same as Case 1 except the tape head moves right $+1$ to tape square $k+1$. Case 3 means that the machine reaches a unique halting state $\mathcal{H}$ and the program execution halts.

A configuration $(q,k,T)$ is called a halting configuration if machine execution starts at $(q,k,T)$, and after a finite number of execution steps, the machine reaches the halting state. A configuration $(q,k,T)$ is periodic if after the execution of $n$ computational steps of the Turing machine, the new configuration $(r,j,S)$ has the same state and the same tape contents. This means $r = q$ and the same tape contents means $T(x+k) = S(x+j)$ for every integer $x$. In this case, $(q,k,T)$ is a periodic configuration. Thus, all periodic configurations are immortal. A Turing machine is called *periodic* if it has at least one periodic configuration or it only has halting configurations.

The state cycle is a notion of recurrence for the Turing machine. A state cycle is a non-halting execution sequence of input commands $(q_0,a_0) \to (q_1,a_1) \to \cdots \to (q_{N-1},a_{N-1}) \to (q_N,a_N)$, such that $q_0 = q_N$ and where pair $(q,\alpha)$ in $\eta(q,\alpha)$ is called an input command. A state cycle is called a prime state cycle if it contains no proper state subcycles. A consecutive repeating state cycle is a state cycle that repeats itself twice, where the second repeat immediately follows the first: $(q_0,a_0) \to (q_1,a_1) \to \cdots \to (q_{N-1},a_{N-1}) \to (q_0,a_0) \to (q_1,a_1) \to \cdots \to (q_{N-1},a_{N-1})$.

In theorem 22.2, a *periodic point theorem that holds for any Turing machine* is proved: a consecutive repeating state cycle uniquely determines a periodic configuration of the Turing machine; and vice versa, a periodic configuration uniquely determines a consecutive repeating state cycle. Thus, to search for a periodic configuration, a procedure may look for consecutive repeating state cycles.

In definition 22.12, the prime directed edge is defined. The prime directed edge represents, over a window of execution, a Turing machine program visiting only one state twice and visiting the other states one time or not at all. A prime directed edge contains a prime state cycle. Pattern matching determines how two prime directed edges are glued together. When prime directed edges are glued together, this is called link matching. The link matching is used to build prime directed edge sequences. Based on $|Q|$ and $|A|$, an upper bound on the number of prime directed edges is computed. The prime directed edge sequences cover all Turing program execution possibilities. As a consequence, the number of prime directed edges is a useful measure of the Turing machine complexity. For a given Turing machine, a procedure for finding all prime directed edges is shown. After this, a periodic point search procedure is described whereby prime directed edges are link matched together to form prime directed edge sequences. This procedure searches for consecutive repeating state cycles inside the edge sequences.

In [9], Kurka conjectured that any Turing machine that has no halting configurations has a periodic configuration. In [2], Blondel et al demonstrated that some Turing machines have only aperiodic immortal configurations. Furthermore, Blondel et al. showed in [2] that determining whether a given counter machine has a periodic orbit in configuration space is undecidable.

In this context, there are two main results. First, that each consecutive repeating state cycle determines a unique periodic point in a Turing Machine and vice versa. Second, using this first result, if a Turing machine is periodic, then the search procedure can determine whether the machine only has halting configurations or, if not, then this procedure finds a periodic configuration of this machine. Furthermore, the procedure demonstrated can be used to find periodic configurations of any length when they exist. Finally, the mathematical tools developed here – e.g. the prime directed edge, the edge substitution operator, link matching, and prime directed edge sequences – can be used to study the aperiodic immortal configurations, which could help better understand the dynamics of the Halting problem [12].

## 22.2 Turing Machines & Periodic Configurations

The Turing Machine is defined here so that its program $\eta$ is explicitly represented as a function.

**Definition 22.1.** *Turing Machine*
A Turing machine is a triple $(Q, A, \eta)$ where

1. $Q$ is a finite set of states that does not contain a unique halt state $\mathcal{H}$.
2. The machine execution starts in an initial state s and s lies in $Q$.
3. $A$ is a finite set of alphabet symbols that are read from and written to the tape.
4. $-1$ and $+1$ represent advancing the tape head to the left or right square, respectively.
5. $\eta : Q \times A \to Q \times A \times \{-1, +1\} \cup \{\mathcal{H}\} \times A \times \{0\}$ is a function. $\eta$ acts as the program for the Turing machine. For each $q$ in $Q$ and $\alpha$ in $A$, $\eta(q, \alpha) = (r, \beta, x)$ describes how machine $(Q, A, \eta)$ executes one computational step. When in state $q$ and scanning alphabet symbol $\alpha$ on the tape:
   (a) Machine $(Q, A, \eta)$ changes to state $r$.
   (b) Machine $(Q, A, \eta)$ rewrites alphabet symbol $\alpha$ as symbol $\beta$ on the tape.
   (c) If $x = -1$, then machine $(Q, A, \eta)$ moves its tape head one square to the left on the tape and is subsequently scanning the symbol in this square.

(d) If $x = +1$, then machine $(Q, A, \eta)$ moves its tape head one square to the right on the tape and is subsequently scanning the symbol in this square.

(e) If $r = \mathcal{H}$, machine $(Q, A, \eta)$ reaches halting state $\mathcal{H}$ and halts.

**Definition 22.2.** *Turing Machine tape*
The Turing machine tape $T$ is represented as a function $T : Z \to A$ where $Z$ denotes the integers. The tape $T$ is $M$-bounded if there exists a bound $M > 0$ such that $T(k) = T(j)$ whenever $|k|, |j| \geq M$.

The Turing machine definitions in [5], [12] assume the initial tape, before program execution begins, is $M$-bounded and the tape contains only blank symbols, denoted here as #, outside the bound. In this paper, the tape is not assumed to be $M$-bounded, unless this is explicitly stated for a particular case. The symbol on the $k$th square of the tape is $T(k)$.

**Definition 22.3.** *Turing Machine Configuration with tape head location*
Let $(Q, A, \eta)$ be a Turing machine with tape T. A configuration is an element of the set $C = (Q \cup \{\mathcal{H}\}) \times Z \times \{T : T \text{ is tape with range } A\}$. If $(q, k, T)$ is a configuration, then $k$ is called the tape head location.

Consider the configuration $(p, 2, \ldots \#\#\alpha\underline{\beta}\#\# \ldots)$. The 1st coordinate indicates that the Turing machine is in state $p$. The 2nd coordinate indicates that its tape head is currently scanning tape square 2, denoted as $T(2)$. The 3rd coordinate indicates that tape square 1 contains symbol $\alpha$, tape square 2 contains symbol $\beta$, and all other tape squares contain the # symbol. Sometimes a periodic configuration $p = (q, k, T)$ is called a periodic point or immortal periodic point.

**Definition 22.4.** *Computational Period and Hyperbolic Degree*
Consider immortal periodic point $p$. If the machine starts its execution at point $p$, then the minimal number of computational steps, denoted $C(p)$, for the machine to return to point $p$ is called the computational period of $p$. Observe that $C(p) = R + L$ where $R$ and $L$ denote the number of right and left tape head moves respectively. Define the hyperbolic degree of $p$ as $m(p) = R - L$. If $m \neq 0$, the periodic point is called *hyperbolic*. Otherwise, $p$ is called non-hyperbolic.

The computational period is motivated by the classical dynamical systems definition of the period of a point $p$ in $X$ for an autonomous (e.g. [6]) dynamical system $f : X \to X$ where $X$ is a topological space, $f$ is a function and $m$ is the minimal positive integer such that $f^m(p) = p$ . The hyperbolic degree is analogous to the classical dynamical systems definition of a hyperbolic periodic point of $f : X \to X$ when $X$ is a manifold and $f$ is differentiable.

A *pattern* W is a finite sequence of alphabet symbols chosen from A. In other words, $W : \{0, 1, \ldots, n-1\} \to A$. The length of $W = n$ and is denoted as $|W| = n$. The $k$th element of the pattern $W$ is denoted as $W(k)$ or $w_k$. Thus, pattern $W$ is sometimes explicitly expressed as $w_0 w_1 \ldots w_{n-1}$. $S$ is a *subpattern* of $W$ if $S = w_j w_{j+1} \ldots w_{k-1} w_k$ for some $j$ and $k$ satisfying $0 \leq j \leq k \leq n-1$ and the length of $S = k - j + 1$. A pattern represents a finite sequence of the tape.

The expression $[4,\overline{12}\underline{1}\overline{212}]$ represents the point $p$ where the machine is in state 4; the tape head is located at the underlined 1; the tape to the right of the tape head contains the periodic pattern 212 212... ; and the tape to the left of the tape head contains the periodic pattern ...12 12 12.

*Example 22.1.* *Non-Hyperbolic Periodic Point*

The state set is $Q = \{q,r\}$ and the alphabet set is $A = \{1,2\}$. The halting state is $\mathcal{H}$. $\eta$ is defined below.

$\eta(q,1) = (\mathcal{H},1,0)$ $\qquad$ $\eta(q,2) = (r,2,-1)$
$\eta(r,1) = (q,2,+1)$ $\qquad$ $\eta(r,2) = (q,1,+1)$

Consider program execution steps: $[r,x\underline{2}2y] \mapsto [q,x1\underline{2}y] \mapsto [r,x\underline{1}2y] \mapsto [q,x2\underline{2}y] \mapsto [r,x\underline{2}2y]$, where $x$ is any infinite left tape sequence chosen from $A$ and $y$ is any infinite right tape sequence. The tape head moves for this non-hyperbolic immortal periodic point are $[+1,-1,+1,-1]$. All points of the form $p = [r,x\underline{2}2y]$ are non-hyperbolic periodic points with period 4.

*Example 22.2.* *Hyperbolic Periodic Point*

The state set is $Q = \{q,r,s,t,u,v,w,x\}$ and the alphabet set is $A = \{1,2\}$. $\eta$ is defined below.

$\eta(q,1) = (r,1,+1)$ $\qquad$ $\eta(q,2) = (\mathcal{H},2,0)$
$\eta(r,1) = (\mathcal{H},1,0)$ $\qquad$ $\eta(r,2) = (s,2,+1)$
$\eta(s,1) = (t,1,+1)$ $\qquad$ $\eta(s,2) = (\mathcal{H},2,0)$
$\eta(t,1) = (\mathcal{H},1,0)$ $\qquad$ $\eta(t,2) = (u,2,+1)$
$\eta(u,1) = (\mathcal{H},1,0)$ $\qquad$ $\eta(u,2) = (v,1,+1)$
$\eta(v,1) = (\mathcal{H},1,0)$ $\qquad$ $\eta(v,2) = (w,2,+1)$
$\eta(w,1) = (\mathcal{H},1,0)$ $\qquad$ $\eta(w,2) = (x,1,-1)$
$\eta(x,1) = (\mathcal{H},1,0)$ $\qquad$ $\eta(x,2) = (q,2,+1)$

The point $p = [q,\overline{12}\ \underline{1}\ \overline{212222}]$ is an immortal periodic point with computational period 8 and hyperbolic degree 6.

**Definition 22.5.** *Window of Execution*
Consider the next $N$ computational steps of a Turing machine. The window of execution, denoted as $[\ell,\mu]$ or $[\ell,\ell+1,\ldots,\mu]$, is the sequence of integers representing the tape squares that the tape head visited during these $N$ computational steps. The length of the window of execution is $\mu - \ell + 1$ which is also the number of distinct tape squares visited by the tape head during these $N$ steps. To express the window of execution for the next $n$ computational steps, the lower and upper bounds are expressed as a function of $n$: $[\ell(n),\mu(n)]$. If $j \leq k$, then $[\ell(j),\mu(j)] \subseteq [\ell(k),\mu(k)]$ which follows from the definition.

The purpose of the window of execution is to describe only the portion of the tape that the tape head visits during the next $N$ computational steps. This is useful because all tape squares outside the window of execution remain unchanged during those $N$ steps. Since the tape squares may be renumbered without changing the results of the machine execution, for convenience it is often assumed that the machine starts execution at tape square 0. In example 22.2, during the next 8 computational steps – that is, one cycle of the immortal periodic point – the window of execution is [0, 6] and its length is 7.

## 22.3 State Cycles

This section introduces state cycles and consecutive repeating state cycles. Subsequently, a proof shows that a consecutive repeating state cycle determines a unique periodic point and a periodic point determines a unique consecutive repeating state cycle.

**Definition 22.6.** *State Cycle*
Consider $N$ execution steps of Turing Machine $(Q, A, \eta)$. After each execution step, the machine is in some state $q_k$ and the tape head is pointing to some alphabet symbol $a_k$. Relabel the indices of the states and the alphabet symbols if necessary and assume the machine has not halted after $N$ execution steps. This execution sequence of input commands is $(q_0, a_0) \mapsto (q_1, a_1) \mapsto \ldots \mapsto (q_{N-1}, a_{N-1}) \mapsto (q_N, a_N)$, where each pair $(q_k, a_k)$ executed as $\eta(q_k, a_k)$ is called an input command. A *state cycle* is a non-halting execution sequence of input commands such that the first and last input command in the sequence have the same state: $(q_k, a_k) \mapsto (q_{k+1}, a_{k+1}) \mapsto \ldots \mapsto (q_{N-1}, a_{N-1}) \mapsto (q_k, a_k)$. The length of this state cycle equals the number of input commands minus one. A state cycle is called a prime state cycle if it contains no proper state subcycles. For a prime state cycle, the length of the cycle equals the number of distinct states in the sequence. For example, $(2,0) \mapsto (3,1) \mapsto (4,0) \mapsto (2,1)$ is called a prime 3-state cycle because it has length 3 and also 3 distinct states $\{2,3,4\}$.

*Remark 22.1.* *Any prime state cycle has length* $\leq |Q|$

*Proof.* This follows from the pigeonhole principle and the definition of a prime state cycle.

*Remark 22.2.* *Any state cycle contains a prime state cycle*

*Proof.* Relabeling if necessary let $\zeta(q_1, q_1) = (q_1, a_1) \mapsto \ldots \mapsto (q_n, a_n) \mapsto (q_1, a_{n+1})$ be a state cycle. If $q_1$ is the only state visited twice, then the proof is completed. Otherwise, define $\ell = \min\{|\zeta(q_k, q_k)| : \zeta(q_k, q_k) \text{ is a subcycle of } \zeta(q_1, q_1)\}$. Then $\ell$ exists because $\zeta(q_1, q_1)$ is a subcycle of $\zeta(q_1, q_1)$. Claim: Any state cycle $\zeta(q_j, q_j)$ with $|\zeta(q_j, q_j)| = \ell$ must be a prime state cycle. Suppose not. Then there is a state $r \neq q_j$ that is visited twice in the state cycle $\zeta(q_j, q_j)$. But then $\zeta(q_r, q_r)$ is a cycle with length less than $\ell$ which contradicts $\ell$'s definition.

**Definition 22.7.** *Consecutive repeating state cycle*
If machine $(Q, A, \eta)$ during program execution repeats a state cycle two consecutive times, $(q_1, b_1) \mapsto \ldots \mapsto (q_n, b_n) \mapsto (q_1, b_1) \mapsto \ldots \mapsto (q_n, b_n) \mapsto (q_1, b_1)$, then $(Q, A, \eta)$ has a consecutive repeating state cycle.

**Theorem 22.1.** *Each periodic point determines a unique consecutive repeating state cycle*

*Proof.* Suppose $p$ is an immortal periodic point with period $n$. Let the input command sequence $(q_1, b_1) \mapsto \ldots \mapsto (q_n, b_n) \mapsto (q_{n+1}, b_{n+1})$ denote the first $n$ input commands that are executed. Since $p$ has period $n$, $(q_1, b_1) = (q_{n+1}, b_{n+1})$. Thus, the first $n$ steps are a state cycle $(q_1, b_1) \mapsto \ldots \mapsto (q_n, b_n) \mapsto (q_1, b_1)$. Since the $n+1$ computational step corresponds to applying $\eta$ to $p$, the window of execution is identical for the next $n$ steps as it was for the first $n$ steps. Thus, the next $n$ steps have an input command sequence that is identical as the first $n$ steps. Thus, the sequence of input commands for $2n$ steps is $(q_1, b_1) \mapsto \ldots \mapsto (q_n, b_n) \mapsto (q_1, b_1) \mapsto \ldots \mapsto (q_n, b_n) \mapsto (q_1, b_1)$.

**Theorem 22.2.** *Each consecutive repeating state cycle determines a unique periodic point*

*Proof.* Suppose Turing machine $(Q, A, \eta)$ begins or resumes execution at some tape square and repeats a state cycle two consecutive times denoted as $(q_1, b_1) \mapsto \ldots \mapsto (q_n, b_n) \mapsto (q_1, b_1) \mapsto \ldots \mapsto (q_n, b_n) \mapsto (q_1, b_1)$. Let $s_k$ denote the tape square just before input command $(q_k, b_k)$ is executed the first time where $1 \leq k \leq n$. Let $t_k$ denote the tape square just before input command $(q_k, b_k)$ is executed the second time where $1 \leq k \leq n$.

Thus, just before input command $(q_1, b_1)$ is executed a second time, the window of execution for the first state cycle is $I_n = \cup_{j=1}^{n+1}\{s_j\}$ where $s_{n+1} = t_1$. The window of execution for the second repetition of the state cycle is $J_n = \cup_{j=1}^{n+1}\{t_j\}$, where $t_{n+1} = t_n + t_1 - s_n$.

Furthermore, observe that the window of execution for the computational steps 1 thru $k$ is $I_k = \cup_{j=1}^{k+1}\{s_j\}$ where the tape head is located at tape square $s_{k+1}$ after input command $(q_k, b_k)$ is executed the first time. Also, observe that the window of execution for the computational steps $n+1$ thru $n+k$ is $J_k = \cup_{j=1}^{k+1}\{t_j\}$ where the tape head is located at tape square $t_{k+1}$ after the input command $(q_k, b_k)$ is executed the second time in the second repeating cycle.

Next a useful notation represents the tape patterns for each computational step. Then the proof is completed using induction. Let $V_1$ denote the tape pattern, which is the sequence of alphabet symbols in the tape squares over the window of execution $I_n$, just before input command $(q_1, b_1)$ is executed the first time. Thus, $V_1(s_1) = b_1$. Let $V_k$ denote the tape pattern, which is the sequence of alphabet symbols in the tape squares over the window of execution $I_n$, just before input command $(q_k, b_k)$ is executed the first time. Thus, $V_k(s_k) = b_k$.

Let $W_1$ denote the tape pattern, which is the sequence of alphabet symbols in the tape squares over the window of execution $J_n$, just before input command $(q_1, b_1)$ is executed the second time. Thus, $W_1(t_1) = b_1$. Let $W_k$ denote the tape pattern, which is the sequence of alphabet symbols in the tape squares over the window of execution $J_n$, just before input command $(q_k, b_k)$ is executed the second time. Thus, $W_k(t_k) = b_k$.

Using induction, it is shown that $V_1$ on window of execution $I_n$ equals $W_1$ on window of execution $J_n$. This completes the proof.

**Base Case.** Since $(q_1, b_1)$ is the input command before computational step 1 and $(q_1, b_1)$ is the input command before computational step $n+1$, then $V_1(s_1) = b_1 = W_1(t_1)$. Thus, $V_1$ restricted to window of execution $I_1$ equals $W_1$ restricted to window of execution $J_1$.

$\eta(q_1, b_1) = (q_2, a_1, x)$ for some $a_1$ in $A$ and where $x = -1$ or $+1$.

Case $x = +1$. **Right Tape Head Move**

$$\begin{array}{llll} & s_1\ s_2 & & t_1\ t_2 \\ V_1 \ldots & \underline{b}_1\ b_2 \ldots \qquad & W_1 \ldots & \underline{b}_1\ b_2 \ldots \\ V_2 \ldots & a_1\ \underline{b}_2 \ldots \qquad & W_2 \ldots & a_1\ \underline{b}_2 \ldots \end{array}$$

Then $s_2 = s_1 + 1$ and $t_2 = t_1 + 1$ and $V_1(s_2) = b_2 = W_1(t_2)$. It has already been observed that $V_1(s_1) = b_1 = W_1(t_1)$. Thus, $V_1$ restricted to the window of execution $I_2$ equals $W_1$ restricted to the window of execution $J_2$ . Furthermore, the tape head is at $s_1$ just before computational step 1 and input command $(q_1, b_1)$ is executed; the tape head is at $t_1$ just before computational step $n+1$ and input command $(q_1, b_1)$ is executed. Also, $V_2(s_1) = a_1 = W_2(t_1)$ and $V_2(s_2) = b_2 = W_2(t_2)$. Thus, $V_2$ restricted to the window of execution $I_2$ equals $W_2$ restricted to the window of execution $J_2$. Furthermore, the tape head is at $s_2$ just before computational step 2 and input command $(q_2, b_2)$ is executed; the tape head is at $t_2$ just before computational step $n+2$ and input command $(q_2, b_2)$ is executed.

Case $x = -1$. **Left Tape Head Move**

$$\begin{array}{llll} & s_2\ s_1 & & t_2\ t_1 \\ V_1 \ldots & b_2\ \underline{b}_1 \ldots \qquad & W_1 \ldots & b_2\ \underline{b}_1 \ldots \\ V_2 \ldots & \underline{b}_2\ a_1 \ldots \qquad & W_2 \ldots & \underline{b}_2\ a_1 \ldots \end{array}$$

Then $s_2 = s_1 - 1$ and $t_2 = t_1 - 1$ and $V_1(s_2) = b_2 = W_1(t_2)$. And $V_1(s_1) = b_1 = W_1(t_1)$. Thus, $V_1$ restricted to the window of execution $I_2$ equals $W_1$ restricted to the window of execution $J_2$. Furthermore, the tape head is at $s_1$ just before computational step 1 and input command $(q_1, b_1)$ is executed; the tape head is at $t_1$ just before computational step $n+1$ and input command $(q_1, b_1)$ is executed. Also, $V_2(s_1) = a_1 = W_2(t_1)$ and $V_2(s_2) = b_2 = W_2(t_2)$. Thus, $V_2$ restricted to the window of execution $I_2$ equals $W_2$ restricted to the window of execution $J_2$. Furthermore, the tape head is at $s_2$ just before computational step 2 and input command $(q_2, b_2)$ is executed; the tape head is at $t_2$ just before computational step $n+2$ and input command $(q_2, b_2)$ is executed. This completes the base case of induction.

**Induction Hypothesis.** Suppose that for the $1, 2, \ldots, k-1$ computational steps and the corresponding $n+1, n+2, \ldots, n+k-1$ steps that for every $i$ with $1 \le i \le k$:

1. $V_1$ restricted to the window of execution $I_i$ equals $W_1$ restricted to the window of execution $J_i$; and for each remaining $p$ where $p \le i$, $V_p$ restricted to the window of execution $I_i$ equals $W_p$ restricted to the window of execution $J_i$.

2. Furthermore, the tape head is at $s_i$ just before computational step $i$ and input command $(q_i, b_i)$ is executed; the tape head is at $t_i$ just before step $n+i$ and input command $(q_i, b_i)$ is executed.

**Induction Step.** Since $(q_k, b_k)$ is the input command before computational step $k$ and before computational step $n+k$, then $V_k(s_k) = b_k = W_k(t_k)$.

$\eta(q_k, b_k) = (q_{k+1}, a_k, x)$ for some $a_k$ in $A$ and $x = -1$ or $+1$.

Case $x = +1$. **Right Tape Head Move** for computational steps $k$ and $n+k$.

$$\begin{array}{llllll} & & s_k \; s_{k+1} & & & t_k \; t_{k+1} \\ V_k & \ldots & \underline{b_k} \; b_{k+1} \; \ldots & W_k & \ldots & \underline{b_k} \; b_{k+1} \; \ldots \\ V_{k+1} & \ldots & a_k \; \underline{b_{k+1}} \; \ldots & W_{k+1} & \ldots & a_k \; \underline{b_{k+1}} \; \ldots \end{array}$$

By the inductive hypothesis $V_k$ restricted to window of execution $I_k$ equals $W_k$ restricted to window of execution $J_k$ and the only change to the tape and tape head after executing $\eta(q_k, b_k) = (q_{k+1}, a_k, +1)$ for the steps $k$ and $n+k$ is that $V_{k+1}(s_k) = a_k = W_{k+1}(t_k)$ and $V_{k+1}(s_{k+1}) = b_{k+1} = W_{k+1}(t_{k+1})$ and that the tape heads move right to $s_{k+1}$ and $t_{k+1}$ respectively. Thus, $V_{k+1}$ restricted to window of execution $I_{k+1}$ equals $W_{k+1}$ restricted to window of execution $J_{k+1}$. And for each $j$ satisfying $1 \le j \le k$, then $V_j$ restricted to window of execution $I_{k+1}$ equals $W_j$ restricted to window of execution $J_{k+1}$.

Case $x = -1$. **Left Tape Head Move** for computational steps $k$ and $n+k$.

$$\begin{array}{llllll} & & s_{k+1} \; s_k & & & t_{k+1} \; t_k \\ V_k & \ldots & b_{k+1} \; \underline{b_k} \; \ldots & W_k & \ldots & b_{k+1} \; \underline{b_k} \; \ldots \\ V_{k+1} & \ldots & \underline{b_{k+1}} \; a_k \; \ldots & W_{k+1} & \ldots & \underline{b_{k+1}} \; a_k \; \ldots \end{array}$$

By the inductive hypothesis $V_k$ restricted to window of execution $I_k$ equals $W_k$ restricted to window of execution $J_k$ and the only change to the tape and tape head after executing $\eta(q_k, b_k) = (q_{k+1}, a_k, -1)$ for the steps $k$ and $n+k$ is that $V_{k+1}(s_k) = a_k = W_{k+1}(t_k)$ and $V_{k+1}(s_{k+1}) = b_{k+1} = W_{k+1}(t_{k+1})$ and that the tape heads move left to $s_{k+1}$ and $t_{k+1}$ respectively. Thus, $V_{k+1}$ restricted to window of execution $I_{k+1}$ equals $W_{k+1}$ restricted to window of execution $J_{k+1}$. And for each $j$ satisfying $1 \le j \le k$, then $V_j$ restricted to window of execution $I_{k+1}$ equals $W_j$ restricted to window of execution $J_{k+1}$.

## 22.4 Prime Directed Edge Sequences

Edge pattern substitution is the mathematical operation used to link match prime directed edges. Prime directed edges are link matched to construct prime directed edge sequences. The notion of an overlap match expresses how a part or all of one tape pattern may match part or all of another tape pattern.

**Definition 22.8.** *Overlap Matching & Intersection Patterns*
Let $V$ and $W$ be patterns. $(V, s)$ overlap matches $(W, t)$ if and only if $V(s+c) = W(t+c)$ for each integer $c$ satisfying $-\ell \le c \le \mu$ such that $\ell = \min\{s, t\}$ and $\mu = \min\{|V| - 1 - s, |W| - 1 - t\}$ where $0 \le s < |V|$ and $0 \le t < |W|$. The index

$s$ is called the head of pattern $V$ and the index $t$ is called the head of pattern $W$. If $V$ is also a subpattern, then $(V,s)$ submatches $(W,t)$. If $(V,s)$ overlap matches $(W,t)$, then define the intersection pattern $I$ with head $u = \ell$ as $(I,u) = (V,s) \cap (W,t)$, where $I(c) = V(c+s-\ell)$ for every integer $c$ satisfying $0 \leq c \leq (\ell + \mu)$ where $\ell = \min\{s,t\}$ and $\mu = \min\{|V|-1-s, |W|-1-t\}$.

**Definition 22.9.** *Edge Pattern Substitution Operator*
Given patterns $V = v_0 v_1 \ldots v_n$ and $W = w_0 w_1 \ldots w_n$ with heads $s,t$ satisfying $0 \leq s,t \leq n$ and pattern $P = p_0 p_1 \ldots p_m$ with head $u$ satisfying $0 \leq u \leq m$. If $(P,u)$ overlap matches $(V,s)$, define the edge pattern substitution operator $\oplus$ as $E = (P,u) \oplus [(V,s) \Rightarrow (W,t)]$ according to the four different cases **A**, **B**, **C** and **D**.

**Case A** $u > s$ and $m - u > n - s$

$$E(k) = \begin{array}{ll} W(k+s-u) & \text{if } u-s \leq k \leq u+n-s \\ P(k) & \text{if } 0 \leq k < u-s \text{ or } u+n-s < k \leq m \end{array}$$

The head of $E$ is $u + t - s$ and $|E| = m + 1$.

$$\begin{array}{llllllll} p_0\ p_1 \cdots & p_{u-s} & \cdots & \underline{p_u} & \cdots & p_{u+n-s} & \cdots & p_m \\ & v_0 & \cdots & \underline{v_s} & \cdots & v_n & & \\ & w_0 & \cdots & w_s & \cdots & w_n & & \end{array}$$

**Case B** $u > s$ and $m - u \leq n - s$

$$E(k) = \begin{array}{ll} W(k+s-u) & \text{if } u-s \leq k \leq n+u-s \\ P(k) & \text{if } 0 \leq k < u-s \end{array}$$

The head of $E$ is $u + t - s$ and $|E| = n + u - s + 1$.

$$\begin{array}{llllllll} p_0\ p_1 \cdots & p_{u-s} & \cdots & \underline{p_u} & \cdots & p_m & & \\ & v_0 & \cdots & \underline{v_s} & \cdots & v_{s+m-u} & \cdots & v_n \\ & w_0 & \cdots & w_s & \cdots & w_{s+m-u} & \cdots & w_n \end{array}$$

**Case C** $u \leq s$ and $m - u \leq n - s$

$E(k) = W(k)$ when $0 \leq k \leq n$. The head of $E$ is $t$ and $|E| = |W| = n + 1$.

$$\begin{array}{llllllll} & p_0 & \cdots & \underline{p_u} & \cdots & p_m & & \\ v_0 \cdots & v_{s-u} & \cdots & \underline{v_s} & \cdots & v_{s+m-u} & \cdots & v_n \\ w_0 \cdots & w_{s-u} & \cdots & w_s & \cdots & w_{s+m-u} & \cdots & w_n \end{array}$$

**Case D** $u \leq s$ and $m - u > n - s$

$$E(k) = \begin{array}{ll} P(k+u-s) & \text{if } n < k \leq m+s-u \\ W(k) & \text{if } 0 \leq k \leq n \end{array}$$

The head of $E$ is $t$ and $|E| = m + s - u + 1$.

$$\begin{array}{llllllll} & p_0 & \cdots & \underline{p_u} & \cdots & p_{u+n-s} & \cdots & p_m \\ v_0 \cdots & v_{s-u} & \cdots & \underline{v_s} & \cdots & v_n & & \\ w_0 \cdots & w_{s-u} & \cdots & w_s & \cdots & w_n & & \end{array}$$

Set pattern $P = \underline{0}101\ 110$. Set pattern $V = 11\ \underline{0}101$. Set pattern $W = 01\ 00\underline{1}0$. Then $(P,0)$ *overlap matches* $(V,2)$. Thus, the *edge pattern substitution operation* is well-defined so $E = (P,0) \oplus [(V,2) \Rightarrow (W,4)] = 01\ 00\underline{1}0\ 110$. The head or index of pattern $E = 4$. Furthermore, $(P,4)$ overlap matches $(V,0)$. Thus, $F = (P,4) \oplus [(V,0) \Rightarrow (W,4)] = 0101\ 0100\underline{1}0$. The index of pattern $F = u + t - s = 4 + 4 - 0 = 8$.

**Definition 22.10.** *Execution node for* $(Q, A, \eta)$
An execution node is a triplet $\mathfrak{N} = [q, w_0 w_1 \dots w_n, t]$ for some state $q$ in $Q$ where $w_0 w_1 \dots w_n$ is a pattern of $n+1$ alphabet symbols each in $A$ such that $t$ is a non-negative integer satisfying $0 \le t \le n$. Intuitively, $w_0 w_1 \dots w_n$ is the pattern of alphabet symbols on $n+1$ consecutive tape squares and $t$ represents the location of the tape head.

**Definition 22.11.** *Halting Execution Node*
Suppose $[q, v_0 v_1 \dots v_n, s]$ is an execution node and over the next $|Q|$ computational steps a prime state cycle is not found. In other words, a prime directed edge is not generated. Then the Turing machine execution halted in $|Q|$ or less steps. Let $W$ be a pattern such that $(W,t)$ submatches $(V,s)$ and $W$ spans the window of execution until execution halts. Define the halting node as $H = [q, W, t]$.

**Definition 22.12.** *Prime directed edge*
A prime head execution node $\mathfrak{H} = [q, v_0 v_1 \dots v_n, s]$ and prime tail execution node $\mathfrak{T} = [r, w_0 w_1 \dots w_n, t]$ are called a prime directed edge iff all of the following hold:

1. When Turing machine $(Q, A, \eta)$ starts execution, it is in state $q$ and the tape head is located at tape square $s$. For each $j$ satisfying $0 \le j \le n$ tape square $j$ contains symbol $v_j$. In other words, the initial tape pattern is $v_0 v_1 \dots \underline{v_s} \dots v_n$.
2. During the next $N$ computational steps, state $r$ is visited twice and all other states in $Q$ are visited at most once. In other words, the corresponding sequence of input commands during the $N$ computational steps executed contains only one prime state cycle.
3. After $N$ computational steps, where $1 \le N \le |Q|$, the machine is in state $r$. The tape head is located at tape square $t$. For each $j$ satisfying $0 \le j \le n$ tape square $j$ contains symbol $w_j$. The tape pattern after the $N$ computational steps is $w_0 w_1 \dots \underline{w_t} \dots w_n$.
4. The window of execution for these $N$ computational steps is $[0, n]$.

A prime directed edge is denoted as $[q, v_0 v_1 \dots v_n, s] \Rightarrow [r, w_0 w_1 \dots w_n, t]$ or $\mathfrak{H} \Rightarrow \mathfrak{T}$. The number of computational steps $N$ is denoted as $|\mathfrak{H} \Rightarrow \mathfrak{T}|$.

**Definition 22.13.** *Overlap matching of a node to a prime head node*
Execution node $\mathfrak{N}$ overlap matches head execution node $\mathfrak{H}$ iff the following hold:

1. $\mathfrak{N} = [r, w_0 w_1 \dots w_n, t]$ is an execution node satisfying $0 \le t \le n$.
2. $\mathfrak{H} = [q, v_0 v_1 \dots v_m, s]$ is a prime head node satisfying $0 \le s \le m$.
3. State $q$ = State $r$.
4. Pattern $(W,t)$ overlap matches $(V,s)$, where $W = w_0 w_1 \dots w_n$ and $V = v_0 v_1 \dots v_m$.

**Lemma 22.1.** *Overlap matching prime head nodes are equal*

If $\mathfrak{H}_j = [q,P,u]$ and $\mathfrak{H}_k = [q,V,s]$ are prime head nodes and they overlap match, then they are equal. Distinct prime directed edges have prime head nodes that do not overlap match.

*Proof.* From the definition of prime edge, $0 \leq u \leq |P|$ and $0 \leq s \leq |V|$. Let $(I,m) = (P,u) \cap (V,s)$ where $m = \min\{s,u\}$. Suppose the same machine begins execution on tape $I$ with tape head at $m$ in state $q$. If $s = u$ and $|P| = |V|$, then the proof is complete. Otherwise, $s \neq u$ or $|P| \neq |V|$ or both. $\mathfrak{H}_j$ has a window of execution $[0,|P|-1]$ and $\mathfrak{H}_k$ has window of execution $[0,|V|-1]$. Let the $i$th step be the first time that the tape head exits finite tape $I$. This means the machine would execute the same machine instructions with respect to $\mathfrak{H}_j$ and $\mathfrak{H}_k$ up to the $i$th step, so on the $i$th step, $\mathfrak{H}_j$ and $\mathfrak{H}_k$ must execute the same instruction. Since it exits tape I at the $i$th step, this would imply that either pattern $P$ or $V$ are exited at the $i$th step. This contradicts either that $[0,|P|-1]$ is the window of execution for $\mathfrak{H}_j$ or $[0,|V|-1]$ is the window of execution for $\mathfrak{H}_k$.

**Theorem 22.3.** *The number of prime directed edges is* $\leq |Q|^2|A|^{|Q|+1}$

*Proof.* From lemma 22.1, each prime head node determines a unique prime directed edge so an upper bound for head nodes provides an upper bound for the prime directed edges. Consider prime head node $[q,V,s]$. There are $|Q|$ choices for the state $q$. Any pattern that represents the window of execution has length $\leq |Q|+1$.

Furthermore, lemma 22.1 implies, for any pattern $P$ where $(V,s)$ submatches $(P,t)$, then the resultant pattern is the same since $V$ spans the window of execution. Thus, $|A|^{|Q|+1}$ is an upper bound for the number of different patterns $V$.

There are two choices for $s$ in a $|Q|+1$ length pattern because the maximum number of execution steps is $|Q|$ (i.e., the tape head move sequence is either $|Q|$ consecutive left tape head moves or $|Q|$ right tape head moves). Thus, $|Q|$ is an upper bound for the number of choices for $s$ unless $|Q| = 1$. The bound holds in the trivial case that $|Q| = 1$. Thus, there are at most $|Q|^2|A|^{|Q|+1}$ prime directed edges.

$|Q|^2|A|^{|Q|+1}$ is not a strict upper bound. Procedure 2 describes an algorithm for finding all the prime directed edges of a Turing machine, which also provides the number of prime directed edges.

**Definition 22.14.** *Edge Node Substitution Operator*
Let $\mathfrak{H} \Rightarrow \mathfrak{T}$ be a prime directed edge with prime head node $\mathfrak{H} = [q,v_0v_1\dots v_n,s]$ and tail node $\mathfrak{T} = [r,w_0w_1\dots w_n,t]$. If execution node $\mathfrak{N} = [q,p_0p_1\dots p_m,u]$ overlap matches $\mathfrak{H}$, then the edge pattern substitution operator in definition 22.9 induces a new execution node $\mathfrak{N} \oplus (\mathfrak{H} \Rightarrow \mathfrak{T}) = [r,(P,u) \oplus [(V,s) \Rightarrow (W,t)],k]$ with head $k = u+t-s$ if $u > s$ and head $k = t$ if $u \leq s$ such that $0 \leq s,t \leq n$ and $0 \leq u \leq m$ and patterns $V = v_0v_1\dots v_n$ and $W = w_0w_1\dots w_n$ and $P = p_0p_1\dots p_m$.

Let $\mathfrak{P} = \{\mathfrak{H}_1 \Rightarrow \mathfrak{T}_1,\dots,\mathfrak{H}_k \Rightarrow \mathfrak{T}_k,\dots,\mathfrak{H}_N \Rightarrow \mathfrak{T}_N\}$ denote the finite set of prime directed edges for $(Q,A,\eta)$. The number of prime directed edges in $\mathfrak{P}$ is $|\mathfrak{P}|$ and is called the *prime directed edge (PE) complexity* of $(Q, A, \eta)$. As defined in 22.14, the link matching step compares two tape patterns, one from execution node $\mathfrak{N}_k$ and the other from head node $\mathfrak{H}_{k+1}$. If the tape patterns overlap match and the state of $\mathfrak{N}_k$ equals the state of $\mathfrak{H}_{k+1}$, then the edge node substitution operation is well-defined and is used to glue prime directed edge $\mathfrak{H}_{k+1} \Rightarrow \mathfrak{T}_{k+1}$ to execution node $\mathfrak{N}_k$. In other words, $\mathfrak{N}_{k+1} = \mathfrak{N}_k \oplus (\mathfrak{H}_{k+1} \Rightarrow \mathfrak{T}_{k+1})$ is computed.

**Definition 22.15.** *Prime directed edge sequence and Link Matching*
A prime directed edge sequence is defined inductively. Each element is a coordinate pair, where the first element is a prime directed edge, the second element is an execution node and each element is expressed as $(\mathfrak{H}_k \Rightarrow \mathfrak{T}_k, \mathfrak{N}_k)$. The first element of a prime directed edge sequence is $(\mathfrak{H}_1 \Rightarrow \mathfrak{T}_1, \mathfrak{N}_1)$ where $\mathfrak{N}_1 = \mathfrak{T}_1$, and $\mathfrak{H}_1 \Rightarrow \mathfrak{T}_1$ is some prime directed edge in $\mathfrak{P}$. For simplicity in this definition, the indices in $\mathfrak{P}$ are relabeled if necessary so the first element has indices equal to 1. If $\mathfrak{N}_1$ overlap matches some non-halting prime head node $\mathfrak{H}_2$, the second element of the prime directed edge sequence is $(\mathfrak{H}_2 \Rightarrow \mathfrak{T}_2, \mathfrak{N}_2)$ where $\mathfrak{N}_2 = \mathfrak{N}_1 \oplus (\mathfrak{H}_2 \Rightarrow \mathfrak{T}_2)$. This is called a link match step. Otherwise, $\mathfrak{N}_1$ overlap matches a halting node, as defined in 22.11. In this case, the prime directed edge sequence terminates and this is called a halting match step. This is expressed as $[(\mathfrak{H}_1 \Rightarrow \mathfrak{T}_1, \mathfrak{T}_1), \mathcal{H}]$.

If the first $k-1$ steps are link match steps, then the edge sequence up to the *kth* element is inductively defined as $[(\mathfrak{H}_1 \Rightarrow \mathfrak{T}_1, \mathfrak{N}_1), (\mathfrak{H}_2 \Rightarrow \mathfrak{T}_2, \mathfrak{N}_2), \ldots,$ $(\mathfrak{H}_k \Rightarrow \mathfrak{T}_k, \mathfrak{N}_k)]$ where $\mathfrak{N}_j$ overlap matches prime head node $\mathfrak{H}_{j+1}$ and $\mathfrak{N}_{j+1} = \mathfrak{N}_j \oplus (\mathfrak{H}_{j+1} \Rightarrow \mathfrak{T}_{j+1})$ for each j satisfying $0 \le j < k$.

## 22.5 Search Procedure for Periodic Points

This section demonstrates how to search for any periodic point – when they exist – by looking for consecutive repeating state cycles inside a prime directed edge sequence. This procedure is useful because it does not search over a tape pattern that has already been previously examined. This is due to Lemma 22.1 that proves if two head nodes overlap match in their respective prime directed edges, then the prime directed edges are equal.

Although there are other methods to look for periodic points, this search procedure demonstrates a broader approach because each aperiodic immortal configuration corresponds to a unique prime directed edge sequence. This is relevant because the non-trivial recurrence of the $M$-bounded aperiodic immortal configurations is integral to the undecidability of the Halting problem.

Link matching is a computational operation used to construct prime directed edge sequences. The following example link matches two prime directed edges.

*Example 22.3.* *Link matching prime directed edges*

State set $Q = \{q, r, s, t, u\}$. Alphabet set $A = \{0, 1\}$. Program $\eta$ is defined below.

$$\begin{array}{ll} \eta(q,0) = (r,1,+1) & \eta(q,1) = (r,1,-1) \\ \eta(r,0) = (\mathcal{H},1,0) & \eta(r,1) = (s,0,+1) \\ \eta(s,0) = (t,0,+1) & \eta(s,1) = (\mathcal{H},1,0) \\ \eta(t,0) = (u,1,-1) & \eta(t,1) = (q,0,+1) \\ \eta(u,0) = (q,1,-1) & \eta(u,1) = (t,0,+1) \end{array}$$

The execution steps of $[u,0\underline{0}010,1] \Rightarrow [q,1000\underline{0},4]$ are shown in table 1.

**Table 22.1** Prime Directed Edge $[u,0\underline{0}010,1] \Rightarrow [q,1000\underline{0},4]$

| STATE | TAPE | HEAD | COMMAND |
|---|---|---|---|
| $u$ | $0\underline{0}010$ | 1 | $\eta(u,0) = (q,1,-1)$ |
| $q$ | $\underline{0}1010$ | 0 | $\eta(q,0) = (r,1,+1)$ |
| $r$ | $1\underline{1}010$ | 1 | $\eta(r,1) = (s,0,+1)$ |
| $s$ | $10\underline{0}10$ | 2 | $\eta(s,0) = (t,0,+1)$ |
| $t$ | $100\underline{1}0$ | 3 | $\eta(t,1) = (q,0,+1)$ |
| $q$ | $1000\underline{0}$ | 4 | |

The execution steps of $[q,\underline{0}1010,0] \Rightarrow [q,1000\underline{0},4]$ are shown in table 2.

**Table 22.2** Prime Directed Edge $[q,\underline{0}1010,0] \Rightarrow [q,1000\underline{0},4]$

| STATE | TAPE | HEAD | COMMAND |
|---|---|---|---|
| $q$ | $\underline{0}1010$ | 0 | $\eta(q,0) = (r,1,+1)$ |
| $r$ | $1\underline{1}010$ | 1 | $\eta(r,1) = (s,0,+1)$ |
| $s$ | $10\underline{0}10$ | 2 | $\eta(s,0) = (t,0,+1)$ |
| $t$ | $100\underline{1}0$ | 3 | $\eta(t,1) = (q,0,+1)$ |
| $q$ | $1000\underline{0}$ | 4 | |

Prime edge $[u,0\underline{0}010,1] \Rightarrow [q,1000\underline{0},4]$ can be link matched to prime edge $[q,\underline{0}1010,0] \Rightarrow [q,1000\underline{0},4]$. After link matching, the sequence of input commands is $[(u,0),(q,0),\ (r,1),(s,0),(t,1),(q,0),(r,1),(s,0),(t,1)]$. Observe that $[(q,0),(r,1),\ (s,0),(t,1),\ (q,0),(r,1),\ (s,0),(t,1)]$ is a consecutive repeating state cycle, which corresponds to the periodic configuration $[q,\overline{1000}\ \underline{0}101\ \overline{0101}]$.

**Definition 22.16.** *Prime Input Command Sequence*
Suppose $(q_1,a_1) \mapsto \ldots \mapsto (q_n,a_n)$ is an execution sequence of input commands for $(Q,A,\eta)$. Then $(q_1,a_1) \mapsto \ldots \mapsto (q_n,a_n)$ is a prime input command sequence if $q_n$ is visited twice and all other states in this sequence are visited once. In other words, a prime input command sequence contains exactly one prime state cycle.

**Lemma 22.2.** *Prime Directed Edges* $\Leftrightarrow$ *Prime Input Command Sequences*

Prime directed edges and prime input command sequences are in one-to-one correspondence.

*Proof.* ($\Rightarrow$) Let $\mathfrak{H} \Rightarrow \mathfrak{T}$ be a prime directed edge where $\mathfrak{H} = [q, v_0 v_1 \dots v_n, s]$ and $\mathfrak{T} = [r, w_0 w_1 \dots w_n, t]$. From the definition of a prime directed edge, over the next $N$ computational steps some state $r$ is visited twice, all other states in $Q$ are visited at most once and there is a sequence of input commands $(q, v_s) \mapsto (q_1, a_1) \mapsto \dots (r, a_k) \dots \mapsto (r, w_t)$ corresponding to these $N$ steps. This is a prime input command sequence.

($\Leftarrow$) Let $(q_1, a_1) \mapsto \dots \mapsto (r, a_{N+1})$ be a prime input command sequence with $N$ computational steps. Then $r$ is visited twice and all other states in the sequence are visited only once. Let $v_0 v_1 \dots v_n$ be the initial tape pattern over the window of execution during the $N$ computational steps. Now $a_1 = v_s$ for some $s$. Let $w_0 w_1 \dots w_n$ be the final tape pattern over the window of execution as a result of these $N$ steps. From the definition of the window of execution, the tape head is at some $t$ satisfying $0 \leq t \leq n$ after these $N$ steps. Thus, $[q, v_0 v_1 \dots v_n, s] \Rightarrow [r, w_0 w_1 \dots w_n, t]$ is a prime directed edge.

**Lemma 3.** *Any consecutive repeating state cycle is contained in a composition of one or more prime input command sequences.*

*Proof.* The proof is in the appendix.

*Example 22.4*

This example illustrates the correspondence between prime directed edges and prime input command sequences for machine $(Q, A, \eta)$ where $Q = \{2,3,4\}$, $\mathcal{H}$ is the halting state, $A = \{0,1\}$ and $\eta$ is specified as $\eta(2,0) = (3,1,-1)$, $\eta(2,1) = (4,0,-1)$, $\eta(3,0) = (4,1,+1)$, $\eta(3,1) = (4,0,+1)$, $\eta(4,0) = (\mathcal{H},0,0)$, and $\eta(4,1) = (2,0,+1)$. The correspondence is shown below.

| **Prime Directed Edges** | **Prime Input Command Sequences** |
|---|---|
| $[2, 0\underline{0}0, 1] \Rightarrow [2, 10\underline{0}, 2]$ | $(2,0) \mapsto (3,0) \mapsto (4,1) \mapsto (2,0)$ |
| $[2, 1\underline{0}0, 1] \Rightarrow [2, 00\underline{0}, 2]$ | $(2,0) \mapsto (3,1) \mapsto (4,1) \mapsto (2,0)$ |
| $[2, 1\underline{1}, 1] \Rightarrow [2, 0\underline{0}, 1]$ | $(2,1) \mapsto (4,1) \mapsto (2,0)$ |
| $[2, 0\underline{0}1, 1] \Rightarrow [2, 10\underline{1}, 2]$ | $(2,0) \mapsto (3,0) \mapsto (4,1) \mapsto (2,1)$ |
| $[2, 1\underline{0}1, 1] \Rightarrow [2, 00\underline{1}, 2]$ | $(2,0) \mapsto (3,1) \mapsto (4,1) \mapsto (2,1)$ |
| $[3, \underline{0}10, 0] \Rightarrow [3, 1\underline{0}1, 1]$ | $(3,0) \mapsto (4,1) \mapsto (2,0) \mapsto (3,0)$ |
| $[3, \underline{1}10, 0] \Rightarrow [3, 0\underline{0}1, 1]$ | $(3,1) \mapsto (4,1) \mapsto (2,0) \mapsto (3,0)$ |
| $[4, \underline{1}0, 0] \Rightarrow [4, 1\underline{1}, 1]$ | $(4,1) \mapsto (2,0) \mapsto (3,0) \mapsto (4,1)$ |
| $[4, \underline{1}1, 0] \Rightarrow [4, 0\underline{0}, 1]$ | $(4,1) \mapsto (2,1) \mapsto (4,0)$ |

There are 9 distinct prime directed edges. Observe that $|Q|^2 |A|^{|Q|+1} = 3^2 2^4 = 144$.

**Definition 22.17.** *Edge Sequence contains a consecutive repeating state cycle*
Lemma 22.2 implies that an edge sequence corresponds to a sequence of prime input commands. The expression *an edge sequence contains a consecutive repeating state cycle* means that the corresponding sequence of prime input commands contains a consecutive repeating state cycle.

**Theorem 22.4.** *Any consecutive repeating state cycle is contained in a prime directed edge sequence.*

*Proof.* This follows immediately from definitions 22.15, 22.17 and lemmas 22.2 and 22.3.

**Procedure 1.** *Searching for a consecutive repeating state cycle in a prime directed edge sequence*

Given an edge sequence whose corresponding prime input command sequence $(q_0, a_0) \mapsto (q_1, a_1) \mapsto ... \mapsto (q_N, a_N)$ has length $N$.

Set $n = \frac{N}{2}$ if $N$ is even; otherwise, set $n = \frac{N+1}{2}$ if $N$ is odd.

For each $k$ in $\{0, 1, 2, \ldots, n-1\}$
For each $j$ in $\{0, 1, \ldots, N-2k-1\}$
{
If $(q_j, a_j) \mapsto \ldots \mapsto (q_{j+k}, a_{j+k})$ equals $(q_{j+k+1}, a_{j+k+1}) \mapsto \ldots \mapsto (q_{j+2k+1}, a_{j+2k+1})$
return consecutive repeating state cycle
$(q_j, a_j) \mapsto \ldots (q_{j+k}, a_{j+k}) \ldots (q_{j+2k+1}, a_{j+2k+1})$
}
If the outer for loop is exited without finding a consecutive repeating state cycle return NO consecutive repeating state cycles were found.

**Procedure 2.** *Prime Directed Edge Search Procedure*

Read Turing Machine $(Q, A, \eta)$ as input.
Set $\mathfrak{P} = \emptyset$.
For each non-halting state $q$ in $Q$
For each pattern $a_{-|Q|} \ldots a_{-2}a_{-1}a_0a_1a_2 \ldots a_{|Q|}$ selected from $A^{2|Q|+1}$
{

| | | | | | | | |
|---|---|---|---|---|---|---|---|
| Square | $-\lvert Q\rvert$ | ... | $-1$ | $0$ | $1$ | ... | $\lvert Q\rvert$ |
| Contents | $a_{-\lvert Q\rvert}$ | ... | $a_{-1}$ | $\underline{a_0}$ | $a_1$ | ... | $a_{\lvert Q\rvert}$ |

1. In start state $q$, $T(k) = a_k$ where $-|Q| \leq k \leq |Q|$, and tape head location 0, execute $(Q, A, \eta)$ until one state is visited twice or a halting state is reached. The execution takes $\leq |Q|$ steps.
2. If execution does not halt, set $r$ equal to the state that is first visited twice.
3. Over this window of execution, construct a prime directed edge $\mathfrak{H} \Rightarrow \mathfrak{T}$ where $\mathfrak{H} = [q, v_0v_1 \ldots v_n, s]$, $\mathfrak{T} = [r, w_0w_1 \ldots w_n, t]$ and $0 \leq s, t \leq n \leq |Q|$.
4. Set $\mathfrak{P} = \mathfrak{P} \cup \{\mathfrak{H} \Rightarrow \mathfrak{T}\}$.

}

*Remark 22.3.* *Prime Directed Edge Search Procedure finds all edges*
Procedure 2 finds all prime directed edges of $(Q, A, \eta)$ and all halting nodes.

*Proof.* Let $\mathfrak{H} \Rightarrow \mathfrak{T}$ be a prime directed edge of $(Q, A, \eta)$. Then $\mathfrak{H} \Rightarrow \mathfrak{T}$ has a head node $\mathfrak{H} = [q, v_0 v_1 ... v_n, s]$, for some state $q$ in $Q$, for some tape pattern $v_0 v_1 \ldots v_n$ that lies in $A^{n+1}$, such that $n \leq |Q|$ and $0 \leq s \leq n$. In the outer loop of procedure 2, when $q$ is selected from $Q$ and in the inner loop when the tape pattern $a_{-|Q|} \cdots a_{-2} a_{-1} a_0 a_1 a_2 \ldots a_{|Q|}$ is selected from $A^{2|Q|+1}$ such that $a_{-s} = v_0 \quad \ldots \quad a_{-k} = v_{s-k} \quad \ldots \quad a_0 = v_s \quad \ldots \quad a_k = v_{s+k} \quad \ldots \quad a_{n-s} = v_n$ then the machine execution in procedure 2 will construct prime directed edge $\mathfrak{H} \Rightarrow \mathfrak{T}$.

When the head node is a halting node, the machine execution must halt in at most $|Q|$ steps. Otherwise, it would visit a non-halting state twice and be a non-halting head node. The rest of the argument for this halting node is the same as for the non-halting head node.

To avoid subscripts of a subscript, let $p_j$ and the subscript $p(j)$ represent the same number. $\mathfrak{P} = \{\mathfrak{H}_1 \Rightarrow \mathfrak{T}_1, \ldots, \mathfrak{H}_k \Rightarrow \mathfrak{T}_k, \ldots, \mathfrak{H}_N \Rightarrow \mathfrak{T}_N\}$ is the set of all prime directed edges. $E([p_1], 1)$ is the edge sequence $[(\mathfrak{H}_{p(1)} \Rightarrow \mathfrak{T}_{p(1)}, \mathfrak{N}_{p(1)})]$ of length 1 where $\mathfrak{N}_{p(1)} = \mathfrak{T}_{p(1)}$ and $1 \leq p_1 \leq |\mathfrak{P}|$. Next $E([p_1, p_2], 2)$ is the edge sequence $[(\mathfrak{H}_{p(1)} \Rightarrow \mathfrak{T}_{p(1)}, \mathfrak{N}_{p(1)}), (\mathfrak{H}_{p(2)} \Rightarrow \mathfrak{T}_{p(2)}, \mathfrak{N}_{p(2)})]$ such that $1 \leq p_1, p_2 \leq |\mathfrak{P}|$ and $\mathfrak{N}_{p(2)} = \mathfrak{N}_{p(1)} \oplus (\mathfrak{H}_{p(2)} \Rightarrow \mathfrak{T}_{p(2)})$. In general,
$E([p_1, p_2, \ldots, p_k], k)$ denotes the edge sequence of length $k$ which is $[(\mathfrak{H}_{p(1)} \Rightarrow \mathfrak{T}_{p(1)}, \mathfrak{N}_{p(1)}), (\mathfrak{H}_{p(2)} \Rightarrow \mathfrak{T}_{p(2)}, \mathfrak{N}_{p(2)}), \ldots, (\mathfrak{H}_{p(k)} \Rightarrow \mathfrak{T}_{p(k)}, \mathfrak{N}_{p(k)})]$ where $\mathfrak{N}_{p(j+1)} = \mathfrak{N}_{p(j)} \oplus (\mathfrak{H}_{p(j+1)} \Rightarrow \mathfrak{T}_{p(j+1)})$ for each $j$ satisfying $1 \leq j \leq k-1$ and $1 \leq p(j) \leq |\mathfrak{P}|$.

**Procedure 3.** *Immortal Periodic Point Search Procedure*

Read Turing Machine $(Q, A, \eta)$ as input.

Use procedure 2 to find all prime directed edges $\mathfrak{P}$.

Set $k = 1$. Set $\mathfrak{E}(1) = \{E([1], 1), E([2], 1), \ldots, E([|\mathfrak{P}|], 1)\}$.

While ( $\mathfrak{E}(k) \neq \emptyset$ )
{
  Set $\mathfrak{E}(k+1) = \emptyset$.
  For each $E([p_1, p_2, \ldots, p_k], k)$ in $\mathfrak{E}(k)$
    For each prime directed edge $\mathfrak{H}_j \Rightarrow \mathfrak{T}_j$ in $\mathfrak{P}$
    {
      if $\mathfrak{H}_j \Rightarrow \mathfrak{T}_j$ link matches with $\mathfrak{N}_{p(k)}$ then
      {
        Set $p_{k+1} = j$.
        Set $\mathfrak{E}(k+1) = \mathfrak{E}(k+1) \cup E([p_1, p_2, \ldots, p_k, p_{k+1}], k+1)$.
        If $E([p_1, p_2, \ldots, p_k, p_{k+1}], k+1)$ has a consecutive repeating state cycle then return the consecutive repeating state cycle and exit the while loop.
      }
    }
  $k$ is incremented.
}

If the while loop exits because $\mathfrak{E}(m) = \emptyset$ for some $m$, then return $\emptyset$; in other words, $(Q, A, \eta)$ has only halting configurations.

*Remark 22.4.* $|\mathfrak{E}(k)| \leq |\mathfrak{P}|^k$

**Definition 22.18.** *Periodic Turing Machine*

A Turing machine $(Q, A, \eta)$ that has at least one periodic configuration, whenever it has an immortal configuration is said to be a *periodic* Turing machine.

*Remark 22.5.* If $E([p_1, p_2, \ldots, p_r], r)$ contains a consecutive repeating state cycle, then the corresponding periodic point has period $\leq \frac{1}{2} \sum_{k=1}^{r} |\mathfrak{H}_{p(k)} \Rightarrow \mathfrak{T}_{p(k)}|$.

*Proof.* Theorem 22.2 implies that a consecutive repeating state cycle determines an immortal periodic point. The length of the state cycle equals the period of the periodic point. Further, the number of input commands corresponding to the number of computational steps equals $|\mathfrak{H}_{p(k)} \Rightarrow \mathfrak{T}_{p(k)}|$ in directed edge $\mathfrak{H}_{p(k)} \Rightarrow \mathfrak{T}_{p(k)}$.

**Theorem 22.5.** *When machine $(Q, A, \eta)$ is periodic, procedure 3 terminates in a finite number of steps with either a consecutive repeating state cycle or for some positive integer $J$, then $\mathfrak{E}(J) = \emptyset$, which means all of the configurations of $(Q, A, \eta)$ are halting.*

*Proof.* If $(Q, A, \eta)$ has at least one configuration $(q, k, T)$ that is an immortal, then by definition 22.18, this implies the existence of a periodic point $p$ with some finite period $N$. Thus, from Theorem 22.1, there is a consecutive repeating state cycle that corresponds to the immortal periodic point $p$. Since procedure 3 searches through all possible prime edge sequences of length $k$, a consecutive repeating state cycle will be found that is contained in a prime directed edge sequence with length at most $2N$. Thus, periodic point $p$ of period $N$ will be reached before or while computing $\mathfrak{E}(2N)$.

Otherwise, $(Q, A, \eta)$ does not have any immortal configurations; in other words, for every configuration, $(Q, A, \eta)$ halts in a finite number of steps. Claim: There is a positive integer $J$ such that every edge sequence terminates while executing procedure 3. By reductio absurdum, suppose not. Then there is at least one infinite prime directed edge sequence that exists. This infinite edge sequence corresponds to an immortal configuration, which contradicts that $(Q, A, \eta)$ is a periodic machine.

*Example 22.5.* *A Turing Machine with only aperiodic immortal configurations*

This example is based on work in [2]. The state set $Q = \{a, b, c, d, e, f\}$ and the alphabet set $A = \{0, 1, 2, 3\}$. The halting state is $\mathcal{H}$. In the following table, the Turing program is presented as quintuples $(q, a, r, b, m)$ where $\eta(q, a) = (r, b, m)$.

| | | | |
|---|---|---|---|
| $(a,0,d,1,+1)$ | $(a,1,f,1,+1)$ | $(a,2,f,2,+1)$ | $(a,3,f,3,+1)$ |
| $(b,0,c,1,-1)$ | $(b,1,e,1,-1)$ | $(b,2,e,2,-1)$ | $(b,3,e,3,-1)$ |
| $(c,0,a,2,-1)$ | $(c,1,f,1,+1)$ | $(c,2,f,2,+1)$ | $(c,3,f,3,+1)$ |
| $(d,0,b,2,+1)$ | $(d,1,e,1,-1)$ | $(d,2,e,2,-1)$ | $(d,3,e,3,-1)$ |
| $(e,0,a,3,-1)$ | $(e,1,a,0,-1)$ | $(e,2,d,0,+1)$ | $(e,3,f,0,+1)$ |
| $(f,0,b,3,+1)$ | $(f,1,b,0,+1)$ | $(f,2,c,0,-1)$ | $(f,3,e,0,-1)$ |

Observe that any periodic tape pattern with any non-halting state is immortal. However, none of these configurations are periodic because none are returned to in a finite number of execution steps.

*Remark 22.6.* *Procedure 3 does not halt on aperiodic Turing machines.*

## 22.6 Discussion and Further Work

In [12], Turing presented the Halting problem and proved that the Halting problem is undecidable with a Turing Machine. In [8], Hooper proved that the Turing Immortality problem is undecidable. Both papers assume that every initial machine configuration is $M$-bounded for some finite $M$ (i.e., the tape is bounded by blank symbols before the Turing machine program begins executing). In [1], Berger demonstrated an aperiodic tiling that proved that the tiling problem was undecidable. In light of [2] and the results presented here, the aperiodicity appears to be an integral part of the undecidability.

Procedure 2 finds all the prime directed edges and works for any Turing machine. Furthermore, the construction of the edge sequences via link matching inside Procedure 3 works on any Turing machine; and at the $k$th pass through the outer loop, this construction explores all possible immortal configurations up to an edge sequence length of $k$ prime directed edges. The limitation of procedure 3 is on the aperiodic Turing machines and is due to the exit condition of finding a consecutive repeating state cycle.

Although the consecutive repeating state cycle characterizes any periodic configuration in the Turing machine, a broader notion of recurrence is needed to address the more complex behavior of aperiodic immortal configurations that are initially $M$-bounded. Lemma 22.1 implies that every immortal configuration is contained by a unique prime directed edge sequence, so prime directed edge sequences cover all possible Turing program behaviors. Research that further develops the mathematical notions described here could help better understand the aperiodic immortal configurations, aperiodic Turing machines and perhaps undecidability.

**Acknowledgements.** I would like to thank Tony Lauck, Michael Jones, Kyandoghere Kyamakya, Don Saari and Sandy Zabell for their helpful advice and discussions. I would like to thank Lutz Mueller for creating newLISP, which has been extremely useful in my research. I am extremely grateful for support from my wonderful wife, Joanne Gomez, without whom this work would not have been possible and Haley Arielle Fiske for her inspiration.

## References

1. Berger, R.: The undecidability of the domino problem. Memoirs of the American Mathematical Society (66) (1966)
2. Blondel, V., Cassaigne, J., Nichitiu, C.: On the presence of periodic configurations in Turing machines and in counter machines. Theoretical Computer Science 289(1), 573–590 (2002)
3. Bowen, R.: Periodic points and measures for Axiom A diffeomorphisms. Transactions for American Mathematical Society 154, 377–397 (1971)
4. Brouwer, L.E.J.: Uber Abbildung von Mannigfaltigkeiten. Mathematische Annalen 71, 97–115 (1912)
5. Davis, M.: Computability and Unsolvability. Dover Publications, New York (1982)
6. Fiske, M.S.: Non-autonomous dynamical systems applied to neural computation. PhD thesis, Northwestern University. UMI Microform 9714584 (1996)
7. Hopf, H.: Abbildungsklassen n-dimensionaler Mannigfaltigkeiten. Mathematische Annalen 96, 209–224 (1926)
8. Hooper, P.K.: The Undecidability of the Turing Machine Immortality Problem. Journal of Symbolic Logic 31(2) (1966)
9. Kürka, P.: On Topological Dynamics of Turing Machines. Theoretical Computer Science 174, 203–216 (1997)
10. Mueller, L.: newLISP Language (1999-2012), `http://www.newlisp.org/`
11. Poincare, H.: Sur les courbes definies par une equation differentielle. Oeuvres 1 (1892)
12. Turing, A.M.: On computable numbers, with an application to the Entscheidungsproblem. Proc. London Math. Soc. ser. 2 42(Parts 3 and 4), 230–265 (1936); [Turing, 1937a] A correction, ibid. 43, 544–546 (1937)

## Appendix

Using the same notation as Theorem 22.2, let $V_1$ denote the initial tape pattern, which is the sequence of alphabet symbols in the tape squares over the window of execution of the prime input command sequence, just before the first input command $(q_1, a_1)$ in the sequence is executed. Let $s_1$ denote the location of the tape head, $V_1(s_1) = a_1$. Let $V_k$ denote the tape pattern just before the kth input command $(q_k, a_k)$ in the sequence is executed and let $s_k$ denote the location of the tape head, $V_k(s_k) = a_k$.

**Definition 22.19.** *Composition of Prime Input Command Sequences*
Let $(q_1, a_1) \mapsto \ldots \mapsto (q_n, a_n)$ and $(r_1, b_1) \mapsto \ldots \mapsto (r_m, b_m)$ be prime input command sequences where $V_k$ denotes the tape pattern just before the $k$th input command $(q_k, a_k)$ with tape head at $s_k$ with respect to $V_k$. $W_k$ denotes the tape pattern just before the $k$th input command $(r_k, b_k)$ with tape head at $t_k$ with respect to $W_k$. Suppose $(V_n, s_n)$ overlap matches with $(W_1, t_1)$ and $q_n = r_1$. Then $(q_n, a_n) = (r_1, b_1)$. The composition of these two prime input command sequences is defined as $(q_1, a_1) \mapsto \ldots \mapsto (q_n, a_n) \mapsto (r_2, b_2) \mapsto \ldots \mapsto (r_m, b_m)$. The composition is undefined if $(V_n, s_n)$ and $(W_1, t_1)$ do not overlap match or $q_n \neq r_1$. If $(q_1, a_1) \mapsto \ldots \mapsto (q_n, a_n) \mapsto (q_1, b_1)$ is a prime state cycle, then it is also prime input command sequence.

*Example 22.6.* Directed Partition procedure

Start with finite sequence ( 0 4 2 3 4 1 3 0 1 2 0 4 2 3 4 1 3 0 1 2 ).

The partition steps are shown below.

1. ( (0 4 2 3) 4 1 3 0 1 2 0 4 2 3 4 1 3 0 1 2). 4 lies in (0 4 2 3).
2. ( (0 4 2 3) (4 1 3 0) 1 2 0 4 2 3 4 1 3 0 1 2). 1 lies in (4 1 3 0).
3. ( (0 4 2 3) (4 1 3 0) (1 2 0 4) 2 3 4 1 3 0 1 2). 2 lies in (1 2 0 4).
4. ( (0 4 2 3) (4 1 3 0) (1 2 0 4) (2 3 4 1) 3 0 1 2). 3 lies in (2 3 4 1).
5. ( (0 4 2 3) (4 1 3 0) (1 2 0 4) (2 3 4 1) (3 0 1 2) ). 0 lies in (3 0 1 2).

**Definition 22.20.** *Tuples*

A tuple is a finite sequence of objects denoted as $(\sigma_1, \sigma_2, \ldots, \sigma_m)$. The length of the tuple is the number of objects in the sequence denoted as $|(\sigma_1, \sigma_2, \ldots, \sigma_m)| = m$. For our purposes, the objects of the tuple may be states, input commands or natural numbers. (3) is a tuple of length one. (1, 4, 5, 6) is a tuple of length four. Sometimes the commas will be omitted as in the previous example. (4 6 0 1 2 3) is a tuple of length six. The 4 is called the first object in tuple (4 6 0 1 2 3). 1 is called a member of tuple (4 6 0 1 2 3).

**Definition 22.21.** *Tuple of Tuples*

A tuple of tuples is of the form $(w_1, w_2, \ldots, w_n)$ where each $w_k$ may have a different length. An example of a tuple of tuples is ( (3), (1, 4, 5, 6), (4, 5, 6) ). The commas are omitted in this example ( (0 8 2 3) (1 7 5 7) (5 5 6) ).

**Definition 22.22.** *Directed Partition of a Sequence*

A directed partition is a tuple of tuples $(w_1, w_2, \ldots, w_n)$ satisfying Rule A and B.

Rule A. No object $\sigma$ occurs in any element tuple $w_k$ more than once.

Rule B. If $w_k$ and $w_{k+1}$ are consecutive tuples, then the first object in tuple $w_{k+1}$ is a member of tuple $w_k$.

*Example 22.7.* Directed Partition Examples

The five examples illustrate element and partition tuples and Rule A and Rule B.

( (0 8 2 3) (8 7 5 4) (5 0 6) ) is an example of a directed partition.

( (0 8 2 3) (8 7 5 4) (5 0 6) ) is sometimes called a partition tuple.

(0 8 2 3) is the first element tuple. The first object in this element tuple is 0.

Element tuple (8 0 5 7 0 3) violates Rule A because object 0 occurs twice.

( (0 8 2 3) (1 7 5 4) (5 0 6) ) violates Rule B since 1 does not lie in tuple (0 8 2 3).

**Definition 22.23.** *Consecutive Repeating Sequence and Extensions*
A consecutive repeating sequence is a sequence $(x_1, x_2, \ldots, x_n, \ldots, x_{2n})$ of length $2n$ for some positive integer $n$ such that $x_k = x_{n+k}$ for each $k$ satisfying $1 \leq k \leq n$. An extension sequence is the same consecutive repeating sequence for the first $2n$ elements $(x_1, x_2, \ldots, x_n, \ldots, x_{2n} \ldots x_{2n+m})$ such that $x_k = x_{2n+k}$ for each $k$ satisfying $1 \leq k \leq m$. A minimal extension sequence is an extension sequence $(x_1, \ldots, x_n, \ldots x_{2n+m})$ where $m$ is the minimum positive number such that there is one element in $x_{2n}, x_{2n+1}, \ldots, x_{2n+m}$ that occurs more than once. Thus, $x_{2n+k} = x_{2n+m}$ for some $k$ satisfying $0 \leq k < m$.

For example, the sequence $S = (4234130120\,4234130120)$ is a consecutive repeating sequence and $\overline{S} = (4234130120\,4234130120\,42341)$ is an extension sequence. $\overline{S}$ contains consecutive repeating sequence $S$ .

**Definition 22.24.** *Directed partition extension with last tuple satisfying Rule B*
Suppose $(x_1 \ldots x_n \ldots x_{2n}, x_{2n+1}, \ldots x_{2n+m})$ is an extension of consecutive repeating sequence $(x_1 \ldots\ x_n \ldots x_{2n})$. Then $(w_1, w_2, \ldots, w_r)$ is a directed partition extension if it is a directed partition of the extension: The last tuple $w_r$ satisfies Rule B if $x_{2n+m}$ is the last object in tuple $w_r$ and $x_{m+1}$ lies in tuple $w_r$.

For example, the extension $(4\;2\;3\;4\;1\;3\;0\;1\;2\;0\;4\;2\;3\;4\;1\;3\;0\;1\;2\;0\;4\;2\;3)$ has directed partition extension $((4\;2\;3)\;(4\;1\;3\;0)\;(1\;2\;0\;4)\;(2\;3\;4\;1)\;(3\;0\;1\;2)\;(0\;4\;2\;3))$ and the last tuple satisfies Rule B since 4 lies in $(0\;4\;2\;3)$.

**Procedure 4.** *Directed Partition procedure*
*This procedure converts a finite sequence into a directed partition.*

Given a finite sequence $(x_1 \ldots x_n)$ of objects.
Initialize element tuple $w_1$ to the empty tuple ( ).
Initialize partition tuple $P$ to the empty tuple ( ).
For each element $x_k$ in sequence $(x_1\ \ldots\ x_n)$
{
 if $x_k$ is a member of the current element tuple $w_r$
 {
  Append element tuple $w_r$ to the end of partition tuple $P$ so that $P = (w_1 \ldots w_r)$.
  Initialize current element tuple $w_{r+1} = (x_k)$.
 }
 else update $w_r$ by appending $x_k$ to end of element tuple $w_r$.
}

The result is the current partition tuple $P$ after element $x_n$ is examined in the loop. Observe that the tail of elements from $(x_1 \ldots x_n)$ with no repeated elements will not lie in the last element tuple of the final result $P$.

**Procedure 5.** *Directed Partition Procedure implemented in newLISP* `<www.newlisp.org>`

*The function findpartition converts any finite sequence represented as a list into a directed partition.*

```
(define (addobject  etuple object)
 (if (member object etuple)
      nil
      (append etuple  (list  object))
))

(define (findpartition seq)
  (let(
          (partition  '())
          (etuple  '())
          (testadd nil)
       )
    (dolist (object seq)
      (set 'testadd (addobject etuple object))
      (if testadd
         (set 'etuple testadd)
         (begin
           (set 'partition (append partition (list etuple)))
           (set 'etuple (list object))
         )
    ))
  partition
))
```

```
> (set 'seq '(4 2 3 4 1 3 0 1 2 0 4 2 3 4 1 3 0 1 2 0 4 2 3 4))

> (findpartition  seq)
((4 2 3) (4 1 3 0) (1 2 0 4) (2 3 4 1) (3 0 1 2) (0 4 2 3))
```

4 lies in the last tuple `(0 4 2 3)`.

*Remark 22.7.* Every Consecutive Repeating Sequence has an extension sequence with a directed partition such that the last tuple satisfies the Rule B property.

*Proof.* As defined in 22.23, extend consecutive repeating sequence $(x_1 \dots x_n \dots x_{2n})$ to the extension sequence $(x_1 \dots x_n \dots x_{2n}, x_{2n+1}, \dots x_{2n+m})$ such that $m$ is the minimum positive number such that there is one element in $x_{2n}, x_{2n+1}, \dots x_{2n+m}$ that occurs more than once. Thus, $x_{2n+k} = x_{2n+m}$ for some $k$ satisfying $0 \leq k < m$.

Apply procedure 4 to $\overline{S} = (x_1 \dots x_n \dots x_{2n}, x_{2n+1}, \dots x_{2n+m})$. Then the resulting partition tuple $P$ extends at least until element $x_{2n}$ and the last tuple in $P$ satisfies rule B. If the partition tuple $P$ is mapped back to the underlying sequence of elements, then it is an extension sequence since it reaches element $x_{2n}$.

**Lemma 22.3.** *Any consecutive repeating state cycle is contained in a composition of one or more prime input command sequences.*

*Proof.* Let $\sigma = [(q_1,a_1) \mapsto \ldots \mapsto (q_n,a_n)(q_1,a_1) \mapsto \ldots \mapsto (q_n,a_n)]$ be a consecutive repeating cycle. Procedure 4 and remark 22.7 show that this sequence of consecutive repeating input commands may be extended to a minimal extension sequence $[(q_1,a_1) \mapsto \ldots \mapsto (q_n,a_n) \mapsto (q_1,a_1) \mapsto \ldots \mapsto (q_n,a_n) \mapsto (q_1,a_1) \mapsto \ldots \mapsto (q_m,a_m)]$.

For simplicity, let $v_k$ denote input command $(q_k,a_k)$. Apply procedure 4 to $(v_1 \ldots v_n v_1 \ldots v_n v_1 \ldots v_m)$ so that the result is the partition tuple $P = (w_1 \ldots w_r)$. Then the sequence of element tuples in $P$ represent a composition of one or more prime input command sequences.

Rules A and B imply that for consecutive tuples $w_k = (v_{k(1)} v_{k(2)} \cdots v_{k(m)})$ and $w_{k+1} = (v_{(k+1)(1)} v_{(k+1)(2)} \cdots v_{(k+1)(m)})$ , then $(q_{k(1)},a_{k(1)}) \mapsto \ldots \mapsto (q_{k(m)},a_{k(m)}) \mapsto (q_{(k+1)(1)},a_{(k+1)(1)})$ is a prime input command sequence. Remark 22.7 implies that the last tuple $w_r$ corresponds to a prime input command sequence and that the consecutive repeating state cycle is contained in the partition $P$ mapped back to the sequence of input commands.

**Definition 22.25.** *Finite sequence rotation*

Let $(x_0 x_1 \ldots x_n)$ be a finite sequence. A $k$-rotation is the resulting sequence $(x_k x_{k+1} \cdots\ x_n x_0\ x_1 \ldots x_{k-1})$. The 3-rotation of (8 7 3 4 5) is (3 4 5 8 7).

**Definition 22.26.** *Rotating a state-symbol cycle*

Let $(q_1,a_1) \mapsto \ldots \mapsto (q_n,a_n) \mapsto (q_1,b_1)$ be a state cycle. This state cycle is called a state-symbol cycle if $a_1 = b_1$. A rotation of this state-symbol cycle is the state cycle $(q_k,a_k) \mapsto \ldots, (q_n,a_n) \mapsto (q_1,a_1) \mapsto \ldots (q_k,b_k)$ for some $k$ satisfying $0 \le k \le n$. In this case, the state-symbol cycle has been rotated by $k-1$ steps.

**Lemma 22.4.** *Any consecutive repeating rotated state cycle generated from a consecutive repeating state cycle induces the same immortal periodic orbit.*

*Proof.* Let $p$ be the immortal periodic point induced by this consecutive repeating state cycle. Rotating this state cycle by $k$ steps corresponds to starting at periodic machine configuration $p$ and executing the Turing machine $k$ steps.

**Procedure 6.** *A newLISP [10] function that searches for a consecutive repeating sequence.*

```
(define  (findpatternrepeats   plength   seq)
  (let (
        (k 0)
        (maxk (- (length seq) (+ plength plength)) )
        (pattern nil)
        (repeatpair nil)
        (norepeats true)
      )

    (while (and (<= k maxk) norepeats)
      (set 'pattern (slice seq k plength))
```

```
            (if (= pattern (slice seq (+ k plength) plength))
                (begin
                  (set 'repeatpair (list pattern k))
                  (set 'norepeats false) )
            )
            (set 'k (+ k 1))
          )
          repeatpair
    ))

    (define (findrepeats seq)
      (let (
             (plength 1)
             (maxplength (/ (length seq) 2) )
             (repeatpair nil)
            )
          (while (and (<= plength maxplength) (not repeatpair))
             (set 'repeatpair (findpatternrepeats plength seq))
             (set 'plength (+ plength 1))
          )
          repeatpair
    ))

(set 'seq1 '(3 5 7 2 3 5 7 11 5 7 ) )
(set 'seq2 '(3 5 7 2 3 5 7 11 5 7 11 2 4 6 8 ) )
(set 'seq3 '(1 2 0 2 1 0 2 0 1 2 0 2 1 0 1 2 1 0 2 1 2 0
             2 1 0 1 2 0 2 1 2 0 1 2 1 0 1 2 0 1 0 1) )

>  (findrepeats seq1)
nil

>  (findrepeats seq2)
((5 7 11) 5)

>  (findrepeats seq3)
((0 1) 38)
```

# Author Index

Printed by Publishers' Graphics LLC